Undercooled Alloy Phases

COVER PHOTO
Microstructure of Nb-25 atm% Ge following undercooling of 530 K. Solute trapping of Ge occurred because of the large degree of undercooling, but recalescence caused precipitation of excess solute in the form of Nb_5Ge_3.
(See page 225)

Undercooled Alloy Phases

Proceedings of the 1986 Hume-Rothery Memorial Symposium, which was organized by the TMS Committee on Alloy Phases, and was held in New Orleans, Louisiana, March 2-6, 1986 at the 115th Annual Meeting of TMS-AIME.

Edited by

E.W. Collings
Battelle Memorial Institute
Columbus, Ohio

C.C. Koch
North Carolina State University
Raleigh, North Carolina

A Publication of The Metallurgical Society, Inc.

Library of Congress Cataloging-in-Publication Data

Hume-Rothery Memorial Symposium (1986 : New Orleans, La.)
Undercooled alloy phases.

Includes bibliographies and index.
1. Alloys - Congresses. **2.** Phase rule and equilibrium - Congresses. **3.** Supercooling - Congresses. **I.** Collings, E.W. **II.** Koch, C.C. **III.** Metallurgical Society Alloy Phases Committee. **IV.** Metallurgical Society. Meeting (115th : 1986 : New Orleans, La.) **V.** Title.
TN689.2.H86 1986 669'.94 86-33243
ISBN 0-87339-059-8

A Publication of The Metallurgical Society, Inc.
420 Commonwealth Drive
Warrendale, Pennsylvania 15086
(412) 776-9000

Printed in the United States of America.
Library of Congress Catalogue Number 86-33243
ISBN NUMBER 0-87339-0598

Table of Contents

PART III: UNDERCOOLED SOLIDIFICATION-- THEORY AND EXPERIMENT

Preface

This book presents the Proceedings of the 1986 Hume-Rothery Memorial Symposium which was held in New Orleans, Louisiana, March 2-6, 1986 at the 115th Annual Meeting of TMS-AIME. The Symposium was organized by the TMS Committee on Alloy Phases in honor of Professor David Turnbull, the recipient of the Hume-Rothery Award for 1986. The subject of the Symposium, "Undercooled Alloy Phases", is one of the research areas in which Professor Turnbull has been a pioneer and in which he remains active. He led off the Symposium with a talk entitled, "Undercoolability and the Exposure of Metastable States". A brief biography of Professor Turnbull is included in this volume.

The purpose of the Symposium was to bring focus on the phenomenon of undercooling and its influence on the formation of metastable phases. It has become recognized that undercooling is central to the metastable structures formed by the various rapid solidification methods. Therefore, it is a topic of intense scientific interest as well as technological importance.

The book has been organized into three sections. Part I, Undercooled Phases and Microstructures, includes Professor Turnbull's paper and ten other papers which focus on undercooling and metastability in specific alloy systems. Part II, Techniques of Undercooling and Metastable Phase Formation, contains five papers which describe research with various experimental methods ranging from pulsed laser quenching to solidification in long drop tubes. Part III, Undercooled Solidification - Theory and Experiment, includes seven papers which address the theory of undercooled solidification behavior as well as experimental studies of this phenomenon. All of the papers were given a critical peer review. The editors wish to thank the participants for performing the double tasks of being both authors and reviewers. We wish to give special thanks to the Session Chairmen, Dr. D. E. Polk, Office of Naval Research, and Professor J. H. Perepezko, University of Wisconsin.

E. W. Collings
Battelle Columbus Laboratories
Columbus, Ohio

C. C. Koch
North Carolina State University
Raleigh, North Carolina

DAVID TURNBULL BIOGRAPHY

David Turnbull was born (February 18, 1915) on a farm near Kewanee, Illinois. He received his B.S. from Monmouth College (Illinois) in 1936, and his Ph.D. in physical chemistry at the University of Illinois in 1939. In 1958, he received an Honorary Sc.D. from Monmouth College, and in 1962, an Honorary M.A. from Harvard University.

From 1939 to 1946, Turnbull was on the faculty of The Case Institute of Technology. He was a research scientist at the General Electric Research Laboratory from 1946 to 1962, where he served as manager of the Chemical Metallurgy Section from 1950 to 1958. He spent the academic year 1957-1958 on leave at the Cavendish Laboratory, Cambridge University. He was adjunct Professor of Metallurgy at Rensselaer Polytechnic Institute from 1954 to 1962. In 1962, he joined the faculty of Harvard University as Gordon McKay Professor of Applied Physics. He became professor emeritus in 1985.

He was Institute of Metals Division (AIME) Lecturer, 1961; Robert S. Williams (MIT) Lecturer, 1969; Campbell Memorial (ASM) Lecturer, 1980; and Van Horn Lecturer (Case Western Reserve University), 1981. He was awarded the Materials Research Society von Hippel Prize and the Acta Metallurgica Gold Metal in 1979, the American Physical Society International Prize for New Materials in 1983, The Metallurgical Society of AIME Hume-Rothery Award and the Japan Prize in 1986.

He is a member of the National Academy of Sciences and a fellow of the American Academy of Arts and Sciences, the American Association for the Advancement of Science, the Metallurgical Society of AIME, the American Society for Metals and the American Physical Society. He is co-editor, with Henry Ehrenreich and Frederick Seitz, of the Solid State Physics series.

Turnbull has made crucial contributions to the understanding of the kinetics of phase transformations in condensed systems. With his collaborators at General Electric he formulated the classical nucleation theory for crystallization and solid-state reactions, and tested it quantitatively. His demonstration that liquid metals could be undercooled by as much as a third of their absolute melting temperature showed conclusively the fundamental difference in short-range order between liquids and crystals, and was the basis of Frank's important proposal of icosahedral order in the liquid state.

In collaboration with M. H. Cohen, Turnbull developed the free volume theory for atomic transport in simple liquids, and predicted that, if crystallization could be avoided, any liquid could be made into a glass. This was soon confirmed by Duwez's demonstration that an amorphous metal could be obtained by rapid quenching of the melt. Turnbull, with his student, H. S. Chen, then demonstrated that these materials were true glasses, in that they exhibited the same manifestations of a liquid-glass transition as the well-known silicate or organic glasses. With his group at Harvard he has continued to make contributions to the study of the properties and transformations of metallic glasses.

He has contributed to the study of diffusion in crystalline materials with the first quantitative characterization of short-circuit diffusion; his proposal, with F. C. Frank, of the dissociative mechanism to explain the diffusion of noble metals in elemental semiconductors; and the work of his Harvard group on the fast diffusion of noble metals in polyvalent metal matrices.

In the field of crystal growth he has made contributions to the understanding of the spiral growth mechanism, cellular precipitation, and, most recently, the mechanisms of ultra-fast crystal growth induced by irradiation of a solid surface by a short laser pulse.

PART I: Undercooled Phases and Microstructures

UNDERCOOLABILITY AND THE EXPOSURE OF METASTABLE STRUCTURES

David Turnbull

Division of Applied Sciences
Harvard University
Cambridge, Ma 02138

The thermodynamic conditions for exposing metastable states are established by de-energizing a material, as in melt quenching, condensation, etc. When these conditions are met, metastable states may appear because they are kinetically preferred. The understanding of these kinetic preferences is reviewed. Formation of metastable structures by melt quenching requires high resistance to crystal nucleation to deep undercooling through the "labile" regime between the liquidus and configurational "freezing" temperature, T_g. Further, the growth of the most stable phase must be thermally activated and so quench suppressible if the metastable state is to persist at $T < T_g$. In metal systems, selective alloying is often crucial to the provision of the nucleation and/or growth resistance requisite to forming various metastable structures. These alloying effects are surveyed and interpreted.

I. Introduction

The general requirements for the appearance and persistence of metastable structures might be summarized as follows:

(1) thermodynamic exposure of the structure

(2) kinetic preference for the structure over the more stable ones

(3) kinetic entrapment of the structure.

Metastability is exposed thermodynamically through energization — by melting, dissolution, irradiation or cold working — followed by de-energization effected by quenching or condensation. During de-energization a metastable structure may form and be trapped kinetically provided evolution to more stable structures must be thermally activated; this proviso, e.g., would not, in general, be met by a structure susceptible to a martensite transition.

In accord with the focus of this symposium the emphasis of my paper is on the exposure of metastability by melt quenching and the effects of alloying on the emergence of metastable structures. First the understanding of kinetic preferences for metastable structures is discussed. Then the undercoolability of liquid metals and its interpretation are reviewed. Finally the effects of impurity admixture on the formation of metastable alloy structures are surveyed and interpreted.

II. Kinetic Preferences for Metastable Structures

De-energization may present thermodynamic options for evolution of the de-energized structure into metastable, or even unstable, states as well as to the most stable state. The rate of transition to a particular state is generally some product function of a kinetic coefficient, p_i, with a thermodynamic factor $f(\Delta G_i)$. Clearly for a metastable structure to appear the p_i determining its formation rate must substantially exceed that for the more stable state.

The general experience on kinetic preference in structural evolution was summarized by the author in earlier publications (1,2). It indicates that metastable states very often are kinetically favored; indeed, so often that Ostwald proposed as a general rule that a system will evolve through the succession, in the order of decreasing free energy, of available metastable states. There are exceptions to Ostwald's rule as stated. The author has noted (1,2) that *the more general tendency in structural evolution, following either de-energization or energization, appears to be toward formation of the available state with entropy closest to that of the initial one.* There seems to be no theoretical proof for a step entropy transformation sequence and there are some apparent exceptions to it. However, the processes most favored kinetically are likely to be those (2) requiring the smallest changes in the positional and motional correlations of the atoms in the evolving group and it is such processes that should be attended by the least entropy changes (2). For example, according to this argument we should expect that amorphous rather than the thermodynamically more stable crystalline films should deposit from the vapor on amorphous substrates held at $T \to 0°K$. Indeed, amorphous rather than crystalline films are formed in both physical and computer simulated condensation of four coordinated elements at low temperature. Also it has been shown (3,4) that the amorphous semiconducting phase, a-sc, of Si grows faster than the crystalline phase, c-sc, into undercooled liquid Si, ℓm, when a thermal gradient, $(\text{grad } T)_i$, sufficient in magnitude to depress the interfacial temperature, T_i, to a level well below that of the metastable a-sc/ℓm equilibrium, is imposed at the c-sc/ℓm interface.

Uniform hard spheres also form an amorphous (5,6), rather than a crystalline, phase in computer simulated condensation at 0°K. In contrast, pure metals generally deposit from their vapors in polycrystalline forms on cold substrates (7). However, it is likely that the number densities of heterophase nucleants on the substrates were high and there is evidence that the movement of the crystal/amorphous interface in pure monatomic systems is not quench suppressible (8-10).

III. Undercoolability: Nucleation Limited

In metals, the velocity of the crystal-melt interface, u, becomes easily measurable at minute departures from equilibrium, as specified by the interfacial undercooling, ΔT_i, and it increases rapidly with further increases in ΔT_i. Indeed, it is well-known that when the imposed thermal gradients are of the usual magnitudes, u can be calculated by supposing that it is controlled entirely by the rate of impurity and/or heat transport, thereby ignoring the kinetic resistance due to the configurational rearrangement at the interface. It follows that *formation of metastable states from liquid metals requires that the liquid be highly resistant to crystal nucleation to deep undercooling through a "labile" regime in which crystallization to the stable state, once initiated, proceeds with extreme rapidity.*

It is now generally accepted that liquid metals do, indeed, exhibit extraordinarily high resistance to crystal nucleation to deep undercooling, provided heterophase nucleants are isolated by dispersion or deactivated by fluxing. The major current issues are (1) what are the undercoolings at the onset of *measurable* homogeneous nucleation. These undercoolings, scaled with the liquidus temperature, T_ℓ, will be labeled ΔT_r^0 ($= \Delta T^0/T_\ell$)? (2) has the occurrence of homogeneous crystal nucleation in any liquid metal been demonstrated? (3) what are the microscopic origins of the high resistance of liquid metals to homogeneous nucleation? These issues are discussed in turn.

A. Undercooling at Onset of Detectable Nucleation

It is convenient to distinguish the liquid metals (group A) which crystallize to closed packed or b.c.c. structures from those (group B) which crystallize to structures in which the nearest neighbor coordination number is less than 8. Metals such as Hg or In which crystallize to somewhat distorted close-packed structures might be placed in an intermediate category (A'). Using dispersion and fluxing techniques the author and his associates demonstrated in the early 1950's that ΔT_r^0 was at least $\sim$0.18 for most of the A metals (11) and $\sim$0.32 for Hg (12). At the time the large magnitudes of these undercoolings were quite unexpected because of the then prevalent view that the atomic short range order (SRO) in liquid metals was essentially similar to that in close-packed crystals. Less surprising were the results of Vonnegut (13), the author, and others that the ΔT_r^0 in group B liquids also were 0.15-0.20, at least.

Later investigators using similar techniques, found onset undercoolings greater than 0.3 for some group B metals and between 0.2 and 0.3 for several group A metals. More particularly, Bosio *et al.* (14) reported ΔT_r^0 approaching 0.5 for certain emulsions of Ga droplets. Perepezko and coworkers (15) obtained ΔT_r^0 in the range of 0.3 to 0.4 for droplet emulsions of a number of other group B metals. Recently Devaud and the author (16) observed that some 0.3-0.5 mm diameter Ge droplets, fluxed in molten dehydrated B_2O_3, exhibited $\Delta T_r^0 \approx 0.35$. Ovsienko (17), Gomersall *et al.* (18) reported ΔT_r^0 ranging between 0.23 and 0.27 for droplets of the group A metals, Fe, Co, Ni. Similar results for metals in the A group are described in some of the papers of this symposium. Table 1 summarizes some of the more recent data on onset undercooling.

These results demonstrate that liquid metals are indeed extraordinarily resistant to homogeneous crystal nucleation to deep undercooling into the labile regime. According to classical nucleation theory ΔT_r^0 should depend upon cooling rate and specimen size. However, the existing evidence indicates, in accord with the theory, that this dependence is rather weak so that, with heterophase nucleants eliminated, large masses of liquid metals will sustain

Table 1 - Reported scaled undercoolings, ΔT_r^0, (excepting for Hg uncorrected for differences in cooling rate) at measured nucleation onset for several pure metals.

Metal	Structure Type	ΔT_r^0	Specimen Diameter	References
Fe	A	0.23 ± 0.01	~ 0.7 cm	18
Co	A	0.27 ± 0.01	60-200 μm	17
Ni	A	0.27 ± 0.01	80-200 μm	17
In	A'	0.26	2-5 μm	15
Hg	A'	0.33 - 0.35	2-5 μm	12,26
Ga	B	0.50 ± 0.02	2-5 μm	14
Ge	B	0.35 ± 0.02	0.05 cm	16
Sn	B	0.37	2-5 μm	15
Bi	B	0.42	2-5 μm	15

undercooling of at least 0.15 to 0.20 for indefinite periods without crystallizing. At such undercoolings a considerable variety of important metastabilities – structural, compositional or morphological – can arise.

B. Occurrence of Homogeneous Nucleation

It is generally presumed that the undercoolings at the onset of measurable homogeneous nucleation must equal or exceed the maximum reported experimental values, *properly adjusted for cooling rate and specimen size*. However, there are conditions, as explained by the author in an earlier publication (19), under which this presumption may not be correct. First, it is evident that, when comparing results of different investigators, proper allowance must be made for experimental uncertainties in the calibration of T^0, the temperature at nucleation onset, relative to the liquidus, T_ℓ. However, apart from calibrational disparities there are other reasons why there might be some apparent or actual specimen-to-specimen dispersion of ΔT_r^0 values at homogeneous nucleation.

An apparent dispersion may arise between various emulsions of droplets coated by solid films, because substantial hydrostatic stresses can develop in the smaller droplets (diameter < 2-3 microns) as they are undercooled, due to the differential volume response of the liquid relative to its coating film. This effect can lead to hydrostatic stresses which at T^0 are quite different from those at T_ℓ. Consequently, the apparent ΔT_r^0 could be displaced considerably from that which would be exhibited by the unstressed liquid. The analysis of Spaepen and the author (20) indicated that this displacement could be as large as + 10% for small $A\ell$ droplets coated by thin films of their oxide. In view of the high mechanical strength of solids in thin film form and the great resistance of liquid metals to cavitation, the possible importance of such stress effects in experiments on small droplets cannot be lightly dismissed. These effects might be assessed by precise evaluation of the thermal expansion, $\Delta\alpha$, and heat capacity, ΔC_p, and their temperature dependences, of the liquid relative to the crystal. If, as the droplets are undercooled, the stresses remain negligible $\Delta\alpha(T)$ and $\Delta C_p(T)$ should correspond, at least near T_ℓ, to the extrapolated bulk metal values of these coefficients.

An actual dispersion of ΔT_r^0 for homogeneous nucleation can result from specimen-to-specimen variation in the concentrations of those *homophase* impurities which have a lower standard free energy of solution in the crystalline than in the liquid phase of the metals (19). Such impurities, even in amounts of order of a few at ppm, can, without measurably displacing T_ℓ, reduce ΔT_r^0 considerably from its pure metal value. This "chemical" effect on nucleation is distinct from that of the *heterophase* impurities which lead to heterogeneous nucleation. The reversible work of forming a nucleus which includes a single impurity atom or molecule may be expressed as (19):

$$W_i^* = 4\pi r^{*2}\sigma + \frac{4}{3}\pi r^{*3}\Delta G_v + \Delta g_i^0 - v_i\Delta G_v - kT\,\ell n(n_i v^*) \qquad (1)$$

$$r^* \cong -2\sigma/\Delta G_v$$

so:

$$W_i^* \cong \frac{16\pi\sigma^3}{3(\Delta G_v)^2} + \Delta g_i^0 - v_i\Delta G_v - kT\,\ell n(n_i v^*) \qquad (2)$$

where:

- σ = crystal-melt interfacial tension
- ΔG_v = Gibbs free energy of crystallization/volume
- Δg_i^0 = standard free energy of solution/atom impurity in crystal minus that in liquid phase
- v_i = volume/atom of impurity
- n_i = impurity atoms/volume
- v^* = volume of a single nucleus.

The last term in W_i^* comes from the entropy of localization, $k\,\ell n(n_i v^*)$, of the impurity atom within the nucleus. Adapting the approximate kinetic analysis of Fisher and the author (21), the frequency/volume, I_i, of forming nuclei including an impurity atom becomes:

$$I_i \cong n_1 k_1 (n_i v^*) \exp\left\{-\frac{(16\pi\sigma^3/3\Delta G_v^2) + \Delta g_i^0 - v_i \Delta G_v}{kT}\right\} \quad (3)$$

where n_1 = number of metal atoms/volume, k_1 = frequency of atomic transfer at the nucleus-melt interface. At zero impurity content the nucleation frequency is:

$$I \cong n_1 k_1 \exp\left[\frac{-16\pi\sigma^3}{3(\Delta G_v)^2 kT}\right] . \quad (4)$$

If $\Delta g_i^0 < 0$ it follows that homogeneous nucleation will be facilitated by inclusion of an impurity atom in the nucleus. The nucleation probability would be still proportional to specimen volume and the value of σ would be virtually identical to that deduced from experiments on an impurity-free liquid. The author has shown that (19), with plausibly chosen Δg_i^0 values, ΔT_r^0 might be decreased 5-10% by the presence of a few ppm impurity. A pronounced chemical effect on the nucleation of polymer crystals from their melt was, indeed, demonstrated by Legras Mercier, and Nield (22).

At steady state, isothermal homogeneous nucleation should be stochastic in volume and time. This condition requires that in droplet emulsions the nucleation probability, k_d, in a droplet of diameter, d, be proportional to Id^3, where I is the nucleation frequency/volume-time. If the undercooling is chosen so that the time for crystal growth is negligible compared with that for nucleation k_d becomes, virtually, the crystallization frequency. Then the isothermal kinetics of crystallization of droplet emulsions can be analyzed to ascertain the dependence of k_d on d. In one method (a) developed by the author (12) and described in detail by Pound (23), the size dispersion of the droplet population is exploited to discriminate between the possibilities. However, a more conclusive discrimination is achieved by method (b) in which $k_{\bar{d}}$, where $\bar{d}$ is the diameter of the droplet of average volume, are compared for separate emulsions with substantially differing $\bar{d}$; then $k_{\bar{d}_1}/k_{\bar{d}_2} = (\bar{d}_1/\bar{d}_2)^3$ if nucleation is homogeneous.

The author's measurements of the isothermal rates of solidification of Hg-Laurate (Hg-L) coated Hg droplet dispersions (12), as well as those with Cormia (24) on n-hepta- and n-octadecane dispersions, fulfilled both tests (a) and (b) for volume nucleation. Thus at a given undercooling the rates were satisfactorily described by a single nucleation frequency/volume. The onset undercooling of Hg in these experiments was $\sim 78°$ ($\Delta T_r^0 \approx 0.33$). Somewhat higher onset undercoolings in differently prepared Hg dispersions were reported by Monge (25) and by Perepezko and Rasmussen (26). When adjusted to the cooling rate of the author's experiments the Monge result is about 3° and that of Perepezko and Rasmussen about 4-5° greater than obtained by the author. These differences are not large and could reflect only the temperature calibration uncertainties in the separate studies. Alternatively, they could be due to small "chemical" effects (19) as discussed in the foregoing. Thus the evidence that homogeneous nucleation was exhibited in the author's experiments on HgL coated Hg droplets seems quite conclusive.

The temperature dependence of I in the author's experiments on Hg was satisfactorily described by the equation derived from classical nucleation theory but with a prefactor about 10^7 times too large (12), when σ and the entropy of crystallization/volume, ΔS_v, were assumed constant. It was shown elsewhere (19) that this disparity could result from the chemical effect noted above. Alternatively, it might, as the author suggested earlier (12), arise from temperature variations of ΔS_v and σ. The temperature variation of $\Delta S_v(T)$ in the undercooled regime can be estimated roughly from the heat capacity, $C_p(T)$, measurements of Busey and Giauque (27) (B-G) on crystalline and liquid Hg in conjunction with those of Perepezko and Rasmussen (26) (P-R) on undercooled Hg droplets. Actually the temperature coefficient of C_p at T_m reported by P-R disagrees in sign, as well as magnitude, with that of B-G on bulk liquid Hg. However, when the P-R $C_p(T)$ is grafted on to the B-G $C_p(T)$ extrapolated to 30° undercooling, the calculated $\Delta S_v(T)$ at 80° undercooling, $T = 154°K$, gives $\Delta S_v(154) = 0.92\ \Delta S_v(T_m)$ compared with $\Delta S_v(154) =$ $= 0.93\ \Delta S_v$ obtained by extrapolating the B-G $C_p(T)$ to $\Delta T = 80$. Putting the $\Delta S_v(T)$ calculated from the composite $C_p(T)$ into the classical expression for I, with the theoretical prefactor, the Hg data can then be fit within experimental error, with an interfacial tension increasing by 0.10 ergs/cm^2-deg with temperature. Over the temperature regime of the I measurements the fitting σ could be described by:

$$\sigma = 11.9 + 0.10T \text{ ergs/cm}^2$$

where T is the absolute temperature. This result is in rough agreement with that of a calculation of Mijazawa and Pound (28) assuming $\Delta C_p \sim 0$. The positive sign and approximate magnitude of $d\sigma/dT$ are consistent with Spaepen's negentropic model for the crystal-melt interfacial tension.

Mijazawa and Pound (28) measured the isothermal rates of solidification of two dispersions of Ga droplets which exhibited onset undercoolings $\sim 100°$, well below the maximum $\sim 150°$ claimed by Bosio *et al.* (14). Their data on the separate dispersions seemed to meet test (a) for volume nucleation. However, according to the author's calculation they failed to meet test (b). Thus, at

$\Delta T = 100°$ the solidification time constant $(1/k_{\bar{d}})$ for their first dispersion, in which $\bar{d} = 5$ μm, was only 3.5 times greater than that for their second dispersion in which $\bar{d}$ (= 21 μm) was about four times larger. Thus their $k_{\bar{d}}$ appeared to be proportional to $\bar{d}$, rather than to $\bar{d}^3$, which would have been required by test (b) for volume nucleation. The origin of such an unusual dependence of $k_{\bar{d}}$ on $\bar{d}$ is not clear.

C. Origins of Nucleation Resistance of Monatomic Liquids

From the foregoing it is clear that the homogeneous nucleation frequency in pure liquid metals is immeasurably small to scaled undercoolings in excess of 0.2. This large resistance to nucleation is qualitatively accounted for by two complementary theoretical approaches. Macroscopically the nucleation resistance indicates, in terms of simple nucleation theory, a scaled crystal-melt interfacial tension at the *temperature of measurable nucleation onset* of about 0.55 for the type A metals and 0.60 for Hg. This scaled tension is defined as:

$$\alpha = v^{2/3}\sigma/\Delta h_f$$

where v is the volume/atom and Δh_f is the heat of fusion/atom.

In metals the volume changes on solidification are small and at constant density the energy is quite insensitive to the precise atomic siting. From this it appears that a crystal-melt interface, *without a density deficit* should be formable at little energy cost. However, depending on liquid structure, formation of such an interface may require localization, with consequent entropy loss, of atomic positions in the liquid adjacent to the interface (29). As Spaepen (30) showed, such a localization negentropy can make a major contribution to α.

Microscopically the high nucleation resistance of liquid metals implies that their crystallization must be attended by some essential reconstruction of the atomic short range order (SRO). Frank (31) suggested that this reconstruction might be from a predominantly icosahedral (ICH) SRO in the liquid to the close packed crystalline (CPC) SRO. In support of this suggestion he showed that if the interatomic interaction potential is of the Lennard-Jones type, the energy of a 13 atom cluster is actually lower in the ICH than in either of the CPC configurations. The ICH configuration is a polytetrahedral (PTH) one and includes no half-octahedra, which are essential components of CPC structures. It has also been established that the energetic preference for PTH over CPC configurations extends well beyond 13 to several tens of atoms clusters and that the density in these PTH clusters is higher than in

CPC ones (32). In the dense random packed (DRP) structure of uniform hard spheres it is found that tetrahedral configurations predominate while there is a conspicuous paucity, relative to CPC, of half-octahedral ones (5,6,33). The appearance of a polytetrahedral type structure in random packing suggests that it is entropically favored and that it may be the most likely prototypical model structure for monatomic liquids. The entropy of a condensed system should increase with the range of local density fluctuations permissible at a given energy. Because of the high density of polytetrahedral clusters this range should be substantially greater in random packed structures than in those where the CPC type of SRO predominates.

Thus this microscopic theory for the nucleation resistance requires that atoms in the liquid be positioned randomly within the repulsive constraints. These atoms would then have to be localized sharply to form a crystal-melt interface with no density deficit. Spaepen and Meyer (34) calculated that the contribution to the scaled tension, α, associated with this localization at the (111) face of an f.c.c. crystal is ~ 0.86 at T_m. This value of α should be an upper limit and it is, indeed, well above those indicated by the crystal nucleation kinetics. *Either the negentropy or the excess energy of the interface could be made to vanish by particular choices of the interfacial configuration but they cannot be made to vanish simultaneously.*

D. Alloying Effects

There have been several studies of the dependence of the temperature $T^0(x)$, at measurable nucleation onset, on melt composition. They generally indicate (35) that in the labile regime $T^0(x)$ roughly parallels the liquidus, $T_\ell(x)$. Thompson and Spaepen (36) showed that this behavior can be explained by supposing that:

(1) The scaled crystal-melt tension, α, as in Spaepen's model, is mainly topological in origin and only weakly dependent on x.

(2) The crystal nucleus is always in contact with melt of average composition; this supposition is the analog of the "mean field approximation."

If $T_\ell(x)$ falls sharply, the scaled glass temperature, $T_{rg}(x) = T_g(x)/T_\ell(x)$, may increase markedly as x increases since $T_g(x)$ usually is much less strongly dependent than $T_\ell(x)$ on x. In those alloy systems, where the glass transition is realized, the kinetic resistance to interface transfer is expected to scale roughly as that to flow, so that $1/k_i \propto e^{a/(T_r - T'_{rg})}$, the Vogel-Fulcher factor in flow resistance; T'_{rg} is a temperature somewhat below, but proportional to the measured T_{rg}. Then as $T_{rg}(x)$ increases the nucleation frequency at a given scaled undercooling sharply decreases. This effect has been calculated, with varying assumptions in several earlier papers (10,29,37). For example, when α is taken to be constant and equated to 0.6, the value indicated by the kinetics of nucleation in undercooled Hg, such calculations show that, while T_r^0 varies little with T_{rg} between 0 and 0.6, the maximum frequency of nucleation, I_{max}, sharply decreases, and the temperature, T_r^{max}, at which it occurs increases as T_{rg} increases. Further, for $T_{rg} > 2/3$ I_{max} should be depressed to immeasurably low levels, $< 10^{-5}$ $cm^{-3}sec^{-1}$. A corollary result is that, with heterophase nucleants removed, the minimum quench rate needed for glass formation should decrease markedly as T_{rg} increases. Experience is in, at least, qualitative accord with this expectation. More particularly, cooling rates of $-\dot{T} \sim 10^6$ $deg\text{-}sec^{-1}$ are generally required to quench those alloys, e.g. Au_4Si or Fe_4B, which exhibit $T_{rg} < 0.5$ to glasses. In contrast, the minimum $-\dot{T}$ for glass formation by the alloys, $Pd_{82}Si_{18}(T_{rg} \approx 0.60)$, $Pd_{77.5}Cu_6Si_{16.5}(T_{rg} \approx 0.63)$, and $Ni_{40}Pd_{40}P_{20}(T_{rg} \approx 2/3)$ are, respectively (38,39,40), 10^{-3}, $< 10^{-2}$ and < 1 $deg\text{-}sec^{-1}$.

Greer (41) found that, after partial crystallization by reheating, glassy specimens of $Fe_4B(T_{rg} \sim 0.5)$ contained $\sim 10^{12}$ crystallites/cm^3, randomly sited and oriented. From this result, in conjunction with thermal analyses of the crystallization, he concluded that copious homogeneous nucleation must have occurred in the alloy, during the quench, at rates consistent with those calculated from the simple theory with $\alpha \sim 0.5$ to 0.6. These and similar observations, while not precise enough for quantitative evaluation of I, show conclusively that homogeneous crystal nucleation does occur in liquid metals.

Experiments of the author and colleagues (40) have shown that the molten alloy $Ni_{40}Pd_{40}P_{20}$ can be slow cooled at $-\dot{T} \sim 1$ $deg\text{-}sec^{-1}$, under a dehydrated B_2O_3 flux, to a glass in virtual bulk form, minimum dimension = 1 cm, and then reheated (42), without crystallization, to the liquidus at $\dot{T} < 2.5$ $deg\text{-}sec^{-1}$. This behavior indicates, in rough agreement with the foregoing calculations, that I_{max} in this alloy cannot have exceeded 10^{-1} $cm^{-3}sec^{-1}$. The relatively high $T_{rg}(\approx 2/3)$ of $Ni_{40}Pd_{40}P_{20}$ apparently reflects its exceptionally low liquidus, $T_\ell \sim 913°K$; its T_g, $\sim 605°K$, is in the range of those exhibited by other Ni- and Pd-metalloid glasses. It is interesting that the ternary liquidus is displaced from that of the binary $Ni_{50}Pd_{50}$ T_ℓ, $\sim 1500°K$, by roughly the same amount as the Ni_4P T_ℓ, $\sim 1155°K$, is displaced from T_m of pure Ni.

IV. Undercoolability: Growth Limited

From the foregoing it is clear that most liquid metals, when nucleant free, can sustain scaled undercooling of at least 0.2 to 0.3 for indefinite periods without crystallizing. However, in rapid solidification processing the liquid metal is often spread quickly and thinly over a cool crystalline substrate which serves as a massive heat sink. Solidification will then begin at and propagate from the substrate. The question arises what will be the undercooling, ΔT_i, at the moving solidication front. It will be this undercooling which provides the potential for metastable structure formation in such processing. The kinetics of such plane front solidification and the resultant interfacial undercooling are reviewed in the following sections.

E. Pure Metals

At steady state the velocity, u, of a planar crystal-melt interface in a pure metal may be related to the interface transfer frequency, k_i, by the equation (43):

$$u = - fk_i\lambda \frac{\Delta G_i}{RT_i} \tag{5}$$

where

λ = interface displacement per transfer

f = fraction of interface sites available for attachment

T_i = interface temperature

ΔG_i = the free energy of crystallization/gm. atom at T_i; it is assumed that $RT_i >> \Delta G_i$ and $\Delta G_i = -\Delta T_i \Delta S_m$ where ΔS_m = the gram atomic entropy of melting and ΔT_i = displacement of the *interface* temperature from equilibrium

u also may be expressed in terms of the rate of transport of the heat of crystallization from the interface, as follows:

$$u \cong \kappa \frac{(\text{grad } T)_i \bar{V}}{\Delta H_m} \tag{6}$$

where

κ = the thermal conductivity of the solid

$(\text{grad } T)_i$ = thermal gradient at the interface

$\bar{V}$ = gram atomic volume.

It is assumed that the heat is carried away through the solid and that the melt temperature is uniform.

Equations (5) and (6) may be combined to give the displacement of the interface temperature from equilibrium:

$$\Delta T_i \cong \frac{\kappa(\text{grad } T)_i}{fk_i\lambda} \frac{\bar{V}RT_iT_m}{\Delta H_m^2} \tag{7}$$

where $\Delta S_m = \Delta H_m/T_m$.

We have noted that in pure metal melts the interface transfer step in growth is extremely rapid, so rapid that, with imposed $(\text{grad } T)_i$ of the usual magnitude, ΔT_i is very small and growth is considered to be heat transport "limited". Indeed, there is evidence, summarized elsewhere (8-10, 43), that the interface transfer is collision limited, occurring with virtually no thermal activation. Then $fk_i = \nu_c$, the collision frequency of melt atoms with the interface and

$$u \cong -\nu_c\lambda \frac{\Delta G_i}{RT_i} = -u_s \frac{\Delta G_i}{RT_i} \tag{8}$$

where $u_s = \nu_c\lambda$ the approximate sound speed in the liquid metal, of order 3 to 5 kilometers/sec. Under these conditions growth will not be quench suppressible.

It is instructive to estimate the maximum undercooling which might be reached at the interface of a pure metal crystal growing into its melt. Assume, for example (2), the following typical assignments:

$u_s = 4$ km/sec, $\kappa = 0.2$ cal/cm^2 sec.deg., $\Delta S_m = 2$ cals/gm-atom deg. and $\Delta H_m/\bar{V} = 300$ cals/cm^3; then (2):

$$\Delta T_i/T_m \sim 1 \times 10^{-9}(\text{grad } T)_i \quad . \tag{9}$$

The thermal gradients reached after irradiation for a few picoseconds by high energy laser pulses are among the highest realizable. These are estimated to be $\sim 10^8$ deg/cm which would give a maximum undercooling (2) $\Delta T_i/T_m \sim 0.10$. This undercooling would be sufficient to activate certain heterophase nucleants, if present, but entirely insufficient for the occurrence of appreciable homogeneous nucleation in advance of the interface or metallic glass formation. Thus in the absence of nucleants it is expected that in such processing pure metal crystals will simply regrow epitaxially into their melts (2). This expectation is born out by experiments on pulsed irradiation of such crystals, for example, those of Lin and Spaepen (44) on iron.

F. Alloys

If the melt is alloyed, impurity redistribution, either by partitioning between melt and crystal or by local reordering without partition, may be a thermodynamic requirement for the advance of the crystallization front (8,45, 46,2). The growth speed must then fulfill the condition (2):

$$u = -\frac{D_i}{c_i} (\text{grad } c)_i \tag{10}$$

where

c_i = impurity concentration in the melt at the interface, i.
$(\text{grad } c)_i$ = concentration gradient in the melt at i.
D_i = melt diffusivity at i.

Under these conditions the upper limiting interface velocity is $u \sim D_i/\lambda$. In the labile regime where $D_i \sim 10^{-5}$ cm^{2}sec^{-1} and with $\lambda \sim 10^{-8}$ cm, this velocity is 10 meters/sec, which is about 2.5 orders of magnitude below that (3-5 kilometers/sec) of collision limited growth. Since impurity transport in the melt (and glass) is thermally activated, the interface movement should be quench suppressible *so long as impurity redistribution is thermodynamically required.* It follows that such a redistribution requirement is a necessary, though not sufficient condition, for metallic glass formation in the presence of crystal nuclei.

Partitionless growth at an impurity atom fraction x becomes possible when the interface temperature falls below that, $T_0(x)$, at which the free energies of melt and crystal of identical composition are equal. In this growth mode impurity may be trapped in the crystal at a higher chemical potential than it has in the melt (47). In many alloy systems $T_0(x)$ for one structure or another will always exceed 0°K, which means that any at x the alloy will be more stable in some *homogeneous* crystalline than in an amorphous solid form. In contrast, there are other systems where the $T_0(x)$ for both terminal structures go to 0°K, e.g. at x_0 and x_0' on nonintersecting courses so that at any composition between x_0 and x_0' the alloy at 0°K will be more stable as an amorphous solid than in any *homogeneous* crystalline form. Note that the thermodynamic conditions which, with the homogeneity constraint, stabilize amorphous forms will also be reflected by equilibrium liquidus temperatures, $T_\ell(x)$, exhibiting marked negative deviations from ideality and leading to abnormally deep eutectic temperatures (46). Boettinger (45) proposed that required partitioning in crystallization is a necessary condition for metal glass formation. However, certain metal-metalloid alloys [e.g. $Fe_{95}B_5$, Ni_3P] which clearly crystallize without partition do form and exist as glasses (48,49). This behavior shows that even if partitioning is not required, the need for local impurity reordering can enforce metal glass formation. When local impurity reordering is not demanded, partitionless crystallization in alloys may be, as in pure metals, only limited by collisions. It could then proceed without thermal activation and so be quench insuppressible.

When growth is restrained by the necessity of impurity redistribution, the possibilities for thermodynamic exposure and kinetic trapping of metastable states are greatly enlarged (2). Under these conditions the maximum undercooling at an advancing crystal-melt interface can be estimated roughly (2) by substituting D_i/λ for $fk_i\lambda$ in equation (7). If such an estimate is correct the interface temperature can be reduced to a level well below that at homogeneous nucleation onset with readily realizable thermal gradients. Then homogeneous nucleation may occur in advance of the interface, thereby producing a fine-grained equiaxed structure in the solidified alloy. If $T_r \gtrsim 0.5$ the thermal gradients realizable in RSP are generally sufficient to depress T_i to levels below T_g so that a metallic glass forms. If the homogeneous nucleation onset temperature is well above T_g there may form in this quench a large number of crystal nuclei which would regrow upon annealing to produce a fine equiaxed grain morphology.

V. Formation of Metastable Structures From Undercooled Liquids

G. Topological Metastability

The appearance of metastable crystals in undercooled liquids generally results from preferred nucleation, either homogeneous or heterogeneous. Obviously, a necessary condition for their formation is that the melt be undercooled to a temperature considerably below the metastable equilibrium points. Thus the formation of metastable phases in *pure* melts is much more likely when the melt is isolated, as when fluxed or in the form of suspended droplets, than when rapidly quenched on a crystalline substrate where the realizable undercooling is much smaller. Some examples of metastable crystal formation in deeply undercooled droplets are cited in what follows.

Several studies have shown (50,51), beginning with that of Cech (50), that the b.c.c. nucleates in preference to the f.c.c. phase in droplets of certain Fe-Ni alloys which were deeply undercooled to temperatures where the f.c.c. structure would have been thermodynamically preferred. Bosio *et al.* (14) found that simple metastable structures nucleate in preference to the less symmetric stable one in deeply undercooled Ga droplets. These observations suggest that the interfacial tension, σ, in the nucleation barrier, must be considerably lower for the metastable than for the stable phase against the melt. According to the negentropic model for σ this condition might be fulfilled if the entropy of fusion of the metastable phase was less than that of the stable phase. There appear to be insufficient thermal data to test this possibility for the Fe alloys and the metastable phases of Ga. However, it is generally found that the entropy of a metal is higher in the b.c.c. than in the f.c.c. structure and also in its more symmetric structures.

It is difficult to impose high cooling rates on suspended droplets. Consequently the undercooling of such droplets is unlikely to lead to glass formation unless T_{rg} is close to T_r^0. When T_{rg} falls well below T_r^0 the interfacial thermal gradients and undercoolings needed for glass formation are much more readily achieved, when impurity redistribution must attend crystal growth, by quenching thin molten films on solid substrates than by cooling droplet suspensions.

Recently it was reported (52), and fully confirmed by numerous investigators, that a phase with icosahedral point group symmetry but exhibiting a sharp diffraction pattern forms in preference to the more stable orthorhombic one in the rapid melt quenching of certain molten Aℓ alloys. Later, Poon *et al.* (53) found that such an icosahedral phase also forms in preference to the more stable one when a $Pd_{60}U_{20}Si_{20}$ glass is reheated. The appearance of the icosahedral phase seems to be due to preferential nucleation. Some unpublished observations of Shechtman (54) indicated an icosahedral nucleation frequency $> 10^{20}$ $cm^{-3}sec^{-1}$ within small rapidly quenched spheres of $A\ell_6Mn$, which is not far below the upper limiting nucleation rate permitted by classical nucleation theory. It might be supposed that the short-range order and entropy of the liquid and glassy phases of these alloys are closer to those of the icosahedral than of the stable phases. However, the thermal and structural data are still insufficient for testing this supposition.

H. Compositional Metastability

The thermodynamic constraints on supersaturated solution formation and solute trapping during solidification were described by Baker and Cahn (47). The actual extent of solute inclusion will be kinetically determined by the interplay of the interface transfer and impurity transport processes. This complex problem, including that of heat transport, has been surveyed and analyzed by several investigators (55-57). Complete solute trapping can occur when the interface temperature is depressed to a level below $T_0(x)$. Such trapping often is kinetically favored since the limiting interface speed may be 3-5 kilometers/sec when interface transfer controlled and only D_i/λ,

~ 10 meters/sec, when impurity transport controlled. The analyses generally indicate (55) that as u is increased by the imposition of steeper thermal gradients the impurity distribution ratio, c_i^s/c_i^ℓ, at the interface increases rather little from its equilibrium value until u approaches D_i/λ. With further increase of u c_i^s/c_i^ℓ rises rapidly but continuously to a level approaching 1, corresponding to complete trapping.

I. Morphological Metastability

Here consideration is limited to the formation of polycrystalline structures from undercooled liquids.

We have noted that with imposed thermal gradients of the normal magnitude ΔT_i is small during growth of metal crystals in their undercooled melts. Under these conditions when a single nucleus forms the melt quickly recalesces and further nucleation is thereby cut off. Thus the as solidified specimen would be a single crystal. This expectation is generally realized but if nucleation occurs at sufficiently deep undercooling, under some conditions, the as solidified grain structure is polycrystalline. In particular, Walker found that when bulk melts of Ni were seeded at various undercooling, ΔT_r, the as solidified structures were essentially monocrystalline (grain size ~ 2 cm) when $\Delta T_r \lesssim 0.10$ but equiaxed polycrystalline (grain size ~ 0.01 cm) when $\Delta T_r \gtrsim 0.10$. Similar transitions in grain structure with initial undercooling in droplets of Ag were reported by Powell (59), in Fe alloys by Kattamis *et al.* (60), and in Ge droplets, under a B_2O_3 flux, by Devaud and the author (16,61). The grain size transition in Ge (61) occurred at $\Delta T_r \gtrsim 0.24$; as ΔT_r at nucleation increased beyond 0.24 the as crystallized grain size became progressively finer, reaching the 10-20 μm range at $\Delta T_r \gtrsim 0.33$.

The origin of this morphological change is still unclear. The observations of Devaud and the author (61) suggested that in Ge it results from dendrite breakup. At large undercooling the primary dendrite structure should be quite fragile and, as Flemings (62) suggested, the convection currents in the melt might break it up by a "melt-back" mechanism. Alternatively, it has been suggested that the rapidly moving dendrite from the primary nucleus may perturb the surrounding melt sufficiently to trigger "dynamic" nucleation (58,63), possibly resulting from the opening and sudden collapse of microcavities. The ultimate grain refinement that can be achieved by increasing the undercooling beyond the transition level is limited by the grain coarsening which occurs as the metal cools.

We have noted that in the rapid solidification of pure liquid metal thin films on crystalline substrates, after crystallization starts from the substrate, the undercooling at the interface will generally be insufficient for further nucleation. All nucleation would have occurred heterogeneously on the substrate and a columnar structure would result. However, when growth must be thermally activated, the undercooling in advance of the interface can be high enough, following imposition of large thermal gradients, for copious homogeneous nucleation to occur with resultant fine grain sizes. Here the quench rate can be sufficient to largely suppress grain coarsening following solidification.

Acknowledgement: I did not know Professor Hume-Rothery personally but I take this occasion to acknowledge that, when the emphasis in my research was shifting from Physical Chemistry to Materials Science, I was especially inspired and instructed by his book on "The Metallic State."

This paper was prepared while I was Bernd T. Matthias Visiting Scholar at the Center for Materials Science (CMS), Los Alamos National Laboratory. I am especially grateful to Dr. Ricardo Schwarz of CMS for critically reviewing my manuscript and to the Laboratory for providing a stimulating environment for thought. My research at Harvard University in the subject areas of this paper was supported in part by grants from NASA NAS8-35416 and ONR N00014-85-K-0023.

References

1. D. Turnbull, Met. Trans. 12A, 695 (1981).
2. D. Turnbull, "Proc. 2nd Israel Mats. Eng. Conf." (ed. A. Grill and S.I. Rokhlin), pp. 1-10, Ben Gurion University, Beersheva (1984).
3. P.H. Bucksbaum and J. Bokor, Phys. Rev. Lett. 53, 182 (1984).
4. D. Turnbull, Mats. Res. Soc. Symp. Proc. (eds. H. Kurz, G.L. Olson and J.M. Poate), in press.
5. C.H. Bennett, J. Appl. Phys. 43, 2727 (1972).
6. J.D. Bernal, Nature 185, 68 (1960).
7. (a) W. Buckel and R. Hilsch, Z. Phys. 138, 109 (1954);
   (b) R. Hilsch, "Non-Cryst. Solids," (ed. V.C. Fréchette), 348, Wiley, N.Y. (1960).
8. (a) D. Turnbull, "Solidification", pp. 1-22, Am. Soc. Metals, Metals Park, Ohio (1970);
   (b) J. Physique 35-C-4, pp. 1-9 (1974).
9. D. Turnbull and B.G. Bagley, "Treatise on Solid State Chemistry," 5 513, Plenum, N.Y. (1975).
10. F. Spaepen and D. Turnbull, "Rapidly Quenched Metals," 2nd Int. Conf. (ed. N.J. Grant and B.C. Giessen), pp. 205-229, MIT Press, Cambridge, MA (1976).

11. D. Turnbull and R.E. Cech, J. Appl. Phys. 21, 804 (1950).

12. D. Turnbull, J. Chem. Phys. 20, 411 (1952).

13. B. Vonnegut, J. Colloid Sci. 3, 563 (1948).

14. (a) L. Bosio, A. Defrain, M. Erney and E. Epelboin, Mem. Sci. Revue Metall. 58, 43 (1961);
(b) L. Bosio and A. Defrain, J. Chim. Phys. 61, 859 (1964);
(c) L. Bosio, A. Defrain and M. Dupont, Ibid. 68, 542 (1971).

15. J.H. Perepezko and J.S. Paik, "Rapidly Solidified Amorphous and Crystalline Alloys" (eds. B.H. Kear, B.C. Giessen and M. Cohen), 49, North-Holland, Amsterdam (1982).

16. G. Devaud and D. Turnbull, Mats. Res. Soc. Symp. Proc. (eds. G.S. Cargill, III, F. Spaepen and K.N. Tu), in press (1986).

17. D.E. Ovsienko, V.P. Maslov and V.P. Kostyuchenko, Kristallografiya 16, 405 (1971); see also A.A. Chernov, Ann. Revs. Mats. Sci. 3, 397 (1973).

18. D.W. Gomersall, S.Y. Shiraishi and R.G. Ward, J. Austral. Inst. Metals 10, 220 (1965).

19. D. Turnbull, Progress in Materials Science (eds. J.W. Christian, P. Haasen and T.B. Massalski), Chalmers Anniversary Volume, 269, Pergamon (1981).

20. F. Spaepen and D. Turnbull, Scripta Met. 13, 149 (1979).

21. D. Turnbull and J.C. Fisher, J. Chem. Phys. 17, 71 (1949).

22. R. Legras, J.P. Mercier and E. Nield, Nature 304, 432 (1983).

23. G.M. Pound, "Liquid Metals and Solidification," pp. 87-105, Am. Soc. Metals, Metals Park, Ohio (1958).

24. D. Turnbull and R.L. Cormia, J. Chem. Phys. 34, 820 (1961).

25. P. Monge, Colloque International sur les propriétes des États Metastable, Pau-France, Resumés, Univ. of Bordeaux, 41 (1968).

26. J.H. Perepezko and D.H. Rasmussen, Met. Trans. 9A, 1490 (1978).

27. R.H. Busey and W.F. Giague, J. Phys. Chem. 75, 8061 (1953).

28. Y. Miyazawa and G.M. Pound, J. Cryst. Growth 23, 45 (1974).

29. D. Turnbull, "Physics of Non-Cryst. Solids" (ed. J.W. Prins), 41, North-Holland, Amsterdam (1964).

30. F. Spaepen, Acta Met. 23, 729 (1975).

31. F.C. Frank, Proc. Roy. Soc. 215A, 43 (1952).

32. M.P. Hoare and P. Pal, Adv. Phys. 24, 643 (1975).

33. J.L. Finney, Proc. Roy. Soc. 319A, 497 (1970).

34. F. Spaepen and R.B. Meyer, Scripta Met. 10, 257 (1976).

35. J.H. Perepezko, Proc. 2nd Int. Conf. Rapid Solidication Processing (eds. R. Mehrabian, B.H. Kear and M. Cohen), 56 Claitore, Baton Rouge, LA (1980).

36. C.V. Thompson and F. Spaepen, (a) Mats. Res. Soc. Proc. (ed. G.E. Rindone) 9, 603 (1982); (b) Acta Met. 31, 2021 (1983).

37. D. Turnbull, Contemp. Phys. 10, 473 (1969).

38. A.J. Drehman and D. Turnbull, Scripta Met. 15, 543 (1981).

39. A.J. Drehman, A.L. Greer and D. Turnbull, Appl. Phys. Lett. 41, 716 (1982).

40. H.W. Kui, A.L. Greer and D. Turnbull, Appl. Phys. Lett. 45, 615 (1984).

41. A.L. Greer, Acta Met. 30, 171 (1982).

42. H.W. Kui and D. Turnbull, Appl. Phys. Lett. 47, 796 (1985).

43. F. Spaepen and D. Turnbull, (a) "Laser Annealing of Semiconductors" (eds. J.M. Poate and J.W. Mayer), pp. 15-42, Academic Press, N.Y. (1982). (b) Ann. Rev. Phys. Chem. 35, pp. 241-263 (1984).

44. C.J. Lin and F. Spaepen, "Chemistry and Physics of Rapidly Solidified Materials" (ed. B.J. Berkowitz and R.O. Scattergood), 273, T.M.S.-AIME N.Y. (1983).

45. W.J. Boettinger, "Proc. 4th Int. Conf. on Rapidly Quenched Metals" (ed. T. Masumoto and K. Suzuki) 1, pp. 99-102, Japan Inst. Metals, Sendai (1982).

46. D. Turnbull, "Proc. EPRI-Acta Met. Workshop on Amorphous Metals and Semiconductors" (ed. R. Jaffee), in press (1986).

47. J.C. Baker and J.W. Cahn, (a) Acta Met. 17, 575 (1969); (b) "Solidification," pp. 23-58, Am. Soc. Metals, Metals Park, Ohio (1970).

48. C.J. Lin and F. Spaepen, Appl. Phys. Lett. 41, 716 (1982).

49. B.G. Bagley and D. Turnbull, Acta Met. 18, 857 (1970); see Figs. 5 and 7.

50. R.E. Cech, Trans. AIME 206, 585 (1956).

51. T.F. Kelly, M. Cohen and J.B. van der Sande, Met. Trans. 15A, 819 (1984).

52. D. Shechtman, I. Blech, D. Gratias, and J.W. Cahn, Phys. Rev. Letters 53, 1951 (1984).

53. S.J. Poon, A.J. Drehman and K.R. Lawless, Phys. Rev. Letters 55, 2324 (1985).

54. D. Shechtman, private communication.

55. M.J. Aziz (a) J. Appl. Phys. 53, 1158 (1982); (b) Appl. Phys. Letters 43, 552 (1985).

56. J.W. Cahn, S.R. Coriell and W.J. Boettinger, "Laser and Electron Beam Processing of Materials" (ed. C.W. White and P.S. Peercy), pp. 89-103, Academic Press, N.Y. (1980).

57. K.A. Jackson, G.H. Gilmer and H.J. Leamy, "Laser and Electron Beam Processing of Materials" (ed. C.W. White and P.S. Peercy), pp. 104-110, Academic Press, N.Y. (1980).

58. J.L. Walker, (a) "Liquid Metals and Solidification," pp. 319-336, Am. Soc. Metals, Metals Park, Ohio (1958).
(b) Phys. Chem of Process Metallurgy, Pt. 2, 845, Interscience, N.Y. (1961).
(c) B. Chalmers, "Principles of Solidification," pp. 114-15, 122-24, Wiley, N.Y. (1964).

59. G.L.F. Powell, J. Australian Inst. Metals 10, 223 (1965).

60. (a) T.Z. Kattamis and M.C. Flemings, Trans. Met. Soc. A.I.M.E. 236 1523 (1966).
(b) M.C. Flemings and Y. Shiohora, Mats. Sci. and Eng. 65, 157 (1984).

61. G. Devaud and D. Turnbull, to be published.

62. M.C. Flemings, private communication.

63. B. Chalmers, "Liquids: Structure, Properties, Solid Interactions" (ed. T.J. Hughel), pp. 308-325, Elsevier, Amsterdam (1965).

EFFECTS OF SUPERCOOLING ON SOLUTE DISTRIBUTION AND MICROSTRUCTURES

A. MUNITZ* AND G. J. ABBASCHIAN**

* Visiting Scientist from Nuclear Research Center, Negev, Beer Sheva, Israel
** Professor

Department of Materials Science and Engineering
University of Florida
Gainesville, Florida 32611

After a brief literature review on the effects of supercooling on the solidification microstructure, solute distribution and grain size of alloys, our recent experimental results of solute distribution profiles in highly supercooled binary alloys are discussed. A prominent effect of supercooling below the solidus is to cause solute-rich coring due to a partitionless solidification during recalescence. An X-ray microanalysis technique was used to determine the fraction of the partitionless solidification as a function of supercooling. The fraction was found to be considerably less than that predicted based on adiabatic-uniform temperature recalescence. The results are discussed based on a non-isothermal recalescence model which also includes interfacial kinetics and curvature effects.

Introduction

In the past few years there has been a great deal of interest in modifying alloy microstructure and segregation profiles by various rapid solidification processes such as powder atomization, melt spinning, and laser surface melting. These processes are generally based upon increasing the rate of heat removal from the solidifying material by improving the heat transfer coefficient to the surroundings and/or by increasing the surface to volume ratio of the material being solidified. Because of heat transfer requirements, rapidly solidified materials generally have limited thickness, as in the case of melt spinning or laser surface melting, or are in the form of fine particulates. On the other hand, rapid solidification in the bulk form can be achieved by supercooling the liquid prior to nucleation, whereby a substantial portion of the heat of fusion is absorbed by the supercooled liquid itself. For a supercooled liquid, the solidification process occurs non-isothermally and with high interfacial velocities (at least at the outset of solidification). These conditions can result in microsegregation profiles which are substantially different from those of the conventionally solidified materials. For the conventionally solidified materials, the center of the dendrites generally contain lower amounts of solute (if $k < 1$). However, this may not be the case for the supercooled structures. Since the initial solidification may take place without partitioning (1,2). In addition, the supercooling may affect the grain size (3-6), change the solidification morphology from dendritic to non-dendritic (6-8), or cause formation of alternate phases (9).

In the first part of this paper, literature illustrating the effect of supercooling on grain size, microstructure, and solute distribution in the solid will be reviewed. In the second part, the results of our current experiments on the microstructure and solute distribution profiles in highly supercooled Fe-Ni alloys will be presented. Finally, the results will be discussed on the basis of the current theories of supercooled melt solidification.

The Effect of Supercooling on Grain Size and Microstructure

Fine grained structures generally have improved low temperature strength, reduced microsegregation, reduced anisotropy, improved heat treatment response and improved workability. One method of controlling the grain size is by melt supercooling prior to solidification. The results of several investigations on the effect of supercooling on grain size of various materials are summarized in Figure 1.

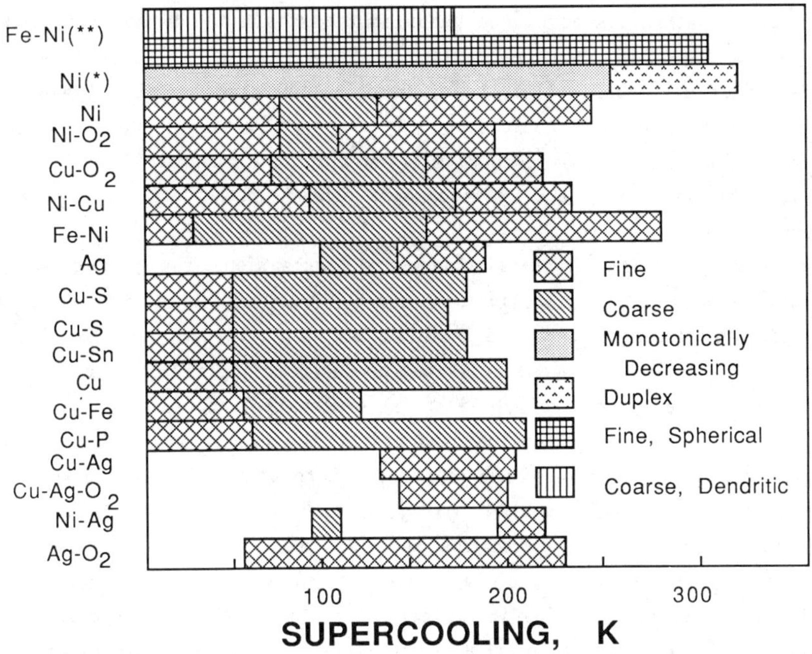

Fig. 1 - Grain size of various metals and alloys as a function of supercooling for nucleation. Maximum recorded supercooling by each investigator is shown by the length of each bar (From (4), *(10), **(6)).

The investigations, with the exception of Fe-Ni (6) and Ni (10) shown on the top of the chart, consisted of supercooling of the liquid surrounded by a layer of molten slag or fused silica. Large supercoolings have been achieved with this arrangement, since the heterogenous nucleating effect of the crucible walls is eliminated (11). On the other hand, for Fe-Ni and Ni, supercooling was achieved by melting and solidifying the samples in a containerless manner using an electromagnetic levitation apparatus (6,10). In most metals and alloys studied, a fine grained structure is observed at limited supercooling, followed by a range where coarse grains are formed. The latter may or may not be followed by another range of fine grains at larger supercoolings. The results of Ag-O_2 system show fine grains in the intermediate supercooling range (12). The authors attribute the grain refinement to the recrystallization following solidification. However, the refinement may also be due to the effect of gas evolution during solidification.

Several mechanisms have been proposed to explain the behavior of grain size as a function of supercooling. These include: (1) Selective activation of different nucleants and denucleation process, (2) Cavitation shrinkage causing dynamic nucleation, (3) Recrystallization due to stresses and/or dislocations, (4) Dendrite remelting and coarsening, and (5) Dendrite fragmentation and coarsening. These mechanisms are briefly discussed below.

According to the classical theory of nucleation, the radius of the critical nucleus and the activation energy for nucleation decrease with increasing supercooling. As a result, smaller clusters and non-potent heterogeneous nucleants become activated as supercooling is increased. While the former should theoretically lead to an exponential decrease in grain size as a function of the supercooling, the latter may result in an abrupt grain refinement. This is because each type of heterogeneous nucleant presents a characteristic nucleation barrier which depends, among several factors, on the lattice disregistry between the nucleant and the nucleus. Thus, with increased supercooling, additional abundant nucleants may become active (5,13). If abundant nucleating sites with low lattice disregistry are available, fine grains may be produced at low supercoolings. Only when these potent sites are absent (denucleation), the supercooling can be pushed to a point where an other abundant source of nucleation becomes active, again resulting in fine grains.

Walker (11) studied the effect of supercooling on the grain size of pure nickel, and reported a drastic reduction in the grain size at a supercooling in the range 150 to 175 K. In these experiments, the melt was isolated from the crucible walls by a layer of fused silica glass, thereby eliminating the heterogeneous nucleation effect of the container walls. Walker also measured the amplitude of sound waves produced during solidification, and found a good correlation between the maximum in the signal and the supercooling corresponding to the abrupt change in the nickel grain size, Figure 2. It was proposed that the grain refinement is caused by a cavitation mechanism, resulting from a rapid flow required to feed the solidification shrinkage. When the cavities collapse, the pressure on the remaining liquid causes enhanced nucleation. Calculations show that this is a possible mechanism for enhanced grain refinement (14).

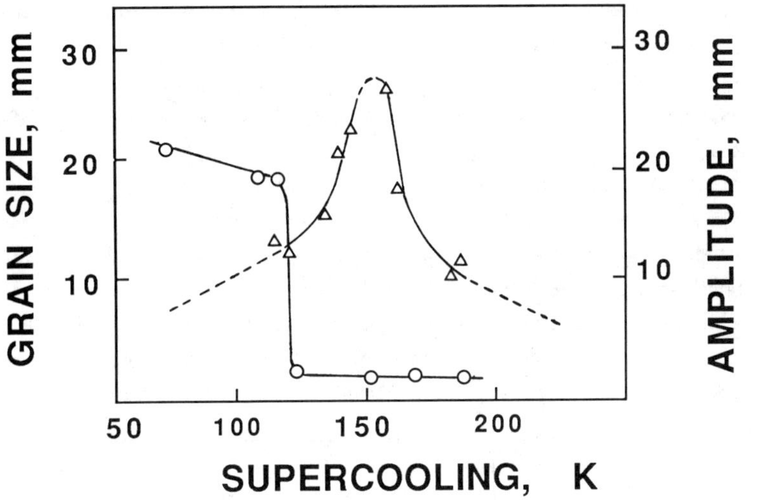

Fig. 2 - Grain diameter and amplitude of the initial pressure pulse as a function of supercooling (11).

A number of other studies have also shown an abrupt reduction in grain size at a critical supercooling (7,8,15-17). The value of the supercooling is shown to be about 175 K for a variety of iron and nickel based alloys. In addition to the abrupt grain refinement for Fe-25 w/o Ni, there is a decided change in the subgrain morphology (7,8). At supercooling less than 175 K, the structure is usually dendritic within the grains, while at supercoolings larger than 175 K, the microstructure exhibits the so called "spherical morphology". It was found that at supercoolings below the critical values, only the dendrite arm spacing, not the grain size, was refined. Above the critical supercooling, however, each grain consisted of a single spherical element, and any subsequent reduction of the element size resulted in grain size reduction. Remelting of secondary dendrite arms and coarsening have been proposed as the mechanism for the changes in the grain size and microstructure as a function of supercooling. Based on this mechanism, the partial remelting and disintegration of dendrites occur during recalescence of alloys supercooled beyond the critical value because the local melting temperature of the root of dendrites is lower than that of the tip. Furthermore, because of the high surface to volume ratio of the spherical elements, extensive coarsening takes place during solidification, which eventually controls the final grain size. As will be shown later, the room temperature grain size might not necessarily be the same as that at the end of solidification because of solid state grain growth. This factor must be taken into consideration in relating the grain size to the supercooling.

Another explanation for the abrupt grain reduction for pure materials as well as for alloys is based on recrystallization in the solid state (12,18). The central idea is that internal stresses accumulated during solidification from a supercooled state can reach a critical level to cause plastic flow and recrystallization. The higher the supercooling, the faster is the solidification rate, and consequently the greater is the accumulation of internal stresses. Therefore, the point at which abrupt grain reduction takes place is where the accumulated internal stresses reach a sufficiently high level to cause plastic deformation and subsequent recrystallization. Measurements of the microstress as a function of supercooling (17), Figure 3, tend to confirm the above mentioned proposition, though no explanation is given for the decrease of the stress level beyond the critical supercooling.

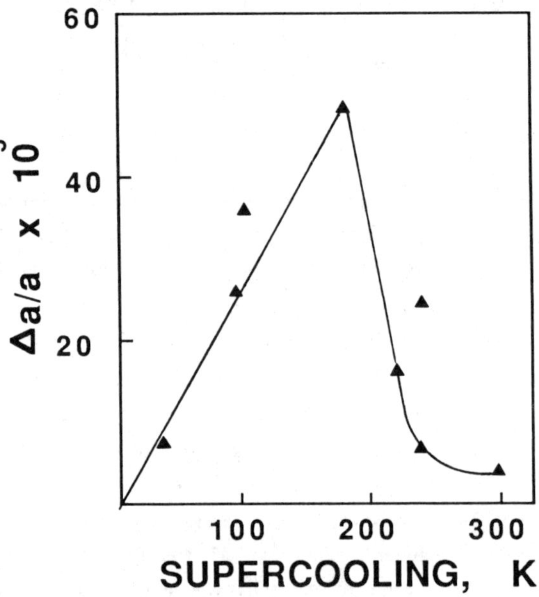

Fig. 3 - Microstress as a function of supercooling in Ni-3.7 w/o Si (17).

In contrast with the above investigations, recent experiments (10) using electromagnetic levitation of high purity Ni (99.99 w/o) show that the grain size of Ni decreases monotonically with increasing supercooling up to 340 K, without an abrupt grain refinement around 150 K. The relationship between grain size and the degree of supercooling is shown in Figure 4.

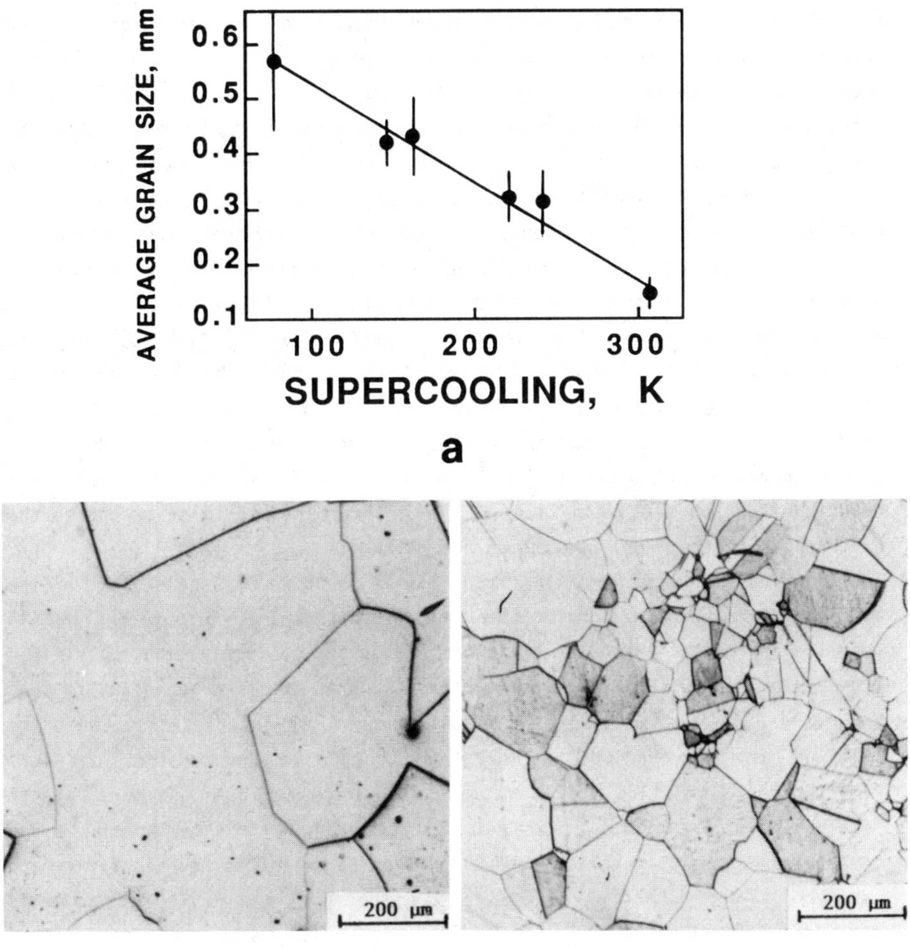

Fig. 4 - (a) The average grain size as a function of supercooling for pure nickel (b) and (c) optical micrographs demonstrating the microstructure after supercooling of 100 and 305 K, respectively. Samples subsequently quenched in water (10).

Examination of the quenched samples at supercoolings up to 340 K shows the appearance of twins in samples quenched from supercoolings in excess of 225 K. Concurrent with the formation of twins at this range, is the appearance of fine grains about one tenth the size of the matrix grains. No abrupt grain size reduction was observed around 175 K. The difference between these results and those of the previous experiments (11) may be attributed in part to the differences in experimental techniques. All the experiments reporting the abrupt grain size reduction were conducted in alumina crucibles covered by a protective glass slag. The levitation experiments, on the other hand, are containerless where the supercooled liquid is solidified by quenching in water. It is possible that the stresses arising from the interaction of the solidifying metal with the container wall, compounded with the inherent solidification stresses, may be responsible for the abrupt grain size transitions. By eliminating the effects of the container walls, the levitation results may be more indicative of the inherent solidification characteristics of supercooled nickel. It should be noted that since fluid flow during solidification can also have a drastic effect on the grain size (6), the differences may also be due to the differences in the magnitude and nature of the flow.

Another, and probably the most likely mechanism responsible for the observed dependence of grain size and microstructure on supercooling is dendrite fragmentation caused by fluid flow during solidification. The fluid flow can originate from different sources, such as the flow necessary to feed the solidification shrinkage, or resulting from the interaction between the melt and the surroundings. Also, the higher the supercooling, the more vigorous the fluid flow due to faster interface velocities and more intense recalescence. The grain size reduction, according to this mechanism, is due to a change in solidification morphology from dendritic to non-dendritic, and as such is not directly related to the supercooling. This mechanism is discussed in more detail later in conjunction with the results of supercooling of Fe-Ni binary alloys.

Experimental

Iron-nickel samples were melted in an electromagnetic levitation apparatus, supercooled, and quenched, as described in detail elsewhere (2,6,10). Briefly, high purity iron (99.98 w/o) and nickel (99.99 w/o) were weighed according to the desired composition (25 or 33.3 w/o Ni) and arc melted in an argon atmosphere. The arc melted specimens (weighing approximately 1.5 g.) were placed in the levitation melting apparatus through which high purity argon, helium, or helium + 2% hydrogen was flowing. The specimens were melted and superheated to about 300 K above the liquidus. The molten samples were cooled by increasing the gas flow rate until nucleation and recalescence was observed after which the sample was remelted. The cooling and heating cycle was repeated several times to record the maximum supercooling attainable with the sample. When the liquid attained a desired supercooling, the levitation power was cut off, allowing the supercooled sample to fall through a diaphragm into a quenching medium which was located beneath the levitation chamber. The quenching medium was either water at room temperature, or molten lead held at approximately 650 K. Some of the samples were splat cooled in a splat-cooling apparatus using copper plattens.

The specimens were sliced and prepared for metallographic and scanning electron microscopic examination. For the former, the samples were electropolished and heavily etched (more than 20 seconds) with 6% Nital, while for the latter, they were very lightly etched (less than 5 seconds) in order to reveal the dendrite element boundaries while keeping geometrical effects during compositional analysis to a minimum. X-ray microanalysis was performed in a scanning electron microscope (SEM) with EDAX capabilities at 25 kV. Ni and Fe compositional line profiles were taken across the dendrite element width at 5 to 15 locations within the sample. Each line profile consisted of 20 points, 1 μm apart; the counts at each point were taken for 200 seconds. The raw data were corrected with a standard ZAF (atomic number-Z, absorption-A, and fluorescence-F, including background as well as dead time corrections) computer program (19). With this procedure the expected relative compositional resolution is approximately $\pm$ 0.1 w/o.

Results

Microstructural Analysis

The microstructure of Fe-Ni alloys as a function of supercooling can be classified in three major categories: dendritic, spherical element, and mixed morphology, as shown in Figure 5. At supercoolings less that 175 K, all three morphologies are frequently observed, while at supercoolings exceeding this value, the spherical morphology tends to predominate. In addition, other microstructural morphologies are also observed at supercoolings higher than 175 K. An example is shown in Figure 6 for a sample which was supercooled 220 K and quenched in molten lead. In many parts of this specimen a "rosette" type structure can be observed, having a radiating fan appearance with elongated cells emanating from central hubs (20). The light particles within the cells shown in Figure 6c represent partitionless solidification of the alloy during recalescence, as discussed later.

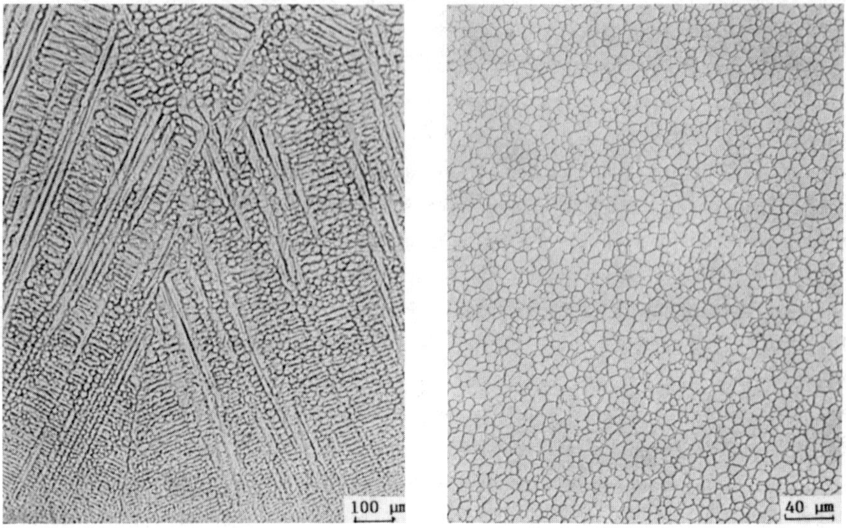

a

b

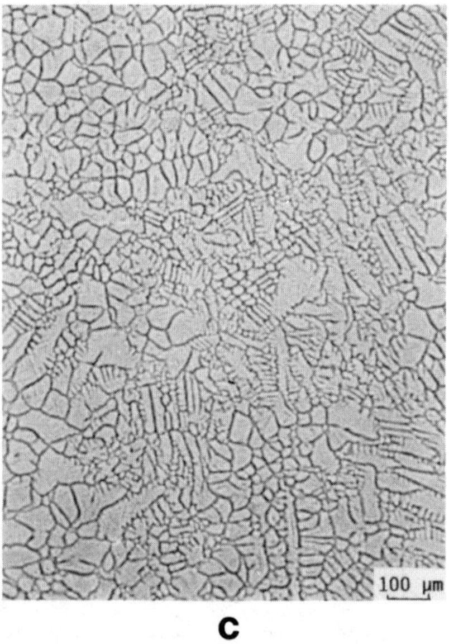

c

Fig. 5 - Optical micrographs demonstrating three types of morphologies in Fe-25 w/o Ni alloy quenched in water: (a) dendritic, (b) spherical, and (c) mixed (2,6)

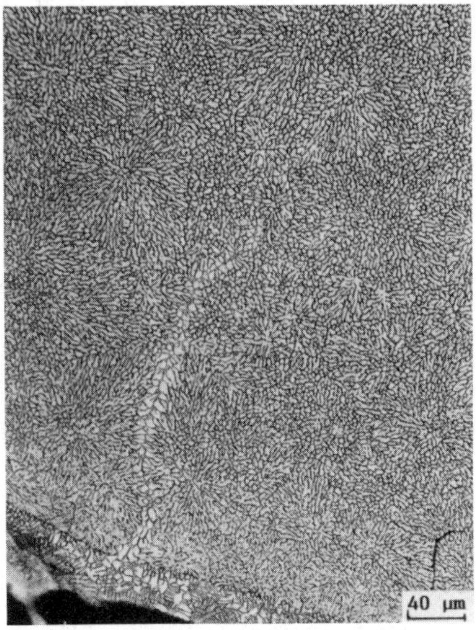

a

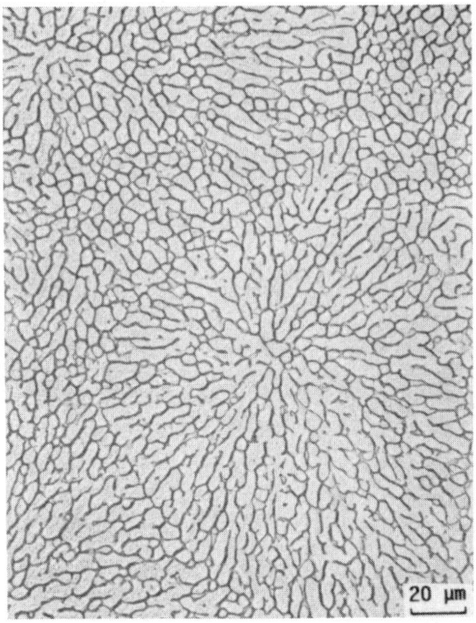

b

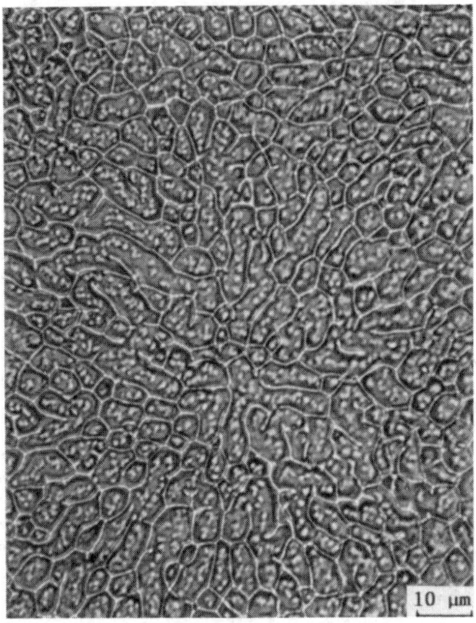

c

Fig. 6 - "Rosette" type microstructure of a sample supercooled 220 K and quenched in molten lead.

The effect of supercooling on the dendrite arm and spherical cell spacings is shown in Figure 7.

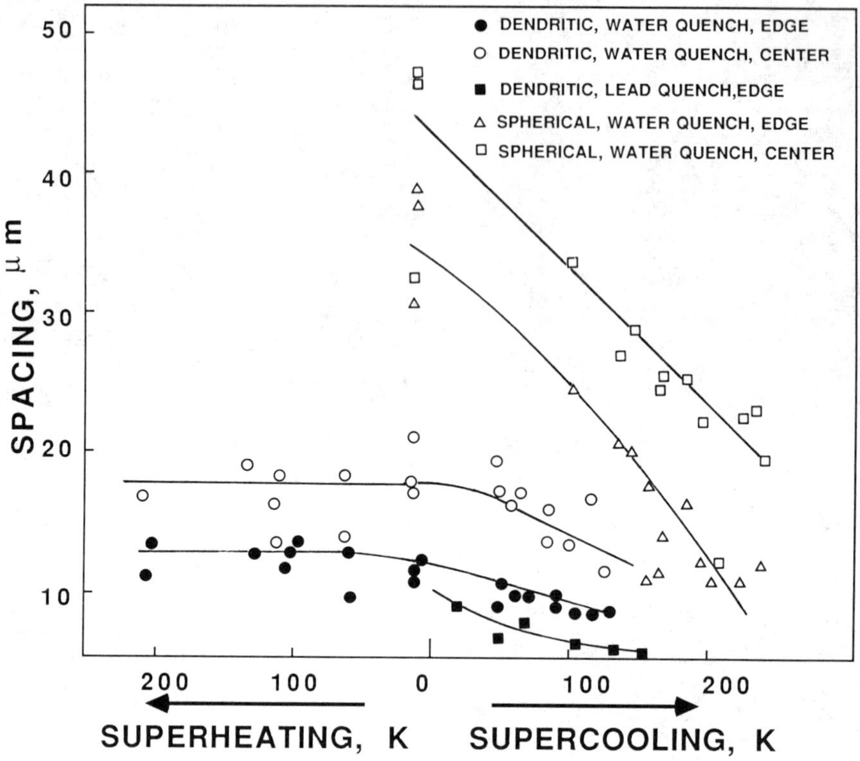

Fig. 7 - The average spacing of spherical elements and dendrite arms as a function of quenching temperature.

The figure also shows dendrite arm spacings for liquid samples which were quenched with up to 200 K superheating above the liquidus. As can be seen, the spacings depend on the supercooling, the quenching medium, and the position within the samples. The superheat does not seem to have an appreciable effect on the spacing. The effect of supercooling on the spacing is very pronounced for the spherical elements. The edge of the samples generally exhibited a finer spacing than that in the center of the samples. Calculations based on the DAS versus cooling rate equations show that for a water quenched sample with 150 K supercooling, the cooling rate at the edge and center of the samples are about 2040 and 480 K/s, respectively (6). For samples supercooled by 100 K, the cooling rates are respectively 1400 and 330 K/s. Furthermore, at any given supercooling, the spherical elements are much coarser than the dendrites. This indicates enhanced coarsening of the spherical elements during the solidification, possibly because of the three dimensional mass transport and stirring in the mushy zone.

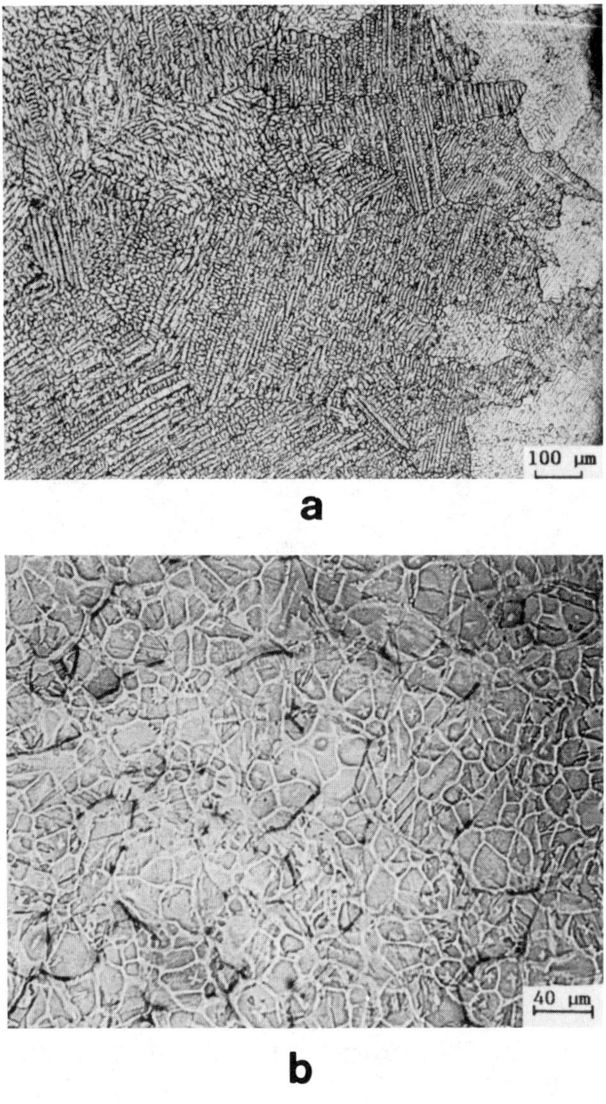

Fig. 8 - Optical micrographs of (a) dendritic and (b) spherical element illustrating the grain and subgrain structure of Fe-25 w/o Ni alloys (6).

The grain size as well as subgrain microstructures of the as-quenched samples are shown in Figure 8. For the dendritic morphology, the grains have jagged and serrated boundaries, as compared with smooth and slightly curved boundaries for the spherical morphology. Furthermore, for the latter morphology each grain consists of many spherical elements, indicating that considerable grain growth has taken place. This is evidenced in Figure 8b, where spherical elements belong to two neighboring grains. This clearly indicates that the final grain size is not necessarily the same as that formed just at the end of local solidification. On the other hand, there is no evidence of appreciable grain growth in samples with dendritic morphology. The average grain size as a function of supercooling and superheating for both morphologies are shown in Figure 9.

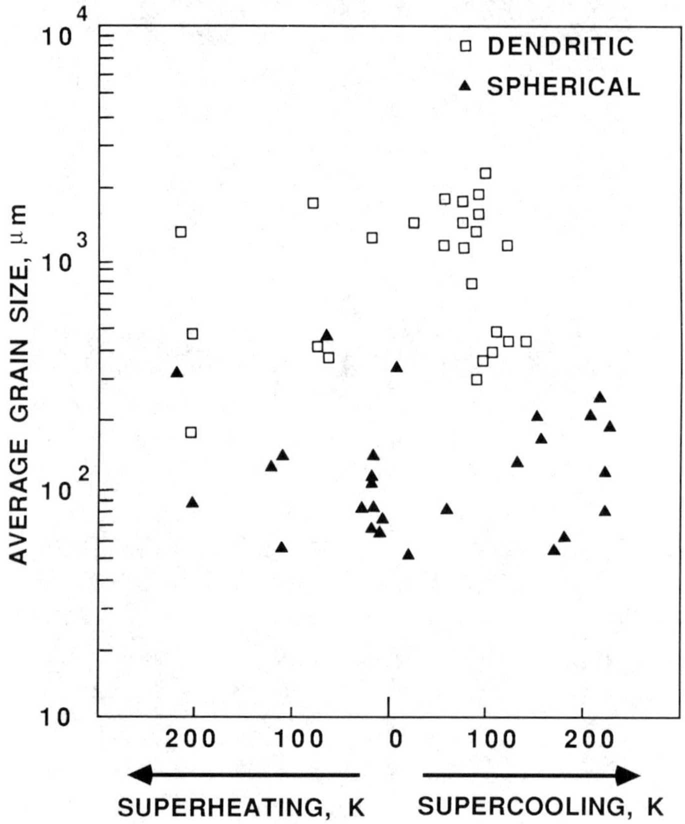

Fig. 9 - Average grain size as a function of temperature for dendritic and spherical microstructures (6).

The average grains associated with the dendritic morphology are approximately an order of magnitude larger than those with the spherical subgrain structure. However, within each morphology, the supercooling appears to have no appreciable effect on the grain size.

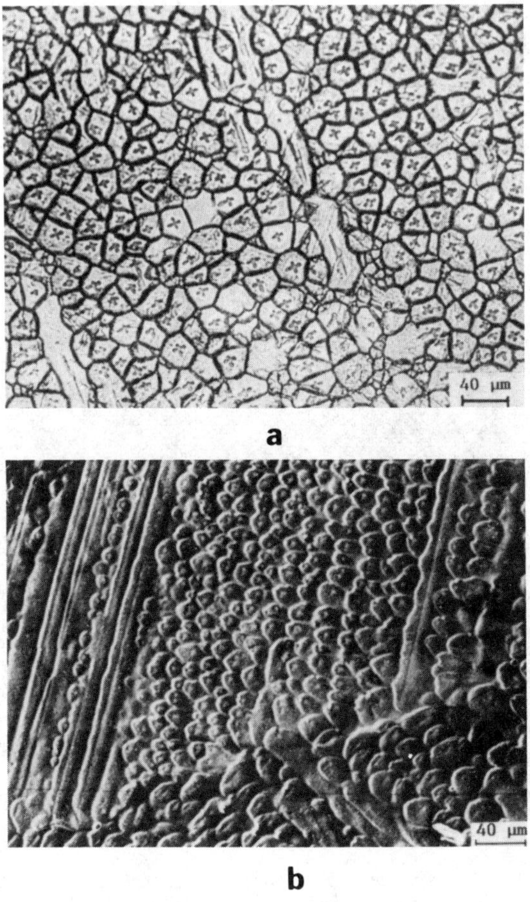

Fig. 10 - (a) Spherical and (b) dendritic morphologies showing solute-rich crosses seen as dark regions in the center of each element, surrounded by a solute-poor region which gradually darkens towards the edge. (a) Fe-25 w/o Ni quenched in water after 20 K supercooling. (b) Fe-33.3 w/o Ni quenched in water after 130 K supercooling.

Compositional Analysis

Further metallographic examination using longer etching times (from 30 to 200 seconds) reveals qualitatively the segregation profile within each dendrite arm or spherical element. As shown in Figure 10a, there is a cross shaped region in the center, surrounded by a lightly etched region which gradually darkens towards the circumference of the spherical element. As we shall see later the crosses are solute (Ni) rich area formed as a result of partitionless solidification associated with a supercooled melt, the surrounding solute-poor region marks the beginning of partitioning. The cell boundaries are very heavily etched, indicating high solute concentration. The back scattered SEM image in Figure 10b shows similar solute-rich cores in the center of dendrites. Figure 11 shows a metallographic cross-section of Fe-25 w/o Ni supercooled to 100 K and splat cooled during recalescence.

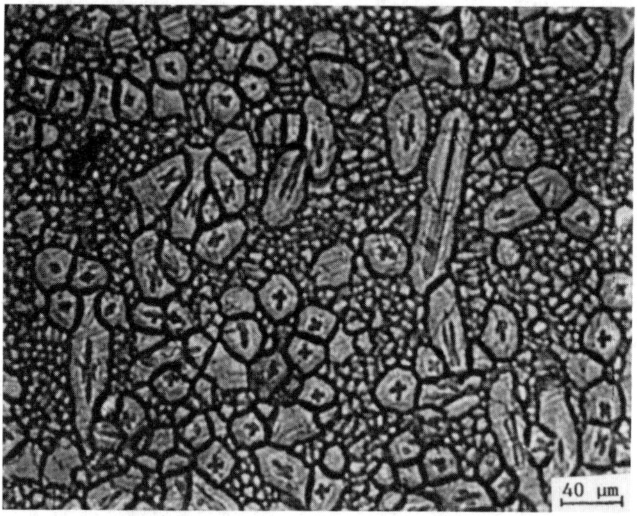

Fig. 11 - Optical micrograph of Fe-25 w/o Ni splat cooled after a supercooling of 130 K prior to nucleation.

The fraction of solid which formed during recalescence of the supercooled melt shows the same microstructural characteristics discussed before, i.e., solute-rich core in the center surrounded by a solute depleted zone with increasing solute content towards the circumference. The fraction of liquid remaining after recalescence solidified rapidly in the splat cooling apparatus, resulting in the surrounding refined microstructure.

The Ni concentration profile across a spherical element for a 33.3 w/o Ni sample supercooled 60 K and quenched in water is shown in Figure 12.

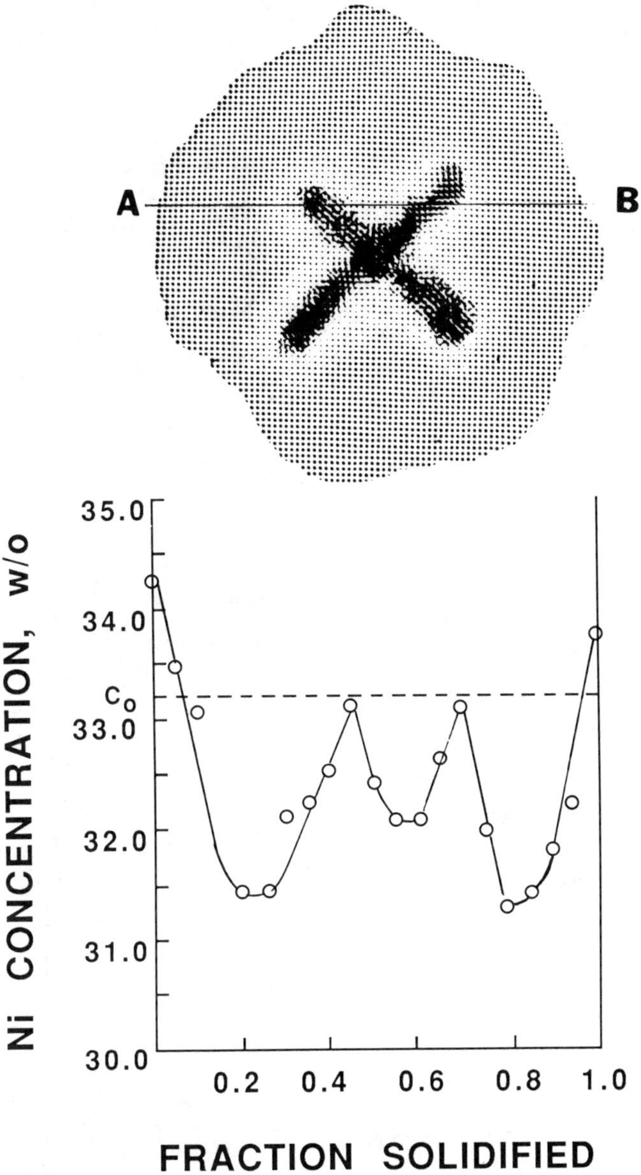

Fig. 12 - Compositional line profile of Ni concentration across a dendrite arm for a sample supercooled 60 K and quenched in water.

The profile is along the line AB shown in the schematic illustration of the element in the upper portion of this figure. The line profiles for samples solidified with 0, 30, 60, 90, and 130 K supercoolings are compared in Figure 13.

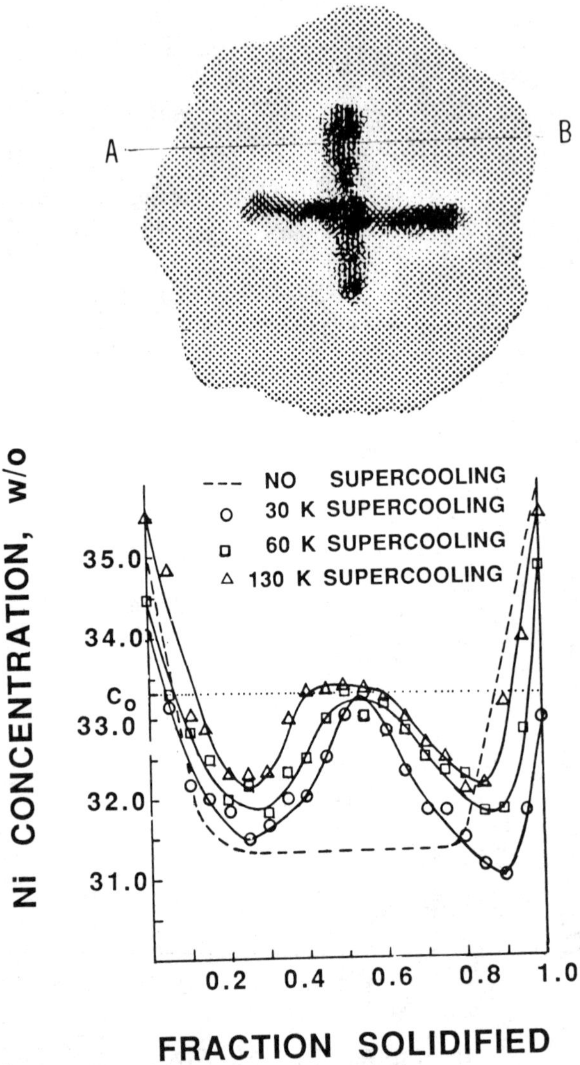

Fig. 13 - Compositional line profiles of Ni concentration along line Ab for four samples with 0, 30, 60, and 130 K supercoolings.

For the sample which solidified with no supercooling, the minimum in concentration (-31.3 w/o Ni) occurs at the central part of the element, and occupies about 2/3 of the concentration line profile. Beyond this region, the concentration is seen to increase monotonically toward the periphery. On the other hand, for supercooled samples, a central region with a composition equal to that of the original alloys (C_o) was detected. This region is followed by a region where the concentration decreases to a minimum, then gradually increases towards the periphery. The width of the central core region (cross thickness) increases with increasing supercooling, as summarized in Table I. The transition from solute-rich at the core to the minimum concentration is relatively sharp for the sample with 130 K supercooling, but becomes shallower as supercooling is decreased. The minimum solute concentration also increases with supercooling, e.g. for the sample solidified with 30 K supercooling, the composition is very close to the minimum in the sample solidified with no supercooling.

Table I: A comparison between the experimental and theoretical fraction of solid with C_o composition.

Supercooling K	Thickness of Cross, μm	Volume Fraction Solidified Segregationless,%		
		Experimental	Theoretical Based on T_s	Theoretical Based on T_o
30	1.2	6 $\pm$ 2	10	11
60	2.2	11 $\pm$ 2	16	18
90	3.4	17 $\pm$ 2	23	24
130	4.2	21 $\pm$ 2	32	34

Note: T_s denotes the solidus temperature and T_o denotes the mid-rib temperature at Composition C_o

Discussion

The microstructural analysis of levitation melted and supercooled Fe-Ni samples show that at supercoolings less than 175 K, the solidification morphology can be spherical, dendritic, or a mixture of the two. At supercoolings larger than about 175 K, the microstructure predominantly consists of the spherical morphology, although rosette type morphology is also observed. The results also show that the average grain size of Fe-Ni samples solidified without a container is almost independent of the amount of superheating or supercooling up to 270 K (as long as the subgrain morphology remains the same), as shown in Figure 9. The grain size, however, is strongly dependent on the solidification morphology; i.e. dendritic vs. spherical. The grains associated with the latter morphology are approximately an order of magnitude finer than those of the former morphology. The previously reported (8) abrupt grain size reduction at 175 K, may be explained on the basis of a change in the solidification morphology from dendritic to spherical.

The spherical morphology is believed to be mainly the result of dendrite fragmentation during the initial stages of solidification. The non-dendritic structure could result from (1) remelting of dendrites during recalescence above T_s, (2) coarsening and remelting at the dendrite roots, and/or (3) dendrite fragmentation caused by fluid flow during solidification. The extent of fragmentation by the first two sources depends on the supercooling and cooling rate. The fragmentation by fluid flow, on the other hand, will depend on the thermal conditions as well as the experimental set up used. In the levitation experiments, the flow can be caused by thermal or solutal convection, solidification shrinkage, drag force by the gas stream, and/or the interaction of the metal with the quenching medium.

The levitation experiments show both spherical and dendritic morphologies in samples superheated or supercooled less than 175 K. This is in contrast with the results of the previous investigators who noted only dendritic morphology below this critical supercooling. The difference is believed to be due to the differences in the nature and magnitude of the fluid flow between the two experimental set ups. At supercoolings larger than 175 K, the flow driven by external sources (interaction with the surroundings) is likely to be of lesser importance compared with the flow driven by internal sources (local thermal and solutal convection, and solidification shrinkage), such that fragmentation takes place regardless of the experimental conditions. Many other examples are given in the literature where externally induced fluid flow has been utilized to break up the dendrites and refine the grain size (21). Non-dendritic morphology in Al 4 w/o Cu that solidified with only a few degrees of supercooling has also been reported (5).

The measured compositional profiles as well as the metallographic observations clearly indicate the existence of a solute-rich core within the dendrite elements. The core is surrounded by a region of lower concentration beyond which the concentration increases to a maximum at the cell boundaries. Numerous theoretical and experimental studies have been done to relate solute distribution to the amount of supercooling prior to nucleation (e.g.,22,23). The most widely used model is based on the relationship between the heat of fusion, heat capacities, and the supercooling. The model indicates that for a supercooled liquid, the fraction that solidifies in a segregationless manner is approximately:

$$f_s^o = \frac{C_p}{H_f} (T_p - T_n) \tag{1}$$

where C_p is the average heat capacity of the solid and liquid, H_f is the heat of fusion, T_n is the nucleation temperature, and T_p is the temperature at which partitioning takes place. The partitioning temperature can be as low as the solidus temperature of the alloy, or as high as the mid-rib (T_o) temperature of the alloy (24). As the temperature increases above T_p, partitioning takes place and solidification continues with the solid decreasing in solute until the maximum recalescence temperature is reached. Beyond this point the temperature decreases and solidification continues in a manner similar to conventional solidification processes, i.e. with no supercooling. The solidification path for supercooled samples is shown schematically in the appropriate portion of Fe-Ni phase diagram, Figure 14. The equations relating the composition of the solid, C_s, to the fraction solid, f_s, during and after recalescence are given by (23):

$$C_s = C_o + \frac{(f_s - f_s^o)H_f}{C_p m_s} \quad \text{for } f_s^o < f_s < f_{sr} \tag{2}$$

$$C_s = C_{sr}\left[1 - \frac{(f_s - f_{sr})}{(1 - f_{sr})}\right]^{k-1} \quad \text{for } f_{sr} < f_s < 1 \tag{3}$$

where C_o is the initial alloy composition, C_{sr} is the solid composition at the maximum recalescence temperature, f_{sr} is the fraction solid at maximum recalescence, m_s is the solidus slope, k is the equilibrium distribution coefficient, and other quantities are defined before. The amount of segregationless solidification can be calculated from Equation (1) if the partitioning temperature was known. However, by setting T_p equal to T_s or T_o, the lower and upper limits for the fraction of the partitionless zone can be calculated, as given in Table I. It should be noted that the supercoolings given in this table correspond to the value at which the supercooled liquid was dropped from the levitation coil. Abbaschian and Flemings (6) have shown that the sample loses an additional 25 K before it impinges the quenching water. Because of this, the calculations are based on the measured supercooling plus 25 K. The fact that the samples may supercool further after immersion in the water has not been taken into consideration.

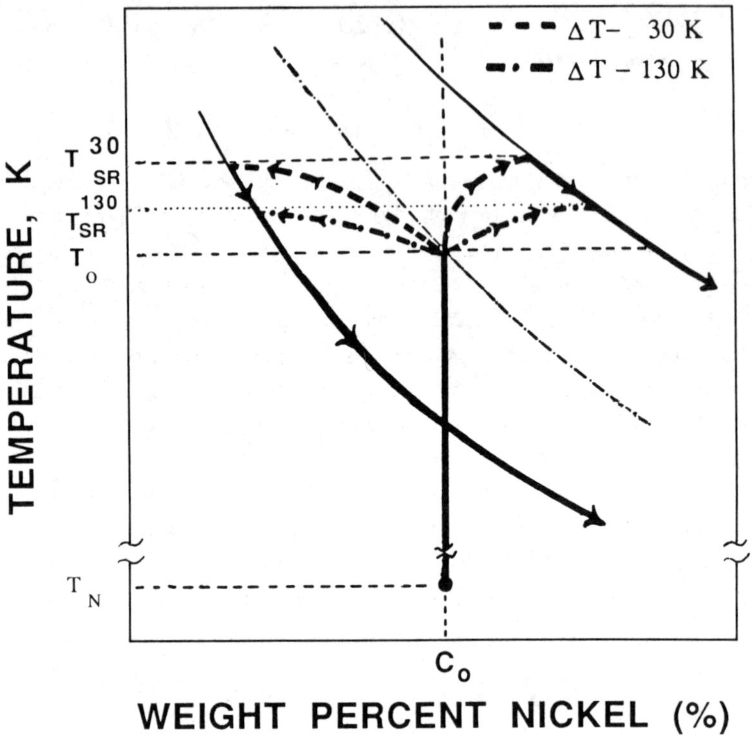

Fig. 14 - Schematic partial phase diagram of the Fe-Ni system showing the solidification path of an alloy with composition C_o nucleating at T_n.

The experimental and the calculated values given in Table I indicate the same trend, i.e. an increase in the fraction solidified in a segregationless manner with increased supercooling. However, the calculations generally tend to overestimate the amount of segregationless solidification. The disagreement is believed to be due to use of several simplifying assumptions in applying Equation 1 (25). These include: (i) adiabatic solidification during recalescence, (ii) negligible solid state diffusion, (iii) no remelting, (iv) no curvature effects, and (v) uniform temperature throughout the sample during recalescence. Because of the short recalescence times, the effect of non-adiabatic conditions on the amount of segregationless solidification is expected to be negligible. For example, Ewasko et al. (26) have found that the recalescence time for Ni 32.5 w/o Sn supercooled 190 K is about 10 ms. Furthermore, the non-adiabatic conditions can only increase the extent of partitionless solidification, thereby increasing the difference between the experimental and calculated results. The assumption of negligible solid state diffusion is also nonrealistic; however, because of the rapid solidification rates (of the order of 10^3 K/s) and slow solid state diffusion rates, its contribution is likely to be small.

The curvature at the tip of the thermal dendrites, which form before partitioning, depresses the liquidus and solidus temperatures, as well as T_o, leading to partitioning at lower temperatures. According to the Gibbs-Thomson effect, the temperature depression depends on the radius of the curvature. The temperature depression is estimated to be in the range 3 to 30 K for 80 K supercooling (25). Such a temperature depression can drastically reduce the extent of the partitionless solidification, bringing the calculated values closer to the experimental results.

The non-uniformity of the temperature within the sample during recalescence is another major factor that reduces the extent of partitionless solidification. This is because the solid-liquid interface temperature during growth into a supercooled liquid is generally higher than that at any other part of the sample. As such, partitioning occurs when the interface temperature, rather than the average temperature of the sample, reaches T_p. The assumption of temperature uniformity within the sample is approximately valid only when the growth process is kinetically controlled. When the growth is heat transfer controlled, the interface temperature will be appreciably higher than the bulk, and may eventually reach the melting point, even though the bulk liquid remains supercooled. Whether the growth is kinetically or thermally controlled depends, among several factors, on the material, interface supercooling, and size of the growing solid. The effect of solid radius on the interface velocity is shown in Figure 15. According to Peteves et al. (25), for a given interfacial supercooling, when the solid radius is small, the kinetically controlled rate is slower than the heat transfer controlled rate. The situation is reversed at higher radii of curvature, when the process becomes thermally controlled. The transition from kinetically to thermally controlled growth occurs at a larger radius of curvature as the supercooling is increased. It should be noted that during growth of a nucleus, the growth will initially be interfacially controlled until the radius increases to the demarcation line shown in Figure 15. Beyond this radius, the growth follows the demarcation line, and the process becomes coupled, i.e. the interfacial velocity and temperature must satisfy both heat transfer and kinetic requirements.

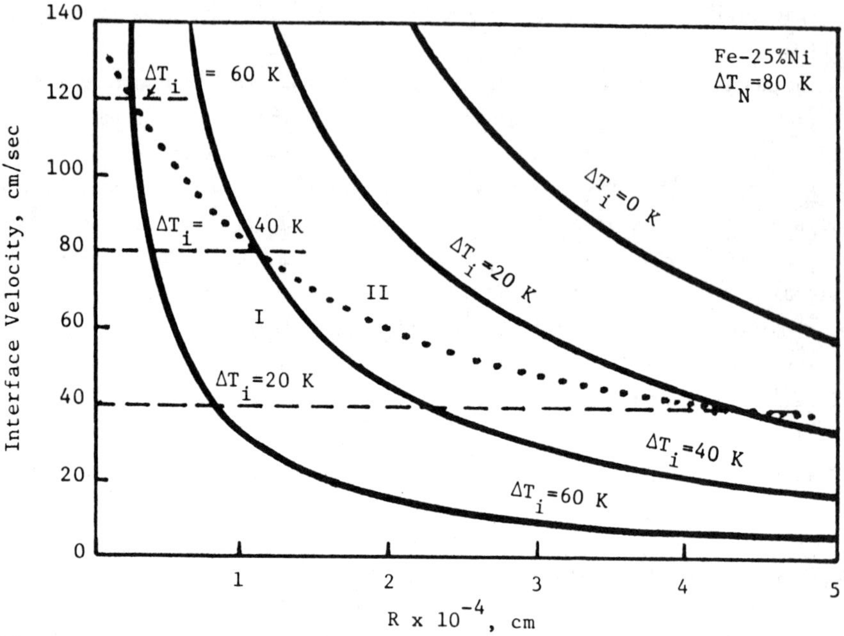

Fig. 15 - Interface velocity as a function of the radius of solid, growing at various constant interfacial supercoolings: solid and dashed curves show the heat flow and kinetically controlled growth, respectively. The dotted curve is the demarcation line between the two growth modes (25).

The measured compositional profiles, Figure 13, also show that the minimum concentration increases with increasing the supercooling. This trend could be due to a reduction in the maximum recalescence temperature as the supercooling is increased. It may also be due to solute trapping because of the rapid solidification rates during recalescence (27). Since the solidification rate is related to the supercooling, more pronounced solute trapping is expected to occur at higher supercoolings. It should be noted that the above mentioned trend in the level of the minimum composition could not be due to the solid state back diffusion because, as recently shown for Al-Cu and Al-Si alloys (28), back diffusion is expected to have an appreciable effect only at much lower solidification rates. Furthermore, back diffusion will tend to increase the minimum concentration, and as such will have an effect opposite of the observed behavior.

Summary

The literature on the effect of supercooling on the grain size was briefly reviewed. For supercooled Fe-Ni alloys, it is shown that the abrupt grain size reduction is more likely to be due to the change in the solidification morphology from dendritic to spherical, rather than being due solely to the supercooling. The latter morphology is believed to result for dendrite fragmentation caused by fluid flow. The effect of supercooling on the solute segregation profile was presented and discussed. The fraction of segregationless solidification was shown to be considerably less than that calculated based on adiabatic-uniform temperature recalescence. The discrepancy can be attributed to the non-uniformity of the temperature within the sample during recalescence, and to the curvature effect at the tip of the thermal dendrites.

Acknowledgement

The support of this research by the Office of Naval Research is gratefully acknowledged. The Scientific Officer of the program was Dr. Bruce A. MacDonald. Special thanks are due to Professor M. C. Flemings for his valuable discussions and contributions.

References

1. W. J. Boettinger, S. R. Coriel, and R. F. Sekerka, Proc. of Rapid Solidification Processing and Technologies, R. Mehrabian Ed., Gaithersburg, Maryland, December 6-8, 1982, p. 45.
2. D. D. McDevitt and G. J. Abbaschian, "Chemistry and Physics of Rapidly Solidified Materials," B. J. Berkowitz and R. O. Scattergood, Eds., St. Louis, Missouri, The Metallurgical Society of AIME, October 26-27, 1982, p. 49.
3. R. T. Southin and G. M. Weston, J. Australian Institute of Metals, 1974, Vol. 19, p. 93.
4. L. M. Hogan, "Physics of Materials," D. W. Borland, L. M. Clarebrough, and A. J. M. Moore, Eds., University of Melbourne, Australia, 1979, p. 111.
5. L. F. Mondelfo, "Grain Refinement in Castings and Welds," G. J. Abbaschian and S. A. David, Eds., The Metallurgical Society of AIME, 1982, p. 3.
6. G. J. Abbaschian and M. C. Flemings, Met. Trans. A, 1983, Vol. 14A, p. 1147.

7. T. Z. Kattamis and M. C. Flemings, Trans. TMS-AIME, 1966, Vol. 236, p. 1523.
8. T. Z. Kattamis and M. C. Flemings, Trans. AFS., 1967, Vol. 75, p. 191.
9. T. F. Kelly and J. B. Vander Sande, ibid 2, p. 35.
10. G. E. Amaya, J. A. Patchett, and G. J. Abbaschian, ibid 5, p. 51.
11. J. L. Walker, "Physical Chemistry of Process Metallurgy," G. R. St. Pierre, Ed., Interscience Publications, New York, 1961, p. 845.
12. G. L. F. Powell and L. M. Hogan, Trans. Met. Soc. of AIME, 1969, Vol. 245, p. 407.
13. P. B. Crosley, A. W. Douglas, and L. F. Mondolfo, "Solidification of Metals," Iron and Steel Inst. Publ. No. 110, London, 1968, p. 10.
14. G. Horvay, Proc. 4th Inter. Congr. of Applied Mechanisms (ASME), 1962, p. 1315.
15. S. Skolianos, P. S. Liu, and T. Z. Kattamis, ibid 5, p. 97.
16. B. L. Jones and G. M. Weston, J. Australian Institute of Metals, 1970, Vol. 15, p. 167.
17. D. E. Ovsienko B. B. Maslow, and B. N. Dneprenko, Russ. Met., 1979, Vol. 4, p. 142.
18. G. L. F. Powell and L. M. Hogan, Trans. Met. Soc. AIME, 1968, Vol. 242, p. 2133.
19. S. J. B. Reed, Electron Microprobe Analysis, Cambridge University Press, Cambridge, 1977, pp. 175-197.
20. D. D. McDevitt and G. J. Abbaschian, Microstructural Science, 1983, Vol. 11, p. 125.
21. J. A. Patchett and G. J. Abbaschian, Met. Trans. B, 1985, Vol. 16B, p. 505.
22. T. Z. Kattamis, Z. Metallkunde, 1970, p. 856-860.
23. T. Z. Kattamis, Ph.D. Thesis, MIT, 1965.
24. J. C. Baker and J. W. Cahn, "Thermodynamics of Solidification in Solidification," ASM, Metals Park, Ohio, 1971, p. 23.
25. S. D. Peteves, D. D. McDevitt, and G. J. Abbaschian, in Rapid Solidification, Principles and Technology III, R. Mehrabian, Ed., NBS Internal Report, Gaithersburg, Maryland, 1982 p. 121.
26. R. Ewasko, S. M. Thesis, Department of Materials Science and Engineering, MIT, 1982.
27. M. Aziz, in these Proceedings.
28. J. A. Sarreal and G. J. Abbaschian, "The Effect of Solidification Rate on Microstructure," to appear in Metallurgical Transactions.

A METASTABLE EUTECTIC REACTION IN THE Au-Sn SYSTEM

K.N.Ishihara,
H.Gohchi
and
P.H.Shingu
Department of Metal Science and Technology,
Kyoto University,
Kyoto 606, Japan

Four metastable phases were identified by solidifying Au-Sn alloys of 29, 31 and 35 at% Sn, three by the droplet undercooling method and two by the gun quenching method. Out of these four metastable phases only the γ phase, which is a Hume-Rothery compound with e/a ratio of 1.60, was produced by both droplet and gun quenching methods. One of these metastable phases melts by a metastable eutectic reaction, 46K below the stable eutectic temperature.

Introduction

As in the case of the cementite(Fe_3C) phase in the Fe-C system, metastable phases usually emerge when a high temperature phase is cooled rapidly. Hence the metastable phase diagrams are often misunderstood as representing the non-equilibrium phases which can only appear by rapid cooling because of their faster growth rates relative to those of the stable phases. However, recent studies revealed that many metals and alloys can be solidified into metastable solid phases by very slow cooling when enough undercooling is given(1-5). We have shown, in a previous paper(6), the necessary condition for the formation of a metastable phase in preference to the stable phase.

Moreover, it is often found that these metastable phases which are formed by slow cooling have the same crystal structures as those formed by rapid quenching(2,5). The advantage of the slow cooling method for investigating the metastable phase diagrams is that the formation and decomposition (melting) temperature of metastable phases can be determined relatively easily by this technique. In the paper previous to this one, a metastable peritectic reaction in the Sn-Pb system was reported by the slow cooling and heating of fine droplets emulsified in silicone oil. In this paper, a metastable eutectic reaction found in Au-Sn system will be reported. An examination of the metallographic morphology difference between stably and metastably solidified structure will also be given.

Experimental Methods

Samples of three different compositions, Au-29at%Sn, Au-31at%Sn and Au-35at%Sn, were melted in silicone oil and emulsified by ultrasonic vibration. The formation and decomposition of phases by cooling and successive reheating were studied by the differential thermal analysis(DTA) and by energy dispersive X-ray diffractometry. The details of these techniques are described in a previous paper(7). For the rapid solidification experiment, the gun method operated in vacuum was used(8).

Results

The emulsion samples solidified in the temperature range of 493-443K corresponding to an undercooling of more than 60K below the eutectic temperature of ζ and the AuSn phase, 553K. On heating, three endothermal peaks are observed for all three different composition samples. The DTA

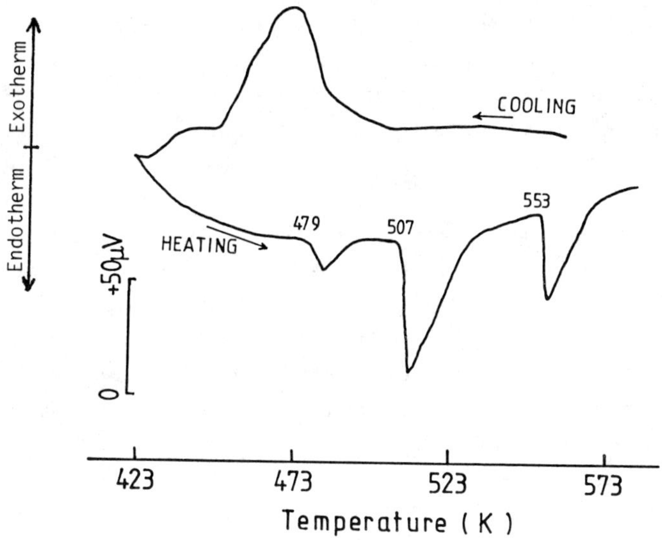

Figure 1 - DTA thermogram on heating and cooling for the sample Au-31at%Sn.

Table 1 - The results of the measurements of X-ray diffractometry. γ phase is identified as γ brass structure with a=0.98nm which is the Hume-Rothery compound. M_H phase is indexed as tetragonal with a=0.585nm and c=0.310nm. The structure of M_L was not identified.

interplaner spacing d (nm)	metastable phases (hkl)		
0.2821			M_L
0.2749		M_H(101)	
0.2718			M_L
0.2672			M_L
0.2619	γ(321)		
0.2486		M_H(111)	
0.2450			M_L
0.2310	γ(330)		
0.2191	γ(420)		
0.2118		M_H(201)	
0.2071		M_H(220)	
0.2036			M_L
0.2023			M_L
0.1922	γ(431)		
0.1849		M_H(310)	
0.1454			M_L

thermogram for the Au-31at%Sn sample is shown in Fig.1. The highest temperature peak corresponds to the eutectic temperature, 553K. The two lower temperature peaks correspond to the melting of two different metastable phases which will be termed M_L and M_H hereafter in this paper.

The X-ray diffractometry at room temperature revealed the interplaner spacing, d, of three metastable phases as shown in Table 1. The interplaner spacing of γ phase match those of the rapid quenched metastable phase reported by Giessen(9). The result of the rapid quenching by the gun method in the present work confirmed the formation of the γ phase and another metastable phase termed X phase by Giessen(9). Neither the M_H nor M_L phase found by the droplet undercooling method was found in the gun quenched samples. Thermal analysis of the gun quenched sample only showed a broad endothermal peak around 473K.

In order to confirm the nature of two endothermal peaks in Fig.1 which are ascribed to the melting of the metastable M_L and M_H phases, X-ray diffractometry was performed at room temperature, 494K, and 524K. The result clearly showed the disappearance, due to melting, of diffraction peaks from the M_L phase at 494K and from the M_H phase at 524K.

It was also found that the diffraction intensity from the stable ζ phase is much weaker at 524K than at 494K. The diffraction intensities from the stable AuSn phase were the same at each of three temperatures, Fig.2. This fact indicates that when the M_H phase melts to become an undercooled liquid, partial melting of stable ζ phase also occurs. From the equilibrium phase diagrams, for the sample compositions used in the present work, it is seen that the ζ phase does not solidify as a primary phase, hence ζ phase which co-exists with M_H phase, should be the product of a metastable eutectic reaction. The metastable eutectic melting of ζ and M_H phases can be clearly visualized by the energy dispersive X-ray diffractometry performed while continuously heating the sample, as shown in Fig.3.

Discussion

It is of interest to compare the metastable phases obtained by static undercooling and rapid quenching, respectively. In previous papers, we have reported the production of the same metastable phase in the case of an $In_{50}Bi_{50}$ alloy(5) and the different metastable phase in the case of Pb-Sn alloys(7). In the case of Au-Sn alloys, γ phase, which is a Hume-Rothery phase with an e/a ratio of 1.60, is produced commonly by both methods, while the M_L and M_H phases are produced only by the droplet method and the X phase only by rapid quenching. Further experiments are needed to explain such

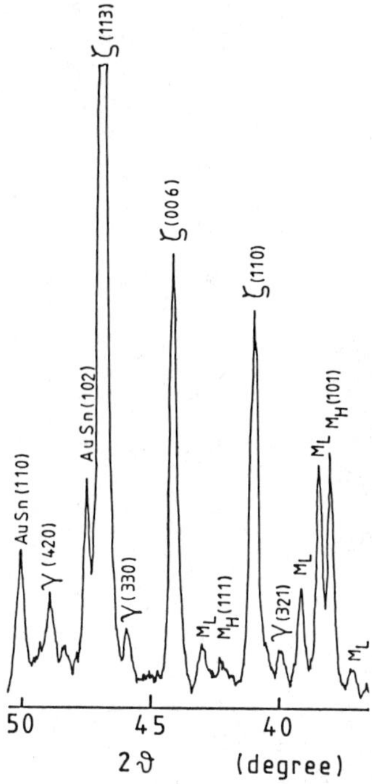

(A)

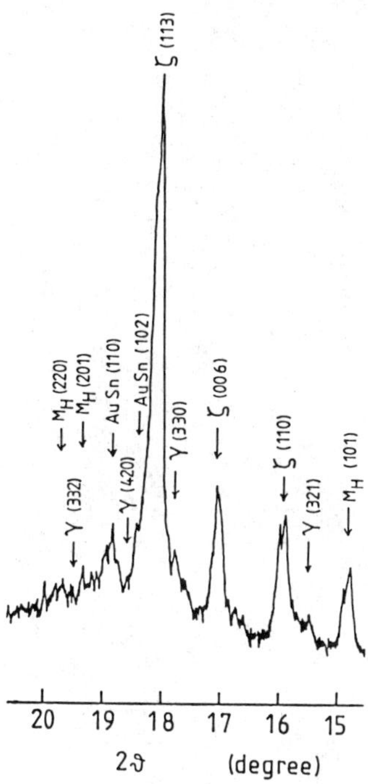

(B)

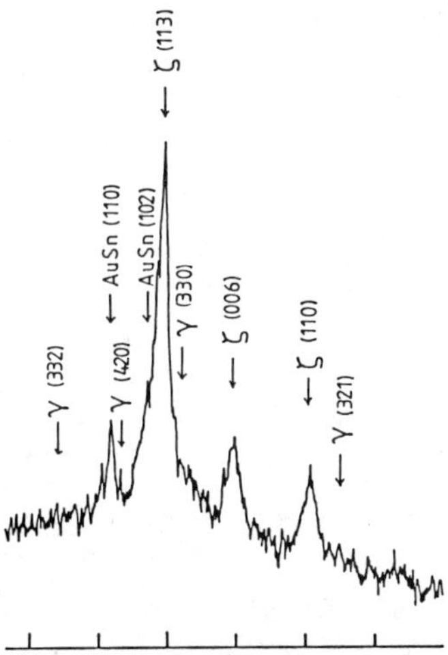

(C)

Figure 2 - The results of X-ray measurements of a Au-31at%Sn sample at different temperatures: (A) room temperature, (B) 494K and (C) 524K. For curves (B) and (C), the sample was first cooled down to 349K and then heated up and held at given temperatures during measurements. CoK_α radiation was used for (A) and MoK_α radiation was used for (B) and (C).

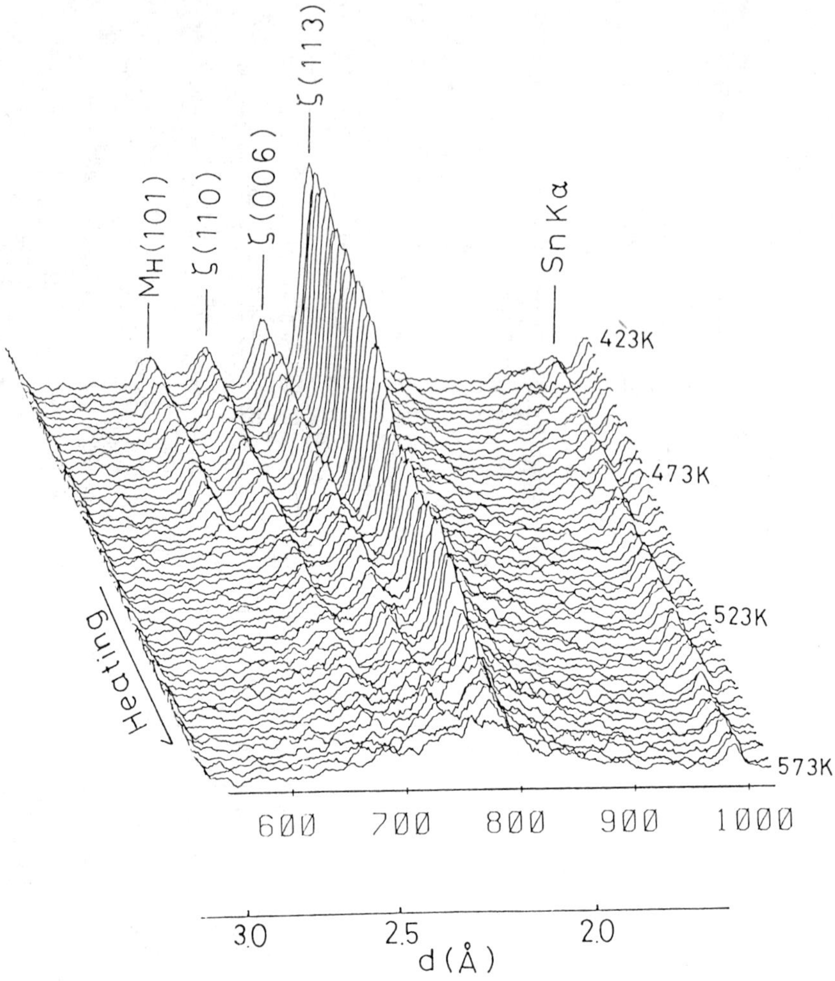

Figure 3 - A series of energy dispersive X-ray diffraction curves for an Au-31at%Sn sample. The sample was heated up at a rate of 0.54K/min. Note that the diffraction peak for the ζ phase decreases as the peak for M_H phase disappears at 507K indicating the metastable eutectic reaction.

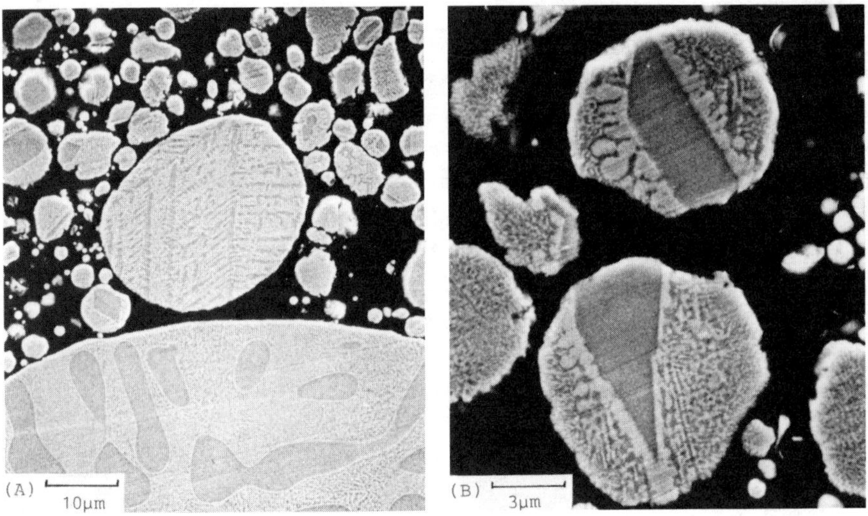

Figure 4 - The microstructure of solidified droplets of Au-35at%Sn sample. Note that in (A) the size of the non-faceted AuSn primary dendrites are much finer in the droplet of about 20μm diam. than those in the large particle seen in the lower part of the photo. Faceted AuSn primary crystals are seen in the smaller particles. (B) Shows an enlarged view of the faceted AuSn primary crystals in smaller droplets. The composition of both non-faceted and faceted primary phases was confirmed by the electron probe microanalyzer to be Au-50at%Sn.

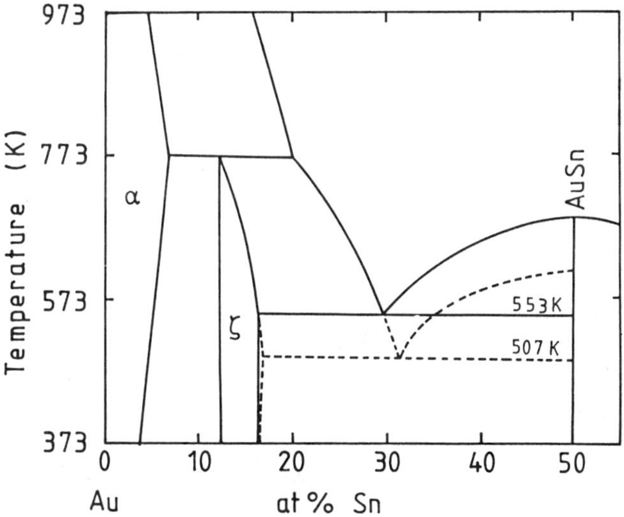

Figure 5 - The estimated metastable phase diagram of Au-Sn system superimposed on equilibrium phase diagram.

differences and coincidences.

In the present work, the amount of metastable phases γ and M_L obtained by the droplet method is small relative to the metastable M_H and the stable ζ and AuSn phases. Hence only the metastable phase diagram relating M_H, ζ and AuSn is obtainable with any degree of certainty. The metallographic microstructure examination of solidified droplets of 35at%Sn samples revealed the primary AuSn phase and fine eutectic structure shown in Fig. 4(A). Electron probe microanalysis showed the composition of the primary phase as AuSn and the average composition of eutectic to be about Au-32at%Sn. In Figure 4(A), it is to be noted that for a droplet of about 20μm diam. (seen in the center) the AuSn primary is dendritic and finely branched. However for the smaller size droplets, the primary phases exhibit rather coarse and faceted morphology as shown in Fig.4(B). The composition of these faceted primary phases, according to electron probe microanalysis, was about Au-50at%Sn.

The size of these primary crystals is too small for structural study by X-ray diffractometry. But the change in morphology from non-faceted to faceted suggests that these primary crystals are of different crystallographic structure. It is thus tentatively concluded that the metastable M_H phase has the composition Au-50%atSn.

From these considerations, an outline of the metastable eutectic phase diagram involving stable ζ and metastable M_H phases may be drawn as in Fig.5.

Conclusions

1. Four metastable phases were identified by solidifying Au-Sn alloys of 29, 31, and 35at%Sn; three by the droplet undercooling method and two by the gun quenching method.

2. By the droplet undercooling method a metastable phase with tetragonal structure, termed as M_H phase, was produced in nearly equal proportion to the stable phase. The M_H phase melts by the metastable eutectic reaction with the stable ζ phase at 507K, 46K below the stable eutectic temperature between ζ and AuSn phase.

3. An outline of the metastable phase diagram involving the M_H, ζ and liquid phases is given.

References

1. A.Defrain, "États Metastables du Gallium Surfusion et Polymorphisme," J. Chim. Phys., 74 (1977), 851-862.

2. J.H.Perepezko, "Crystallization of Undercooled Liquid Droplets," Rapid Solidification Processing. Principles and Technologies II, ed. R.Mehrabian, B.H.Kear and M.Cohen (Claitor's Publishing Division, Baton Rouge, LA, 1980), 56-67.

3. K.P.Cooper, I.E.Anderson and J.H.Perepezko, "Crystallization of Undercooled Pb-Sn Alloys," Proc. 4th Int. Conf. on Rapidly Quenched Metals, ed. T.Masumoto and K.Suzuki (Japan Institute Metals, Sendai, 1983), 107-110.

4. J.H.Perepezko and I.E.Anderson, "Metastable Phase Formation in Undercooled Liquids," Synthesis and Properties of Metastable Phases, ed. E.S.Machlin and T.W.Rowland (Metallurgical Society of AIME, Warrendale, PA, 1980), 31-61.

5. K.N.Ishihara, K.Mori and P.H.Shingu, "Formation of a High Pressure Phase in In-Bi Alloys Solidified from Highly Undercooled Liquid at Atmospheric Pressure," Proc. 5th Int. Conf. on Rapidly Quenched Metals, ed. S.Steeb and H.Warlimont (Elsevier Science Publishers B.V., Amsterdam, 1985), 55-58.

6. K.N.Ishihara, M.Maeda and P.H.Shingu, "The Nucleation of Metastable Phases from Undercooled Liquids," Acta Metall., 33 (1985) 2113-2117.

7. M.Mori, K.N.Ishihara and P.H.Shingu, "Metastable Phase Diagram of the Pb-Sn System," Mater. Sci. Eng. 78 (1986) 157-162.

8. P.H.Shingu et al., "An Amorphous Phase in a Splat-Cooled Fe-3.8wt%C Alloy," Scripta Metall. 8 (1974) 1317-1320.

9. B.Giessen, "A Metastable -Brass Phase in the Gold-Tin System and a Note on Non-Equilibrium Hume-Rothery Phases," Z. Metallkd. 59 (1968) 805-809.

UNDERCOOLED PHASES IN ALUMINUM-IRON ALLOYS

C. M. Adam, V. R. V. Ramanan and D. J. Skinner

Metals and Ceramics Laboratory
Allied Corporation
Morristown, NJ 07960

There have been numerous investigations of metastable alloy phases in aluminum-iron alloys. In 1955 Scheil and Masuda investigated metastable microstructures in several hyper-eutectic alloys by quenching into liquid sodium. Recent progress in rapid solidification technology has produced new phases variously described as quasi-crystals, "O" phase, and icosahedral phases, and the undercoolings associated with the formation of this type of metastable phases can now be assessed with reasonable accuracy. This paper will discuss the various metastable phases found in these aluminum-iron alloys and their associated degree of undercooling.

Introduction

Dix [1] in 1925 described the major morphological features of aluminum-iron binary alloys containing up to 9 wt.% Fe and accurately established the binary diagram. He reported that a fine rod eutectic structure without primary phases was obtained by chill casting 2.7 and 3.5 wt.% Fe alloys into a graphite mold. Similar microstructures were obtained by Scheil and Masuda [2] in 1955 for 5-6 wt.% Fe alloy droplets quenched into liquid sodium. It is now obvious that Dix, and Scheil and Masuda, induced metastable eutectic solidification by using faster cooling rates than those responsible for the equilibrium Al-Al_3Fe eutectic structures. Based on later crystallographic work [3,4], it has been postulated that the eutectic structure shown by Dix, and Scheil and Masuda, were examples of Al-Al_6Fe eutectic colonies formed by metastable eutectic solidification. Adam and Hogan [5,6] based on their experimental data postulate an Al-Al_6Fe metastable binary diagram, with the metastable eutectic at 3.5 wt.% Fe and 923 K. Jones [7] examined splat cooled alloys containing up to 39 wt.% Fe, of which 8 wt.% Fe alloys were claimed to be representative, and in which previously unreported microstructures were obtained. These microstructures consisted of optically irresolvable, featureless areas, called Zone A by Jones, and very fine primary Al_6Fe intimately associated with dendritic Al-Al_6Fe eutectic colony structures, referred to as Zone B. Subsequent transmission electron microscopy on the Zone A areas [8] showed the structure to consist of two phases; super-saturated aluminum and a metastable cubic iron-aluminum phase with lattice parameter a = 0.36 nm, although more recent microscopy [9,10] has raised the possibility that the intermetallic phase is amorphous in sufficiently rapidly quenched Al-8Fe-2Mo ternary alloys. The arrangement of the two phases within the Zone A structure has many of the classical features of a eutectic solidification structure resulting from mutual short-range lateral diffusion at a strongly coupled solid-liquid interface, and has therefore been described as a microeutectic structure [11].

Further complications in phase identification arise due to the effect of relatively rapid solidification rates either altering the morphologies of the above intermetallics or, which is more likely, favoring the formation of further metastable phases. Al_mFe [12] and Al_xFe [13,14] are two such intermetallics that have been found recently in commercial, dilute Al-Fe-Si alloys. This has led Young and Clyne [13] to postulate the following correlation between the rate of cooling and the Al-Fe phases which precipitate:

$$<1\ Ks^{-1} \rightarrow Al_3Fe;\ 3\ to\ 12\ Ks^{-1} \rightarrow Al_6Fe;$$

$$0.7\ to\ 6\ Ks^{-1} \rightarrow Al_xFe;\ and,\ >12\ Ks^{-1} \rightarrow Al_mFe.$$

As the cooling rate is increased much in excess of these values a further Al-Fe phase appears that possesses icosahedral ($m\bar{3}\bar{5}$) symmetry. This phase along with similar ones found in Al-Mn, Al-Cr [15] and Al-Fe-V [16], is regarded as a quasi-crystalline solid, possessing all the characteristics

of a crystalline solid, i.e., sharp X-ray and electron diffraction peaks, but showing no translational symmetry.

This paper reports on the phases present in binary Al-Fe alloys having been jet or planar flow cast and the levels of undercooling that are necessary to form these phases.

Experiments

Al-Fe binary alloys containing 8 to 35 wt.% Fe were rapidly solidified as 25-40 μm thick ribbons by conventional free-jet melt spinning, and by planar flow casting. Details of these procedures, and the heat-transfer considerations involved, have been recently reviewed [17]. Table I lists the compositions investigated. The as-solidified ribbons were examined by transmission electron microscopy in a Philips EM400T STEM, and by differential scanning calorimetry using a Perkin-Elmer DSC-2C, following standard experimental procedures. Several alloys were subjected to both isochronal and isothermal heat-treatments to elucidate the kinetics of the transformations of the solidified phases.

Table I

The list of the binary Al-Fe alloys studied during this investigation.

Sample No.	Fe Content (wt.%)
1	7.9
2	11.9
3	15.0
4	17.8
5	21.0
6	24.6
7	29.5
8	34.2

Results and Discussion

Microstructural Development

Depending on the alloy content and the attained solidification velocity, the microstructure varied from sample to sample. However, the solidification microstructure observed can be broadly classified into three types, the salient features of which are illustrated in Figures 1 through 3.

Alloys containing 8 to about 15 wt.% Fe solidify with a microeutectic microstructure (Figure 1), typical of published Zone A structures referred to above. This structure results from plane-front solidification with some lateral solute rejection [18]. Based on the inter-rod spacings of 50-100 nm, the analyses of Jackson and Hunt [19] and Trivedi [20] show that the local growth velocity attained for this microstructural type is in the range of 0.1 to 0.5 ms^{-1}. There is considerable uncertainty regarding the composition and crystallographic nature of the intermetallic phase found in these alloys. Burden and Jones [21] presented electron diffraction data indexed as a simple cubic AlFe structure, while crystallographic data obtained during this investigation places the composition closer to Al_mFe (see below).

In alloys containing 15-22 wt.% Fe quite different microstructures are produced (Figure 2), representative of independently nucleating spherical primary particles referred to as "O" phases [17], and possessing electron diffraction data typical of the icosahedral phases. In the richest alloys of this group approximately 95% of the microstructure consists of the spherical, icosahedral, "O" phase, while the more dilute alloys (15-20 wt.% Fe) contain 25-30% of the microeutectic structures. Quantitative analytical microscopy confirmed the iron content of the "O" phase at about 20 wt.% (~11 at.%), which is somewhat in disagreement with the composition found in the analogous Al-Mn system which has been found by various means to be approximately Al_4Mn [22,23]. In both Al-Fe and Al-Mn systems the solute content of the icosahedral, "O" phase is independent of the alloy chemistry, implying that the structure results from diffusionless solidification. A simple solute diffusivity calculation for a diffusionless transformation [17,24] yields 1-2 ms^{-1} for the probable solidification velocity.

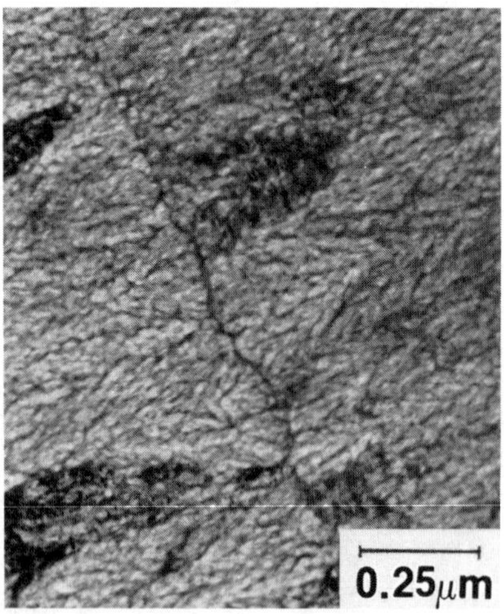

Fig. 1 - Micrograph illustrating the typical microeutectic structure in low iron containing alloys.

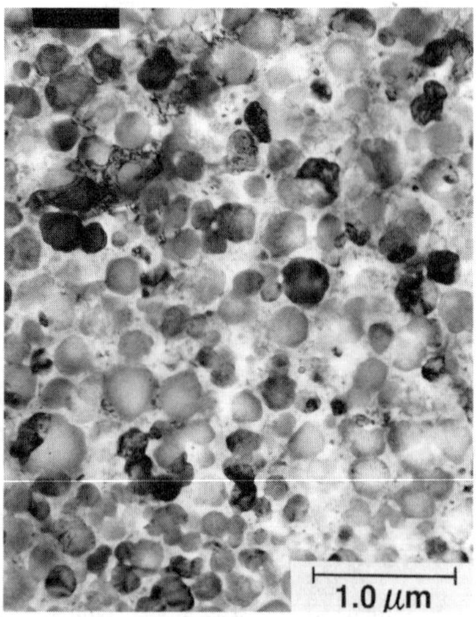

Fig. 2 - Micrograph from the 21.0 wt.% Fe alloy showing an abundance of the "O" phase.

Alloys containing 25-35 wt.% Fe invariably contained primary Al_3Fe (Figure 3), showing a radial array of heavily twinned Al_3Fe branches. This growth morphology is quite different from coarse Al_3Fe primary particles found in conventionally cast alloys, although the individual radiating branches showed the characteristic (001) twinning and (010) growth axis crystallography.

Fig. 3 - Micrograph from the 24.6 wt.% Fe alloy showing the Al_3Fe phase.

Fig. 4 - Micrograph from the 15.0 wt.% Fe jet-cast alloy showing the phase identified as Al_mFe. (See text.)

The presence of the Al_mFe phase (m≈4.4) was also found (see Figure 4) in the Al-Fe alloys over a wide compositional range. This identification follows from the fact that the electron diffraction data from this phase cannot be indexed as due to Al_3Fe (or Al_6Fe) and is indicative of a tetragonal phase. Further, energy dispersive X-ray analysis yields a composition close to Al_mFe. Finally, from the published data, this is the primary phase to be expected under these rapidly solidified conditions.

Differential Scanning Calorimetry

The alloys were subjected to DSC scans at 20 K/min. from 450 to 950 K. As expected, a strong endothermic reaction at (929 ±1) K is noted in all the alloys and corresponds to the eutectic temperature in the equilibrium phase diagram. However, major differences in the solid state transformation behavior of these alloys exist at lower temperatures. Figure 5 displays this data on selected jet-cast alloys. The exothermic reactions observed in the 17.8 and 21.0 wt.% Fe alloys are noteworthy here. The observed shifts in peak temperatures between DSC scans at varying heating rates (5 to 80 K/min.) [25] on the 21.0 wt.% Fe alloy indicate an activation energy of ~2 eV for each transformation.

It is a reasonable assumption that information revealed during reheating of a rapidly quenched alloy is directy related to the non-equilibrium phase diagram through which the alloy has been quenched. Unsuppressed nucleation, during quenching, of any phases allowed by the non-equilibrium kinetics will manifest in the as-solidified microstructure and, via transformations in the solid state. In the ultimate limit of process conditions, nucleation of all phases may be suppressed; a metallic glass is the result. The exothermic peaks of Figure 5 thus point to the existence of a non-equilibrium phase boundary in the Al-Fe diagram which spans the composition range 18-21 wt.% Fe. Alternatively, the extent of undercooling achieved in the jet-casting

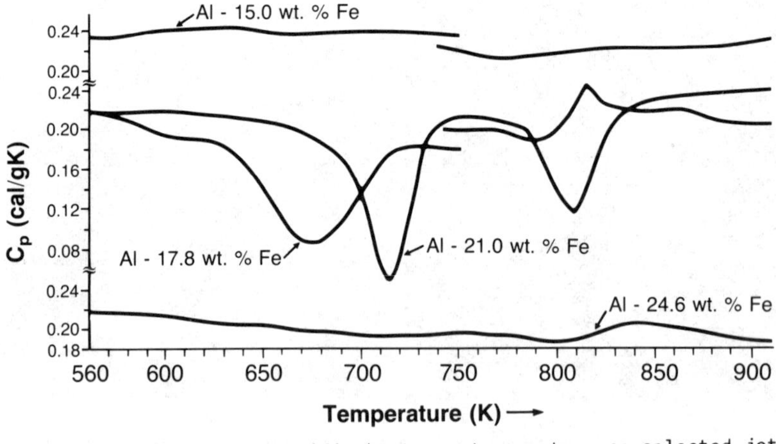

Fig. 5 - DSC data on specific heat vs. temperature on selected jet-cast Al-Fe binary alloys.

process is such that nucleation of all metastable phases, except the one represented by the first, or lowest temperature exotherm in the DSC trace, has been suppressed in the 17.8 and 21.0 wt.% Fe alloys. If this phase boundary were to extend to other compositions, the absence of the exotherms in the 15.0 and 24.6 wt.% Fe alloys implies insufficient undercooling. Confirmation of this is obtained from a planar flow cast (PFC) ribbon of the 15.0 wt.% Fe alloy; as is well known, planar flow casting allows larger undercooling [17]. Figure 6 shows the DSC data on this PFC ribbon, where the exothermic reactions are clearly seen: the second exotherm resides in the high temperature shoulder of the trace.

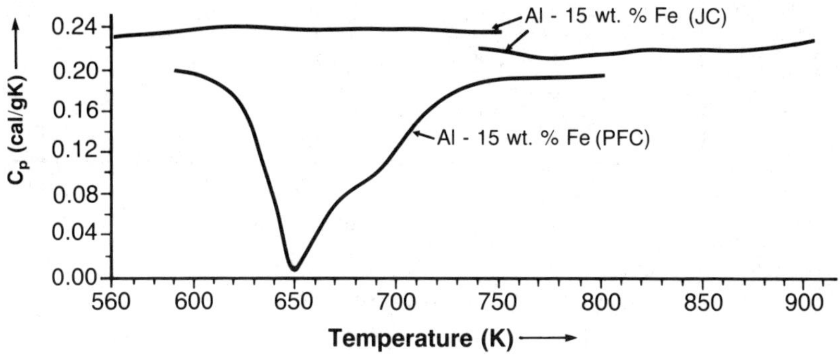

Fig. 6 - Comparison of DSC data on jet-cast and planar flow cast ribbons of the 15.0 wt.% Fe alloy.

One of the most striking results from the DSC data is the large level of undercooling achieved in the alloy casts. The onset temperature for the first exotherm in the 21.0 wt.% Fe alloy is ~690 K, indicating that the alloy has been undercooled by at least 610 K (allowing for recalescence) below the liquidus temperature (~1300 K). Since the activation energy for this exothermic reaction is ~2 eV, it is expected that at lower levels of undercooling the phase transformation represented by the exotherm would proceed rapidly, and the observed, as-solidified microstructure would not be substantially "O" phase. (As confirmation, it has been found that following a 50 s anneal at 720 K the first exotherm in this alloy is eliminated; the second exotherm vanishes after an anneal of about an hour at the same temperature.) Similarly, the data from the 17.8 wt.% Fe alloy indicate an undercooling of ~640 K below the liquidus.

Since the superheat temperature above the liquidus was a constant for all the compositions cast, it is a reasonable assumption that the undercool temperature parallels the liquidus in the equilibrium phase diagram. For the jet-casting process, this temperature is estimated to be ~650 K below the liquidus, per the arguments advanced above. Further, the data from the 15.0 wt.% Fe alloy indicate that the PFC process achieves, roughly, an additional 50 K in undercooling.

The importance of undercooling is underscored in the data of Figure 7.

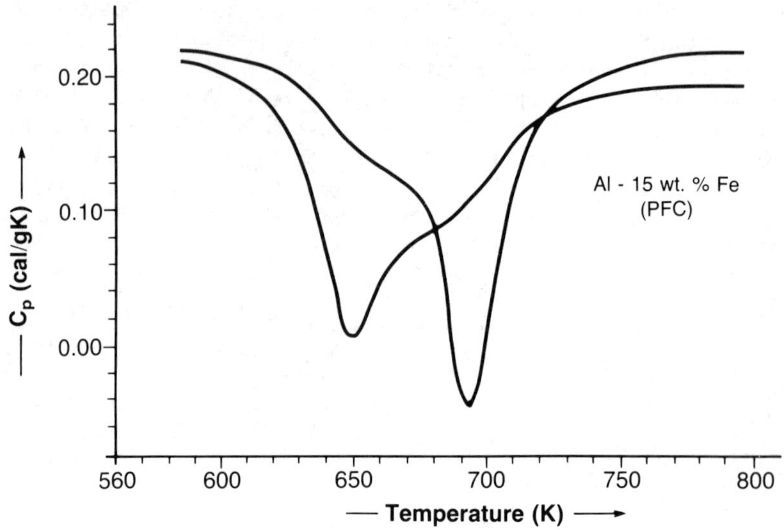

Fig. 7 - Comparison of DSC data on planar flow cast ribbons of the 15.0 wt.% Fe alloy with different degrees of undercooling.

Here, data on two different ribbons from the same PFC cast of the 15.0 wt.% Fe alloy are presented. It should be noted that the shoulder in either trace is located at the peak of the other trace, so that the same transformations are being seen. Alloys with Fe ⩾15 wt.% are especially difficult to planar flow cast into good, uniform ribbon. Figure 7 merely indicates that the degree of undercooling has varied during the cast, nucleating the high temperature phase in one instance (lower undercooling) and the low temperature phase in the other.

A sample of the 21.0 wt.% Fe alloy was scanned at 20 K/min. to 900 K and subsequently cooled at ~300 K/min. to 500 K. A rescan at 20 K/min. on this sample revealed no phase transformation. This experiment suggests that the two exotherms correspond to the transformation of metastable phases to a stable phase which is unaffected by the above mentioned "slow" cooling in the DSC.

The Al-Fe "Non-equilibrium" Phase Diagram

The results from electron microscopy show that the first exotherm in the DSC data corresponds to the transformation of the "O" phase to an intermediate metastable phase. The second exotherm then is a result of the transformation of this resulting intermediate phase, Al_mFe, to the stable phase Al_3Fe, in the alloys with 15-21 wt.% Fe. Therefore, using the temperatures for the onset of these transformations, we are led to postulate the Al-Fe "non-equilibrium" phase diagram as depicted in Figure 8. The subscript "h" for Al_3Fe is used to indicate the transformed stable phase obtained during heating, as opposed to the primary Al_3Fe found in alloys with higher Fe contents.

The observed microstructure in the as-quenched 21.0 wt.% Fe alloy corresponds substantially to the "O" phase. Thus, the "peak" of the "O" phase boundary lies in the vicinity of this composition. Since the "O" phase has not been observed in the 24.6 wt.% Fe alloy, the assumption of an undercooling curve that roughly parallels the liquidus implies a rapid drop of the "O" phase boundary for Fe>21 wt.% as suggested by the dashed line in Figure 8. This, again, is at variance with the observations on the Al-Mn alloys [26].

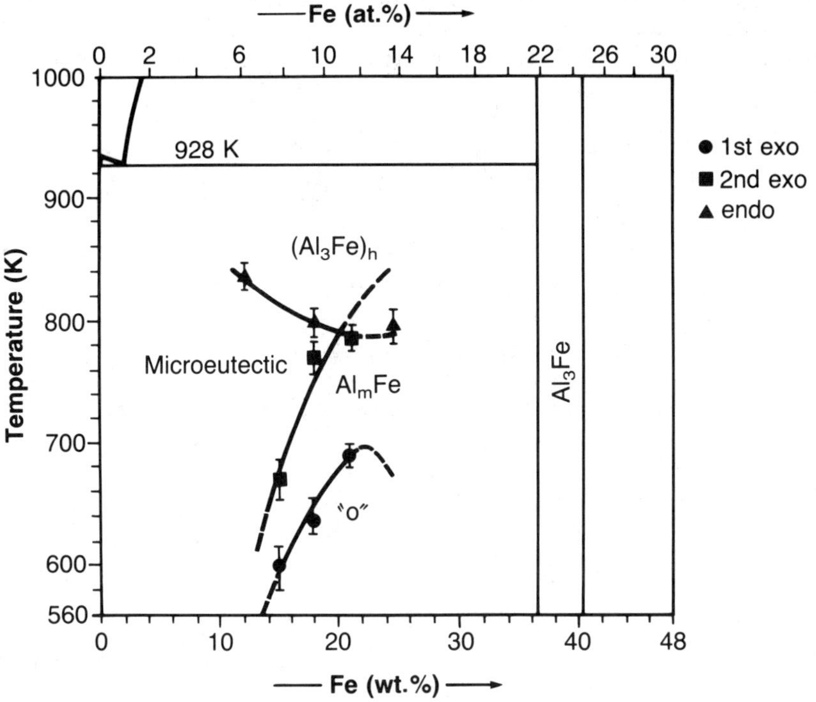

Fig. 8 - The Al-Fe "non-equilibrium" phase diagram suggested by our results.

The endothermic reactions of Figure 5 are also observed in the 11.9 wt.% Fe alloy. The onset (or peak) temperatures for these reactions are not close to any calculated temperatures for solid solution melting (T_0) of Al or Al_6Fe [27]. Since most other conceivable reactions are expected to yield exotherms a dissolution reaction is identified as the most likely cause for the endotherms in these alloys. Therefore, the onset temperatures for the endotherms indicate, it is assumed, the solvus line for the microeutectic structure observed in the low iron containing alloys. As implied in Figure 8, this microeutectic structure is assumed to comprise the α-Al and the Al_mFe phases.

Since there is a large tendency for primary Al_3Fe to precipitate in rapidly quenched alloys with Fe ≥25 wt.%, it is hard to qualify the boundaries of Figure 8 beyond this composition. Work to quantify this figure further is under way. (See note added in proof.)

Conclusions

The undercooled phases in a series of rapidly quenched binary Al-Fe alloys have been studied by electron microscopy and by differential scanning calorimetry. A "non-equilibrium" binary phase diagram has been proposed, based on the combination of results from these studies. The most striking result is the indication of substantial undercoolings, about 650 K below the liquidus, achievable in rapidly quenched alloys using the jet or the planar flow casting processes.

It is a pleasure to acknowledge the technical help provided by Patricia Drew, Elisabeth Musso, and Andrew Townsend during the course of this study.

References

1. E. H. Dix, Proc. ASTM, 25 (1925), 120.

2. E. Scheil and Y. Masuda, Aluminum, 31 (1955), 51.

3. E. H. Hollingworth, G. R. Frank and R. E. Willett, Trans. AIME, 224 (1962), 188.

4. L. K. Walford, Acta Cryst., 18 (1965), 287.

5. C. M. Adam and L. M. Hogan, J. Aust. Inst. Met., 17 (1972), 81.

6. C. M. Adam and L. M. Hogan, Acta Met., 23 (1975), 345.

7. H. Jones, Mater. Sci. Eng., 5 (1969-70), 1.

8. M. H. Jacobs, A. G. Doggett and M. J. Stowell, J. Mat. Sci., 9 (1974), 1631.

9. C. M. Adam and R. G. Bourdeau, Proc. Second Int'l. Conf. on Rapid Solidification Processing: Principles and Technologies, (Claitors, Baton Rouge, LA., 1980), p. 246.

10. J. Zindel, J. T. Stanley, R. D. Field and H. L. Fraser, Materials Research Society Symposia Proceedings, Vol. 28, ed. B. H. Kear and B. C. Giessen, (North-Holland, Amsterdam, 1984), p. 317.

11. C. M. Adam, Rapidly Solidified Amorphous and Crystalline Alloys, (Elsevier, New York, 1982), p. 411.

12. I. Miki, M. Kosuge, and K. Nagahoma, J. Jap. Inst. Metals, 32 (1975), 1.

13. R. M. K. Young and T. W. Clyne, Scripta Met., 15 (1981), 1211.

14. H. Westengen, Z. Metallkunde, 73 (1982), 360.

15. D. Shechtman, I. Black, D. Gratias and J. W. Cahn, Phys. Rev. Lett., 53 (1984), 1951.

16. D. J. Skinner, S. K. Das and C. M. Adam, unpublished work.

17. C. M. Adam and R. E. Lewis, Rapidly Solidified Crystalline Alloys, ed. S. K. Das, B. H. Kear and C. M. Adam, (TMS-AIME, 1986), p. 157.

18. W. Boettinger and J. H. Perepezko, ibid., p. 21.

19. K. Jackson and J. D. Hunt, Trans. AIME, 239 (1967), 864.

20. B. Trivedi, Proc. NATO Advanced Workshop on the Science and Technology of the Undercooled Melt, Amberg, FRG, 1985, in press.

21. M. H. Burden and H. Jones, Metallography, 3 (1970), 307.

22. K. M Krishnan, R. Gronskey and L. E. Tanner, Scripta Met., 20 (1986), 239.

23. K. Kimura, T. Hashimoto, K. Suzuki, M. Ino and S. Takeuchi, J. Phys. Soc. Jap., 54 (1985), 3217.

24. W. Kunz, B. Giovanola and R. Trivedi, Acta Met., 34 (1986), 823.

25. E. A. Marseglia, J. Non-cryst. Solids, 41 (1980), 31.

26. L. Bendersky et al., Symposium on Rapidly Solidified Crystalline Alloys, TMS-AIME Northeast Regional Meeting, Morristown, NJ, 1985, unpublished work.

27. J. L. Murray, Materials Research Society Symposia Proceedings, Vol. 19, ed. L. H. Bennett, R. B. Massalski and B. C. Giessen, (North Holland, Amsterdam, 1983), p. 249.

Note added in proof

Since the time the foregoing manuscript was prepared, alloys with 23 and 26.1 wt.% Fe have been planar flow cast with the attendant greater degrees of undercooling. DSC scans on ribbons from these casts show two exotherms, apart from the endotherm at ~930 K. Fig. 8(a) below shows these additional data. Results on the microstructure from these ribbons are not yet available. However, the onset temperatures for the lowest temperature exotherms from these casts suggest that the "O" phase boundary does not drop off as rapidly as is implied in Fig. 8 for Fe≥21 wt.%.

Recently published data [28] on jet-cast $Al_{86}Fe_{14}$ (25.2 wt.% Fe) ribbons, also included in Fig. 8(a), are seen to be quite consistent with the data reported above. A higher casting speed is reported in Ref. 28 and, therefore, higher cooling rates obtained than in the jet-cast alloys of the present study.

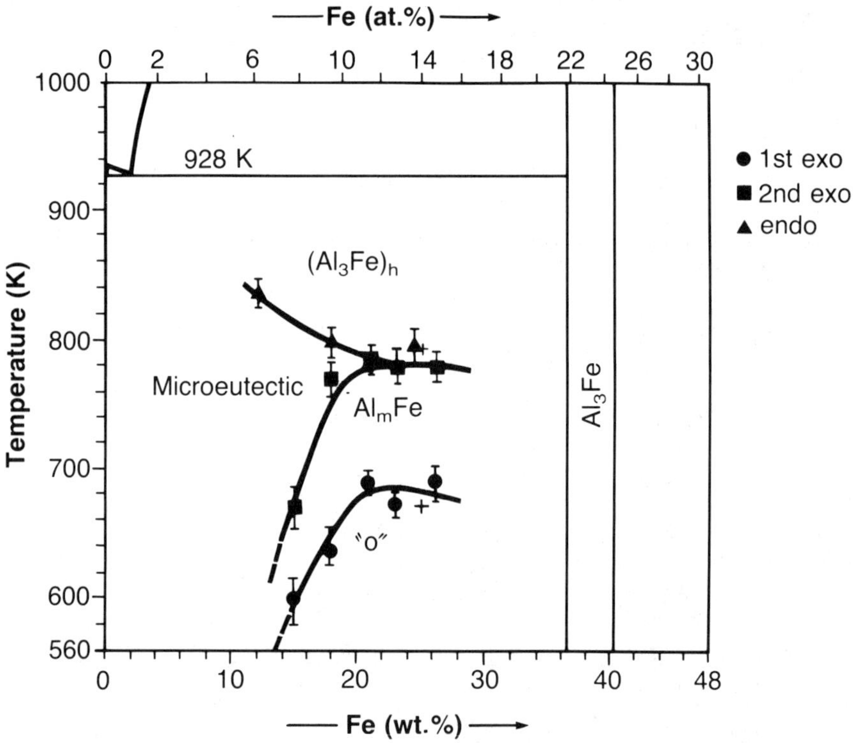

Fig. 8(a) - Modified version of Fig. 8, incorporating additional data referred to above. The "+" symbols represent the exotherms in the DTA data of Ref. 28.

Ref. [28]. R. A. Dunlap and K. Dini, Can. J. Phys. 63 (1985), 1267.

STUDIES OF UNDERCOOLING IN THE IRON-CARBON-SILICON SYSTEM

William Oldfield
Materials Research & Computer Simulation Corp.
Santa Barbara, CA 93110

and

Alan J. Markworth
Battelle Columbus Division
Columbus, Ohio 43201

Studies of the stable and metastable phase diagrams for Fe-C-Si are described. Solidification at varying rates of heat removal, coupled with thermal analysis, was used as a means of inferring the equilibrium eutectic temperatures for the two systems (the stable and metastable eutectics). Heating experiments in the region between the two diagrams were used to show melting of the metastable phase, to confirm the measured temperatures. Finally, the concepts used in this work are related to the more general problem of metastable phase relationships in rapidly cooled materials.

INTRODUCTION

The iron-carbon-silicon system, (cast iron), exhibits a broad range of solidification phenomena. A major factor influencing the solidification process is the existence of two distinct eutectic systems:

$$\text{(a)} \qquad Fe_{\gamma} + C \text{ (graphite)} \qquad (1)$$

$$\text{(b)} \qquad Fe_{\gamma} + Fe_3C \qquad (2)$$

In consequence, there are two distinct phase diagrams, as illustrated by Figure 1. The stable graphite-austenite system is shown by broken lines, while the metastable iron carbide-austenite system, is shown by solid lines.

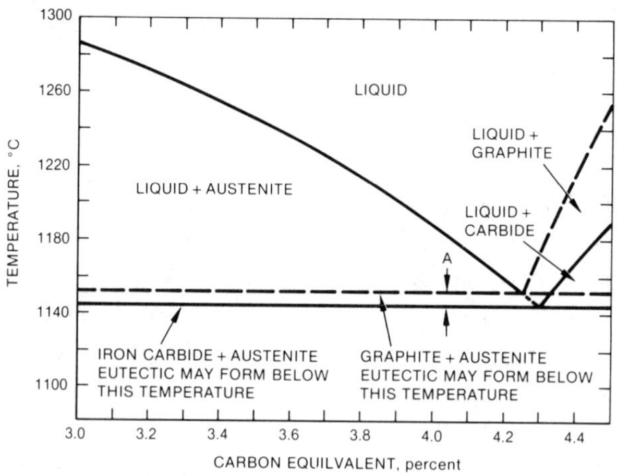

Figure 1. The Fe-C-Si phase diagram, showing both the stable (graphite + austenite) and metastable (iron carbide + austenite) eutectic systems.

The relative position of the two phase diagrams is important, since it helps to determine the structure of the solid, and dominates the metallurgy of commercial cast iron. (For example, the material formed by system (a) is soft and readily machineable, while the product from system (b) is hard and brittle and difficult to machine.) The relative position of the two phase diagrams is strongly influenced by the silicon content of the alloy, which lowers the (b) eutectic temperature. Thus, higher silicon contents are used to promote solidification in the graphite eutectic system. The reverse effect can be produced by other minor constituents such as chromium, which reduce the temperature interval between the two eutectics. Although the experimental work on these studies was done several years ago, it has not, to our knowledge been repeated. It is now put forward because of advances in experimental procedures since the time the work was done. This should permit the procedures to be used with other systems, for which the liquid must be cooled at much higher rates in order to obtain metastable phases.

The phase diagrams form only a small part of the complex solidification phenomena of these alloys. The kinetic phenomena also play a role in determining the gross structure of the solidified alloy. The rate of solidification can be expressed as:

$$\frac{dX}{dt} = f(\Delta T) \tag{3}$$

where X is the local position of the solid-liquid interface at time t, and ΔT is the undercooling. i.e., the difference between the temperature measured at the growing surface, and the appropriate phase equilibrium temperature.

The rate of solidification is much faster, at a specific undercooling, with system (b) than system (a). This may be due in part to the reduced diffusion necessary for Fe_3C-Fe_γ eutectic solidification, and in part to the relatively complex structure of the graphite crystal, which requires a more complex atomic attachment sequence.

STUDIES OF THE PHASE DIAGRAMS

A tie-line in a phase diagram defines the transition between solid and liquid, i.e. the temperature at which solidification can occur if the liquid alloy is cooling, or the temperature of melting if the solid alloy is being heated. Thus, the tie-line can be defined by either the solidification or melting approach.

Solidification Approach

Consider first the situation when cooling liquid alloy through the eutectic equilibrium temperature. When liquid Fe-C-Si alloy of eutectic composition is cooled below the (a) eutectic temperature, the probability of graphite nucleation increases with increasing undercooling. After the eutectic is nucleated, the crystal growth rate increases with undercooling, and the surface area of growing solid also increases, so that the rate of liberation of latent heat of fusion increases rapidly. Eventually, a balance between heat removal and heat liberation might be achieved, where the liberated heat first equals and then exceeds the rate of heat removal, and the temperature of the melt rises. The temperature at which the temperature just stops falling will be termed T_1. Clearly, for a given sample of liquid alloy, this temperature will fall with increasing rate of heat removal. (The concept of a single sample of liquid is important, to avoid differences which might be found between different samples in the the number of nucleation sites.) As the rate of heat removal is progressively increased, T_1 eventually falls below the equilibrium temperature for eutectic system (b). There is now a probability of the metastable eutectic forming, which increases with rate of cooling.

Figure 2 illustrates an experiment which puts the ideas developed in the last paragraph into effect.[1,2] A batch of molten Fe-C-Si alloy was separated into a number of samples which were cooled at different rates. The temperature of each sample was monitored continuously, and the value of T_1 determined. As the the cooling rate increased, T_1 fell steadily. However, the rate of fall of temperature changed sharply, as eutectic (b) solid formed.

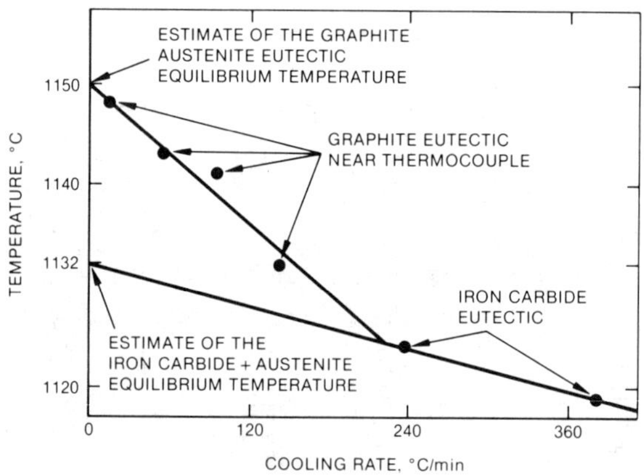

Figure 2. Determination of the metastable eutectic temperature by freezing samples of liquid at increasing cooling rate.

This metastable eutectic has faster kinetics than the graphite eutectic, so that the response to increased cooling rate was more gradual. The extrapolation of this lower line, denoting (b) eutectic, to zero rate of cooling, was taken as the equilibrium eutectic temperature. Figure 3 shows the results of several similar experiments, which reveal the influence of silicon content on the two eutectic temperatures revealed in this manner.[1,3] Figure 4 shows the results of a similar series of experiments which showed the influence of chromium on the eutectic temperatures.[2]

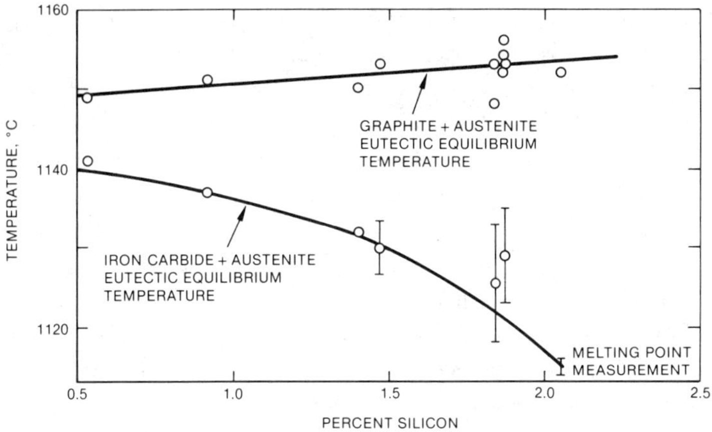

Figure 3. The effect of Si content on the stable and metastable eutectic temperatures.

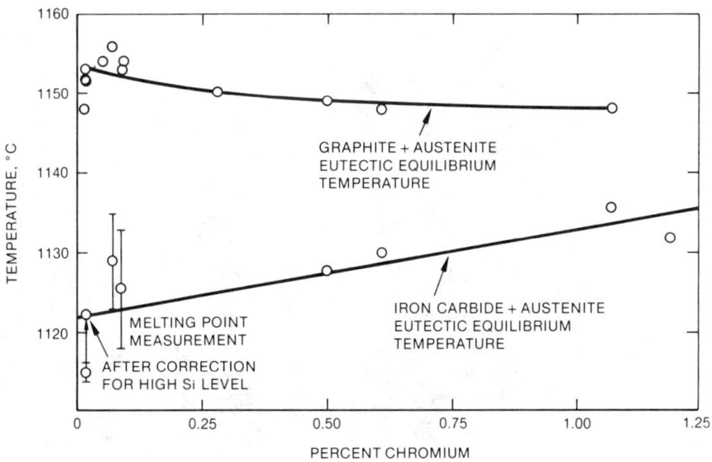

Figure 4. The effect of Cr on the stable and metastable eutectic temperatures.

Melting Experiments

If a sample of the metastable eutectic (b) is heated rapidly to a temperature above its equilibrium temperature, it should melt. When it is heated to a temperature between the two eutectic equilibrium temperatures, it should immediately re-solidify. This provides a means of confirming the results of solidification experiments. (Melting experiments on the metastable eutectic of the Fe-C system have also been reported by Okada et al. [4,5])

Samples of eutectic (b) were prepared by cooling liquid Fe-C-Si alloy at about 400 C/min.. The samples were sliced into thin specimens about the size of 1¢. A bath of molten copper was prepared, thermostatically controlled at temperatures between the eutectics as determined by experiments such as that shown in Figure 2. Samples were placed in the bath for varying periods of time (of the order of 10-60 s), taken out and immediately quenched in water. Water quenched material was readily distinguished from the eutectic (b) solidified at a moderate rate; regions within the specimen exhibited regions of water quenched material after a soaking period of about 20 s (Figure 5). The eutectic had clearly melted. Other regions could be found in the same specimens which contained a fine graphite structure, (Figure 6), which was different from structures formed through decomposition in the solid state, and which was not present in the start material. These are believed to be regions in which eutectic (b) had first melted, and then re-solidified in eutectic system (a). The melting and re-solidification phenomena helped to confirm the results of the solidification studies.

Figure 5. Structure formed by heating metastable eutectic to a temperature between the two eutectic equilibria followed by a water quench. x 600

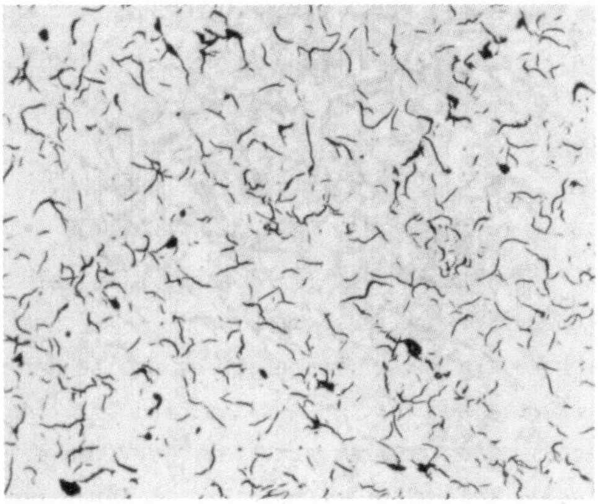

Figure 6. The structure obtained when metastable eutectic is held at temperatures between the two eutectics for longer periods. The liquid initially formed, re-solidifies in the stable system. x 100

EXTENSION OF THE WORK TO OTHER SYSTEMS

Determination of the Stable and Metastable Phase Diagrams

Apart from the prediction of the phase diagrams from thermodynamic considerations, direct measurement of solidification and melting temperatures appears to be a straightforward method for phase determination. However, one particular advantage of the Fe-C-Si system is that it does exhibit structures and kinetics which clearly lend themselves so readily to the experimental approach. Other systems that exhibit similar behavior may be more difficult to study in this way. In most cases, cooling rates will be many orders of magnitude greater than those employed here, making direct measurement of undercooling impossible. However, the temperature of the cooling liquid can be computed as a function of time and position if the cooling rate is known. If this is coupled with a means of determining time or position of solidification, it would allow the solidification temperature to be estimated. By linking this with a means for varying the rate of cooling, a version of Figure 2 could be constructed. Perepezko [6] determined melting temperatures for metastable components in the Sn-Bi system, and used the temperatures to infer features of the metastable phase diagrams.

Similarly, rapidly cooled material could be used in heating experiments to determine the equilibrium temperature of the metastable phase, in much the same manner as the metastable iron carbide-austenite. In this way, the phase diagram for metastable phases produced in splat cooling could be established.

CONCLUSION

Experimental work on the Fe-C-Si stable and metastable eutectic systems is presented as a model for work with other systems which also form metastable phases. The phase transition temperatures can be determined by cooling liquid through the transition at varying rates, and relating a measure of the instant of solidification to the cooling rate. They can also be determined by rapidly heating solid metastable material to temperatures in a range about the equilibrium temperature. Above the equilibrium temperature, melting and possibly re-solidification in the stable system take place and can identified by means of microstructural analysis.

REFERENCES

1. W. Oldfield, "The Solidification of Hypoeutectic Cast Iron" (M.Sc. Thesis Manchester University, 1961), 9-14.

2. W. Oldfield, "The Chilling Mechanism of Chromium in Cast Iron", BCIRA Journal, 9 (1961), 506-518.

3. W. Oldfield, "The Chill-Reducing Mechanism of Silicon in Cast Iron", ibid., 10 (1962), 17-27.

4. A. Okada, H. Miyake and R. Ozaki, J. Japanese Foundrymen's Soc., 53 (1981), 9-14.

5. A. Okada, H. Miyake and R. Ozaki, ibid., 55 (1983), 676-682.

6. J.H. Perepezko, "Proc. 2nd Int. Conf. on Rapid Solidification Processing, Principles, and Technologies II," R. Mehrabian, B.H. Kear, and M. Cohen, eds., (Claitor's, Baton Rouge, 1980), 56-67.

ACKNOWLEDGEMENTS

One of the authors (W.O.) acknowledges the guidance and vision of H. Morrogh, FRS, and thanks the British Cast Iron Research Association for its support at that time.

UNDERCOOLING AND NUCLEATION IN IRON NICKEL ALLOYS

Don H. Rasmussen and Kazi Javed
Department of Chemical Engineering
Clarkson University
Potsdam, NY 13617

Pure iron, pure nickel and iron-nickel alloys have been undercooled a maximum of 0.25 Tm (actual maximum observed undercoolings varied from 440 degrees for pure iron to 420 degrees for a 58 w/o nickel alloy). The dispersion of the preformed and surface treated metal powders in a ceramic matrix has permitted control of the droplet undercooling through surface chemistry. Even though these undercoolings exceed the classical 0.2 Tm level for homogeneous nucleation, the nucleation event is not likely to be homogeneous in origin because of the importance of droplet surface chemistry on determining the undercooling. The droplets crystallized into the FCC structure of the equilibrium alloys. The temperature of nucleation mirrors the liquidus curve in composition dependence. This behavior is explained with a thermodynamic model of nucleation in regular solutions.

Introduction

This study is the beginning of an experimental and theoretical program to understand the undercooling to nucleation in binary and multicomponent alloys. Experimentally, at low undercooling, stable crystal structures nucleate from the liquid and from solution while metastable crystal structures are observed only at ultra high undercoolings. The temperature of nucleation, in either case, is affected by solutes in a "thermodynamic" colligative manner. That is, experimentally, solutes act on the stability of the nucleus in the same manner as they affect bulk liquid-crystal equilibrium. It is a fact that classical nucleation theory [1] is fundamentally kinetic in character. Turnbull and Fisher's application of absolute reaction rate theory [2] transformed classical gas phase liquid-droplet nucleation kinetics into a form applicable to the liquid-to-solid transformation. This treatment retained much of the liquid-drop condensation formalism while replacing the kinetic theory of gases with a more realistic model of atomic interactions in condensed systems. Rasmussen [3] modified the classical analysis of cluster stability in supersaturated systems by incorporating a curvature dependence to the interfacial free energy. Here it is argued that the undercooling to nucleation for binary or multicomponent alloys is described by a model of clustering and nucleation which is formally similar to that used by Turnbull and Fisher [2] but because of the curvature dependence to the interfacial free energy, the undercooling is limited by a "physical spinodal" or a supersaturation at which barrierless cluster growth can take place. The composition dependence of this spinodal limit -- while not the actual magnitude of the undercooling -- is similar to for the observed experimental results in the iron-nickel system.

The clustering model of classical nucleation theory [1] is a development of a metastable state within a supersaturated system by fluctuations of the state of the system about its mean energy, temperature, volume, entropy, order, etc. This metastable state has the property that a local smaller fluctuation can either cause the metastable nucleus to become more stable and begin to grow, or to become less stable, and decompose back into the parent phase milieu. The nucleation process is a random stochastic process where one of a very large number of sequential <u>OR</u> random fluctuations occurs with a large enough amplitude to reach or exceed the metastable state. The probability of this happening is described by the theory of thermodynamic fluctuations of, for example, Landau and Lifshitz [4]. The ability of a liquid to resist crystal nucleation for long periods of time at temperatures near and just below the melting point indicates that the fluctuations about the mean energy etc. of a system under these conditions are normally not large enough for the stabilization of embryos or nuclei, of the crystal phase. Since the size of the mean fluctuation in system properties is directly proportional to temperature, the size of the mean fluctuation will decrease with increasing supersaturation or undercooling. It is therefore not the magnitude of the fluctuation which is limiting nucleation of a crystal from the melt, but a barrier to the decay of this fluctuation into a crystal nucleus. Nucleation becomes possible at high undercooling because the barrier to the decay of the fluctuation into a crystal nucleus decreases or disappears.

The barrier to formation of a crystal nucleus is the generation of an interface between the liquid and a cluster with crystal structure, bonding and symmetry. The interface between the nucleus and the liquid phase that must be formed in the fluctuation is not necessarily like the interface between bulk crystals and bulk liquid -- i.e. like the interface assumed by the classical capillarity approximation.

A bulk crystal will equilibrate to a thermodynamic state which will minimize the total system free energy. Such a state for an equilibrium interface between the crystal and the liquid may require that preferred low energy crystal planes -- those with high planar atom density, crystallographic surface reconstruction, and/or solute or solvent adsorption -- separate the crystal from the liquid or solution. The crystal form which presents the lowest interfacial free energy will be the equilibrium crystal form.

The equilibrium habit will be affected by the actual size of the crystal because the energy of the edges of the crystal and ledges on the faces of the crystal have not been considered as contributing to the planar surface free energies [5]. For very small crystals, the fraction of surface atoms which lie along the edges of the crystal is significant and the difference in the energy of the surface because of the presence of such edges must be acounted for. Some edges have high energy (more liquid like) especially those which are less than 90 degrees in internal angle. Other large interior angle edges have accordingly smaller energies. When the edge energies are added to the surface energies of very small crystals, the sharp 'cusps' in a polar diagram for the surface free energy versus surface orientation begin to disappear and the surface energy becomes more and more isotropic. When the crystalline clusters of sub-nucleus and nucleus sizes are considered, the surface energies are likely to be isotropic and the equilibrium surface of spherical morphology.

The interfacial region between the crystalline body and the random liquid structure is not infinitely thin, the initial clustering process must form a surface volume element as well as a bulk cluster volume of the new phase. The surface volume element contributes less change in enthalpy or internal energy per atom added to the cluster than an atom which is totally transfered to the bulk of the cluster. It has been demonstrated that the internal energy change on moving an atom from the equilibrium interface to the bulk of the crystal can range from 1/4 to 1/2 of the enthalpy of the transformation [6]. The process of addition of corner and edge monomers to a cluster must necessarily condense less enthalpy of transformation than the addition of surface site atoms on solidification. The addition of an atom to a very small cluster will add a higher percentage of surface than bulk to the cluster and a higher percentage of less stable surface that planar surface. Another way of stating this phenomenon is that, for the reverse fusion process, the heat of fusion of a cluster must be an increasing function of cluster size starting from zero for a monomer cluster increasing until an asymptotic limit is reached where the interface can propagate without increase or decrease in surface area per atom. The first atoms added to a growing cluster cannot locate stable fully bonded internal or equilibrium surface sites -- which do not yet exist in or on the cluster. These very small clusters do not either have an asymptotic bulk interior, or an asymptotic macroscopic interface.

These two changes, development of an asymptotic surface energy with the size and shape of a crystal and achievement of an asymptotic enthalpy of transformation with the size of the cluster, oppose one another and, taken together, suggest the following scheme for the formation of a cluster and nucleus. First, it can be assumed that each added monomer contributes a constant macroscopic enthalpy of transformation to the stability of the cluster at the undercooling at which nucleation is observed. Next, for very small clusters of atomic dimensions, the interface is isotropic and the interfacial free energy per unit area is size dependent. The clusters are spherical with a surface energy which increases from zero with increasing size of the cluster. The nucleus may become stable and begin to grow before low index planes with high planar atom density achieve stability relative to an isotropic spherical surface and, for that matter, before internal stacking sequences of the equilbrium lattice are fully determined. The corresponding low index planes appear as facets and decrease the overall surface energy of the cluster below that of a sphere with an isotropic interface. The facets will eventually become dominant and a platelet or polyhedral equilibrium form for the crystal will then propagate until total equilibrium is achieved. The size of the crystal for which a change in morphology is important will be very small indeed. The size dependent effects on cluster structure and stability will be limited to cluster dimensions within the first 10, or so, atomic diameters. By the time a crystal has grown to a microscopically observable size, the faceting or nonfaceting will depend on the equilibrium Wulff construction [7].

The effect of temperature on the interfacial free energy between a nucleus and an undercooled liquid cannot be separated from the Gibbs - Thompson [8] equilibrium melting point depression. When the surface free energy is ascribed a size dependence, this size dependence only pertains at thermodynamic equilibrium, or at the metastable melting point for a crystal of finite size. Alternatively, when the surface free energy is ascribed a temperature dependence, the surface free energy at any one temperature pertains to an equilibrium or metastable equilibrium interface at the metastable melting point for a crystal of finite size. Therefore, the interfacial tension can be considered a function of size or a function of temperature but cannot be considered to be both a function of crystal size and temperature. Therefore the reference states used in certain kinetic theories of nucleation in condensed systems which required impossible positive or negative pressures [9] are inappropriate. In this work, the interfacial tension is considered a function of cluster, or crystal, size only. A temperature dependence will occur only when colligative changes such as bulk pressure or solute independently modify bulk, or macroscopic, phase equilibrium.

Increase in compositional difference between a crystal (or cluster) and its equilibrium (or metastable equilibrium) mother liquid will increase the interfacial free energy [10]. A regular solution model has been developed which predicts this effect. The interfacial free energy between a cluster and the parent phase will be minimized when the cluster has the same composition as the liquid from which it forms. Thus a massive or diffusionless nucleation event has a smaller barrier to nucleation than the nucleation of an equilibrium crystal with widely different composition than the parent alloy. This is quite different behavior from that predicted by the parallel tangent construction.

Nucleation will occur when a larger than average fluctuation decays with the formation of a critically large cluster -- i.e. number of atoms, degree of order, extent of interfacial match to the local liquid lattice -- such that further addition of atoms with crystal-like bonding and organization stabilizes this nucleus. The probability of this happening is small at high subfreezing temperatures but increases more than exponentially with increased undercooling. Eventually, with sufficient undercooling or supersaturation, the increase in the free energy release on transformation will equal and then exceed the product of an increasing surface area per unit volume of cluster and a decreasing surface free energy per unit area of cluster. Beyond this critical undercooling clusters present in the liquid state of smaller size can grow without a surface energy barrier to macroscopic size.

A formal thermodynamic analysis of the dynamic interfacial free energy between a cluster and undercooled liquid and a kinetic analysis of clustering and nucleation in the condensed state has been developed [10]. Homogeneous nucleation is understandable in terms of a simple modification of classical nucleation theory dubbed, nucleation on approach to a physical spinodal.

Prediction of the Concentration Dependence of the Physical Spinodal

The thermodynamics and kinetics of cluster formation in a unary system, according to Rasmussen et. al. [11], is easily extendable to crystal nucleation from binary liquid systems. The appropriate thermodynamic potential applicable to crystal nucleation from a regular solution is the J potential developed by Reiss [12]. The undercooling to nucleation as a function of composition is determined by combining the state property of the J potential and the dynamic interfacial tension between cluster and parent liquid.

For nucleation from solution, it is the dynamic interfacial tension that is required and not the equilibrium interfacial tension because the interface to form during clustering is incapable of relaxing by adsorption to an equilibrium interface. The cluster does not have enough "bulk" to provide the material for this excess solute and the relaxation time for surface equilibration is of the same order of magnitude as the cluster lifetime. The dynamic interfacial tension [8] is given by,

$$\sigma(r)_{dyn} = \{X_1\sigma_o^1 + X_2\sigma_o^2 + (\alpha m/A)(X_2^p - X_2^n)^2\}(1-(\delta/r))^2 \qquad (1)$$

where σ_o^1 and σ_o^2 are the interfacial free energies for isotropic planar interfaces between parent liquid and bulk crystal of components 1 and 2 at their respective triple point temperatures, α is the mixing coefficient for the regular solution, m the number of bonds for each molecule within the interface which attach the surface molecule to a molecule in the interfacial layer on the other side of the surface of tension, A is the surface area of the cluster, X_2^p and X_2^n are the atom fractions solute in the parent alloy and nucleus respectively, δ is the thickness of the interface between crystal and solution and r is the radius of the cluster.

σ_o^1 and σ_o^2, the interfacial free energies for isotropic planar interfaces between the pure crystalline metal and liquid metal, are themselves unknown quantities. For pure single component systems, the isotropic interfacial tensions for liquid-vapor interfaces are equal to the equilibrium interfacial tensions because the hydrodynamic atomic or molecular relaxation times are short compared to the times necessary for the determination of the interfacial tension. For the isotropic, nonfaceted liquid-solid interface, some surface reconstruction may be necessary, especially for high entropy of fusion materials [13]. Whether or not such surface reconstruction can occur faster than the rates of interface movement caused by atom or molecule attachment or dissolution will determine if σ_o^1 will be equal to or larger than σ_{eq}^1. Since even this equilibrium quantity is obtained from the interpretation of experimental nucleation measurements, the corresponding isotropic quantity will hereby be considered as equivalent to it and calculable from a correlation applied to the undercooling and nucleation of all pure materials [14]. σ_o^1 can be related to the undercooling observed for pure metals and is the interfacial free energy necessary to balance to the Gibbs-Kelvin equation at maximum undercooling to the limiting physical spinodal supersaturation. Applying the Gibbs-Kelvin equation to a pure metals at the physical spinodal yields to following equality,

$$\Delta G_{ps} = \Delta S \quad \Delta T = -0.464 \; \sigma_o^1 / \delta \qquad (2)$$

In this equation, both the dynamic interfacial free energy and the interface thickness are unknown. If δ, the interface thickness parameter is assumed to equal the monomer radius, then the dynamic value for the interfacial free energy is calculable from knowledge of the entropy of fusion at the melting point and the undercooling to nucleation near the physical spinodal. Equation (2) becomes,

$$\Delta S_f \quad \Delta T = -0.464 \; V_m \{X_1 \sigma_o^1 + X_2 \sigma_o^2 + (\alpha m/A)(X_2^p - X_2^n)^2\}/\delta \qquad (3)$$

where ΔS_f is the entropy of formation of the crystal, V_m is the molar volume of the crystal, X_i is the mole fraction of component i, σ_o^i is the planar interfacial free energy between the liquid and crystalline phases of pure component i, α is the interaction parameter of the regular solution model, m is the number of nearest neighbors across the interface with which a particular atom interacts, A is the area of the interface per atom and δ is the interface thickness. Assuming that the density of the interfacial region is similar to that of the crystal, $V_m = A\ \delta$. In this equation, ΔS_f is the entropy of formation of a crystal from the liquid where the structure is the equilibrium crystal structure but with composition as a variable. For regular solutions, the liquid and cluster compositions influence the entropy of the transformation in direct proportion to the difference between the liquid and cluster compositions. This compositional dependence is offset by the dependence of the interfacial energy on the square of the difference between the composition between the cluster and the liquid. Thus the undercooling to the physical spinodal will be minimized if the composition of the cluster or nucleus at the physical spinodal is the same as the composition in the parent liquid. For each pure metal in the binary alloy,

$$\Delta S_f \quad \Delta T = -0.464\ V_m\ \sigma_o\ /\ \delta\ . \tag{4}$$

Then, setting $X_2^p = X_2^n$, and substituting equation (4) into equation (3) for each pure metal, yields the following composition dependence for the physical spinodal,

$$\Delta T\ /\ T_m = X_1\ (\Delta T_1\ /\ T_{m1}) + X_2\ (\Delta T_2\ /\ T_{m2}) \tag{5}$$

where $\Delta T_i\ /\ T_{mi}$ are the relative undercoolings for the pure metals, i = 1,2. This simple mixing rule will hold as long as both pure metals crystallize into the same structure. More complex cases will be taken up later.

Experimental Procedure for Measuring Undercooling in Iron-Nickel Alloys

Undercooling behavior of pure nickel, pure iron, and three iron-nickel alloy powders have been examined by using the solid-in-solid suspension technique. A Perkin-Elmer differential thermal analysis (DTA 1700) system along with a 3700 thermal analysis data station (TADS) was employed to establish the melting and nucleation temperatures of the metal and alloy powders.

Pure nickel, pure iron, and the iron-nickel alloy powders were obtained from commercial sources (Alfa products and Johnson Matthey Inc.). The purity of the iron and nickel particles are 99.9% or better, while the alloy particles are about 99+% pure. These commercial powders were prepared by conventional atomization techniques. The pure metals have a size range of 2.5-10 microns and the alloy particles have a size range of 10-50 microns.

The solid-in-solid dispersion technique [15] was used to suspend the commercially available powders and to permit preparation of samples for thermal analysis. The primary objective of the solid-in-solid suspension technique was to disperse the metal particles in a suspending medium such that all the particles were physically seperated from one another.

The suspending medium used in the sample preparation must satisfy two conditions: 1) thermal stability over the temperature range of the experiment and 2) no chemical reaction with the metal powder over the experimental temperature range. A powder that satisfies both these requirements and at the same time are available in very small (.05 micron diameter) size is gamma alumina.

In order to modify the surface oxide, a natural oxide coating which is capable of catalyzing crystal nucleation in the undercooled liquid metal, the metal particles were treated with a number of surfactant reagents. Boric acid and phosphomolybdic acid (PMA) consistently produced maximum undercooling for a given particle size distribution. Continuous use of PMA at high temperature proved to be detrimental to the platinum sample cups of the Perkin Elmer DTA-1700 thermal analysis unit. Therefore, only boric acid was used for these experiments.

The sample was prepared by mixing the metal powder with gamma-alumina suspending powder in a 1:5 ratio. A small amount of the surfactant oxide was dissolved in methanol. This solution was subsequently added to the metal-alumina mixture. This mixture was then sonified in a Branson Cell

Disruptor at 20,000 Hz. Because of the large density difference between the metal and the alumina powder, the metal particles tend to settle at the bottom during the methanol evaporation process. Therefore it was necessary to subsequently mix the suspension mechanically in order to disperse all the metal particles in the alumina powder. A thorough mechanical mixing reduced agglomeration of particles to a minimum. Optical microscopy was used to study such agglomeration. If agglomeration of metal particles was observed additional alumina powder was added and the sample subjected to another mixing routine.

The effect of thermal cycling on the metal powders was determined by scanning electron microscopy, metallography and X-ray diffraction. The particle size distribution was characterized from scanning electron and optical photomicrographs of the sample before and after the thermal analysis. The alloy particles were embedded in epoxy resin (Buehler Ltd.) and fine polished using standard metallographic techniques. The microstructures of the alloy droplets were examined in the SEM before and after the thermal analysis. The alloy powders were also subjected to X-ray diffraction in a Philips diffractometer following thermal analysis.

Results

Various levels of undercooling were observed for a given particle size distribution for different surfactant reagents. Such behavior indicated that the particle surface coating was the primary controlling factor for the undercooling. The relationship between a given surface chemistry and the corresponding degree of undercooling has been established. However, the exact physicochemical nature of the surface coating is yet to be determined. The results shown in table 1 correspond to the maximum observed undercoolings obtained with boric acid as the fluxing oxide generator.

Any significant agglomeration resulted in multiple exothermic peaks showing that a very small fraction of the droplets was initially in contact with more effective nucleants. If additional alumina powder was added to the same sample batch and the more dilute powdered metal mechanically mixed, only one exothermic peak corresponding to the previous maximum undercooling was obtained. The results are explained by percolation theory where a volume fraction of less than 16% must be maintained to prevent significant agglomeration. Agglomerated particles coalesce during the thermal cycling as evident from the multiple nucleation peaks and a microscopically observed change in the particle size distribution. For dispersions of metal powder in alumina where the metal particle make up less than 10% of the total volume, the parlticle size distributions did not change on thermal cycling. Photomicrographs of an as received Fe-42wt%Ni powder and the same post-thermal cycled alloy powder are presented in figure 1.

Table 1. Undercoolings for Iron-Nickel Powders Fluxed with Boric Acid and suspended in Gamma-Alumina powder.

Metal	Drop Size	Tm	Undercooling	
			To Exotherm Peak	Maximum
Nickel [16]	--	1453		470
Nickel [17]	--	1453		452
Nickel	2.5-3 μm	1453	345	370
Nickel	3-7 μm	1453	345	365
Fe-Ni(20:80)*	-100 mesh	1425	360	390
Fe-Ni(50:50)	-100 mesh	1427	365	400
Fe-Ni(58:42)	-325 mesh	1435	370	420
Iron	2-20 μm	1535	390	440
Iron [16]	--	1528		550
Iron [17]	--	1538		535

*wt. ratios

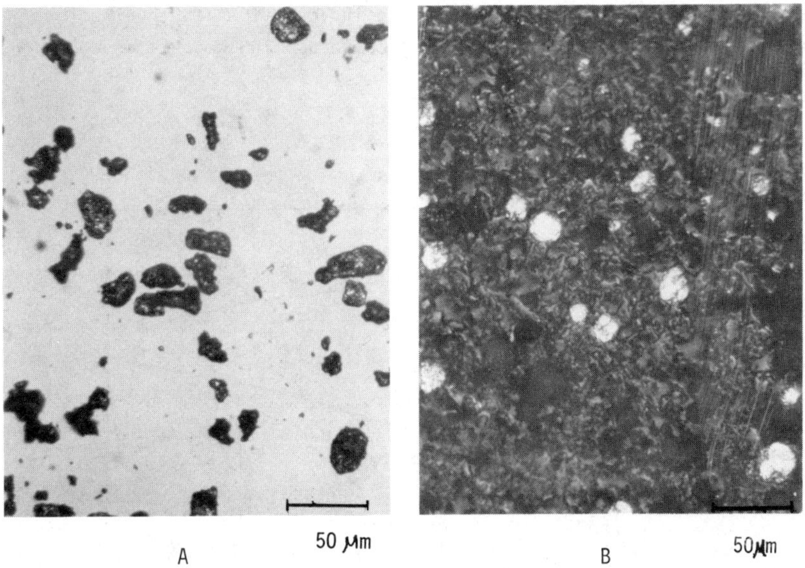

Figure 1. Photomicrographs of as received Fe-42wt%Ni powder, A, and post alumina mixed and thermal cycled sample of same, B.

The effect of particle size on undercooling for iron-nickel are being examined currently. It appears that the nature of the surface coating is the primary controlling factor for undercooling and not the particle size. The iron-nickel alloy particles were 10-50 microns in diameter compared to the pure nickel and iron which were obtained in particle sizes between 2.5 to 10 microns. However, no significant reduction in undercooling was observed for the iron-nickel alloys. This result does not indicate whether the nucleation is homogeneous or heterogeneous in nature. Also the level of undercooling represents only about 25% of the absolute melting temperature.

The effect of the alumina powder, which is in direct contact with the liquid metal droplet, is not known. However, use of different surfactants which resulted in different levels of undercooling with the same alumina suggests that the alumina powder used in this research does not have any significant effect on the nucleation process. A distinct single exothermic peak is obtained for each of the three iron-nickel alloy compositions. Figures 2 and 3 present thermograms for pure Ni and Ni-20w/o Fe respectively. The singular nature of the melting and crystallization peaks indicate that stable FCC solid solutions nucleated at maximum undercooling and melted as expected on reheating. A range of cooling rates, 10 to 50 °C/min, was used to monitor the nucleation events. No significant difference in the degree of undercooling was observed for this range of cooling rates. The composition dependence of the undercoolings indicate that the nucleation temperature scales with the liquidus temperature in composition dependence.

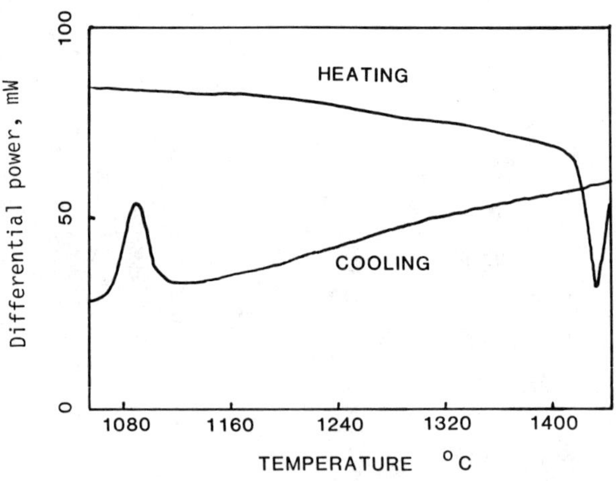

Figure 2. Thermogram for Pure Nickel.

Discussion

The experimental results indicate that the solid-in-solid dispersion process does not substantially limit undercooling of liquid metal droplets. The data presented were obtained on commercially prepared powders and the fluxing of existing surface oxides increased the observable undercooling of the droplets. The thermograms in figures 2 and 3 indicate that all droplets within each sample have been fluxed equally with the result that only one nucleation limited undercooling is observed. Figure 1 shows before and after thermal cycle photographs of the droplet distribution for an alloy powder. The droplet size distributions are substantially the same though the droplet shape has changed during the thermal cycle. The alumina powder has prevented agglomeration of the metal droplets and does not appear to be the limiting influence on undercooling.

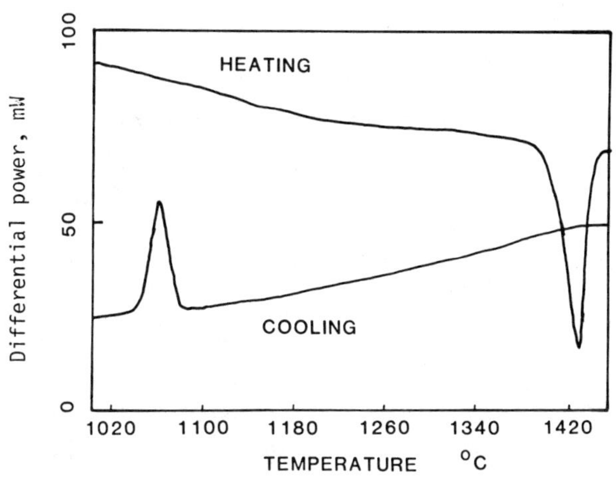

Figure 3. Thermogram for Ni-20wt%Fe.

The undercoolings presented in table 1 include the maximum values recorded by single droplet techniques for pure iron and nickel. These maximum undercoolings are still 33% greater than the undercoolings observed in this study. The main difference between the techniques is the surface preparation method. The fluxing oxide technique used here relies on solubility of surface or contaminant oxides for their removal while gas phase reduction or oxide evaporation has been carried out in the single droplet suspension studies. The pure iron and nickel results were obtained with micron particles on a hot stage microscope [15]. These results have been duplicated with single droplets suspended in a reducing gas environment [16]. Methods to achieve surface oxide reduction within the solid-in-solid distribution are under study.

The alloy powders were too small to give internal segregation as the solid-liquid interface of the solid solution was presumably stable to morphological instability during droplet solidification at high undercooling. The solidification products were all FCC solid solutions and no hint of alternate phase nucleation was observed by X-ray diffraction. These results were to be expected as the compositions which might show BCC soldification structure according to the theory of alternate phase nucleation are substantially higher in iron content at the undercoolings observed in this work [18].

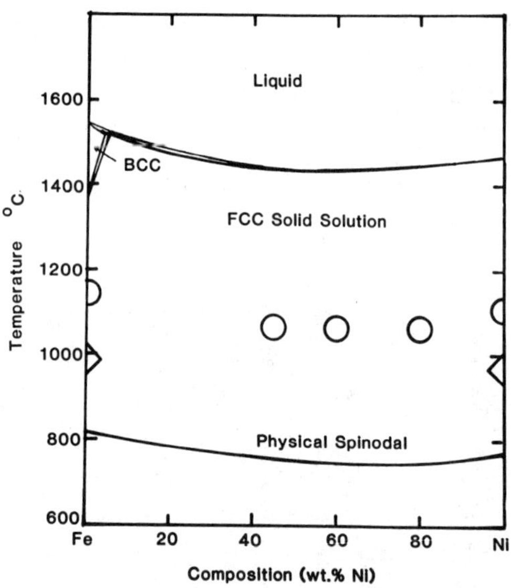

Figure 4. "Phase Diagram" for the Iron-Nickel System containing points of maximum undercooling and the physical spinodal limit to undercooling.

The physical spinodal curve for the iron-nickel system is presented in figure 4 along with the data from table 1. This "phase diagram" demonstrates that the undercoolings observed experimentally occur at relative undercoolings which scale with the liquidus and solidus curves (actually, the scaling would be with the To curve). The temperature at which the physical spinodal is predicted to exist also scales in the same manner with concentration. To calculate the undercooling to the physical spinodal as a function of composition from equation 5 required values for the undercooling to the physical spinodal for the pure elements, iron and nickel. The pure element undercoolings to the physical spinodal limit were obstained from a correlation between this limit and the liquid state stability to be presented elsewhere. It is believed that the parameters selected to quantify this correlation maximize the undercooling to the physical spinodal and overpredict the undercooling which one might observe in a droplet nucleation experiment. The undercooling to the physical spinodal predicted for pure iron and pure nickel are about 0.4 Tm. It is not expected that nucleation studies will in fact achieve this level of undercooling because the barrier to nucleation decreases exponentially on approach to this limit and disappears at the physical spinodal.

Because the nucleation temperatures observed in this study are higher than the maximum undercoolings observe for the pure metals and substantially higher than the predicted spinodal limit, it is believed that heterogeneous nucleation limits the solidification process. This belief will be tested in subsequent measurements of the nucleation rate for narrowly defined droplet distributions.

Acknowledgement

This research program is supported by ARO grant DAAG29-85-K-0196.

References

1. F.F. Abraham, Homogenoeus Nucleation Theory (Academic Press, New York, 1974).

2. D. Turnbull and J.C. Fisher, J. Chem. Phys. 17, 71, 1949.

3. D.H. Rasmussen, Journal of Crystal Growth, 56, 45 (1982).

4. L.D. Landau and E.M. Lifshitz, Statistical Physics, (Addison-Wesley,

Reading Massachusetts, 1969) p. 471.

5. B. Lewis, in Crystal Growth Second Edition, Editor Brian R. Pamplin, (Pergamon Press, Oxford, 1980) p. 23.

6. E. A. Moelwyn-Hughes, Physical Chemistry (Pergamon Press, Oxford, 1965).

7. Lawrence E. Murr, Interfacial Phenomena in Metals and Alloys, (Addison Wesley Publ. Co., Reading, Massachusetts, 1975)pp. 3-10.

8. D.P. Woodruff, The Solid-Liquid Interface (Cambridge University Press, London, 1973)

9. J.L. Katz and Frans Spaepan, Philosophical Magazine, B37, 137, 1978.

10 D.H. Rasmussen, J. Chem. Phys. 85, 2277, 1986.

11. D.H. Rasmussen, M.R. Appleby, G.L. Leedom, S.V. Babu and R. Neumann Journal of Crystal Growth 64, 229 (1983).

12. Howard Reiss, Methods in Thermodynamics (Blaisdell Press, New York, 1965).

13. K.A. Jackson, Solidification and Crystal Growth (MacMillan, New York, 1968).

14. D.H. Rasmussen, in preparation

15. D.H. Rasmussen, K. Javed, M. Appleby, and R. Witowski, Materials Letters, 3, 344, 1985.

16. V.P. Skripov, in 1976 Crystal Growth and Materials, Edited by E. Kaldis and H.J. Scheel, (North Holland, Amsterdam, 1977) p. 348.

17. J. Schade, A. McLean and W.A. Miller, This Conference Proceedings.

18. Thomas F. Kelly, Morris Cohen and John B. Vander Sande, Met. Trans A, 15A, 819-833, 1984.

CRYSTALLIZATION BEHAVIOR OF A MELT-SPUN Fe-Ni BASED STEEL

G. M. Michal and V. Laxmanan
Department of Metallurgy and Materials Science
Case Western Reserve University
Cleveland, OH. 44106

and

T. K. Glasgow
Metal Science Branch
NASA Lewis Research Center
Cleveland, OH. 44135

Whether Fe-Ni based alloys solidify with a BCC or FCC structure has been observed by many investigators to be a stronger function of kinetics and undercooling than strictly free energy minimization. Such behavior has been ovserved in an $Fe_{52.8}$, $Ni_{28.7}$, $Al_{3.4}$, $Ti_{6.1}$, $B_{9.0}$ alloy. The alloy was cast as ribbons ∿45 μm thick using a dual "free-jet" variation of chillblock melt spinning against a Cu wheel. Optical, X-ray and electron analyses of the as-cast and annealed ribbons were performed. A microstructure of at least four layers containing combinations of FCC, BCC, and amorphous phases in differing proportions was observed in the as-cast ribbon. The mid-thickness layer had the most unusual features containing FCC grains ∿75 nm in size encompassing spherulitic regions as large as 15 μm comprised of FCC grains ∿25 nm in size. The crystallization sequence responsible for the as-cast microstructure is discussed in terms of the competition between the formation of BCC and FCC phases as influenced by undercooling, recalescence, and variations in cooling rate experienced by the as-cast ribbon.

Introduction

At large undercoolings Fe-Ni alloy melts have been observed to solidify with a BCC structure even though their most stable state is FCC [1,2]. The same phenomenon of BCC nucleating sooner at large undercoolings than the more thermodynamically stable FCC structure has been observed in 303 and 316 stainless steels [3,4]. The sluggish nucleation behavior of FCC structures has also recently been observed in rapidly solidified Ni-Mo alloys [5]. These observations confirm the early predictions of Turnbull et al. [6,7] that the solid phase nucleated first from a supercool-liquid is not necessarily the most stable phase.

Additions of B to Fe and Fe-Ni alloys can lead to the formation of an amorphous structure, if the alloy is rapidly cooled from the liquid state to room temperature [8-15]. Increasing B levels up to 20 at. pct. reduce the magnitude of the cooling rate required to stabilize the amorphous structure [15]. Recent work has discovered an amorphous phase in Fe-Ni alloys with no B [16]. The alloys were subjected to the extreme cooling conditions and improbability of heterogenerous nucleation of 20 nm liquid droplets produced by electrohydrodynamic atomization. Isothermal aging of such amorphous structures at elevated temperatures causes crystallization to occur. Commonly such crystallization proceeds by the nucleation and growth of spherulites [8-13, 17, 18] which are spherical crystalline bodies of radiating crystal fibers.

In the present study metastable BCC and amorphous structures along with spherulite features were found in a rapidly solidified alloy in the as-cast condition. This alloy was part of a program to develop an austenitic steel for elevated temperature applications by utilizing TiB_2 as a strengthening phase [19]. The alloy contained a relatively large percentage of B, not to induce an amorphous structure, but only as a consequence of trying to obtain a large volume fraction of TiB_2 in the completely processed alloy. A variety of microstructural features were observed across the thickness of as-melt-spun ribbon of the alloy. The formation of this complex microstructure is discussed in terms of the variable cooling rate inherent in a melt-spinning process conducted in vacuum. The vast change in cooling rate between the ribbon when in contact, or out of contact, with the wheel will influence the ribbon undercooling, recalescence, and thus the microstructure developed.

Experimental Procedures

The composition in atomic percent of the alloy investigated was as follows: 28.7 Ni, 3.4 Al, 6.1 Ti, 9.0 B and balance Fe. The alloy was cast as ribbons 45 μm thick by 5 mm wide using a dual "free-jet" variation of chillblock melt spinning against a copper wheel [20]. One ribbon was cast against a lower thermal conductivity 4340 steel wheel to assess the phases formed under less dynamic cooling conditions. Runs were made with approximatly 20 g charges. A wheel speed of 20 m/s was employed. The casting chamber was maintained at a vacuum of about $1x10^{-5}$ torr during each run. The casting process for each run was recorded with a high speed video camera.

To assess the stability of the microstructural features found in the as-cast alloy, sections of ribbon were annealed under a vacuum of $<10^{-5}$ torr for three hours in the temperature range from 500 to 1150°C. The microstructural features present in the ribbons as-cast, and after the various heat treatments, were evaluated by combined optical, X-ray diffractometry and transmission electron microscopy (TEM) analyses. Only X-ray diffractometry was performed on the ribbon cast against the 4340 steel wheel. Specimens for optical microscopy were etched in a solution of 33 acetic acid, 33 nitric acid, 1 hydrofluoric acid, 33 water by volume. TEM foils were prepared by a jet polishing technique in a 95 acetic acid, 5 perchloric acid by volume electrolyte at 12°C.

Results

The casting procedures employed produced continuous ribbons about 10 m in length with very uniform cross sectional dimensions and microstructure features for all but the first meter of their length. Good wetting of the Cu wheel by the molten alloy was observed once this steady-state casting condition was reached on each run. Figure 1 contains optical micrographs of the steady-state portion of as-cast ribbon viewed along its longitudinal and transverse edges. Four distinct layers are shown. At the wheel side surface of the ribbon is a light etching featureless layer about 5 μm thick. Above this is a layer about 30 μm thick that contains etch resistant circular features with a range of sizes. A large (15 μm) one in the right portion of the longitudinal micrograph shown in Figure 1 and a smaller (2 μm) one in the left portion are indicated by arrows. The two orthogonal views of the microstructure in Figure 1 serve as evidence that these circular features are spherical in three dimensions. These features, which will be called spherulites, are embedded in a darker-etched matrix. Above this second layer is a thin third layer about 2.5 μm thick. This layer has the same etch resistant, white, featureless appearance as the spherulites found in the layer below. A fourth layer about 4 μm thick runs along the outer, or free, surface of the ribbon. This layer etched much darker than the third layer and spherulitic regions in the second layer.

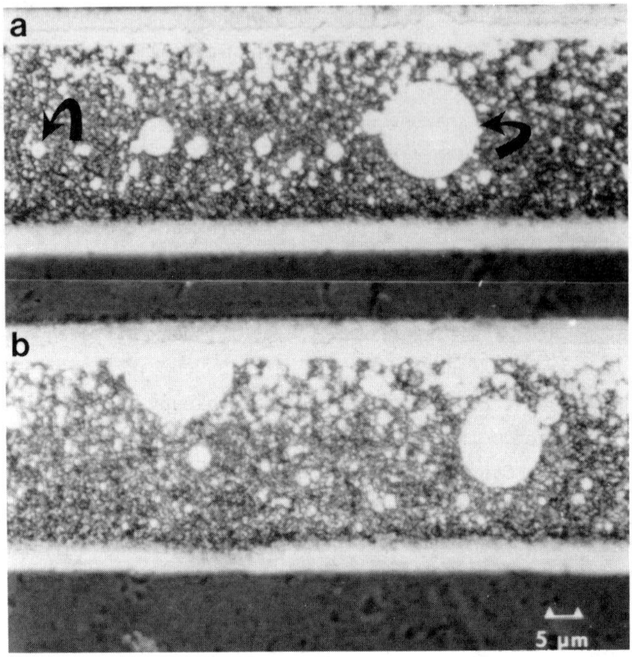

Fig. 1 - Optical micrographs of the "a" longitudinal and "b" transverse edges of the as-cast ribbon. The lower surface contacted the Cu chillblock wheel.

Figure 2 contains two optical micrographs that illustrate additional features found in the as-cast ribbons. Often the interior of a spherulite contained a darker etching region as shown in the center of Figure 2a. Figure 2b illustrates the same situation for spherulites that impinge upon the third layer. This darker etching phase, sometimes found in the center of the spherulites, displays etching characteristics very similar to those of the fourth layer. The micrographs in Figure 2 also show more detail in the fourth layer than those in Figure 1 due to changes in illumination and focus designed to highlight features found in that layer. The fourth layer appears to contain a very fine cellular dendritic structure just resolvable in the micrographs contained in Figure 2.

TEM examination of the mid-thickness region (second layer) of the as-cast structure revealed the presence of a bimodal grain size distribution.

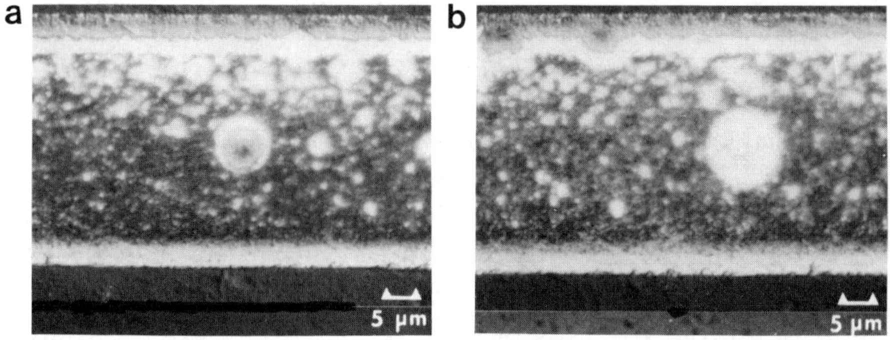

Fig. 2 - Optical micrographs of longitudinal edges of the as-cast ribbon showing a dark etching BCC phase in the center of several spherulites.

Figure 3 illustrates an area of the ribbon that contained fine grains ∿ 25 nm in size. Other areas in the same foil contained larger grains ∿ 75 nm in size as shown in Figure 4. Electron diffraction patterns taken from both types of areas indicated that they both were comprised of only an FCC phase. The relationship between these regions with different grain sizes could not be assessed in this as-cast material because foil thinning was only obtainable in small isolated regions.

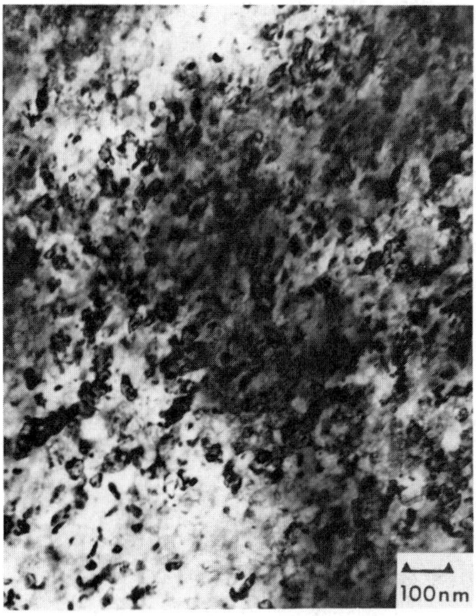

Fig. 3 - A bright field image of a fine grained FCC region in the center of an as-cast ribbon.

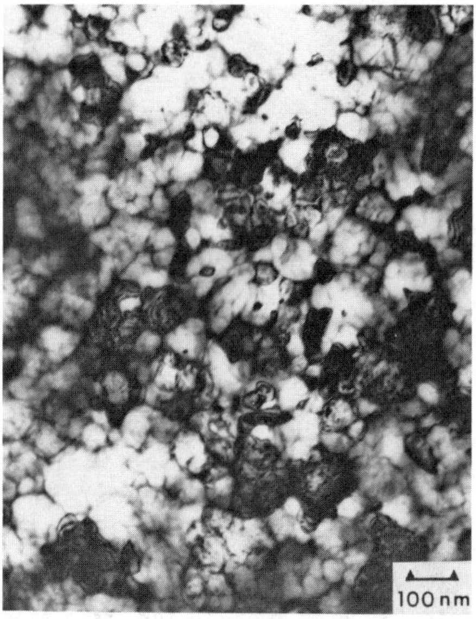

Fig. 4 - A bright field image of a larger grained FCC region in the center of an as-cast ribbon.

However, after annealing at 600°C, the ribbons were able to be thinned with exceptional uniformity. This allowed large areas of a foil to be examined and composite images recorded at magnifications as low as 2,800X. Such magnifications approach those used for optical examination of the ribbons. A typical area imaged in such a fashion is shown in Figure 5. This normal-to-the-surface view of the interior of the ribbon contains two dark circular areas. These are spherulites now observed in the third orthogonal direction. Higher resolution dark field imaging revealed that the spherulitic areas were comprised of grains ∿ 60 nm in size. The surrounding grain size was observed to be ∿ 230 μm. Electron diffraction indicated that the grains in both areas had a FCC structure.

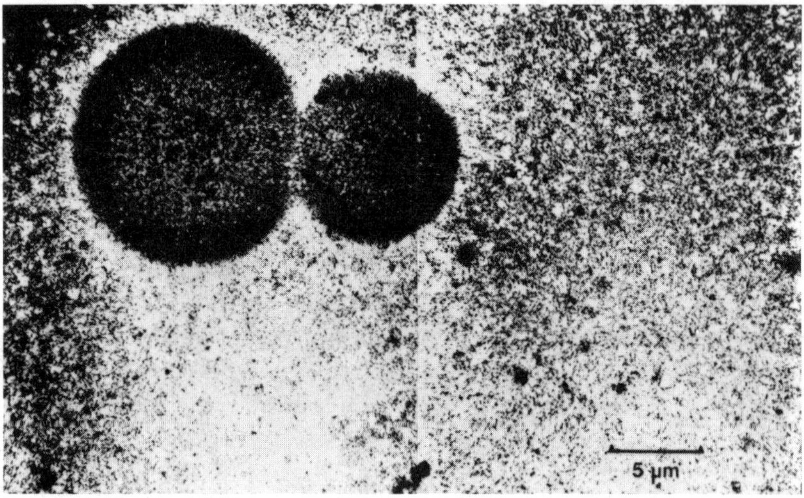

Fig. 5 - A low magnification bright field image of a ribbon that was annealed at 600°C for 3 hrs. Two large spherulitic regions are evident.

The results of X-ray diffractometry scans run on the wheel and free sides of ribbons in the as-cast and annealed-at-600°C conditions are listed in Table I. The wheel side of as-cast ribbons generated a relatively high background count rate and a low count total under the three peaks for each of the FCC and BCC phases. The peak sizes indicate that roughly equal amounts of both phases were present on that side of the wheel. The low peak heights and high background levels measured are taken as evidence for the presence of an amorphous phase in the microstructural layer directly adjacent to the wheel surface. The free side displayed a very different behavior. There was a very dominant BCC-phase present that exhibited a strong 001 texture. This type of strong texture is indicative of a cellular dendritic microstructure in as-cast materials. This X-ray result is consistent with the observation of the fine cellular dendritic structure found in the fourth layer as shown in Figure 2. The penetrating power of the X-rays employed was large enough that the FCC peaks detected on the free side of the as-cast ribbon may have come exlusively from the second and third layers of microstructure.

TABLE I

X-RAY ANALYSIS OF RIBBON SURFACES

		FCC PLANES			BCC PLANES		
		111	002	022	011	002	112
WHEEL SIDE	AS CAST	117	10	14	110	44	21
	600°C/3 HRS.	256	110	83	10	8	5
FREE SIDE	AS CAST	196	272	237	25	18,500	5
	600°C/3 HRS.	64	228	640	8	20	5

The X-ray analysis of the as-cast ribbons was repeated after holding the samples at -196°C for 30 minutes. No change in X-ray counts was observed due to the low temperature excursion.

After the anneal at 600°C the amount of each phase detected from each side of the ribbon changed significantly. On both sides of the ribbon the BCC peak counts fell to essentially a background level. The number of counts in the FCC peaks increased substantially. This indicates that whatever BCC phase was present in the as-cast ribbon was unstable at 600°C and transformed to FCC. As further evidence for the FCC phase being the most stable, X-ray analysis of the ribbon cast against the 4340 steel wheel revealed only FCC peaks even in the as-cast condition.

Discussion

Analyses of the as-cast ribbons revealed the presence of a layer of BCC phase along the ribbons' free surface. Some BCC phase was also detected in the region of the ribbon near the wheel surface. There was also evidence found that the microstructural layer of the ribbon directly adjacent to the wheel surface contained an amorphous phase. To create these metastable phases the ribbon must have experienced substantial undercooling during contact with the Cu chillblock wheel.

Estimates of the degree of undercooling achieved by the ribbon while in contact with the wheel based upon the conditions needed to form the metastable phases have been made in several ways and with a range of results. Simple extrapolation of the T_0 temperature for solidification of a BCC phase with the same composition as the liquid in the Fe-Ni phase diagram indicates that there is some driving force to form the BCC phase with as little as $\sim$ 150 degrees of undercooling at the Fe to Ni ratio of the alloy ribbon. To actually nucleate and grow the BCC phase requires additional undercooling, since the BCC phase must form earlier than the FCC if it is to appear in the as-cast ribbon. A calculated undercooling of about 500 degrees is needed to nucleate the BCC phase in preference to the more stable FCC phase in Fe-Ni alloys with the same Fe to Ni ratio as the alloy ribbon [2]. This number is consistent with the experimentally determined undercooling of 475 degrees needed to observe the formation of a BCC phase from the melt in 316 stainless steel [4].

The observed 475° undercooling may have been greater than that required to initially form some BCC, the excess being necessary to preserve the structure formed. Kelly [4] calculated that BCC could form (with increasing driving force) for undercoolings of 250 degrees or more.

To see if undercoolings of the magnitude just quoted were feasible in the short time the ribbon was in contact with the wheel, a heat transfer calculation was performed. The calculation was based on a model for cooling along a single dimension a slab with both surface and internal resistance to heat transfer [21]. The result of such a calculation for the thermal profile of the ribbon at its moment of last contact with the wheel is shown in Figure 6. The calculation was based upon an initial melt temperature of 1500°C, average temperature of the wheel in contact with the ribbon of 200°C, thermal diffusivity of the ribbon of $6 \times 10^{-6} m^2/s$, an interfacial heat transfer coefficient of $2.4 \times 10^5 W/m^2$-°C and a contact time of 7.5×10^{-4}-s. The calculation showed that the liquid alloy could cool about 600 degrees during its brief contact with the wheel. The computed temperature differential between the wheel and free surfaces was about 100 degrees which converts into an average temperature gradient of 2.2×10^6°C/m. Thus, this simple calculation indicates that the ribbon had the ability to undercool sufficiently that a BCC structure potentially could have formed. The region of the ribbon adjacent to the wheel surface may have undercooled to a much greater extent than the rest of the ribbon and locked into an amorphous phase.

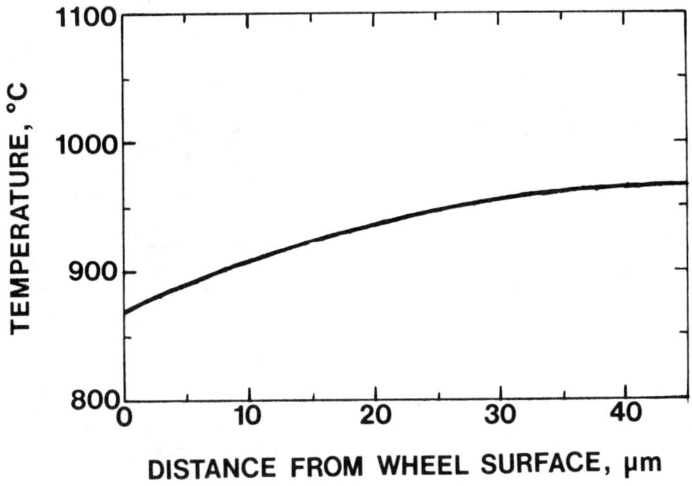

Fig. 6 - The thermal profile across the ribbon at the moment of loss of contact with the Cu chillblock wheel.

The thermal profile calculation performed was highly dependent on the estimated value for the interfacial heat transfer coefficient. The value of $2.4 \times 10^5 W/m^2$-°C used is consistent with the recent work of Huang et al. [22] for melt spinning in a vaccum.

A possible sequence of transformations occurring during the cooling of the alloy ribbon from a liquid to the complex microstructure illustrated in Figures 1 and 2 is schematically represented in Figure 7. The sequence was based upon variable cooling rates experienced by various areas of the ribbon due to two factors. The first was the large difference in heat transfer between the ribbon in-contact and out-of-contact with the wheel. The second was the effect of recalescence which is directly related to the rate that regions of ribbon crystallized.

Figure 7a represents the ribbon at its moment of first contact with the wheel. After completing 7.5×10^{-4}-s of contact with the wheel, the ribbon consisted of undercooled liquid as shown in Figure 7b. All but the free side region of the ribbon may have cooled below its glass transition temperature at this point. The structural rigidity of the ribbon, indicated

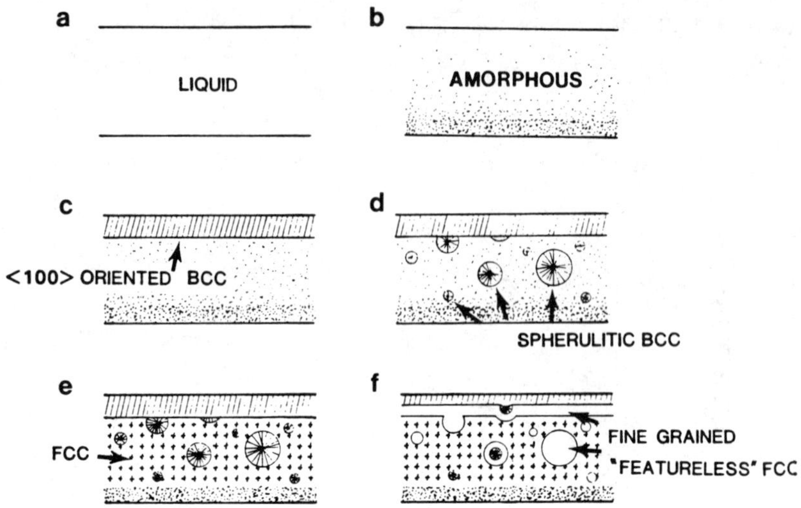

Fig. 7 - A suggested sequence of transformations that occurred to produce the final as-cast microstructure.

by its flatness and constant cross section, serve as evidence that undercooling to a glass transition temperature was achieved while the ribbon was in contact with the wheel. The first speculated crystalization of the ribbon occurred after loss of contact with the wheel. After loss of contact the rate of heat loss is controlled by radiation. A cellular dendritic front of BCC phase was able to nucleate near the "warmer" free-side of the wheel and grow to the free side surface. The microstructure of the ribbon after this event is shown in Figure 7c. Most of the heat from recalescence was extracted through the established temperature gradient towards the wheel side of the ribbon. This caused the interior of the ribbon to gain heat because radiative heat loss from the ribbon is slow. Then over a period of time throughout the interior of the ribbon additional BCC phase was able to nucleate in the glassy phase and grow radially to form spherulites. The ribbon after formation of spherulites is illustrated in Figure 7d. The growth of the spherulites caused further heat to be expelled into the glassy matrix causing its temperature to rise. This caused a net slowing of the cooling rate in the ribbon faciliting the nucleation of the stable FCC phase in the remaining glassy regions. The growth of the FCC phase then consumed the remaining glassy phase in the interior of the ribbon as shown in Figure 7e. Recent work has shown that a Ni-based amorphous alloy containing 9 at. pct. B crystallized after annealing at 750°C into FCC grains with a size comparable to those found between the spherulites [23]. The recalescence from this third crystallization step again caused a slowing of the cooling rate in the interior of the ribbon. Time was then available for the solid state transformation of much of the primary-crystallization BCC phase to a fine-grained, "featureless" FCC phase.

The final microstructure is illustrated in Figure 7f. The BCC phase regions that survived the total cooling sequence to room temperature are found predominantly on the free side surface and in spherulites near the wheel surface. During the solid state crystallization in the ribbon interior, these regions cooled by radiation. This allowed a greater suppression of the solid state transformation to FCC. On occassion BCC phase appears to remain in the center of large spherulites even in the center of the ribbon. The ribbon must have grown too cold before the transformation to FCC could consume all of these spherulites.

The complex sequence of transformations suggested in Figure 7 is consistent with a time-temperature-transformation diagram for the alloy ribbon schematically illustrated in Figure 8. The diagram shows that at large undercoolings the BCC phase forms prior to the FCC phase. However, at longer hold times at those undercoolings the BCC phase will transform to FCC. The arrest of cooling due to relascence from the BCC crystallization would promote this solid state transformation. Figure 8 also indicates that at very low temperatures the FCC phase may form before the BCC phase. Evidence for such a transformation sequence is suggested by crystalization experiments performed on amorphous Fe-Ni-B alloys [8,13,14].

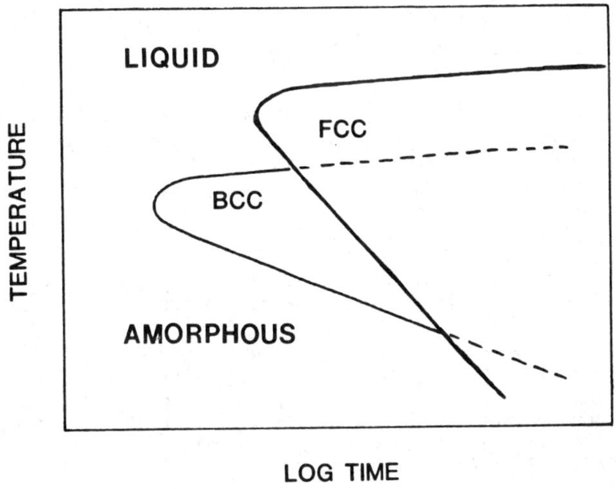

Fig. 8 - A schematic representation of the time-temperature-transformation kinetics for the BCC and FCC phases.

Conclusions

A complex microstructure consisting of four layers was observed in an as-cast Fe-Ni-Al-Ti-B melt spun ribbon cast against a Cu chillblock wheel. The layers contained regions with amorphous and BCC phases that were unstable with respect to a FCC phase.

Sufficient undercooling was achieved while the ribbon was in contact with the Cu chillblock wheel that the BCC and amorphous phases could form. Depending on the local cooling path these metastable phases either survived the much slower cooling rate after loss of contact with the wheel and appear in the as-cast microstructure, or they transformed to the FCC phase. The local cooling path in the ribbon was a function of both position with respect to the wheel and free side surface and the local rate of recalescence.

References

1. R. E. Cech, "Evidence for Solidification of a Metastable Phase in Fe-Ni Alloys", Trans. AIME, 206 (1956) 585-589.

2. Y.W. Kim and T.F. Kelly, "Alternative Crystallization Phases in Sub-Micron Droplets of Fe-Ni Alloys" Proceedings of the 43rd Annual Meeting of the Electron Microscopy Society of America, ed. G.W. Bailey (San Francisco, CA: San Francisco Press, Inc., 1985) 46, 47.

3. T.F. Kelly, "A High Resolution Structural and Compositional Analysis of Rapidly Solidified Strainless Steels" (Sc.D. thesis, Massachusetts Institute of Technology, 1982)

4. D.G. MacIsaac, Y. Shiohara, M.G. Chu and M.C. Flemings, "Structure of Undercooled Iron and Nickel-Based Alloys" Grain Refinement in Castings and Welds eds. G.J. Abbaschian and S.A. David (Warrendale, PA: The Metallurgical Society, 1983) 87-95.

5. S.N. Tewari and T.K. Glasgow, "Undercooled and Rapidly Quenched Ni-Mo Alloys" Undercooled Alloy Phases eds. E.W. Collings and C.C. Koch (Warrendale, PA: The Metallurgical Society, 1986)

6. D. Turnbull, "Formation of Crystal Nuclei in Liquid Metals, J. of Applied Physics, 21 (1950) 1022-1028.

7. J.H. Hollomon and D. Turnbull, "Nucleation" Progress in Metal Physics, Vol. 4, ed. B. Chalmers (London: Pergamon Press, 1953) 333-388.

8. R.S. Tiwari, S. Ranganathan and M. von Heimendahl "TEM of the Kinetics of Crystallization of Metglas R 2826", Z. Metallk., 17 (1981) 563-568.

9. M. von Heimendahl and G. Kuglstatter, "The Activation Energies of Crystallization in the Amorphous Alloy Metglas R 2826A", Journal of Mat. Sci., 16 (1981) 2405-2410.

10. A.L. Greer and J.A. Leake, "The Stability and Transformation of $Fe_{80}B_{20}$" Rapidly Quenched Metals III, Vol. 1, ed B. Cantor (London: The Metals Society, 1978) 299-306.

11. V. Koster and V. Herold "Crystallization of Amorphous $Fe_{80}B_{20}$" Scripta Met., 12 (1978) 75-77.

12. J.L. Walter and S.F. Bartram "Crystallization of Some Amorphous Alloys", Rapidly Quinched Metals III, Vol. 1, ed. B. Cantor (London, The Metals Society, 1978) 307-314.

13. T. Watanabe and M. Scott, "The Crystallization of the Amorphous Alloy $Fe_{40}Ni_{40}P_{14}B_6$," J. of Mat. Sci., 15 (1980) 1131-1139.

14. V. Herold and V. Koster, "The Influence of Metal or Metalloid Exchange on Crystallization of Amorphous Iron-Boron Alloys" Rapidly Quenched Metals III, Vol. 1, ed. B. Cantor (London: The Metals Society, 1978) 281-290.

15. E. Hornbogen and I. Schmidth, "Metastable Crystalline Phases and Glasses in Ternary Fe-C-X Alloys" Rapidly Quinched Metals III, Vol 1 ed. B. Cantor (London: The Metals Society, 1978) 261-264.

16. Y.W. Kim and T.F. Kelly, "A Study of Alternative Crystallization Phases in Sub-Micron Spheres of Iron-Nickel Alloys", Undercooled Alloy Phases, eds. E.W. Collings and C.C. Koch (Warrendale, PA: The Metallurgical Society, 1986)

17. E. Hornbogen, "Combined Reactions", Met. Trans., 10A (1979) 947-972.

18. A.J. Drehman and A.L. Greer, "Kinetics of Crystal Nucleation and Growth in $Pd_{40}Ni_{40}P_{20}$, Glass", Acta Met., 32 (1984) 323-332.

19. G.M. Michal, T.K. Glasgow and T.J. Moore, "Elevated Temperature Strengthening of a Melt Spun Austenitic-Steel by TiB_2", Rapidly Solidified Materials, eds. P.W. Lee and R.S. Carbonara (Metals Park, OH: ASM, 1985) 265-271.

20. R.W. Jech, T.J. Moore, T.K. Glasgow and N.W. Orth, "Rapid Solidification Via Melt Spinning: Equipment and Techniques" J. of Metals, 36 (4) (1984) 41-45.

21. H.J. Carslaw and J.C. Jaeger, Conduction of Heat in Solids, (Oxford: University Press, 1959)

22. S.C. Huang, R.P. Laforce, A.M. Ritter and R.P. Goehner, "Rapid Solidification Characteristics in Melt Spinning a Ni-Base Superalloy" Met. Trans. 16A (1985) 1773-1779.

23. P.A. Scott and J.M. Rigsbee, "Microstructural Analysis of Crystallization Processes in a Ni-Based Amorphous Alloy", Materials Letters 2 (1984) 328-332.

A STUDY OF ALTERNATIVE CRYSTALLIZATION PHASES IN SUB-MICRON SPHERES OF IRON-NICKEL ALLOYS*

Yeon-Wook Kim, Hong-Ming Lin, and Thomas F. Kelly

Department of Metallurgical and Mineral Engineering
University of Wisconsin
Madison, Wisconsin 53706

Three iron-nickel alloys (30, 40, 50 at% Ni) have been processed in vacuum by electrohydrodynamic atomization to produce sub-micron spheres. Transmission electron microscopy is used to determine crystal structure, composition, and microstructure of the solidified particles. The alternative crystallization phase, BCC, is found in the 30 and 40 at% Ni alloys. The 50 at% Ni alloy is found to crystallize as FCC only. Both of these findings are in qualitative agreement with predictions of a model of alternative-crystallization-phase nucleation. An FCC phase is found in some of the 30 and 40 at% Ni alloys and its origin is considered to be due to heterogeneous nucleation. In addition, a very small fraction of the particles are found to have transformed martensitically from fcc to bcc. Furthermore, about one-fourth of the smaller droplets (less than 40 nm diameter) are found to be amorphous. The implications of these findings for classical nucleation theory are discussed.

---

* This work was supported by an award from the International Business Machines Corporation and by a grant from the Department of Energy, Basic Energy Sciences #DE-FG02-85ER45215.

Introduction

When a liquid becomes highly supercooled, the crystal phases that nucleate from it may be any of several alternatives other than the primary crystallization phase. At the temperature at which they nucleate, these alternative phases are not necessarily metastable with respect to the primary crystallization phase. It is possible to have a situation where a crystal phase other than the primary nucleates at a temperature where it is the most stable crystal phase. A distinction should also be made between crystalline solidification and amorphous solidification. Thus, though this type of phase production is commonly referred to as metastable phase formation, the crystalline phases are here called by the more general name of alternative crystallization phases [1,2].

If nucleation catalysts of high catalytic potency are present in the liquid, then nucleation of a crystal phase will occur at small liquid supercooling where the primary crystallization phase is the likely nucleation product. In order to obtain alternative crystallization phases, these potent nucleants must be eliminated. Isolation of the potent heterogeneous nucleants by small droplet processing (atomization) is used in this work to obtain large liquid supercoolings in a large fraction of the liquid volume. If all heterogeneous nucleants are eliminated from a given liquid droplet, then homogeneous nucleation becomes possible.

Electrohydrodynamic atomization (EHD) [3,4] is a process that will produce metal droplets from 2 nm to 50 micrometer diameter. A wire or rod of material is melted at its tip and an electrostatic field draws small droplets from the molten pool. The liquid droplets solidify in free flight through a vacuum at 10^{-4} Pa or better. This size particle can be observed in the transmission electron microscope to determine the crystal structure by electron diffraction, the composition by x-ray analysis, and the microstructure from the image.

At the vacuum levels present in this process, surface films that might grow on the liquid droplets will be minimal but can be present. Also, at 10 nm diameter any individual droplet has a negligible probability of containing a heterogeneous nucleant in its liquid volume.

The iron-nickel alloy system is of interest for this study because it has an fcc and a bcc phase that have changing relative stability with

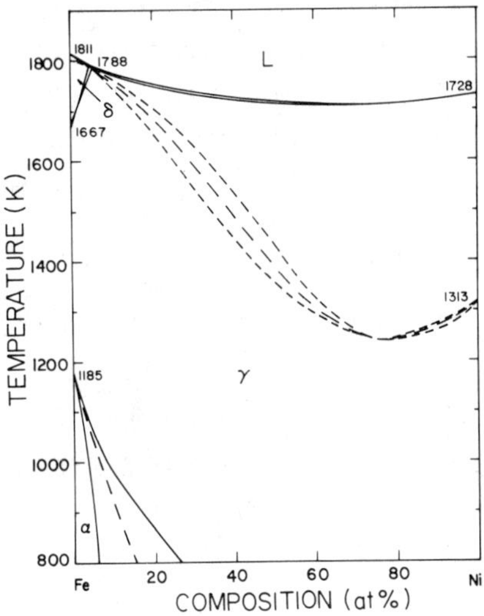

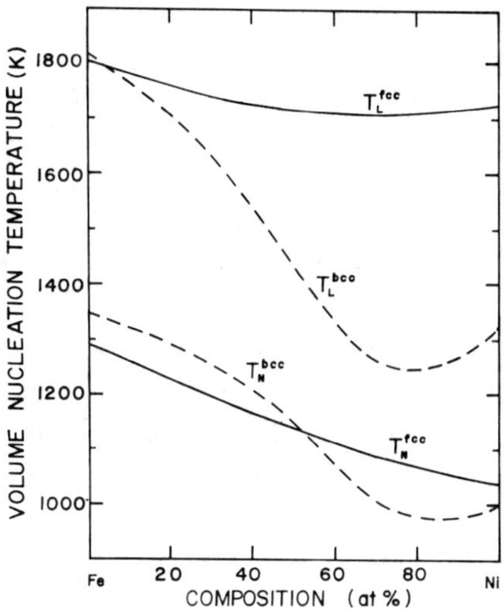

Fig. 1 Iron-nickel phase diagram per Chuang et. al. [5,6] which shows calculated phase boundaries.

Fig. 2 Calculated volume nucleation temperature for the iron-nickel system.

composition and temperature, Figure 1. The iron-nickel system has also been modelled thermodynamically. This allows that the free energy, enthalpy, and temperatures of cystallization may be calculated for both stable and metastable phases. It is thus convenient to systematically study relative nucleation rates of the two phases experimentally and analytically. Analytical models of this nucleation process have been used to predict which phases will be produced under certain experimental conditions. These predictions are being tested by studying the solidification structures of the EHD powder and results of these studies are reported below.

Analytical Model

Kelly and Vander Sande [2] have calculated nucleation probabilities of alternative crystallization phases in the iron-nickel system for homogeneous nucleation in 100 nm droplets. Classical nucleation theory is used to calculate nucleation frequencies as a function of temperature. The total number of nuclei produced is obtained as the integral of the nucleation frequency in time. The free energy change on crystallization for both fcc and bcc at any temperature and composition is obtained from the thermochemical models of Chuang et. al. [5,6] for iron-nickel. The dashed lines in Figure 1 are the calculated liquidus and solidus curves for the bcc phase obtained by Chuang et. al. The liquid/crystal interface energy for both fcc and bcc are taken from the work of Spaepen et. al. [7-9]. At each temperature in the integration, the composition of the crystal phase that results in the maximum nucleation rate is calculated. The nucleation temperature for homogeneous nucleation is determined as the temperature where the integrated number of nuclei reaches unity.

The Kelly and Vander Sande results, which are shown in Figure 2, indicate that the bcc phase has a greater nucleation temperature in these droplets for compositions less than 52 at% Ni. According to these calculations, if iron-nickel alloys are droplet processed by EHD to produce 100 nm spheres, then the bcc phase would be expected to nucleate and grow for compositions less than 52 at% Ni.

Experimental

The iron-nickel alloy powders were prepared from 99.99 wt% iron and nickel stock. The EHD process developed by Phrasor Scientific, Inc. in Duarte, California [3,4] has been used to produce liquid droplets in the range of 20 to 150 nm diameter. Iron-nickel powders of that size are transparent to 100 kV electrons such that they may be imaged in the transmission electron microscope (TEM) without the need for thinning.

The droplets solidify in free-flight through a vacuum chamber operated at about 10^{-4} Pa and impinge on a substrate after solidification is complete. Two substrates that have been used are single crystal rock salt and acetate tape. Rock salt is the preferred substrate since it minimizes hydrocarbon contamination in subsequent observation in the TEM. Specimens for TEM were prepared by depositing a very thin film of carbon onto the substrate and cutting these into 3 mm squares. The substrate was then dissolved in an appropriate solvent: distilled water for the rock salt and acetone vapor for the acetate tape. The carbon film with powder particles was mounted on 1000 mesh TEM grids. By using a very fine mesh size for a supporting grid, it was possible to make the support film very thin which aided observation of the diffraction patterns of the particles.

Analytical electron microscopy was used for determining the crystal structure versus composition results of the solidified particles. The crystal structure can be determined directly by electron diffraction techniques as shown below. The composition of particles was determined by energy dispersive x-ray analysis and was found to be essentially the nominal compositions. Additionally, the diameter of each particle can be determined with great accuracy from the image of the particle.

This work was performed on a Vacuum Generators field-emission HB501 STEM. There are two different operating modes on this instrument that have been used for diffraction and imaging work in this investigation as shown in Figure 3. In normal image mode, all lenses are operational and the scanned beam is focussed to a 0.5 nm spot. Images with better than 0.5 nm resolution are obtained. The convergence of the beam in this mode, however, is on the order of 5 milliradian which is large for electron diffraction work. Because of the large convergence angle of the incident illumination, the diffracted beams appear as discs in the diffraction pattern. Complex patterns with diffracted beams that are close to each other cannot be sorted out in this mode.

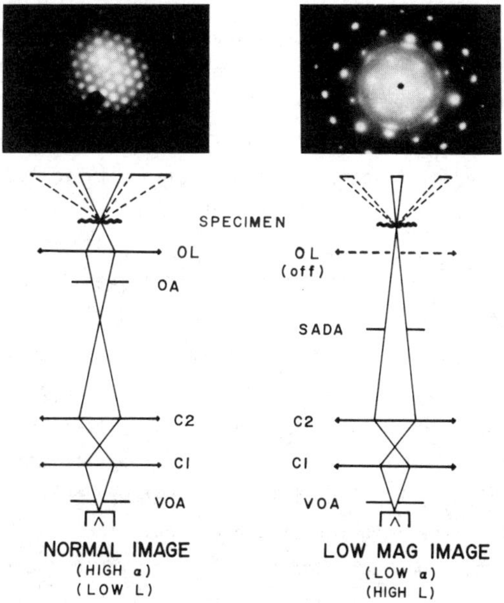

Figure 3 Illustration of the electron diffraction modes used in this study.

It is also possible to operate the instrument with the objective lens turned off to facilitate low magnification imaging. In this latter mode, the scanned beam is focussed to a 10 nm spot with a beam convergence of only 0.5 milliradian. The low beam convergence of this mode is beneficial for observing fine detail in the diffraction information. However, the image resolution of this mode is too poor to resolve microstructural information in the nominally 100 nm diameter particles of this study. Thus, both high-resolution imaging and low-beam-convergence diffraction can be obtained from very small particles (10 nm diameter) but they cannot be obtained simultaneously. Furthermore, though it is possible to obtain both the high-resolution image and a low-beam-convergence diffraction pattern from the same particle, in practice this is not readily accomplished.

Results

An image and corresponding diffraction pattern of a typical solidified droplet are shown in Figures 4 and 5 for fcc and bcc respectively. Typically, the entire particle was surveyed by moving the beam to all areas of the image while observing the diffraction pattern. The solidified particles are almost exclusively single crystal independent of whether they are fcc or bcc. The microstructures are usually relatively free from features and do not show evidence of a large defect content. The microstructures of Figures 4 and 5 are representative of the majority of the particles observed. Furthermore, the tabulated results below represent only the results of analysed diffraction patterns but are typical of a much larger number of structures for which only the image was observed.

A small fraction of the particles of 30 and 40 at% Ni contain microstructural features that have the appearance of twins. An example of this type of microstructure is shown in Figure 6. A diffraction pattern from a particle similar to that of Figure 6 is shown in Figure 7. Since it is not possible to obtain the low-beam convergence pattern in normal image mode, it has not always been possible to clearly identify the patterns with the corresponding micrograph. The diffraction pattern shown in Figure 7 exhibits {112} habit plane twinning of a bcc structure. Another complex diffraction pattern that was observed is shown in Figure 8. Indexing of the pattern indicates that it is consistent with a partial martensitic transformation from fcc to bcc where the relative orientation is consistent with the Nishiyama-Wasserman orientation relationship [10,11]. Additional

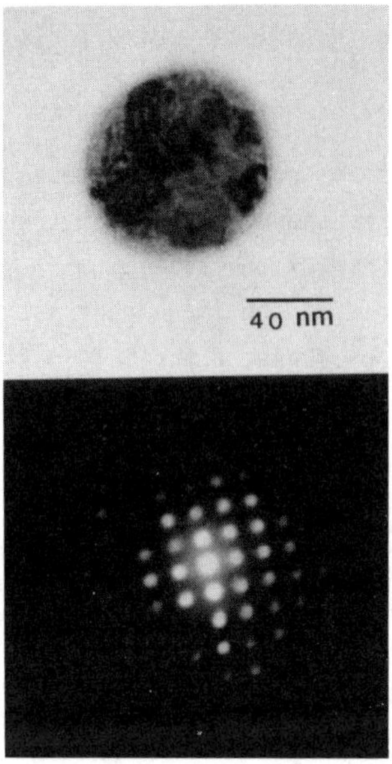

Fig. 4 EHD particle of fcc with high beam convergence diffraction pattern.

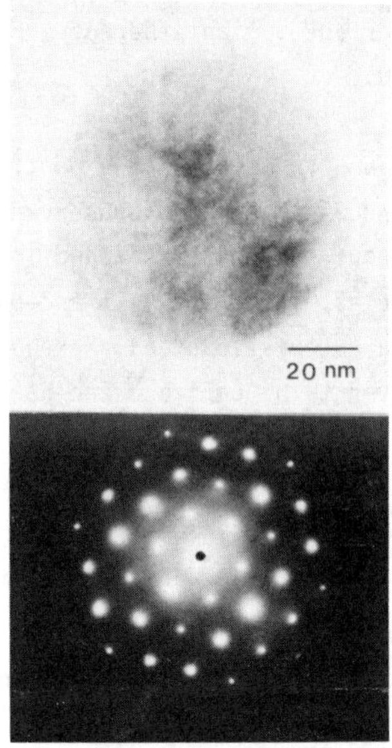

Fig. 5 EHD particle of bcc with low beam convergence diffraction pattern.

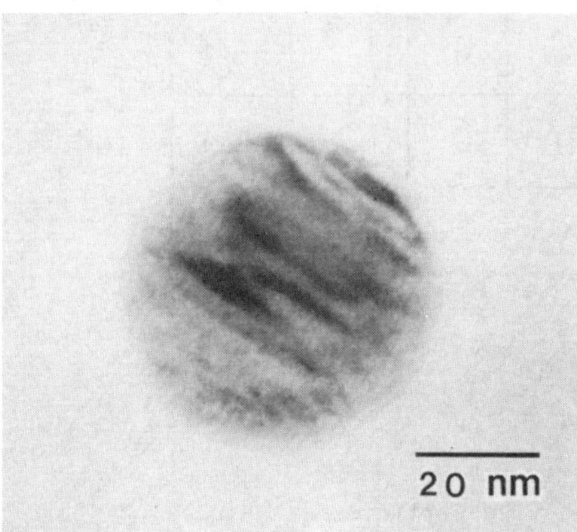

Fig. 6 EHD particle of 30 at% Ni with twinned structure.

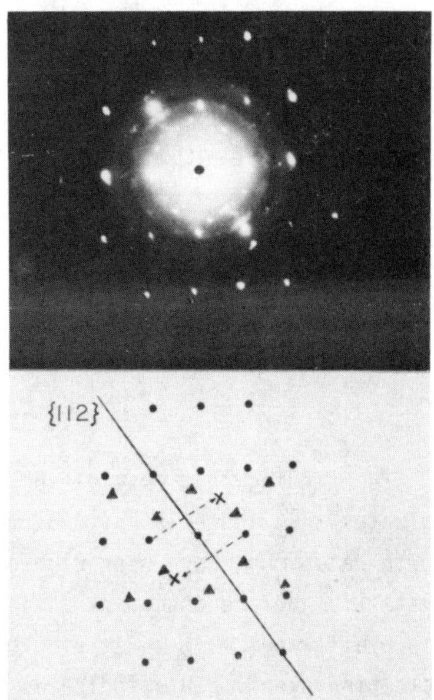

Fig. 7 Diffraction pattern and schematic of a twinned bcc structure with a {112} habit plane.

complex patterns have been observed that cannot be indexed as either fcc, bcc, or any variant thereof. Further work on these latter structures is needed.

In a fraction of the particles smaller than about 40 nm diameter, an amorphous structure is observed. An example is shown in Figure 9. The amorphous structure is found in about one-fourth of the sub-40 nm particles. Note the salt-and-pepper appearance of the image contrast which is characteristic of an amorphous phase. In one case, a particle was observed to be part bcc and part amorphous, Figure 10.

The various results that have been obtained for iron 30, 40, 50 at% nickel alloys are summarized in Table 1 according to the type of structure observed.

Table 1 Summary of microstructures observed.

ALLOY	fcc	bcc	fcc + bcc	complex	amorphous (< 40 nm)
Fe-30 wt % Ni	25	52	2	15	~1/4
Fe-40 wt % Ni	4	19	0	—	~1/4
Fe-50 wt % Ni	13	0	0	—	0

Discussion

The finding that most of the solidified particles are single crystal indicates that there is but one nucleation event per droplet. Crystal growth rates must have been high enough to cause the droplet to fully crystallize before a second nucleation event occured. The droplet of Figure 10 exhibits what must be attributed to sluggish growth after a single nucleation event. This indicates that crystal nucleation rates were never large and the liquid never reached a point where copious nucleation could occur. Though copious nucleation is indicative of homogeneous nucleation, the single nucleation events in the droplets of this study are consistent with either homogeneous or heterogeneous nucleation.

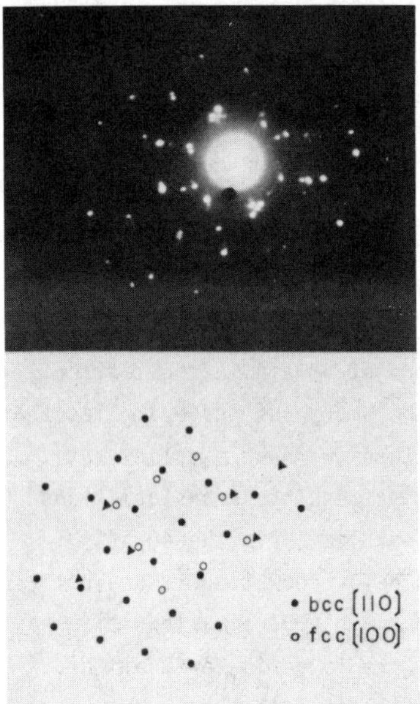

Fig. 8 Diffraction pattern of a structure with the Nishiyama-Wasserman relationship.

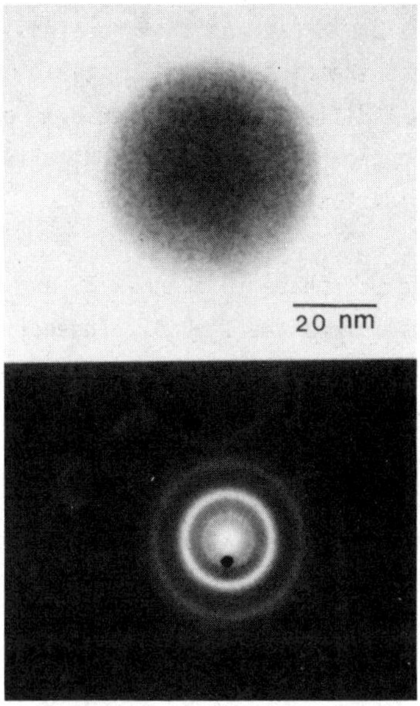

Fig. 9 Amorphous particle and low-beam-convergence diffraction pattern.

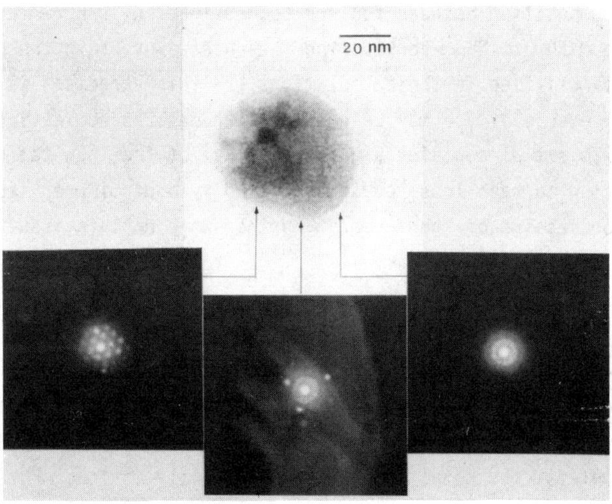

Figure 10 EHD particle of Fe-30 at% Ni that is part bcc, part amorphous.

The fact that Fe-50 at% Ni is exclusively fcc, that the diffraction patterns are clean and show no signs of strain, and that the phase diagram indicates that fcc should be stable to low temperatures implies that the fcc phase solidified from the liquid. For the same latter two reasons, the fcc phase in the Fe-30 and 40 at% Ni alloys also appears to be due to solidification from the liquid.

The bcc phase in the Fe- 30 and 40 at% Ni alloys appears to have solidified from the liquid. Indeed, Figure 10 shows bcc growing from the liquid! The diffraction patterns show no signs of strain in the lattice. Furthermore, the fact that bcc is seen with the amorphous phase implies that it must have been retained to ambient temperatures without a solid state transformation. Based on what is known about martensitic transformations [12] in iron-nickel alloys, the bcc phase is not likely in the 40 at% Ni alloy. The M_s temperature of Fe-30 at% Ni is below room temperature but surface martensite can form. Still, only one example of a martensitic structure was found in Fe-30 at% Ni and none were found in Fe-40 at% Ni. The twinned bcc structure may be due to a fcc particle that fully transformed martensitically. Twinned bcc structures have been observed as products of martensitic transformations [13].

The presence of the fcc phase in the Fe-30 and 40 at% Ni is not in accord with the calculations for homogeneous nucleation shown in Figure 2 which predict bcc nucleation up to 52 at% Ni. Experimentally, the cross-over composition between fcc and bcc nucleation is here found to be between 40 and 50 at% Ni. Considering the uncertainties in the physical property data utilized in the calculations [2] this agreement is quite good. The presence of fcc may be due to heterogeneous nucleation. The fact that the Fe-30 at% Ni contains a greater amount of fcc than the Fe-40 at% Ni yet is expected to have less, could be caused by a dirtier environment during the processing of the former. Further work on this issue is needed.

One of interesting findings is the amorphous phase in Fe-30 and 40 at% Ni alloy powders that are smaller than 40 nm diameter. Clearly, the liquid was supercooled to the point where atomic mobility was very low and neither nucleation nor growth could occur rapidly. Because of the very small size of these droplets, the total nucleation probability is low. Furthermore, because the surface area-to-volume increases as the diameter of a sphere increases, the cooling rate is highest in the smallest droplets (on the order of 10^6 K/s for 100 nm droplet at 1700 K). Thus, if an amorphous structure was to be produced in any droplet, it is most likely that it should be found in the smallest particles.

This effect can be illustrated by calculating the nucleation temperature for homogeneous nucleation as a function of droplet size. This is shown for each of 30, 40, 50, and 60 at% Ni for fcc and bcc in Figures 11 and 12, respectively. If the model were correctly accounting for the temperature dependence of the liquid mobility, the nucleation temperature would drop to impossibly low values at smaller diameters. From these data

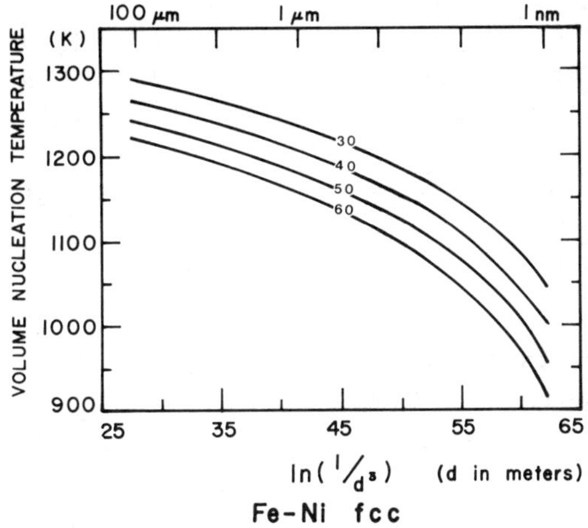

Figure 11 Size dependence of nucleation temperature for fcc.

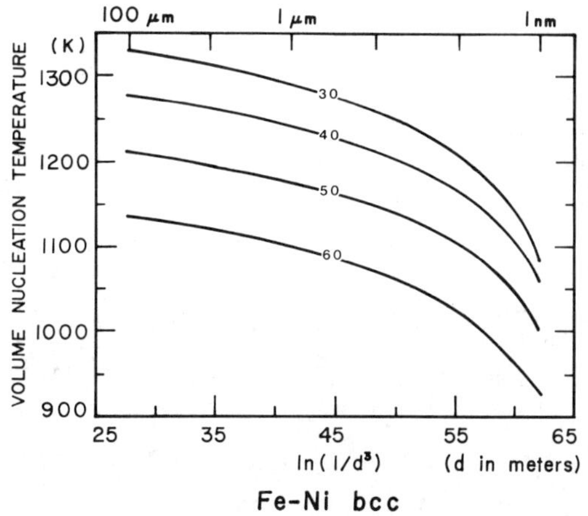

Figure 12 Size dependence of nucleation temperature for bcc.

it appears that the liquid mobility terms in the nucleation expression are not low enough. A reduced glass transition temperature of 0.25 has been used in the calculations and higher values like 0.33 may be more appropriate.

Conclusions

A bcc phase is observed as an alternative crystallization phase in submicron droplets of Fe-30 and 40 at% Ni.

The fcc phase in the Fe-50 at% Ni alloy nucleates and grows from the liquid. The presence of the fcc phase in the submicron droplets of Fe-30 and 40 at% Ni is due to heterogeneous nucleation.

Nucleation rates do not reach the very high values that are associated with copius homogeneous nucleation.

About one fourth of the spheres of Fe-30 and 40 at% Ni that are smaller than 40 nm diameter are amorphous. The amorphous structure is produced as a result of both low nucleation rates and low growth rates at the high liquid supercoolings that obtain in the very small droplets.

The cross-over composition between fcc and bcc nucleation calculated by Kelly and Vander Sande is greater than observed experimentally but the data is consistent with the qualitative trends.

Bibliography

1. T.F. Kelly and J.B. Vander Sande, in Chemistry and Physics of Rapidly Solidified Alloys, TMS-AIME, Warrendale, PA, (1983) pp. 35-47.

2. Thomas F. Kelly and John B. Vander Sande, "Nucleation of Alternative Crystallization Phases", submitted to Acta Metall. (1986).

3. J. Perel, J.F. Mahoney, P. Duwez and B.E. Kalensher, in Rapid Solidification Processing Principles and Technologies II, R. Mehrabian, B.H. Kear and M. Cohen (Eds.), Claitor's Publishing Division, Baton Rouge, LA, (1980), p. 287.

4. J. Perel, J.F. Mahoney, S. Taylor, Z. Shanfield and C. Levi, in Rapid Solidified Amorphous and Crystalline Alloys, B.H. Kear, B.C. Giessen and M. Cohen (Eds.), North Holland, New York, NY, (1982), pp. 131-132.

5. Y.-Y. Chuang, K.-C. Hsieh, and Y.A. Chang, submitted to Metall. Trans. A, (1984).

6. Y.-Y. Chuang, Y. A. Chang, and R. Schmid, submitted to Metall. Trans. A, (1984).

7. F. Spaepen, Acta Metall., (1975) Vol. 23, pp. 729-743.

8. F. Spaepen and R.B. Meyer, Scripta Metall., (1976) Vol. 10, pp. 257-263.

9. C.V. Thompson, Ph.D. Thesis, School of Applied Sciences, Harvard University, Cambridge, MA, (1981).

10. Z. Nishiyama, Sci. Rep. Tohoku Univ., vol. 23 (1934) p. 637., and G. Wasserman, Arch. Fur das Eisenhutt, vol. 16 (1933) p. 647.

11. J. W. Edington, Practical Electron Microscopy in Materials Science, Monograph Two, Philips Technical Library, p. 116.

12. H. M. Ledbetter and R. P. Reed, Mat. Sci. Eng., vol. 5 (1969/70) pp. 341-349.

13. G. Kraus, Acta Metall., vol. 11, (1963) pp. 499-509.

CONTRIBUTION OF COARSENING TO THE ESTABLISHMENT OF GRAIN SIZE IN BULK-UNDERCOOLED Ni-Ag ALLOYS

S. Skolianos, M. Chen and T. Z. Kattamis

Department of Metallurgy
Institute of Materials Science
University of Connecticut
Storrs, Connecticut 06268

The grain size of bulk-undercooled dilute Ni-Ag alloy specimens was studied versus initial undercooling, average cooling rate and solute concentration. Initial undercooling varied between 20 and 250K, cooling rate between 0.14 and 6K/s and silver content between 1 and 3wt%. The grain size of dendritic specimens which were subcritically undercooled ($\Delta T <\simeq 170K$) was practically independent of initial undercooling, cooling rate and silver content. In supercritically undercooled nondendritic specimens ($\Delta T >\simeq 170K$) the grain size decreased with increasing undercooling, cooling rate after recalescence and solute concentration. A coarsening model was introduced which predicts for a given undercooling (220K) the effects of cooling rate and composition on average grain size, grain density and specific grain boundary area. This model assumes that the crystallite size distribution at the maximum recalescence temperature at the onset of cooling can be deduced from the grain size distribution in a specimen which was undercooled 220K and splat-cooled.

Introduction

It was first reported by Walker, who studied the solidification behavior of bulk-undercooled nickel and cobalt melts, that an abrupt grain refinement occurs when the initial undercooling prior to nucleation of the solid exceeds a critical amount (1). In nickel he established an as-cast grain size transition from 0.02 m at no undercooling to 10^{-4} m beyond an initial undercooling of about 175 K and detected the emission of audible noises associated with the rapid freezing. Emission of acoustical disturbances and the drastic grain refinement were also reported and discussed by Glicksman(2) in a theoretical and experimental study of dynamic effects during high speed solidification.

The variation of cast grain size in Fe-25wt%Ni and Ni-25wt%Cu alloys versus undercooling was reported previously (3,4). A substantial grain refinement was observed at and beyond an undercooling of about 170 K. This refinement was accompanied by elimination of dendritic grains and formation of nondendritic grains exhibiting spherical solute coring. Microstructural indications of partial remelting during recalescence and grain multiplication during solidification of highly undercooled melts were discussed elsewhere (5). Assume a binary melt of composition C_0 nucleated at T_N, Figure 1a. During recalescence solidification is partitionless, provided (6,7): 1) The solid-liquid interface temperature is maintained below T_0, which is a necessary condition and; 2) the interface velocity, R, exceeds the speed at which solute atoms can diffuse away from the advancing interface, which is a sufficient condition. This speed is usually given by D_L/a, where D_L is liquid diffusion coefficient and a is interatomic distance of the solidifying phase. The recalescence boundary, which at a given moment is the geometric locus of primary dendrite arm tips spreads out from the point of nucleation as a roughly spherical wave (2,8) which travels at constant velocity R until it approaches the container's walls. This dendrite tip velocity is a function of initial undercooling prior to nucleation, ΔT. During partitionless solidification thermal dendrites of uniform composition C_0 are expected to form. While primary dendrite arms are advancing at constant tip temperature the undercooling of the remaining interdendritic liquid decreases with time, because released latent heat of fusion flows therein. Thus, the growth velocity of secondary dendrite arms gradually decreases until it becomes smaller than D_L/a at a temperature T' and solute partitioning initiates. During recalescence between T' and the maximum recalescence temperature, T_R, at which the remaining liquid is no longer undercooled, solidification is controlled by diffusion of solute. While this solidification takes place the solid of composition C_0 becomes superheated above the equilibrium solidus temperature, T_S. It is then expected to remelt during recalescence on a fine sub-dendrite scale, so that solute diffusion through the solid toward the liquid channels which formed during remelting would take place sufficiently rapidly to relieve superheating and maintain local equilibrium between the solid and the liquid. After T_R has been reached solidification of the remaining liquid is accompanied by loss of heat to the surroundings.

The decrease in interfacial velocity versus fraction solid and initial undercooling was calculated by Levi,et al. (9,10),for undercooled aluminum droplets. These authors analyzed the stages of recalescence and concluded that for a substantial initial undercooling in a first solidification regime, the droplet absorbs most of the released latent heat and external cooling plays a minor role. During the second regime growth is slower and depends on rate of heat extraction through the droplet surface. During recalescence of hypocooled Al-6wt%Si droplets it was speculated that some supersaturated solid behind the interface could have been heated above T_0 , but no remelting had occurred, due to the short times available (10). Remelting after recalescence was suspected in larger, dendritically solidified tin-lead alloy particles, where heat flow conditions were more sluggish (11).

The time variation of the average temperature of a hypoeutectic alloy specimen nucleated at T_N is illustrated in Figure 1b. Remelting leads to dendrite fragmentation which is most likely enhanced by intense fluid flow in the undercooled liquid and formation of pressure waves beyond a critical undercooling (2,12). Coarsening of the dendrite fragments will operate during the very short recalescence stage, the thermal arrest at T_R and especially the long post-recalescence cooling period. Diffusional coarsening was recently studied by Voorhees, et al. (13,14), by solving and applying the multi-particle diffusion problem. These authors found that when the initial particle size distributions were scaled by the average particle radius, they evolved to identical time-independent distributions which were different from those predicted by LSW and became broader and more symmetric with increasing volume fraction of coarsening phase, while ripening kinetics became faster.

It was the main aim of this investigation to establish and interpret the dependence of final nondendritic grain size on cooling rate and, within certain narrow limits, on alloy composition. The investigation was conducted on dilute Ni-Ag alloys, namely Ni-1,2 and 3wt%Ag. In the monotectic Ni-Ag system silver particles precipitate along the grain boundaries late during the solidification process and restrain solid state grain growth.

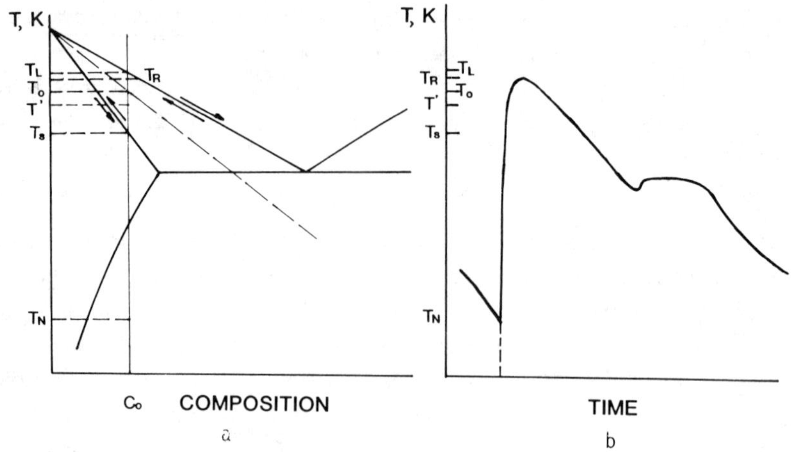

Fig. 1- Schematic phase diagram (a) and time variation of the average temperature of a hypoeutectic alloy specimen nucleated at T_N (b).

Experimental Procedure

Several cylindrical ingots (0.015 m dia. x 0.06 m long) were solidified from melts that were undercooled between 20 and 250 K, following the glass encasement procedure (15). They were allowed to cool at about 0.14 K/s after recalescence. Temperature was monitored with a fine Pt/Pt-10%Rh thermocouple protected in a fused silica tube. Some ingots were surface-nucleated at the desirable temperature with a nickel wire and were internally chilled by plunging a water-cooled copper U-tube in the bulk of the metal (15). Thus, a spectrum of cooling rates after recalescence was established with distance from the chill. The spatial distribution of cooling rates was determined using three coaxial thermocouples with beads situated at different distances from the chill. The microstructures studied corresponded to average cooling rates of 1 and 6 K/s.

Partial remelting experiments were carried out on previously homogenized polycrystalline specimens and the microstructures studied as a function of heating rate in the vicinity of 1714.8 K, the solidus of the alloy at which remelting would normally initiate. Specimens 0.5 to 1.5 gm were heated in a levitation furnace at different rates. Their temperature was monitored using a two-color optical pyrometer coupled with a fast response oscilloscope. In order to preserve the remelting pattern (liquid channels and inclusions) the specimens were quenched when the temperature approached T_S. The power was automatically switched off and the specimen fell into water. In some cases a jet of cold Ar was flushed on the specimen without modifying the levitation power. The very narrow temperature range (4 K) between the liquidus and the solidus made the experiment very difficult. In some specimens the temperature did not reach T_S and no remelting occurred, in most of them the temperature exceeded T_L and they melted completely. However, three of them were heated at 3, 10 and 50 K/s and were successfully partially remelted.

Remelting at high rates of 10^4- 10^5 K/s was studied using wire-drawn alloy, which was annealed, homogenized and "exploded" by passing a high density current through it. The current pulse was generated by a capacitors bank. The heating rate was measured with a high sensitivity two-color pyrometer coupled with a rapid response digital storage oscilloscope.

Results

The variation of grain size versus degree of undercooling and composition is illustrated in Figure 2 for specimens of Ni-1,2 and 3wt%Ag cooled at about 6 K/s. The average cooling rate was approximately the same before nucleation and after recalescence. At and beyond a critical undercooling of about 170K the grain size is drastically reduced (1,3) and grain growth ceases to be dendritic (3,15). The microstructural transition from dendritic to nondendritic growth is illustrated by the inset, Figure 2. Within the composition range studied herein the silver content does not seem to affect the dendritic grain size. However, there is a definite pattern indicating that nondendritic grain size significantly decreases with increasing silver content from 1 to 3wt%.

Secondary dendrite arm spacing was also plotted in Figure 2 versus undercooling for subcritically undercooled specimens. The resulting experimental curve coincides with that of nondendritic grain size versus undercooling. Measurements corresponding to initial undercoolings of about 70K and 150K indicate that for a given cooling rate the dendrite arm spacing is refined with increasing solute concentration, which was well established for non-undercooled alloys (16).

The dependence of grain size on cooling rate for specimens which were nucleated at undercoolings of 150 K and 220 K is shown in Figure 3. The size of nondendritic grains in the specimens which were undercooled 220 K decreased from 55×10^{-6}m for a cooling rate of 0.14 K/s to 25×10^{-6}m for 6 K/s. There was a parallel increase in the grain number density from about 2.8×10^{13} to 1.9×10^{14} m^{-3}, Figure 4.

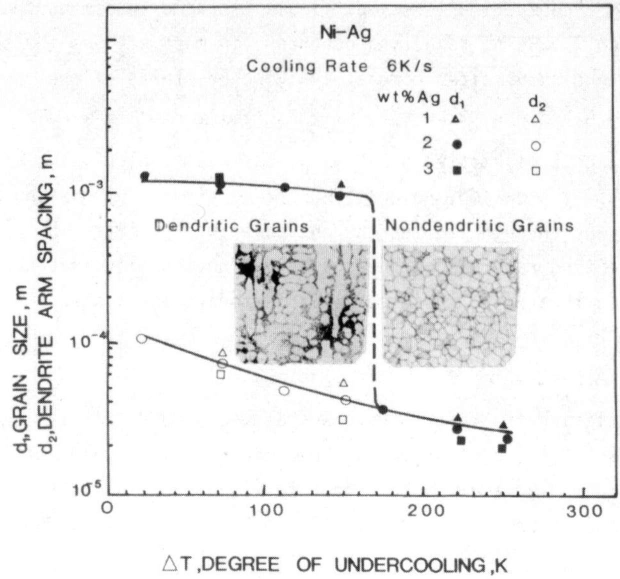

Fig. 2- Dendrite arm spacing in dendritic specimens, and grain size in dendritic and nondendritic specimens versus undercooling and composition. Dilute Ni-Ag alloys cooled at about 6 K/s.

In homogenized Ni-2wt%Ag specimens which were successfully remelted at 3, 10 and 50K/s the liquid appeared at grain boundaries, as well as within the grains in the form of inclusions. The decrease in average liquid inclusion size and spacing with increasing heating rate is illustrated in Figure 5. Partial remelting at higher heating rates was achieved by exploding wires of different diameters with different current densities. Fine wire specimens were exploded at average heating rates between 2.7×10^4 and 10^5K/s. The wires actually remelted completely and broke off at a given location where the diameter was intentionally reduced. In the vicinity of the remelted wire tip partial remelting occurred, region A, Figure 6. Region B resulted from resolidification of liquid which surrounded the tip. Within region A the volume fraction liquid decreased with increasing distance from the remelted wire tip, while the continuous liquid channels which delineate fine subgrains within the original grains degenerated into discrete inclusions. Also, the average channel or inclusion spacing within region A remained practically independent of location. The variation of average liquid inclusion or channel spacing, L(m), versus heating rate, $\dot{T}$ (K/s),

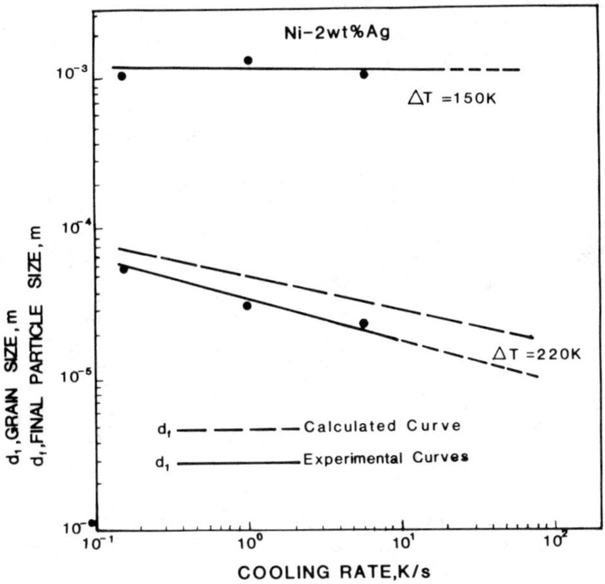

Fig. 3- Grain size in dendritic specimens (undercooled 150 K) and nondendritic specimens (undercooled 220 K) and calculated final grain size in specimens undercooled 220 K versus average cooling rate after recalescence. Ni-2wt%Ag alloy.

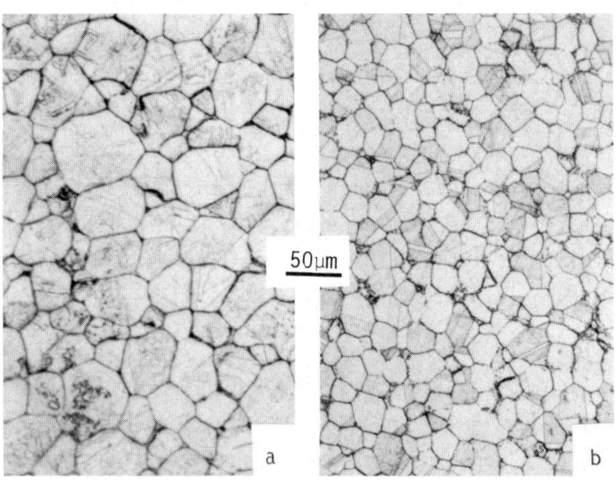

Fig. 4- Photomicrographs of Ni-2wt%Ag specimens undercooled 220 K and cooled at (a) 0.14 and (b) 6 K/s.

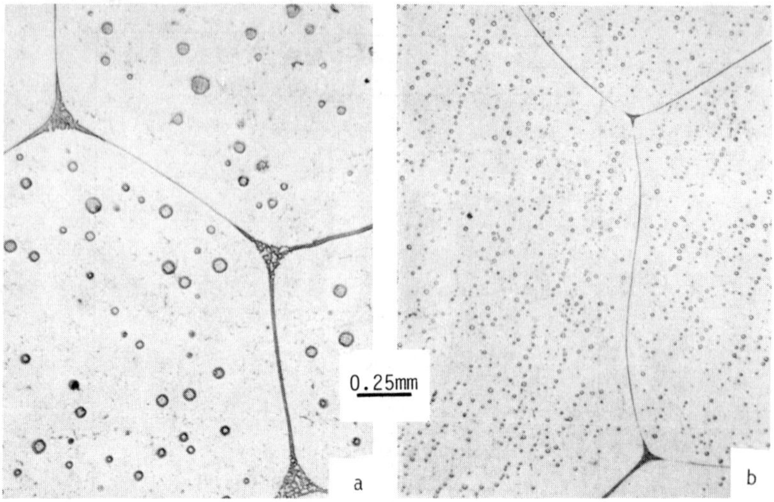

Fig. 5- Photomicrographs of Ni-2wt%Ag alloy specimens which were partially remelted at:(a) 10 K/s and (b) 50 K/s, and quenched.

prevailing at about the monotectic temperature (1708 K) is shown in Figure 7. This variation is closely represented by the relationship: $L^2 = 6.25 \times 10^{-7}/\dot{T}$.

Discussion

The drastic grain refinement which occurs at a critical undercooling of about 170 K was previously investigated (1,2,12). The focal point of this investigation was the dependence of final grain size on cooling rate after recalescence and solute concentration for geometrically identical specimens which were nucleated at the same supercritical undercooling. Previous investigations qualitatively indicated that coarsening is responsible for final grain size in supercritically undercooled alloys (15) and that remelting accompanied by coarsening operates during recalescence (5).

It can be assumed that at the maximum recalescence temperature, T_R, two geometrically identical ingots of a given alloy which were nucleated at the same supercritical undercooling would consist of the same fraction solid and, statistically, the same crystallite size distribution. During subsequent cooling the crystallite size increases because of normal growth during freezing and diffusional coarsening, as well as collisional coalescence, and the crystallite density (number of crystallites per unit volume) decreases.

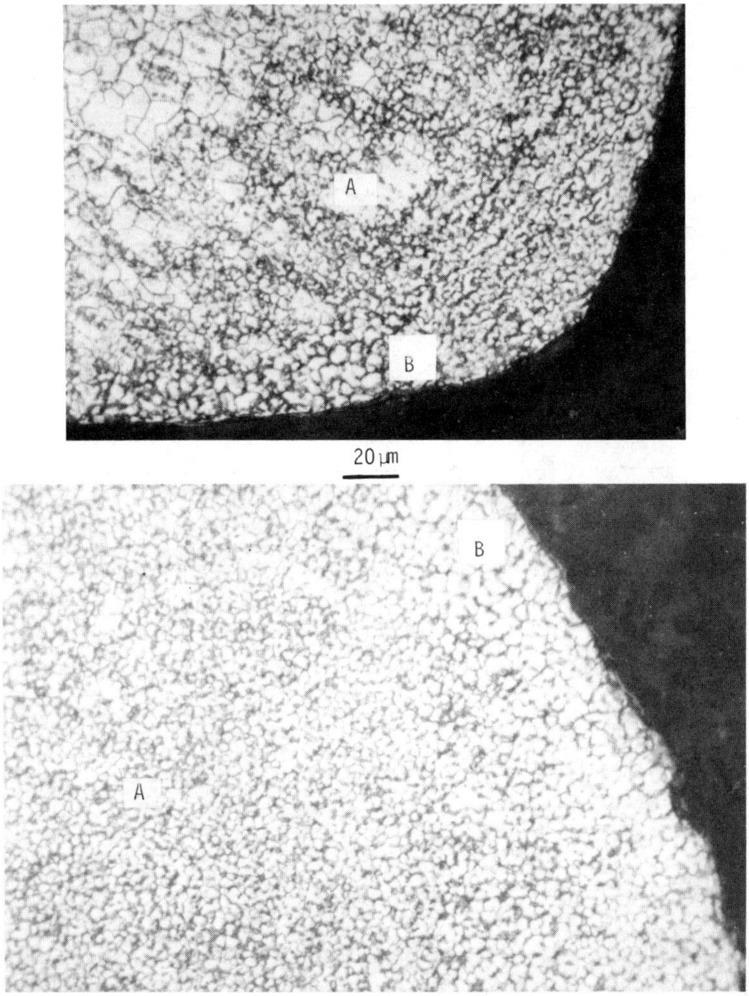

Fig. 6- Photomicrographs of longitudinal sections of exploded wires. Partial remelting occurred in region A close to the wire tip. Complete remelting and resolidification occurred in region B at the tip. Homogenized Ni-2wt%Ag alloy.

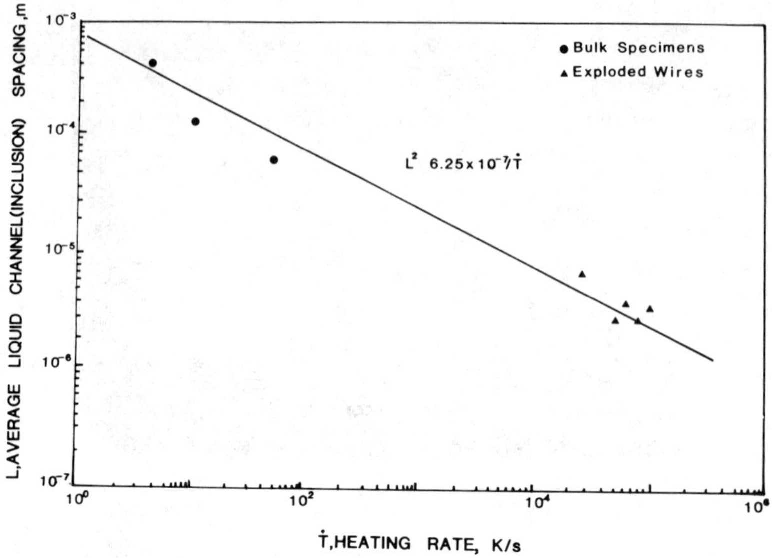

Fig. 7- Average liquid channel and inclusion spacing versus heating rate. Ni-2wt%Ag alloy.

The dependence of final nondendritic grain size and density on cooling rate after recalescence indicates the operation of coarsening which is the only mechanism that can account for the elimination of crystallites during freezing.

In order to establish quantitatively the effect of coarsening during cooling after recalescence it is necessary to know the crystallite size distribution at the maximum recalescence temperature and follow its evolution during cooling as a function of time. Ideally, if the size distribution and shape factors of dendritic fragments resulting from remelting and a precise recalescence curve were available the crystallite size distribution at T_R would have been calculated. Recently, very precise recalescence and cooling curves were experimentally established by Piccone, Wu, Shiohara and Flemings (17) during solidification of levitated and undercooled Ni-25% Sn and Ni-32.5%Sn eutectic alloy droplets. Their temperature measuring system consisted of two silicon photodiodes operating in the near infrared wavelength bands, two amplifiers, a rapid response digital storage oscilloscope interfaced with a computer for calculating the data and a graphics plotter. The response time of this system did not exceed 20 microseconds.

A. Crystallite Size Distribution at T_R

In order to determine the average dendrite segment length, L, which forms during remelting at a heating rate $\dot{T}$ a model previously developed for a binary system (5) was used. This model assumes that as the temperature rises the supersaturation of the solid increases until liquid channels form and that new channels continue to form until, at a given temperature, T, their density is large enough to drain off solute at a rate which keeps the supersaturation constant, below that required for nucleation. A relationship:

$$L^2 = \frac{D \pi^2 \bar{C} S^* |m_s|}{\dot{T}} \qquad [1]$$

was derived, where D is average diffusivity of solute in the solid between the equilibrium solidus T_S and temperature T, $\bar{C}$ is average solute concentration in the alloy, S* is critical supersaturation function and m_s is solidus slope in the binary phase diagram. The experimental points in Figure 7 satisfy a relationship of this type: $L^2 = 6.25x10^{-7}/\dot{T}$.

The temperature measuring system employed in this investigation yielded thermal profiles analogous to that shown in Figure 8 for a Ni-2wt%Ag ingot that was undercooled 220K prior to nucleation of the solid. This figure gives an overall view of recalescence and subsequent cooling. An approximate maximum recalescence rate of $2.2x10^5$K/s was determined using a rapid response oscilloscope. By extrapolating Figure 7 it can be seen that the average dendrite segment length corresponding to this recalescence rate is about $1.2x10^{-6}$ m. The dendrite segmentation is favored by convection in the undercooled liquid (5) and most likely by the generation of pressure waves (2,12). The segments would coarsen during recalescence up to T_R and during subsequent thermal arrest. This coarsening, accelerated by intense fluid flow which is caused by temperature gradients during recalescence is accompanied by an increase in volume fraction solid. It is not possible to calculate the crystallite size distribution at the onset of cooling knowing the average dendrite segment length, if the actual segment size distribution is not available. This distribution was then determined experimentally, as closely as possible to reality. An alloy droplet (1.2 gm) was undercooled in a levitation furnace by about 220K and splat-cooled as previously described for Fe-25wt%Ni alloy (18). Nucleation of the melt occurred when the moving platen touched the falling droplet. A cooling rate in excess of 10^4 K/s was attained. To get a thin specimen, hence a high cooling rate it was essential that platen velocity exceed

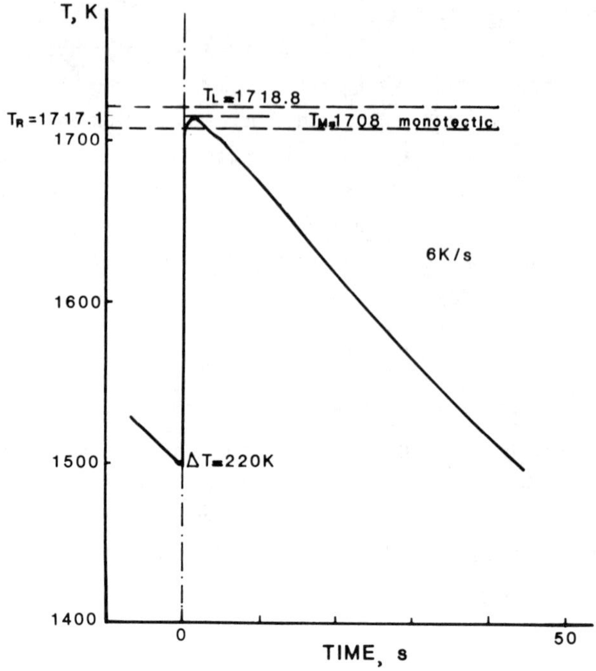

Fig. 8- Thermal profile during solidification of a Ni-2wt%Ag alloy specimen undercooled 220 K.

2.6 m/s, the growth velocity at this undercooling. It was assumed that no substantial coarsening occurred during this fast cooling and hence no crystallites disappeared by dissolution. From the grain size distribution in this specimen, the crystallite size distribution at the onset of cooling was deduced. For this purpose it was assumed that each crystallite received during solidification after recalescence a volume increment proportional to its volume fraction relative to the total volume fraction solid at T_R (19). Following the procedure described elsewhere (20) the fraction solid at the maximum recalescence temperature, f_{SR} and T_E were calculated using expressions (3), (4), (6), (11) and (12) from that analysis (20). The following numerical values were adopted for the various parameters: $\rho c = 5.8 \times 10^6$ J/m^2.K, $\Delta H = 2.35 \times 10^9$ J/m^3, where ρ is density, c is specific heat and ΔH is latent heat of fusion (21),the solidus slope $m_s = -4.95$ K/wt%, Figure 9, the equilibrium partition ratio K = 0.726, $T_L - T_S = 4$K, where T_L is the equilibrium liquidus and $T_L - T_N = 220$K. It was found that $f_{SR} = 0.539$ and $T_R = 1717.1$K. Assume a population of n crystallites at T_R. Let

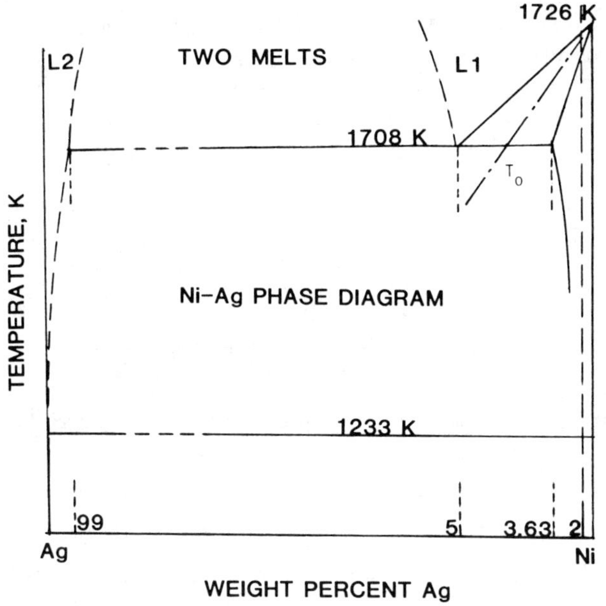

Fig. 9- The Nickel- Silver phase diagram.

$r_i)_R$ and $r_i)_F$ be the radii of a given crystallite at T_R and at completion of solidification, respectively. The previous assumption leads to:

$$\frac{\frac{4}{3}\pi\, r_i^3)_R}{0.539}(1 - 0.539) = \frac{4}{3}\pi[r_i)_F^3 - r_i)_R^3] \qquad [2]$$

Hence, $r_i)_R^3 = 0.539\ r_i)_F^3$. Thus, the initial crystallite size distribution may be calculated by measuring $r_i)_F$ in the splat-cooled specimen. To calculate the evolution of the initial crystallite size distribution during post-recalescence cooling a previous analysis (19) was used. The cooling curve, Figure 8, was subdivided into a large number of steps, each one consisting of an isothermal hold of duration Δt, followed by an instantaneous quenching of a fraction of a degree K. Coarsening is assumed to operate during the isothermal hold. Solidification growth, which results from this instantaneous cooling is assumed to occur at the beginning of the isothermal hold. The radius of particle i, r_i at time $t = \sum_{k=1}^{k} \Delta t_k$, where Δt_k is the duration of the kth time step, is given by:

$$r_i = r_o + \Delta r_{ik})_c + \Delta r_{ik})_g \qquad [3]$$

where $\Delta r_{ik})_c$ and $\Delta r_{ik})_g$ are the changes in r_i occurring during time t due to coarsening and solidification growth, respectively, and r_o is the initial radius. From reference (19):

$$\Delta r_{ik})_c = \lim_{\substack{\Delta t_k \to 0 \\ k \to \infty}} \sum_{t=0}^{t_k} \frac{dr_{ik}}{dt} \Delta t_k \qquad [4]$$

where $\frac{dr_{ik}}{dt} = \frac{dr_i}{dt}$ is given by expression (16) in reference (19) and

$$\Delta r_{ik})_g = \frac{1}{4\pi\rho r_i^2} \frac{M_T}{(1-K)} \frac{(T_M-T_L)^{\frac{1}{1-K}}}{(T_M-T)^{\frac{2-K}{1-K}}} \varepsilon \frac{r_i^2}{\sum_{i=1}^{n} r_i^3} \Delta t_k \qquad [5]$$

where M_T is the total mass of the volume element considered (19), K is the equilibrium partition ratio, T_M is the melting point of the pure Ni and ε is cooling rate. This analysis neglects crystallite coalescence.

Thus, the radii of all spherical crystallites of the system can be calculated at the end of any number of time steps, hence at any time t, by evaluating expression [3] by means of a digital computer. These radii are then adjusted to take into account normal solidification growth by means of expression [5] and the system can be described at any time.

The preceding analysis was applied to specimens cooled at 0.14, 1, and 6 K/s, using as initial crystallite size distribution at T_R that which was deduced from the grain size distribution in the splat-cooled foil. A population of 20 grains was randomly selected and the calculated crystallite radii distribution was: r_1 = 1.8, r_2 = 1.85, r_3 = 1.90, r_4 = 1.90, r_5 = 2.0, r_6 = 2.0, r_7 = 2.10, r_8 = 2.10, r_9 = 2.10, r_{10} = 2.20, r_{11} = 2.20, r_{12} = 2.20, r_{13} = 2.20, r_{14} = 2.20, r_{15} = 2.25, r_{16} = 2.25, r_{17} = 2.25, r_{18} = 2.25, r_{19} = 2.3 and r_{20} = 2.3×10^{-6}m. Other values used were: K = 0.704, $m = m_L$ = -3.6 K/wt%Ag (the liquidus slope in the phase diagram, Figure 9), liquid solute diffusivity D = $10^{-9} m^2/s$, surface tension σ = 0.4 J/m^2, density $\rho = \rho_{Ni}$ = 8.9×10^3 Kg/m^3. From the phase diagram, Figure 9, the concentration of solute in the liquid at temperature T is C_L = (1726-T)/3.6. It was assumed that, as far as coarsening is concerned, solidification is completed at 1708 K, the monotectic temperature. In fact, at T_R, f_{SR} = 0.539. Assuming that Scheil-type solidification prevails during cooling, f_s = 0.983 just above the monotectic temperature and f_s = 0.9997 after completion of the monotectic reaction. The remaining volume fraction liquid of 0.0003, which during further cooling deposits pure silver particles at the boundaries of the nickel-rich particles, was neglected.

Knowing the crystallite radii distribution at the end of each time step it is possible to calculate the specific crystallite surface, S_v (crystallite surface-to-volume ratio) and the remaining number of undissolved crystallites, N, at each moment (19).

$$S_v = \frac{\sum_{i=1}^{20} 4\pi r_i^2}{\sum_{i=1}^{20} \frac{4}{3}\pi r_i^3} \qquad [6]$$

The calculations of average crystallite size, $\bar{d} = 2\bar{r}$, where $\bar{r}$ is average crystallite radius, S_v and N were carried out for three alloy compositions: Ni - 1, 2 and 3wt%Ag and three average cooling rates after recalescence: 0.14, 1 and 6 K/s. Initial values were $\bar{d}_o = 4.11 \times 10^{-6}$m, $S_{v_o} = 1.52 \times 10^6 m^{-1}$ and $N_o = 20$.

The calculated average crystallite sizes versus time during cooling after recalescence are given in Figures 10 and 11 for Ni-1wt%Ag and Ni-3wt%Ag, respectively and for three average cooling rates after recalescence: 0.14, 1 and 6 K/s. In these curves time 0 corresponds to the onset of cooling after recalescence. In all cases it was assumed that undercooling prior to nucleation was 220K. Each curve is terminated at the moment solidification was completed and the corresponding average crystallite size is then average grain size of the solidified specimen. The calculated average grain size versus average cooling rate after recalescence for specimens of Ni-1wt%Ag, Ni-2wt%Ag and Ni-3wt%Ag that were undercooled about 220K prior to nucleation is given in Figure 12. Three experimental points corresponding to an average cooling rate of 6 K/s and to 1, 2 and 3wt%Ag are also shown in the same graph. A more detailed comparison of analytical predictions and experimental results is made in Figure 3, which exhibits calculated average grain size versus cooling rate after recalescence for Ni-2wt%Ag alloy specimens undercooled 220K prior to nucleation and experimental measurements made for cooling rates of 0.14, 1 and 6 K/s. The discrepancy which is considerable at the lowest cooling rate is not surprising in view of the oversimplifying assumptions made and the uncertainty of numerical values adopted for various parameters. Basically, coarsening is responsible for the removal of finer crystallites from the system, which enhances the contribution of growth during cooling, because the solidifying alloy is then deposited on fewer crystallites. The specific crystallite surface area versus time during cooling after recalescence is given in Figure 13 for Ni-2wt%Ag undercooled 220K prior to nucleation and three cooling rates: 0.14, 1 and 6 K/s and in Figure 14

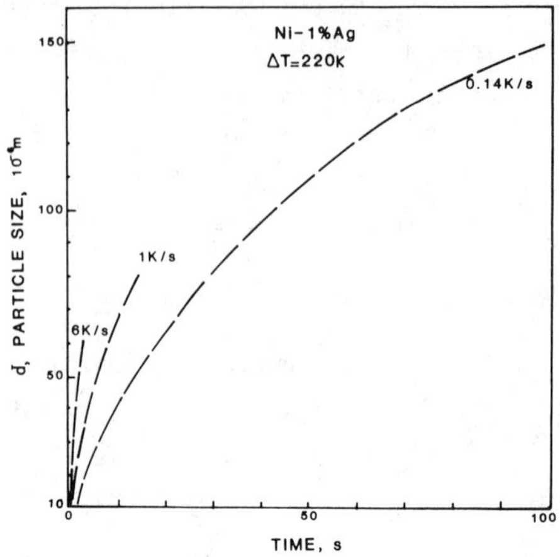

Fig. 10- Calculated average crystallite size versus time during cooling after recalescence for rates of 0.14, 1 and 6 K/s. Time 0 is the onset of cooling after recalescence. Ni-1wt%Ag undercooled 220 K.

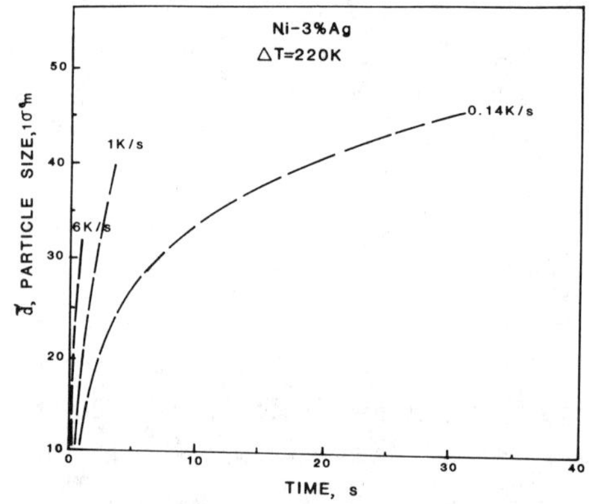

Fig. 11- Calculated average crystallite size versus time during cooling after recalescence for rates of 0.14, 1 and 6 K/s. Time 0 is the onset of cooling after recalescence. Ni-3wt%Ag undercooled 220 K.

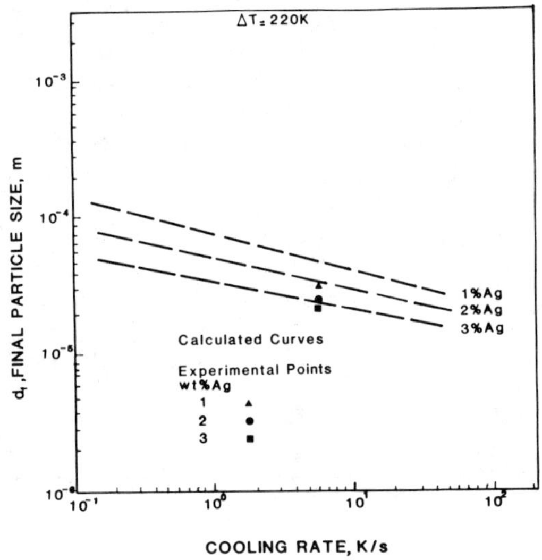

Fig. 12- Average grain size versus average cooling rate after recalescence. Ni-1, 2 and 3wt%Ag specimens undercooled 220K.

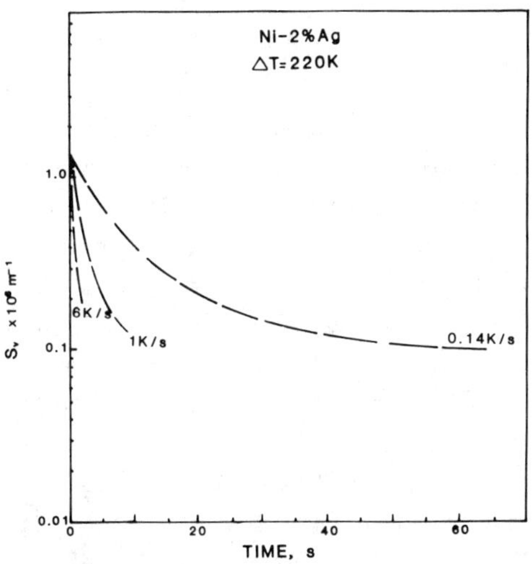

Fig. 13- Specific crystallite surface area versus time during cooling after recalescence for rates of 0.14, 1 and 6 K/s. Ni-2wt%Ag undercooled 220 K.

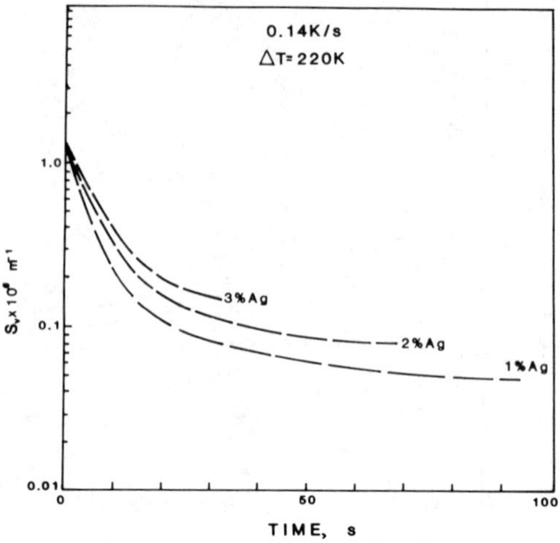

Fig. 14- Specific crystallite surface area versus time during cooling after recalescence for a rate of 0.14 K/s and three compositions: Ni-1, 2 and 3wt%Ag. Specimens undercooled 220 K .

for Ni-1wt%Ag, Ni-2wt%Ag and Ni-3wt%Ag undercooled 220 K prior to nucleation and cooled at 0.14 K/s after recalescence. The terminal points on the curves of Figure 13 correspond to specific grain boundary areas and are plotted versus cooling rate in Figure 15. The experimental points fit the analytical curve quite well. The normalized number of remaining undissolved crystallites from the original population of N_0 = 20 crystallites, N/N_0 is plotted in Figure 16 versus time for Ni-2wt%Ag alloy alloy undercooled 220K prior to nucleation. Figure 17 shows the variation of N/N_0 versus time for Ni-1, 2 and 3wt%Ag alloy specimens undercooled 220K and cooled at 0.14 K/s after recalescence. The terminations of curves in Figure 16 correspond to relative numbers of final grains in the as-solidified specimens with respect to the initial number of 20. The relative number of final grains remaining undissolved versus cooling rate after recalescence was plotted in Figure 18 for Ni-2wt%Ag alloy specimens undercooled 220K prior to nucleation. Discrepancy between experimental points and the analytical curve is reasonable.

Nondendritic grain size is controlled by coarsening kinetics and therefore depends on cooling rate and composition, Figures 3 and 12. For a given undercooling it decreases with increasing cooling rate and alloy concentration. By extrapolating the curves of Figure 12 it can be seen that

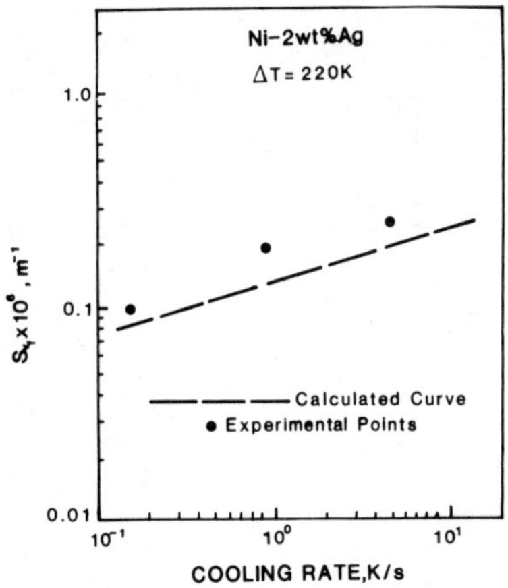

Fig. 15- Specific grain boundary area versus cooling rate. Ni-2wt%Ag alloy specimen undercooled 220 K .

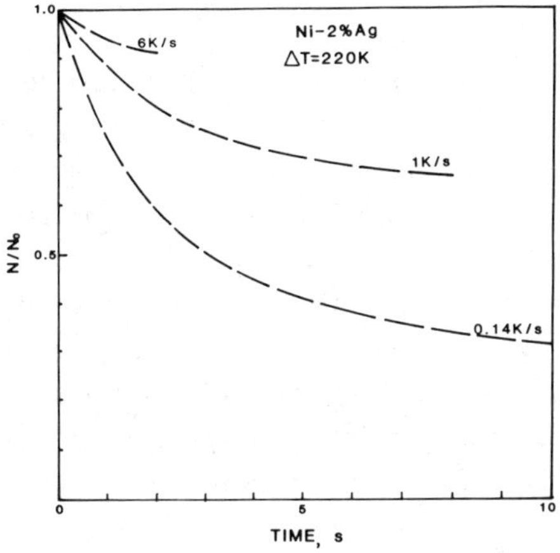

Fig. 16- Relative number of remaining undissolved crystallites versus time during cooling after recalescence for rates of 0.14, 1 and 6 K/s. Ni-2wt%Ag specimens undercooled 220 K .

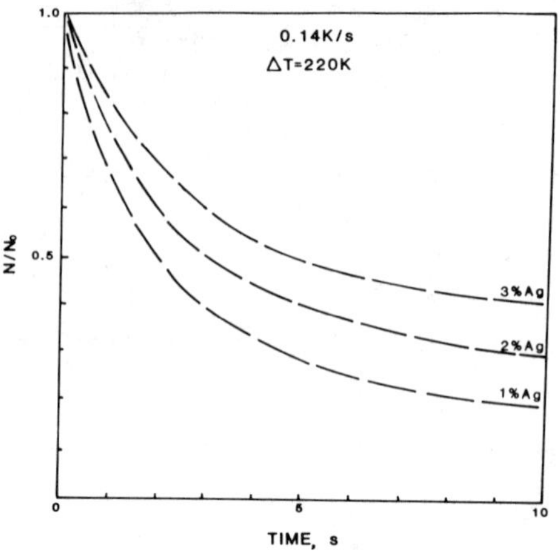

Fig. 17- Relative number of remaining undissolved crystallites versus time during cooling after recalescence for Ni-1, 2 and 3wt%Ag specimens undercooled 220 K and cooled at 0.14 K/s

submicron grain sizes can be reached at cooling rates on the order of 10^6K/s. Dendritic grain size in subcritically undercooled specimens does not depend on cooling rate, Figure 3. In this case coarsening operates between dendrite arms and controls their geometry and spacing. This explains why the dendrite arm spacing curve for dendritic specimens, Figure 2, coincides with that of grain size for nondendritic specimens. Coarsening between dendritic grains operates only and to a limited extent when they are very close to each other. In the literature on rapid solidification there is frequent confusion between grains, subgrains, sections of cellular structures and dendrite elements, which explains frequent erroneous claims that rapid solidification substantially refines cast grain size. Only in processes which allow undercooling prior to nucleation does rapid cooling yield a finer grain size (4). In surface remelting and resolidification using a high power beam growth is epitaxial, no substantial undercooling is possible and grain size can be very coarse. In melt-spun ribbon only in the vicinity of the chill-block where melt undercooling can take place has grain refinement been observed. Finally, during atomization chances for high undercooling at least of the finest particles are excellent and, despite slower cooling than in the other two processes, grain refinement can be significant.

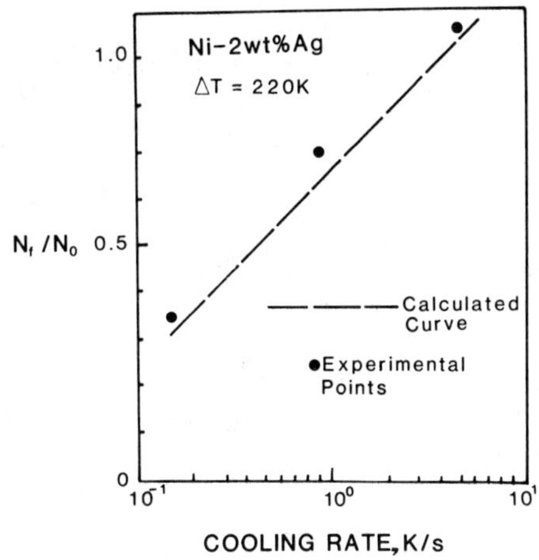

Fig. 18- Relative number of final grains remaining undissolved versus cooling rate after recalescence for Ni-2wt%Ag alloy specimens undercooled 220 K.

Conclusions

1. The dendritic grain size in dilute Ni-Ag alloys is practically independent of undercooling prior to nucleation, cooling rate and composition.
2. The nondendritic grain size in supercritically undercooled alloys depends on undercooling, cooling rate and composition. It decreases with increasing undercooling, cooling rate after recalescence and alloy concentration, whereas the specific grain boundary area increases.
3. A coarsening model was introduced which predicts the effects of cooling rate and composition on average grain size, specific grain boundary area and grain density. Qualitatively the predictions are correct. Quantitatively, there are some discrepancies between them and experimental results.

Acknowledgements

Support of this work by the University of Connecticut Research Foundation is gratefully acknowledged. The wire remelting was carried out at M.I.T. in Professor M. C. Flemings' laboratory. We are indebted to Dr. Y. Shiohara, M.I.T., for many helpful discussions.

References

1. J. L. Walker, in Phys. Chem. of Process Metallurgy, Part 2, G. R. St. Pierre, ed., Interscience Publishers, New York, N.Y., (1961) 845.

2. M. E. Glicksman, Acta Met., 13 (1965) 1231.

3. T. Z. Kattamis and M. C. Flemings, Mod. Casting, 52 (1967) 191.

4. S. Skolianos, P. S. Liu and T. Z. Kattamis, in Grain Refinement in Castings and Welds, G. J. Abbaschian and S. A. David, ed., The Metallurgical Society of AIME, Warrendale, PA (1983) 97.

5. T. Z. Kattamis, J. Cryst. Growth, 34 (1976) 215.

6. J. C. Baker and J. W. Cahn, Solidification, ASM (1971) 23.

7. M. J. Aziz, Appl. Phys. Lett., 43 (1983) 552.

8. G. A. Colligan and B. J. Bayles, Acta Met., 10 (1962) 895.

9. C. G. Levi and R. Mehrabian, Met. Trans. A, 13A (1982) 13 and 221.

10. R. Mehrabian, in Proceedings of the Third Conference on Rapid Solidification Processing, National Bureau of Standards, Dec. 6-8 (1982) 1.

11. J. H. Perepezko, Y. Shiohara, J. S. Paik and M. C. Flemings, in Proceedings of the Third Conference on Rapid Solidification Processing, National Bureau of Standards, Dec. 6-8 (1982) 28.

12. G. Horvay, ASME Proc. Fourth Nat. Congr. Appl. Mech., (1962) 1315.

13. P. W. Voorhees and M. E. Glicksman, Acta Met., 32(1984) 2001 and 2013.

14. M. E. Glicksman and P. W. Voorhees, Met. Trans. A, 15A (1984) 995.

15. T. Z. Kattamis and M. C. Flemings, Trans. TMS-AIME, 236 (1966) 1523.

16. P. W. Peterson, T. Z. Kattamis and A. F. Giamei, Met. Trans. A, 11A (1980) 1059.

17. T. Piccone, Y. Wu, Y. Shiohara and M. C. Flemings, submitted for publication, Met. Trans.

18. T. Z. Kattamis, W. F. Brower and R. Mehrabian, J. Cryst. Growth, 19 (1973) 229.

19. N. J. Whisler and T. Z. Kattamis, J. Mater. Sci., 7 (1972) 888.

20. T. Z. Kattamis, Z. Metallkde, 61 (1970) 856.

21. S. R. Coriell and D. Turnbull, Acta Met., 30 (1982) 2135.

UNDERCOOLED AND RAPIDLY QUENCHED Ni-Mo ALLOYS

S.N. Tewari* and T.K. Glasgow

National Aeronautics and Space Administration
Lewis Research Center
Cleveland, Ohio 44135

Hypoeutectic, eutectic, and hypereutectic nickel-molybdenum alloys have been rapidly solidified by both bulk undercooling and melt spinning techniques. Alloys were undercooled in both electromagnetic levitation and differential thermal analysis equipment. The rate of recalescence depended upon the degree of initial undercooling and the nature (faceted or non-faceted) of the primary nucleating phase. Alloy melts were observed to undercool more in the presence of primary β (NiMo intermetallic) phase than in γ (fcc solid solution) phase. Melt spinning resulted in an extension of molybdenum solid solubility in γ nickel, from 28 to 37.5 at % Mo. Although the microstructures observed by undercooling and melt spinning were similar the microsegregation pattern across the γ dendrites was different. The range of microstructures evolved has been analysed in terms of the nature of the primary phase to nucleate, its subsequent dendritic growth, coarsening and fragmentation, and final solidification of interdendritic liquid.

---

*Senior National Research Council NASA Research Associate; on leave from Defence Metallurgical Research Laboratory, Hyderabad, India.

Introduction

Rapid solidification of metallic alloys frequently results in extended solute solubility, greatly refined microstructures and formation of metastable phases. Chill block melt spinning is an attractive technique for production of thin ribbons which solidify with very high solid-liquid interface velocities (1). These high velocities are obtained by rapid heat extraction to the solid quench wheel. Large interface growth velocities are also obtained when the solid is allowed to grow in an undercooled alloy melt (2). The probability of heterogeneus nucleation at low undercoolings is reduced by using high purity elemental materials, maintaining purity during alloying, and by thermal conditioning of the melt in a suitable slag. When high degrees of undercoolings are attained initial solidification is rapid. Both the chill block melt spinning and bulk undercooling techniques are known to produce similar microstructural changes, such as reduced dendrite spacing (3-4), refined grain size (5-6), formation of metastable phases (7), particulate eutectics (8-9), extended solute solubility (8 and 10), etc. However, very little work has been reported where both the bulk undercooling and melt spinning techniques have been used to investigate and study microstructural and microchemical changes produced in the same binary alloy compositions. In this study hypoeutectic, eutectic, and hypereutectic nickel-molybdenum alloys have been rapidly solidified using the bulk undercooling and melt spinning techniques. Nickel-molybdenum binary alloys form the basis of some of the most important superalloys. This system is also attractive because it has a steep liquidus on the molybdenum rich side of the eutectic composition and a shallow liquidus on the nickel rich side (11). Differential thermal analysis and electromagnetic levitation techniques have been used to study the effect of alloy composition, level of initial undercooling, and nature of the solidifying phases on the microstructure obtained. Microsegregation across γ dendrites has been compared for the melt spun and undercooled alloys. Recalescence behavior after the nucleation of undercooled alloy specimens has been examined together with the resulting alloy microstructure. Melt spun foils have been examined for their range of microstructures produced across the ribbon thickness.

Experimental

Nickel (99.9 percent purity)-molybdenum (99.9 percent) charges with the alloy compositions marked on the Ni-Mo phase diagram (11) shown in Fig. 1 were vacuum induction melted and investment cast to yield 1 cm diameter by 10 cm long alloy bars. Samples for undercooling and melt spinning were obtained from these cylinders machined to 6 mm diameter. Approximately

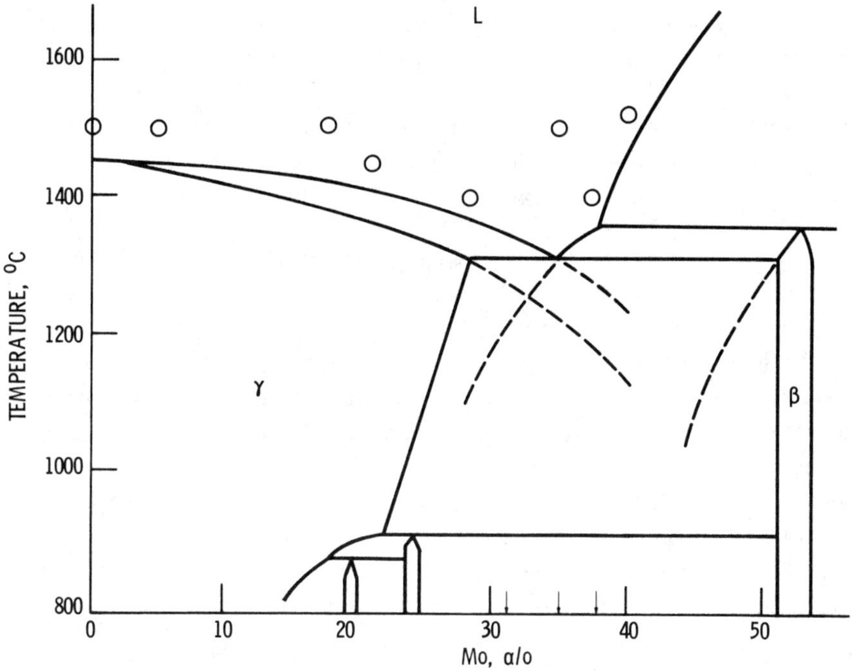

Figure 1. - Nickel-molybdenum phase diagram. ○ shows the compositions and melt temperatures for the Chill Block Melt Spun ribbons. ↓ shows the compositions solidified by bulk undercooling.

3 mm diameter alloy specimens were produced by thermal conditioning treatments carried out in a flowing argon atmosphere in a Pyrex/Vycor slag. These were then undercooled in a differential thermal analysis (DTA) unit at a cooling rate of 10 K/min, and solidified from various levels of undercoolings. The DTA furnace was switched off just after recording the final exothermic (solidification) peak. Through the courtesy of Prof. M.C. Flemings at Massachusetts Institute of Technology several thermally conditioned specimens were undercooled in an electromagnetic levitation (EML) unit. The specimen rested in a bed of Pyrex/Vycor powder in flowing argon. Directly coupled RF power which heated the specimen was turned off to cool the specimen. The specimen temperature was recorded by a two color optical pyrometer. The solidified specimens were metallographically examined and chemically analyzed by electron micro probe analysis.

The melt spun ribbons were cast by the free jet chill block melt spinning process in an apparatus described in Ref. 12. A charge of about 20 g was heated to about 50 K above the liquidus temperature of the alloy and discharged in vacuum on to a 4340 steel wheel finished with 600 grit paper. The ejection pressure used was $8.6x10^4$ N/m^2, the ejection hole was 1 mm in. diameter, and the wheel speed was 20 m/sec. Thin foils for TEM studies were made by electrochemical thinning using an alcohol/butyl cellusolve/ perchloric acid mixture in a twin jet electropolishing equipment. Thin regions near the wheel side or the free surface were obtained by masking the other side using a lacquer.

Results

Differential Thermal Analysis

Figure 2 shows the typical DTA plots obtained for the Ni-35 at % Mo eutectic alloy. Only one melting curve showing the eutectic temperature to be 1593 K is shown in this figure. At small undercoolings two exothermic peaks corresponding to two nucleation events were obtained. At greater undercoolings only one exothermic peak was observed. An examination of the nickel molybdenum phase diagram (Fig. 1) shows that for an alloy of eutectic composition two possibilities exist for the first phase to nucleate. It could be either nickel solid solution (γ) or intermetallic phase (β).

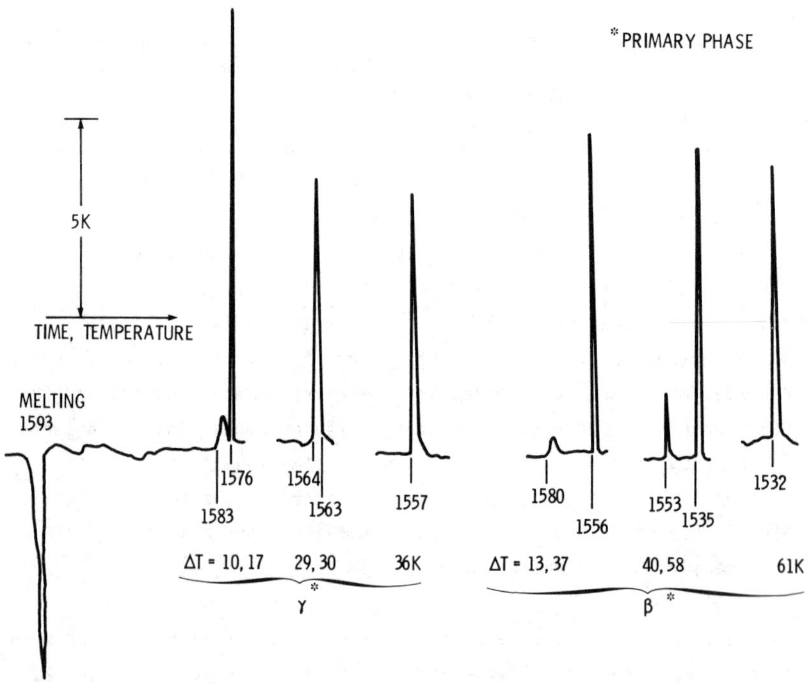

Figure 2. - Differential thermal analysis (DTA) plots for eutectic Ni-Mo alloy.

Microstructural examination of the solidified specimens showed this behavior. The primary phase has been identified for the various DTA plots shown in Fig. 2. The alloy melt can be seen to undercool in the presence of either the γ or β phase before the other phase nucleated. However, as the level of undercooling increased for nucleation of the primary phase, the extent of further melt undercooling for nucleation of the second phase decreased. Only one nucleation event was observed for the greatest levels of undercooling. Metallographic examination of the solidified specimens showed, however, that these specimens also had two nucleation events, which could not be distinguished from each other due to the insufficient response time of the DTA. The alloy melt was observed to undercool much more in the presence of primary "β" phase as compared to primary "γ" phase (Fig. 2).

Figure 3 shows two typical types of microstructures observed in the eutectic alloy samples solidified in the DTA unit. For some of the specimens an orthogonal network of primary γ phase dendrites/or dendrite elements surrounded by γ-β eutectic was observed. The average spacing between these γ dendrite elements was observed to decrease with increasing undercooling, as has also been reported for other alloys (4). For other specimens the microstructure showed the evidence of primary β phase nucleating and growing in a "branched" manner in the melt before nucleation of the γ phase. The microstructure around the β dendrites consisted of nearly spherical γ dendrite elements surrounded by β phase.

Electromagnetic Levitation (EML)

Figure 4 shows typical thermal profiles obtained for the EML specimens. The raw temperature data obtained were quite noisy and had to be filtered in the manner described by Piccone (13) (filtered in both the directions, eight data points in each). The nucleation temperature (T_n) and the maximum recalescence temperature (T_m) are marked in this figure. The recalescence time is very small, a few milliseconds. The maximum racalescence temperature is approximately the same as the eutectic temperature for this specimen. The shape of the recalescence curve is sigmoidal, i.e., the rate of temperature rise goes through a maximum during recalescence. The recalescence time was observed to decrease with increasing undercooling, as has been reported for Ni-Sn alloys (13).

Figure 5 shows two typical microstructures obtained in the specimens solidified in the EML. One type of microstructure showed evidence of primary γ phase nucleating and growing in the melt in a fine dendritic manner, before solidification of rest of the melt. The microstructure consists of γ dendrite elements surrounded by either γ-β eutectic or β phase.

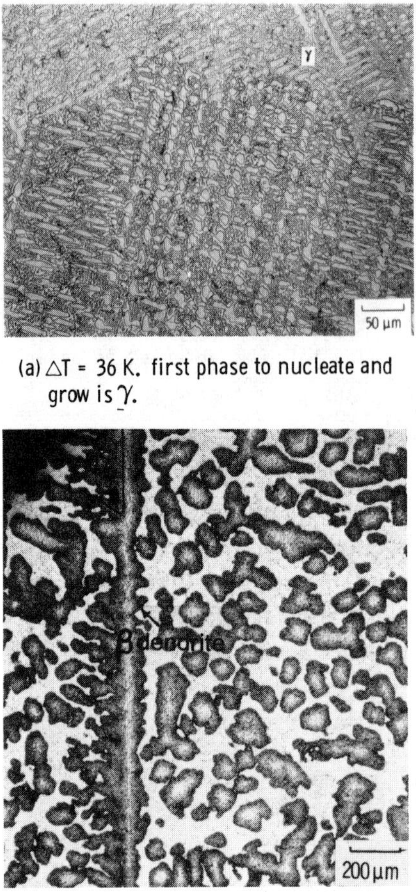

(a) ΔT = 36 K. first phase to nucleate and grow is $\underline{\gamma}$.

(b) ΔT = 114 K, first phase to nucleate and grow is β.

Figure 3. - Microstructure of eutectic alloy solidified in DTA.

The other showed evidence of primary β dendrites nucleating and growing in a nonorthogonal branched manner before rest of the alloy melt solidified. Both types of microstructures contain regions showing a "particulate eutectic" type of structure.

Figure 6 plots the maximum rate of temperature rise during recalescence measured from the thermal profiles for the hypoeutectic, eutectic and hypereutectic alloys, (similar to one shown in Fig. 4) as a function of the initial level of undercooling (alloy liquidus-nucleation temperature). These $\log(\delta T/\delta t)_{max}$ versus $\log(\Delta T)$ plots show a straight line fit with the slopes (m) as marked in the figure. The primary phase to nucleate and

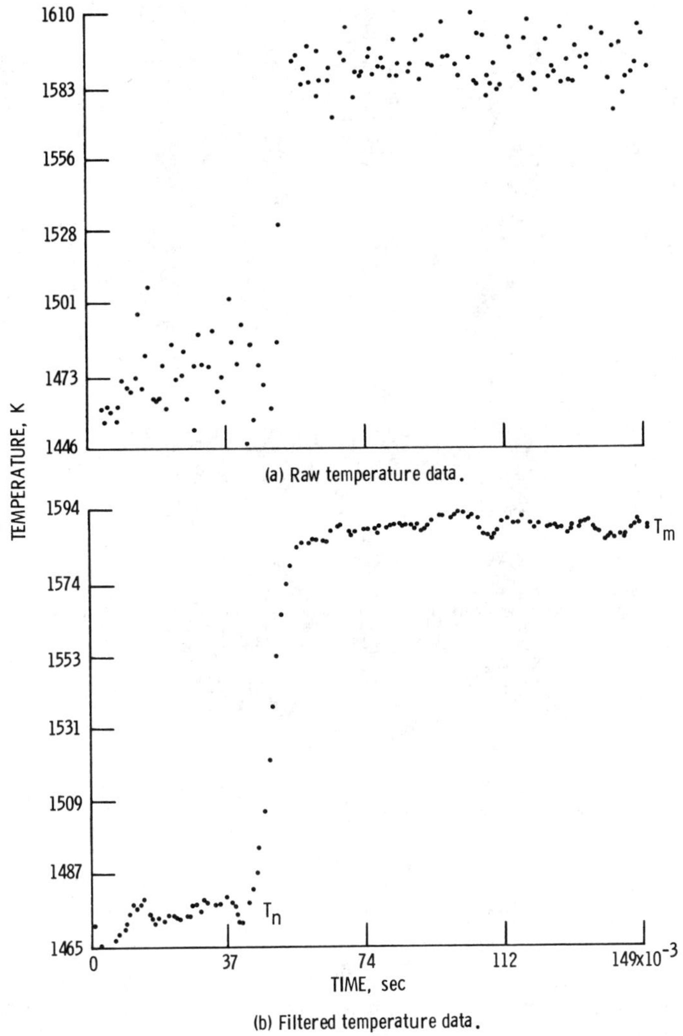

Figure 4. - Typical recalescence behaviour for eutectic alloy solidified from an undercooling (ΔT) of 160 K.

grow was γ for the hypoeutectic alloy, β for the hypereutectic alloy and either γ or β for the eutectic alloy. β did nucleate first for some hypoeutectic samples also. However, these have not been included in this figure.

Chill Block Melt Spinning

Two main features of the melt spun ribbon microstructures are shown in optical micrographs shown in Fig. 7. The alloys with molybdenum contents of 5, 17.5, and 28 at % showed a cellular/dendritic microstructure across

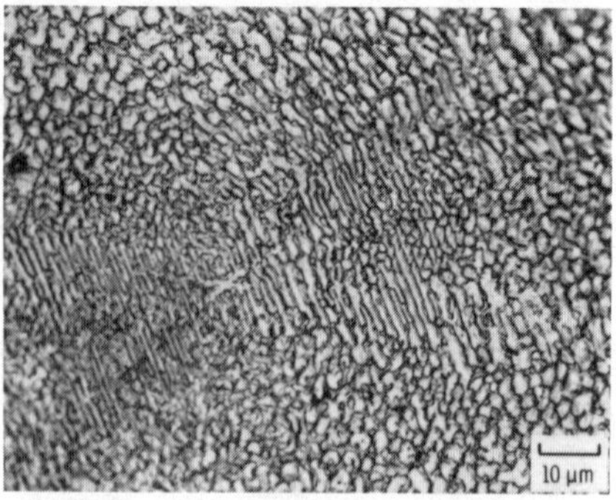

(a) ΔT = 107 K, first phase to nucleate and grow in the melt is γ.

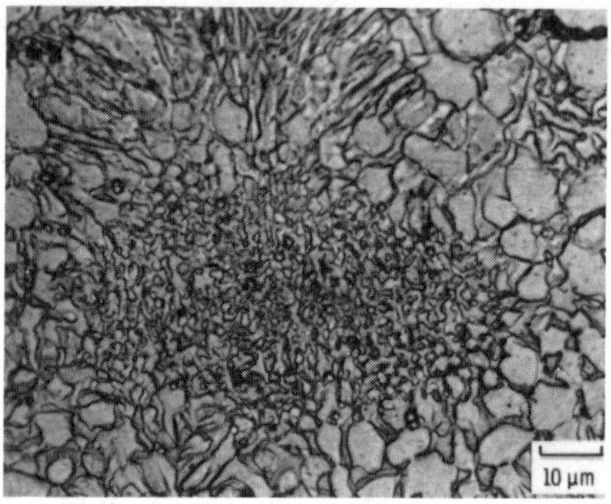

(b) ΔT = 58 K, first phase to nucleate and grow in the alloy melt is β.

Figure 5. - Microstructure of eutectic alloy (Ni-35 at % Mo) solidified in electromagnetic levitation unit.

the ribbon thickness. These cellular/ dendritic features were identified to be due to nickel rich solid solution, fcc(γ) phase. The eutectic alloy (35 at % Mo) also showed these cellular/dendritic microstructures. In addition it showed a thin "featureless" region at optical microscopy resolution near the quench surface. This zone is not of uniform thickness along the ribbon length, most of the ribbon being occupied by the cellular/ dendritic microstructure just described. Increasing molybdenum content of the alloy resulted in increased thickness of this "featureless" zone. This "featureless" zone was identified as the intermetallic, NiMo(β) phase for the eutectic alloy, and the β phase distributed in an amorphous matrix for the hypereutectic alloys.

Figure 8 shows the microstructure of γ phase in various regions of the eutectic alloy ribbon. The microstructure near the wheel surface

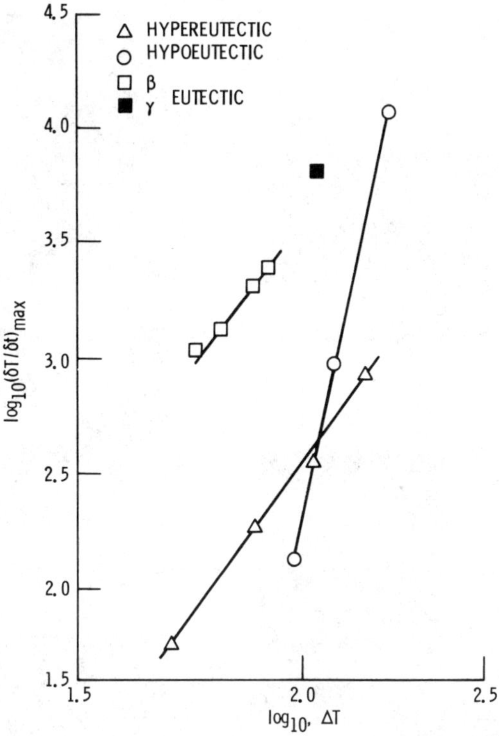

Figure 6. - Dependence of the maximum rate of temperature rise during recalescence $(\delta T/\delta t)$ max on the initial level of undercooling (ΔT).

(Fig. 8 a) consists of only γ grains. The linear features observed in this figure are due to the faulted nature of the supersaturated γ phase (14). There is no other phase observed in this region. The microstructure in the mid-ribbon portion consists of γ cells with the intercellular space occupied by β phase (Fig. 8(b)). The free surface region of the ribbon has a dendritic microstructure, with the interdendritic regions occupied by β phase (Fig. 8(c)). Note that the contrast within the cellular looking features due to the interdendritic β phase observed in the most slowly solidified regions of the ribbon (near free surface) is absent in the mid-ribbon region. This observation confirms that the γ in mid-ribbon region is cellular, whereas the γ near free surface region is dendritic.

Figure 9 shows the lattice parameter of the fcc nickel rich solid solution phase (γ) as a function of composition for the conventionally cast and rapidly solidified nickel-molybdenum alloys. For the conventionally cast material a linear increase in lattice parameter with increasing molybdenum

(a) 28.0 at % Mo.

(b) 35.0 at % Mo.

Figure 7. - Microstructure across the chill block spun ribbons of nickel-molybdenum alloys.

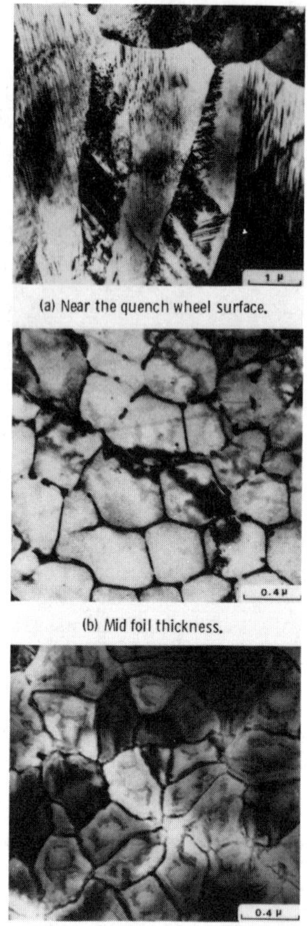

(a) Near the quench wheel surface.

(b) Mid foil thickness.

(c) Near the free surface.

Figure 8. - Microstructure of γ phase formed in various regions of melt spun eutectic alloy ribbon.

content of the γ phase is observed up to 27.5 at % Mo. It remains constant beyond this molybdenum content, as would be expected from the Ni-Mo phase diagram, Fig. 1. A straight line fit up to the molybdenum content of 37.5 at % for the melt spun alloys observed from this figure shows that rapid solidification has resulted in extending the solute solubility from an equilibrium value of 28 to 37.5 at % Mo.

Discussion

Bulk Undercooling Versus Solid-Liquid Interface Temperature Depression

As shown in Fig. 8 the microstructure across the ribbon thickness was not uniform. The fastest quenched region of eutectic alloy showed only

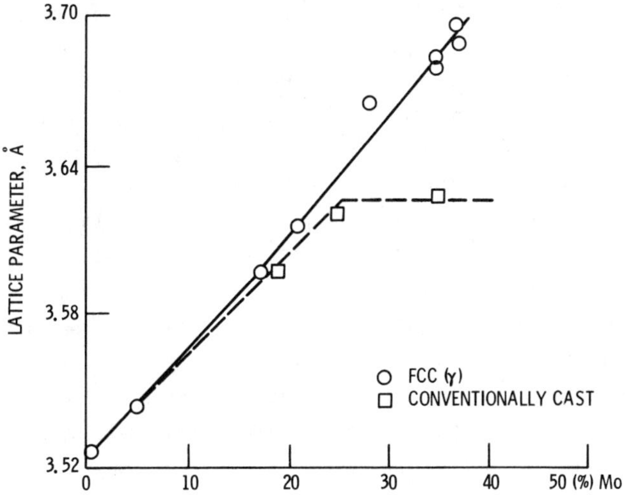

Figure 9. - Lattice parameter of the FCC Nickel γ phase as/a function of composition in conventionally cast and rapidly solidified Ni-Mo alloys.

grains of γ phase in this micrograph. This region therefore must have solidified in a segregationless manner with a stable plane front liquid solid interface. This is possible because of the morphological stability associated with a very high liquid-solid interface velocity in this region (15). Such solidification is thermodynamically only possible at or below 'T_0', the temperature at which solid and liquid of the same composition have equal free energy (16). This observation suggests that solidification in this region occurred with the liquid-solid interface temperature depressed by at least 60 K (assuming that 'T_0' is located midway between the liquidus and solidus temperature). The observed extension of molybdenum solubility from the γ lattice parameter measurements (Fig. 9) also suggests that melt spinning results in solidification occurring in the temperature regimes given by the extended liquidus-solidus.

The microstructure across most of the ribbon thickness however, is cellular/dendritic γ surrounded by β. The microstructure obtained from undercooled alloys, especially the one with the highest cooling rate after nucleation and larger bulk undercooling (e.g. Fig. 5(a), EML specimen with undercooling, ΔT = 107 K) also consists of γ dendrite elements surrounded by β. Rapid solidification by melt spinning and bulk undercooling therefore produces similar microstructures, except that the features are much finer for melt spun alloy. An examination of the microsegregation across the γ dendrites however, shows an important difference between melt spinning and bulk undercooling.

Figure 10 shows the microstructure and microsegregation across γ dendrites in a hypoeutectic alloy. Several solidification paths have been proposed for the solidification of an undercooled alloy melt (17). All of these suggest that the core of γ dendrite would be nickel poor: (First solid to form is of same composition as the melt. During recalescence the solid forming has same composition till either the solidus temperature or 'T_0' is reached. The subsequent solid forming follows the equilibrium solidus till the maximum recalescence temperature is reached. During subsequent cooling further formation of solid follows the solidus composition until the nucleation and growth of eutectic.) The solute profile in the undercooled specimen shows this nickel poor core, Fig. 10(a). However, the melt spun alloy shows the usual segregation pattern with a nickel rich core, Fig. 10(b). This observation suggests that the bulk melt puddle is not undercooled during melt spinning, but only the liquid-solid interface temperature is depressed due to curvature and large growth velocity effects. This observation is however, only strictly valid for the cellular/dendritic region of the melt spun foil. The fastest quenched region showed a segregationless solidification of the γ phase. For this region it is not possible to say whether the solidification occurred in an undercooled melt or not. A similar observation has been reported for a melt spun aluminum-copper alloy with a cellular microstructure (18).

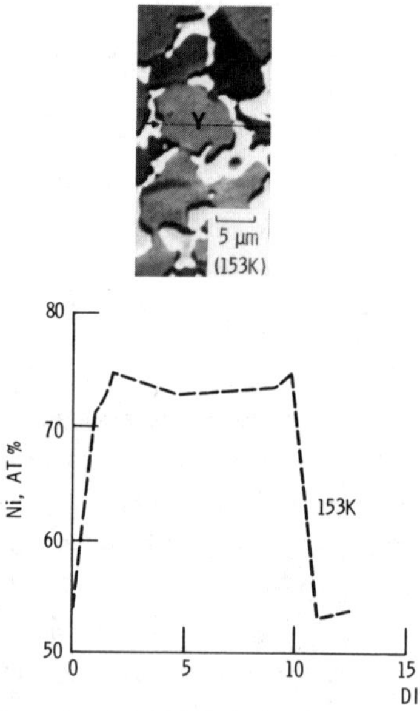

(a) From an undercooling of 153 K (DTA).

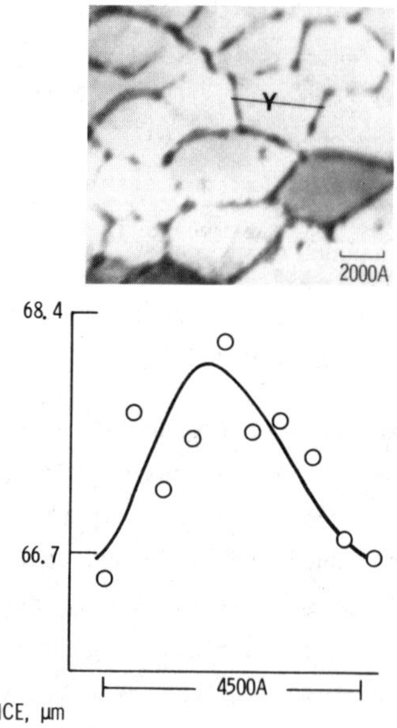

(b) Chill block melt spun alloy.

Figure 10. - Microstructure and microsegregation in solidified hypoeutectic alloy (Ni-31 at % Mo).

Recalescence Behavior

For the EML specimens when the RF power is turned off the specimen cools until the solid nucleates. The latent heat released by rapid growth of the solid in the undercooled melt makes the specimen temperature rise sharply (Fig. 4). At the beginning of the recalescence even though the dendritic growth rate is maximum due to the maximum undercooling in the liquid, the area of the liquid-solid interface is also small. This results in a low rate of solid formation and low rate of temperature rise. As the specimen temperature rises during further recalescence the dendritic growth rate decreases but the liquid-solid interfacial area is greater, resulting in a higher rate of temperature rise. Near the end of recalescence the dendritic growth speed due to the low level of undercooling is least. Such effects should cause the sigmoidal appearance of the recalescence behavior (Fig. 4). The technique of data filtering (eight data points in each direction, Ref. 13) can however, also lead to such a sigmoidal appearance; especially when there are only a small number of data points available during the recalescence.

The temperature sensed by the pyrometers is expected to be a complex function of the specimen size, thermal properties of the undercooled liquid and solid, Pyrex/Vycor environment, undercooling, nature of the growing phase/phases, etc. A rigorous theoretical understanding of the recalescence behavior is therefore very nearly impossible. In order to make some sense from these recalescence data we would make some crude assumptions. Let us <u>assume</u> that the growth rate, R, is proportional to the maximum rate of temperature rise ($\delta T/\delta t_{max}$) observed during the recalescence and that this growth rate corresponds to the initial level of undercooling, ΔT. The dendritic growth speed in an undercooled melt is known to increase with the increasing degree of undercooling in a manner described by, $R(\Delta T)^m$, with the m value between 1.8-3 (2). A plot of $\log(\delta T/\delta t)_{max}$ versus $\log(\Delta T)$ would therefore be expected to give rise to a straight line fit with a slope equal to m. Based on the limited sets of data in Fig. 6 such a straight line fit is observed. This figure gives m equal to 8.6 for the γ dendrites growing as compared to only 2.9 for the β dendrites. The NiMo (β) phase is a faceting type and its growth is expected to be controlled by the solute attachment kinetics. The β phase is therefore expected to grow less rapidly than γ, a nonfaceting phase, for which solute diffusion in the liquid is expected to control the growth rate. This results in a steeper dependence of $(\delta T/\delta t)_{max}$ on ΔT for the γ phase growth as compared to the β phase growth.

Microstructural examination of the melt spun foils also shows that even at the relatively high heat extraction rates encountered during melt spinning the γ phase grows with a much faster growth velocity than β phase. Figure 7(b) shows that the γ phase nucleating on the wheel surface grows much faster than the β phase nucleating on the wheel surface. The gamma phase also spreads laterally much faster. As a result only γ dendrites are observed in most of the eutectic alloy ribbon.

It is also interesting to compare the γ grain sizes for the 28.0 at % Mo alloy near the chill surface with those γ grains nucleated on the β surface in the 35.0 at % Mo alloy (Fig. 7(b)). The grain size for γ grains nucleating in contact with β phase is much larger than those nucleating in contact with the chill block. As shown in Fig. 2 the β phase has been shown to be a poor nucleant for γ in nickel-molybdenum alloy melt, resulting in considerable undercooling of the alloy melt in the presence of β dendrites. The nucleation rate of the λ phase for the same melt temperature is therefore expected to be lower, when it nucleates in contact with β phase, as compared to when it nucleates in contact with the wheel surface. This lower nucleation frequency explains the above men-mentioned grain size variation.

Microstructural Evolution

Figure 11 shows the range of microstructures obtained in the nickel-molybdenum eutectic alloy for various undercoolings and solidification conditions. It also shows schematically the mechanisms responsible for producing these microstructures. As explained earlier either the γ or the β phase can nucleate first from the alloy melt. The solid-liquid interface growing in an undercooled alloy melt is inherently unstable and prone to dendritic breakdown unless the growth velocity exceeds the "absolute stability" limit of Mullins and Sekerka (15). A dendritic network of either γ(orthogonal) or β(nonorthogonal) therefore immediately forms in the melt. During subsequent solidification this network grows, coarsens, fragments and the other phase nucleates and grows. If the first phase to nucleate is γ, two types of microstructures result. If the local solidification conditions are such that the interdendritic spacing is of the same order as interlamelar spacing of the eutectic, a "particulate eutectic" microstructure results, in which γ dendrite elements are surrounded by β phase (Fig. 11(a)). Otherwise the γ interdendritic region is occupied by γ-β eutectic (Fig. 11(b)). If the first phase to nucleate is β, three types of microstructures can evolve. Large undercoolings would prevent

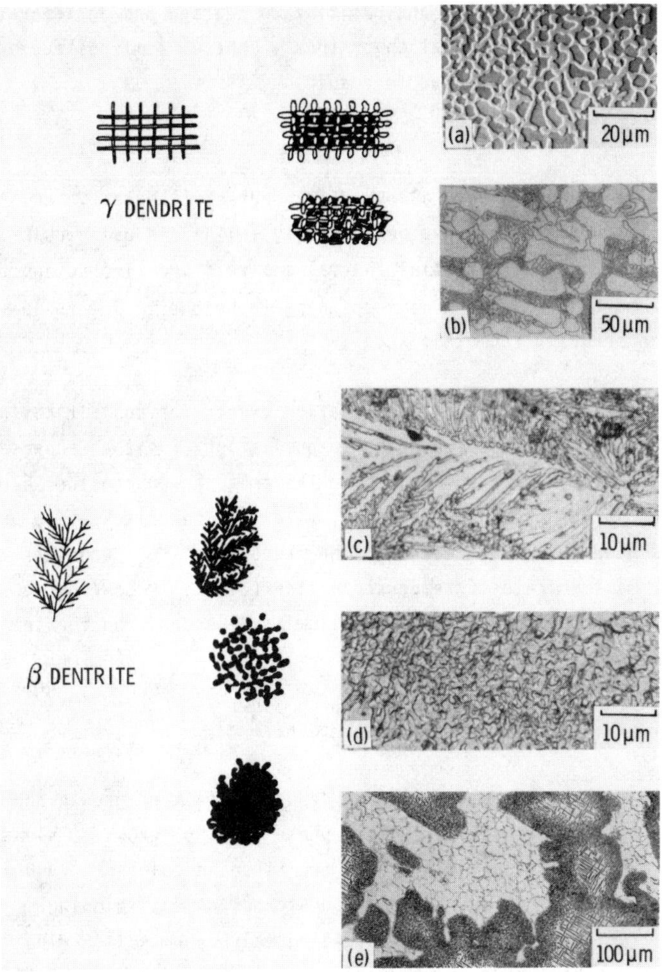

Figure 11. - Range of microstructures formed by solidification of eutectic alloy from an undercooled melt and their evolution. The dark phase in the schematic representations is β.

coarsening of the β dendritic network. The γ phase nucleating subsequently in the β interdendritic regions would produce a "complex regular" type of microstructure which consists of fine β dendritic skeletons on which γ-β "eutectic like" microstructure has formed (Fig. 11(c)). For a lesser degree of undercooling in which the β dendritic network achieves some coarsening and it also undergoes fragmentation of its network before the γ phase nucleates in the interdendritic regions, the microstructure shown in Fig. 11(d) would evolve. This microstructure would contain uniformly distributed γ and β phases ("particulate eutectic"). In case the β phase nucleates first and has time to coarsen and agglomorate considerably before the nucleation of the γ phase, a microstructure shown in Fig. 11(e) would be expected to result.

Conclusions

Nickel-molybdenum binary alloys of hypoeutectic, eutectic and hypereutectic compositions have been rapidly solidified using bulk undercooling by both differential thermal analysis and electromagnetic levitation and by chill block melt spinning techniques. The following conclusions can be drawn from this study,

1. DTA specimens showed two nucleation events for solidification of the undercooled alloy melt corresponding to nucleation of either supersaturated γ(fcc) or hypostoichiometric β(NiMo). The melt showed considerable further undercooling in the presence of either γ or β dendrites. The undercooling was however, much greater in the presence of β dendrites.

2. The maximum rate of temperature rise $(\delta T/\delta t)_{max}$ during recalescence for the EML specimens has the following dependence on the initial level of undercooling (ΔT),

$$(\delta T/\delta t)_{max} \propto (\Delta T)^m,$$

the value of m being 8.6 for the γ dendritic network growth and 2.9 for the β dendritic growth. The faceted phase (β,NiMo) grows at slower rate as compared to the nonfaceted phase (γ,fcc Ni solid solution) even at the very rapid rates of heat extraction encountered in melt spinning.

3. Melt spinning resulted in extending molybdenum solid solubility from a maximum equilibrium value of 28 to 37.5 at %. Rapid solidification by melt spinning does not cause solidification in an undercooled melt puddle. The liquid-solid interface temperature however, may be considerably depressed, as much as 60 K for the eutectic alloy.

4. The nucleation frequency during melt spinning of the molybdenum supersaturated γ grains is much lower in contact with the β phase as compared to that in contact with the wheel surface.

5. A range of microstructures can be formed by solidification of undercooled Ni-Mo alloys. These evolve due to nucleation and growth of dendritic network of either the γ phase (orthogonal network) or β phase (nonorthogonal network) first in the melt and subsequent nucleation and growth of the other phase in the interdendritic regions. When recalescence is sufficient, the particulate eutectic is one such microstructure.

Acknowledgement

Appreciation is expressed to Prof. M.C. Flemings of Massachusetts Institute of Technology and his colleagues, Dr. Y. Shiohara and Y. Wu for making the EML facilities available. Support from the National Research Council is gratefully acknowledged by one of the authors (SNT). Appreciation is expressed to Dr. Hugh R. Gray for his keen interest and continued support.

References

1. S.C. Huang, R.P. Laforce, A.M. Ritter, and R.P. Goehner: Metall. Trans. A, 1985, vol. 16A, pp. 1773-1779.
2. K. Kobayashi, M.I. Kumikawa, and P.H. Shingu: J. Crystal Growth, 1984, vol. 67, pp. 85-90.
3. H.A. Davies, N. Shoji, and D.H. Warrington: in Rapid Solidification Processing: Principles and Technologies, R. Mehrabian, B.H. Kear, and M.Cohen, eds., pp. 153-164, Claitor's Publishing Division, Baton Rouge, LA, 1980.
4. T.Z. Kattamis and M.C. Flemings: Mod. Cast., 1967, vol. 52, pp. 97-104.
5. P.G. Boswell and G.A. Chadwick: Scripta Metall., 1977, vol. 11, pp. 459-465.
6. T.Z. Kattamis: in Fundamentals of Alloy Solidification Applied to Industrial Processes, NASA CP-2337, pp. 15-32, National Aeronautics and Space Administration, 1984.
7. D. Turnbull: Metall. Trans. B, 1981, vol. 12B, pp. 217-230.
8. R. Cheese and B. Cantor: Mater. Sci. Eng., 1980, vol. 45, pp. 83-93.
9. K.P. Cooper, I.E. Anderson, and J.H. Perepezko: Rapidly Quenched Metals IV, vol. 1, T. Masumoto and K. Suzuki, eds., pp. 107-110, The Japan Institute of Metals, Sendai, Japan, 1982.
10. M.G. Chu, Y. Shiohara, and M.C. Flemings: Metall. Trans. A, 1984, vol. 15A, pp. 1303-1310.
11. Metals Handbook, vol. 8, p. 319, American Society for Metals, Metals Park, OH, 1973.
12. R.W. Jech, T.J. Moore, T.K. Glasgow, and N.W. Orth: J. Met., 1984, vol. 36, no. 4, pp. 41-45.
13. T.J. Piccone: Thermal History and Microstructure of Undercooled Nickel-Tin Alloys, M.S. Thesis, Massachusetts Institute of Technology, Sept. 1984.
14. N. Jayaraman and S.N. Tewari: "Fault Structures in Rapidly Quenched Ni-Mo Binary Alloys," (submitted to Metall. Trans.).
15. W.W. Mulins and R.F. Sekerka: J. Appl. Phys., 1964, vol. 35, pp. 441.
16. J.C. Baker and J.W. Cahn: Acta Metall., 1969, vol. 17, pp. 575-578.

17. T.Z. Kattamis: Z. Metallkunde, 1970, vol. 61, pp. 856-860.

18. L.J. Masur and M.C. Flemings: Rapidly Quenched Metals IV, vol. 2, T. Masumoto and K. Suzuki, eds., pp. 1557-1560, The Japan Institute of Metals, Sendai, Japan, 1982.

OCCURRENCE OF METASTABLE PHASES IN BINARY TI-RICH ALLOYS QUENCHED FROM THE MELT

Sung H. Whang

Department of Metallurgy & Materials Science
Polytechnic University
333 Jay Street
Brooklyn, New York 11201

Abstract

Metastable phases that are introduced in binary Ti alloys by rapid quenching include extended solid solutions, the metastable β-phase, the ω-phase, and the glassy phase. The occurrence of these phases was found to be correlated with equilibrium phase diagram features and may be interpreted using two-dimensional maps of various fundamental parameters such as atomic size, heat of formation, electronegativity, and electron density. Good correlations between the metastable phase formation and these parameters were found. There is evidence that the martensite transformation in the binary alloys may be suppressed by rapid quenching from the liquid.

Introduction

Metastable phase formation in Ti alloy systems has been known for some time. Typical phases occurring in the quenched Ti alloy systems include the α', α'', and ω phase. In the past, the studies of these phases were carried out by solid solution treatment followed by quenching in various cooling media, in which the cooling rate ranged from 10^2 to 10^5 $^oC/sec$. The formation of these metastable phases has been well documented [1-6].

By and large, recent rapid solidification processing may be characterized by the rapid quenching of alloys from the molten state. This processing involves very large undercooling [7] as well as very high solidification rate ($\tilde{}10^6$ $^oC/sec$) [8]. To avoid any confusion between the solution treatment and liquid quenching in this paper, rapid quenching means exclusively rapid liquid quenching. In Ti alloys, because of the $\beta \rightarrow \alpha$ transformation, the undercooled liquid experiences more complex phase transformations than rapidly quenched Al alloys. Such $\beta \rightarrow \alpha$ transformations in binary Ti alloy systems is known to be affected by alloying, and probably behaves differently during rapid quenching. The resulting metastable phases, in addition, the α', α'', and ω phases, include glassy phases, and new metastable crystalline phases that can not be produced by the solution treatment process. Some of glass-forming Ti systems have already been reported [9].

Nevertheless, the effects of the rapid solidification on the $\beta \rightarrow \alpha$ transformation, the martensite transformation and the ω phase formation have not been investigated until now. This paper deals with some of these issues through an experimental as well as phenomenological approach.

Experiments

Small alloy buttons of binary Ti alloys were prepared in an arc furnace under an argon atmosphere. Thin foils of the alloys were produced by the hammer-and-anvil technique from the melt. These foils (20 μm thick) were electropolished by the twin jet polisher into thin films for TEM study. The structures of the metastable phases were identified from the TEM micrographs.

Results and Discussions

Based on the alloying response to the $\beta \rightarrow \alpha$ transformation, Ti alloy systems may be conveniently classified into four categories. They are the β-isomorphous type, the eutectoid type, the peritectoid type and the monotectic-peritectic type.

Experimental results on the α-phase extension by rapid quenching (~10^6 K/sec) are tabulated in Table 1, together with the maximum equilibrium solid solubility of α-phase at the invariant reaction temperature. In the β-isomorphous systems, the extended solubility is twice or more the equilibrium value. All the extended compositions fall within the α+β field.

On the other hand, the metastable β-phase is replacing the α+β phase field from the opposite side of the phase diagram. Therefore, it appears clear that the α+β decomposition is mostly suppressed in this type of systems to yield extension of the α-phase as well as the metastable β-phase.

Table 1: Solid Solubility of Various Elements in α-Ti Phase

Type	Alloy System	Maximum Equilibrium Solubility	Extended Solubility	Invariant Reaction Point
β-Isomorphous	Ti-V	3.5 (at. %)	6 (at. %)	
	-Nb	2.5	5	
	-Ta	2.2	7.5	
	-Mo	0.4	2.0	
Eutectoid	Ti-Cr	0.5	5	13.5(at. %)
	-Mn	0.35	2.5	14.5
	-Fe	0.44	2.5	15
	-Co	0.8	2.5	6
	-Ni		5	
	-Cu	1.2	7.5	5.5
	-Si	0.8	6	1.1
	-W	0.2	2.5	
Peritectoid	Ti-C	0.55	10	
	-B	0.43	10	
	-Al	15	15	
	-Ge	2.7	2.5	
	-Sn	9	10	
	-Pb	4.2	5	
Monotectic-Peritectic	Ti-Y	0.1	1	0.2
	-La	1	0.3	1
	-Ce	1	0.6	1

In the eutectoid type systems, rapid quenching leads to an increase in the α-phase solubility of as much as ten times the equilibrium value and to stabilizing the β-phase at room temperature. This means that both the eutectic and eutectoid reactions can be suppressed over a wide range, which makes it possible to form metastable β-phase at room temperature. Nevertheless, the extended α′-phase limit is still much smaller than that of the corresponding eutectoid composition, as shown in Table 1. Instead, by virtue of the eutectoid suppression, metastable β-phase is produced over a wide compositional range.

Figure 1a and 1b show the acicular shaped α-phase in an as-quenched Ti-$Fe_{2.8}$ alloy, and a dispersed ω-phase in the β-phase matrix in an as-quenched Ti-Fe_5 alloy, respectively. The Ti-Fe equilibrium phase diagram and block diagrams for the room temperature phases [10, 11] are shown in Fig. 2, in which hypothetical T_0 lines (T_0 line is the locus of the equal free energy of two phases in the phase diagram) are drawn to demonstrate their relationship to the occurrence of metastable phases. By comparing the two block diagrams, it can be seen that the α+TiFe field of the equilibrium phase diagram converts into metastable α-phase and β-phase fields by rapid quenching. The opening between T_0 (1) and T_0 (2) predicts the occurrence of the metastable β-phase at room temperature, provided a sufficient cooling rate prevents both eutectic and eutectoid decomposition.

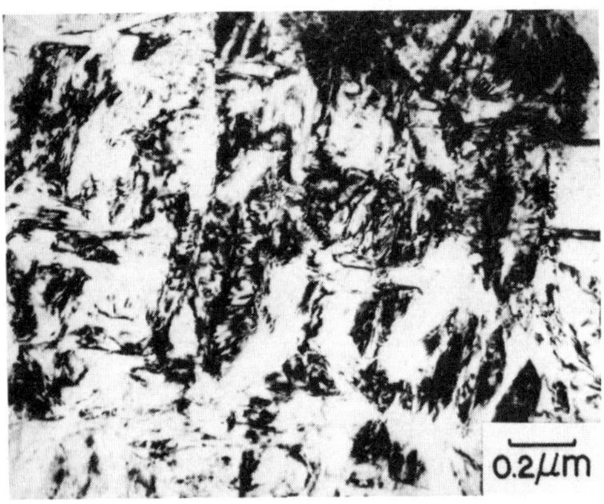

Figure 1

TEM Micrographs of As-Quenched Foils (~20μm thick)

a) $Ti_{97.5}Fe_{2.5}$

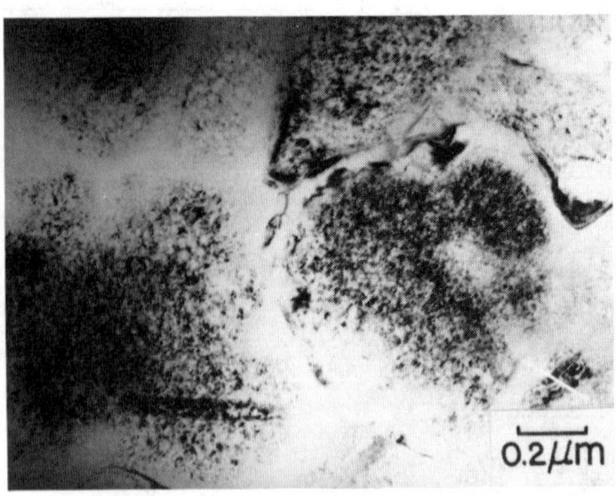

b) $Ti_{95}Fe_5$

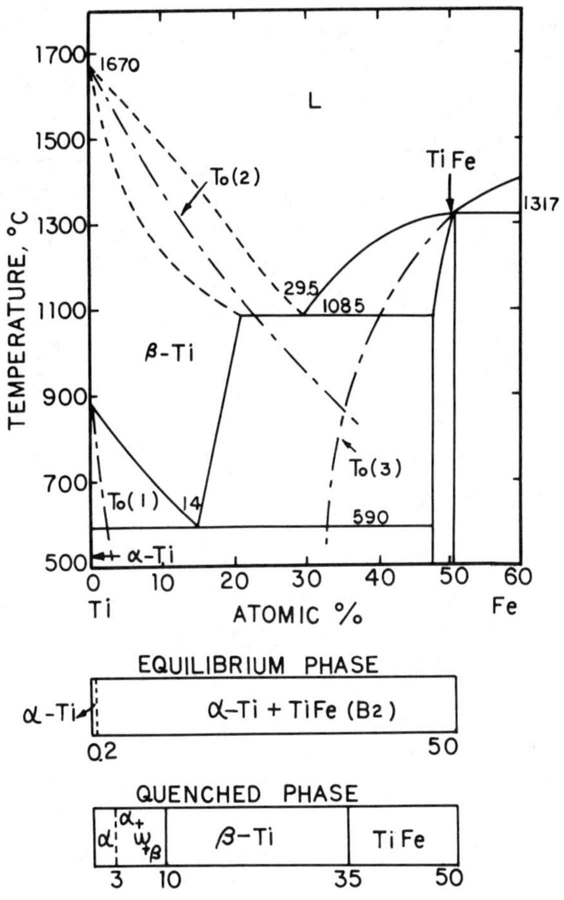

Figure 2

Ti-Fe Phase Diagram Showing A Relationship Between the Quenched Phases and T_0.

For better understanding, schematic T-T-T diagrams for Ti-Fe alloys are provided in Fig. 3a, b and c. When the quench rate exceeds the critical cooling rate (T_c) to avoid the eutectoid transformation, the quenched phase will be the martensitic α-phase as shown in Fig. 3a. The dashed M_{s1} line represents a new martensite starting temperature corresponding to a very high cooling rate. Although such a cooling rate-dependent transformation should not appear in this plot, it shows that this martensite transformation can be suppressed by a very high cooling rate. A detailed discussion will be given in a later section. Hence, this drawing represents the situation in as-quenched Ti-Fe (0-3 at.%). Also, in this case, the ω-phase formation is not possible since the e/a ratio is not favorable. With increasing

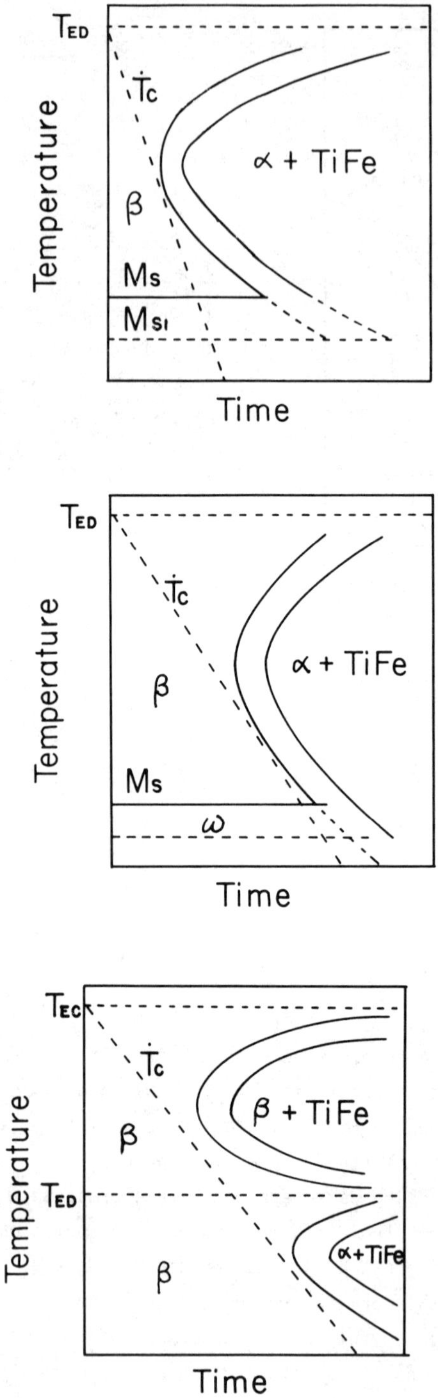

Figure 3

Heuristic T-T-T Diagram for Ti-Fe Alloy Systems During Rapid Quenching.

a) 0-3 at. % Fe

b) 3-10 at. % Fe

c) 10-35 at. % Fe

solute (Fe) concentration, the martensite starting temperature (M_s) decreases, and the ω-phase transformation becomes possible due to the right e/a ratio. However, the martensite transformation can not be completed at this composition. The resulting phases include α', metastable β and ω phases in this range (3-10 at. % Fe) (Fig. 3b). At high solute concentration (10-35 at. %), the deep eutectoid depression facilitates the formation of the metastable β-phase. Both martensite and the ω phase no longer interfere with the metastable β-phase during quenching (Fig. 3c). The as-quenched phase is the metastable β phase (Fig. 3c). Such a three-stage transition is clearly shown in rapidly quenched Ti-Mn alloys ($\alpha' \rightarrow \omega + \alpha' + \beta \rightarrow \beta$): in the first stage at 2.5 at. % Mn the martensitic α-phase is formed (Fig. 4a); in the second stage at 5 at. Mn, the metastable β-phase, the ω-phase and a trace of the α-phase (Fig. 4b); in the third stage at 10 at. % Mn the metastable β-phase formed (Fig. 4c).

The peritectoid type systems show two different tendencies in the solubility extension. In other words, those systems containing small size solute atoms such as C or B show high solubility extension upon rapid quenching, whereas the rest of the peritectoid systems do not show any significant extension in α-phase solubility, although they have high equilibrium solubility. So far, no sign of segregation in the $Ti_{90}C_{10}$ and the $Ti_{90}B_{10}$ system was observed by TEM under the high magnification (100kX), and a significant lattice parameter increase was confirmed by x-ray diffraction in the Ti-B system [12].

Hence, this may be an indication of the fact that C and B are dissolved in the matrix either interstitially or di-substitutionally [13] in such a way that the lattice expansion is possible. Other alloying elements such as Al, Sn, Ge etc. are known as α-phase stabilizers; the $\alpha \rightarrow \beta$ transformation temperature increases with increasing solute concentrations. There is no experimental evidence that rapid quenching stabilizes the β phase down to room temperature.

The monotectic - peritectic type system is commonly found in Ti-rare earth systems. Almost no solubility increase is found in the as-quenched Ti-rare earth system (Y, La, Ce) (Table 1). This may be explained by the equilibrium phase diagram: The characteristic feature of the phase diagram is that the two liquids are immiscible at high temperatures. The heat of formation calculated by Miedema [14, 15] is positive, indicating that the repulsive force is dominant in this system (Fig. 5). No stable liquid is found at the center of this phase diagram. Instead, the separation between a liquid and solid phase exists over a wide temperature range. As a result

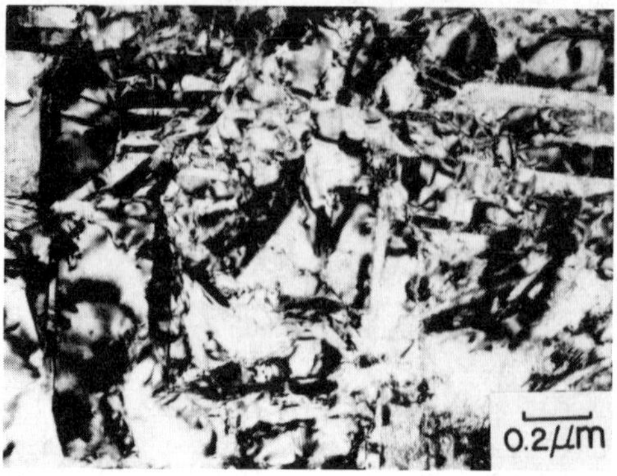

Figure 4

TEM Micrographs of As-Quenched Ti-Mn Alloys (~20μm thick)

a) $Ti_{97.5}Mn_{2.5}$

b) $Ti_{95}Mn_{5}$

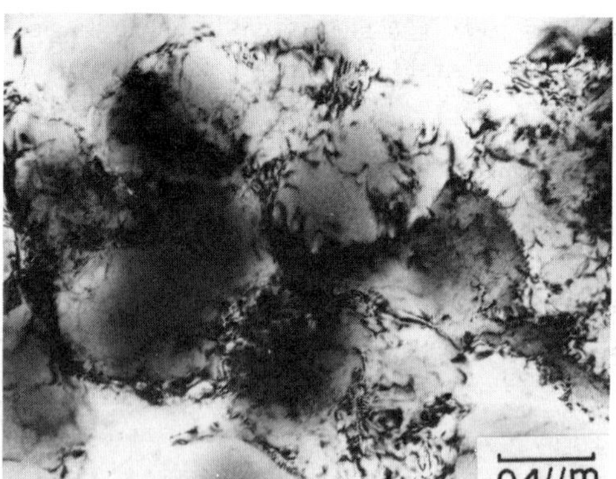

c) $Ti_{90}Mn_{10}$

any recalescence during solidification likely occurs in the two phase field, i.e., the liquid-solid mixed zone, which make it impossible to form an extended solid solution.

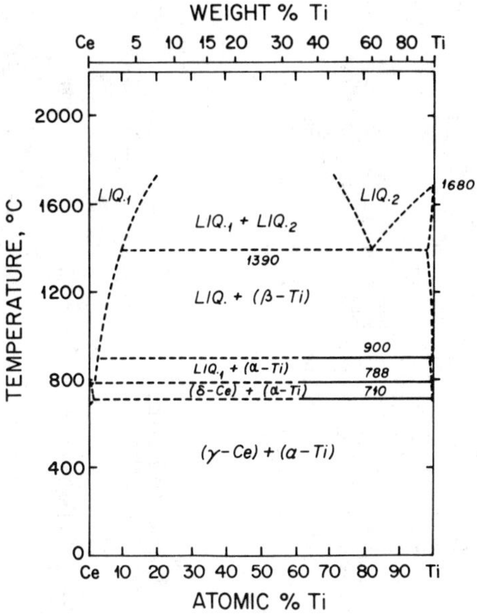

Figure 5

Ti-Ce Phase Diagram

Glassy Phase

Experimental as well as theoretical aspects of glassy phase formation in metal systems are well documented [16, 17, 18, 19, 20].

It has been demonstrated that T_0 is one of the important criteria for predicting glass formation [9]. Also, it is very important to consider the effect of crystal topology on the nucleation and growth to predict metastable phase formation [21]. For example, the liquid phase may be competing with a compound phase having simple structure such as BCC (or CsCl), which is an entropy stabilized phase where growth does not need a diffusional sorting process. As a result, glass forming near such a compound is difficult [22]. An example is shown in Fig. 6 where the liquid phase is competing with the metastable β-phase over the compositional range of 30 at. % Ni or less, and with the Ti_2Ni and TiNi phases for a Ni content larger than 30 at. %. It is interesting to point out that Ti_2Ni with an fcc type structure, can be readily suppressed by rapid quenching, allowing glass formation.

In contrast, the TiNi phase (CsCl) can not be suppressed, so that the glass forming compositional range is confined to a narrow region.

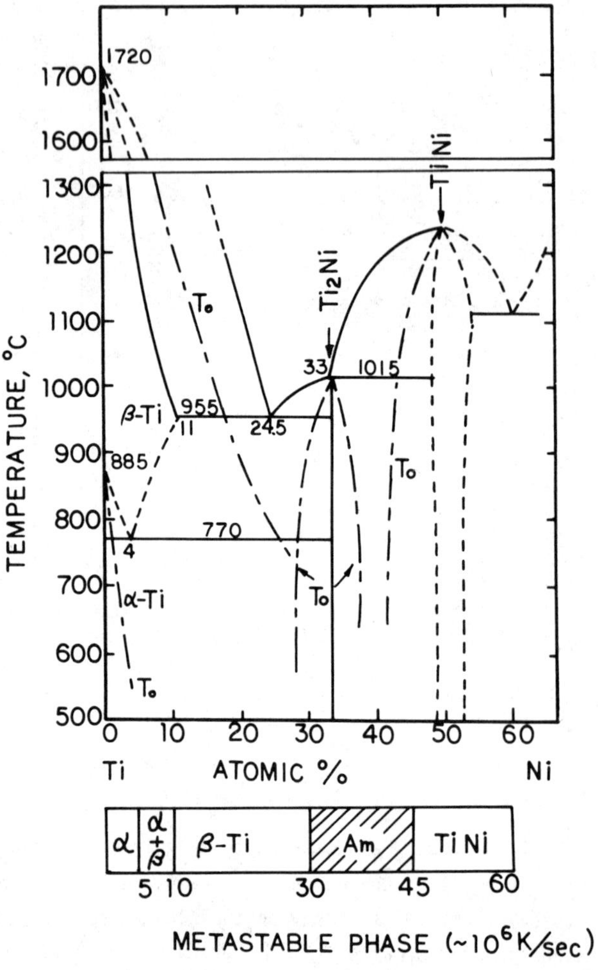

Figure 6

Ti-Ni Phase Diagram Showing Metastable Phase Formation Depending Upon T_0 Line Features

Two Dimensional Map

Following the pioneering work by Hume-Rothery and co-workers [23] on correlating the solid solubility limit with relative atomic size, electronegativity and electron valency between the host and the solute atoms, Darken and Gurry [24] demonstrated that terminal solubility of solute in binary alloy systems (Al, Mg, Ag host) can be identified on a map of atomic radius and electronegativity. Since then, combinations of two parameters such as electronegativity - electron density [25], atomic radius - Miedema's heat of formation of liquid alloy [26], were effectively used in predicting the solid solubility limit for various alloy systems.

Alonso and Simozar [27] have shown that Wigner-Seitz atomic radius and Miedema's heat of formation can be successfully used to predict the solubility of various solutes in Fe, Co, Cu and Ag host systems. Such a solubility prediction utilizing two dimensional maps was also employed for non-equilibrium alloy systems. The first example is that of implanation of solute atoms into a Cu, Fe, Ni, Be and V matrix [28]. The solubility limit of the solutes in these systems were analyzed by the Darken - Gurry map. It has been shown that Hume - Rothery's size criterion of + 15 % is expanded to + 25 % in this plot, while the - 15 % rule remains the same.

In another example, the solubility limit in rapidly solidified (RS) Al alloy systems was studied by the map of atomic radius and heat of formation [29]. No violation of the Hume - Rothery size criterion was found in this case. This discrepancy may be explained from the fact that ion implantation provides a higher cooling rate (~10^{13} ^{o}C/sec) than the liquid quenching (~10^{6} ^{o}C/sec).

Similar two-dimensional maps may be used for rapidly solidified Ti alloys. Firstly, a plot of Wigner - Seitz radius (atomic unit) vs. electronegativity was made in an attempt to predict the α-phase extension, as shown in Fig. 7. The solubility tendency can be arbitrarily divided into three different regions by dashed lines in this plot, i.e., region 1: a solubility greater than 5 at %; region 2: a solubility between 2 at % and 5 at %; region 3: a solubility less than 2 at %. In the plot, the region 1 group is represented by α-phase stabilizing elements, the region 2 group, by β-stabilizing elements and the region 3 group, by immiscible alloying elements. The only disagreement is found in Ti-Ge system. It should be pointed out that the radius values of Ge and Si are not realistic (too large), and do not correspond to values in the stable systems. Also, it is noted that Hume - Rothery's 15 % rule is well obeyed as is the case in rapidly quenched Al alloy systems [29].

Another scheme is to use Miedema's parameters: the electronegativity (ϕ) and the electron density at the cell wall (n_b) can be employed to construct a map (Fig. 8). The separation in this plot is very similar to the plot of atomic radius and electronegativity. Hence the superiority of the Miedema coordinates over the Darken - Gurry coordinates claimed by Chelikowsky [25] was not confirmed.

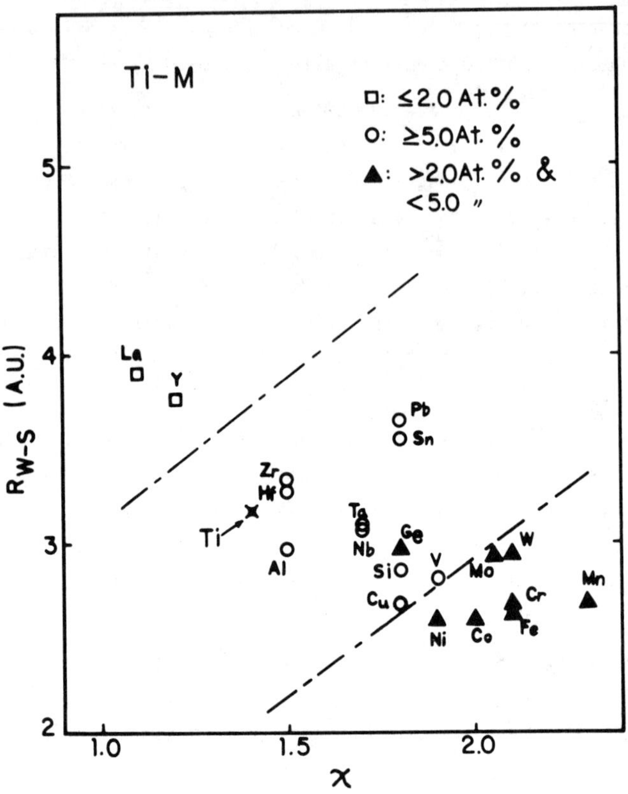

Figure 7

A Plot of the Wigner-Seitz Atomic Radius (R_{W-S}) vs. the Pauling's Electronegativity (x) of a Solute in Binary Ti Alloy Systems

According to Miedema [14], heat of formation of a liquid alloy ΔH_ℓ is given by:

$$\Delta H_\ell = -P(\Delta \phi^*)^2 + Q(\Delta n_b{}^{1/3})^2 - R$$

where P and Q are universal constants, R is a constant to describe the behavior of polyvalent metals with p electrons.

Surprisingly, the plot of the atomic radius and the heat of formation (Fig. 9) yields rather poor results. In particular, the predictions for Ti-V, Ti-Cu and Ti-Ge systems are incorrect. This is quite contrary to the report by Alonso and Simozar [27] that a significant improvement in pre-

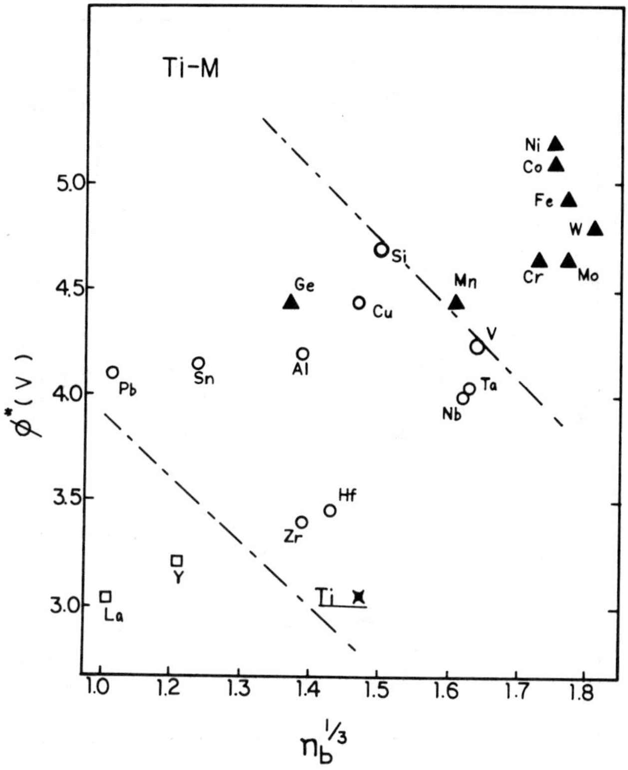

Figure 8

A Plot of Two-Miedema's Parameters: The Electronegativity (ϕ*) vs. The Electron Density (n_b) at the Cell Boundary

dicting the solid solubility was possible from a map of atomic radius vs. heat of formation, compared with any other schemes. It appears that the separation does not improve further by changing the maps, the two coordinate to a three coordinate ones (e.g. R_{W-S}, ϕ* and n_b). Further evidence is needed to clarify these discrepancies.

For the prediction of glass forming systems, the same plot of radius vs. heat of formation was applied to binary Ti alloy systems (Fig. 10). It clearly shows that the atomic size effectively identifies the glass forming systems, i.e., the solute atom should be smaller than the host atom. Also, the glass forming alloys have negative heats of formation as indicated by the dashed line. It should be noted that glass forming systems are located

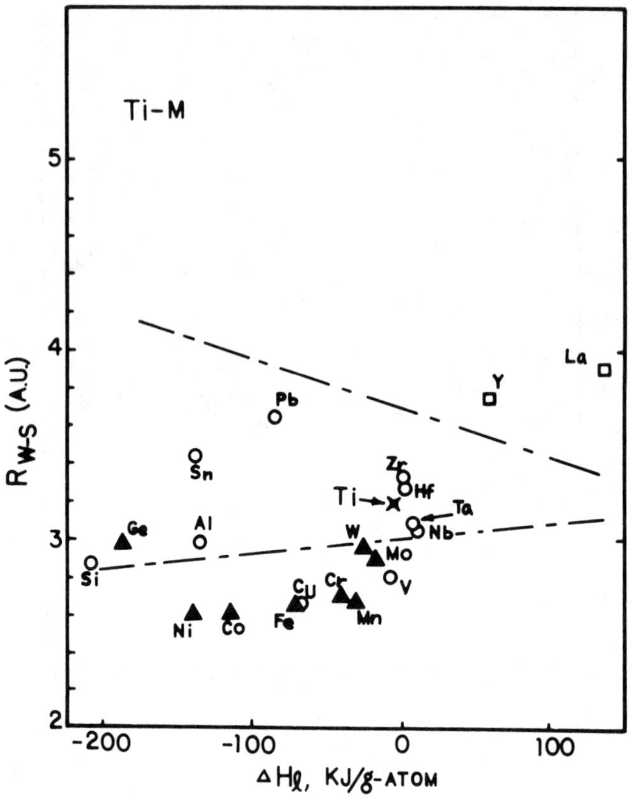

Figure 9

A Plot of the W-S Atomic Radius vs. the Miedema's Heat of Formation for Predicting the Extended α Solid Solubility in Binary Ti Alloy Systems

in the region 2, whereas the highest solubility systems exist in the region 1.

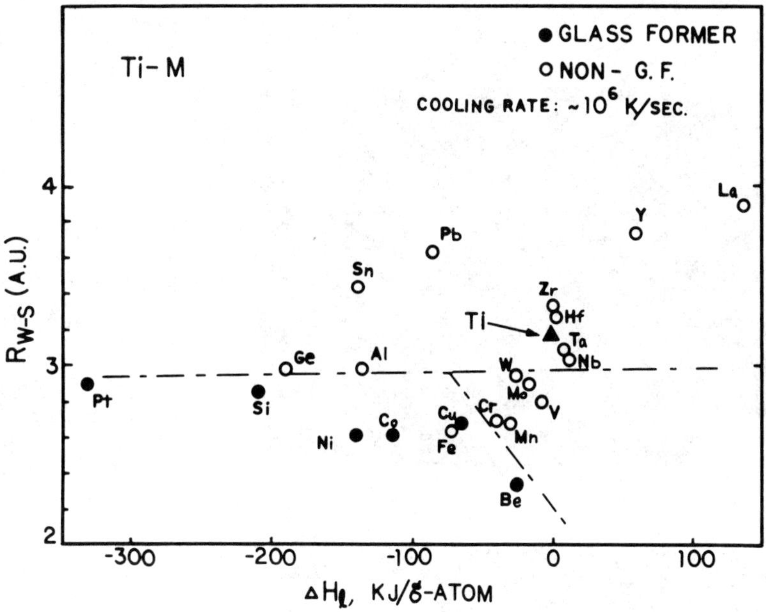

Figure 10

A Plot of the W-S Atomic Radius of a Solute vs. the Heat of Formation for Predicting a Glass Formation in Binary Ti Alloy Systems

Martensite Transformation and ω Phase

Previous studies show that the martensite transformation is independent of the cooling rate beyond the critical rate in binary Ti alloys [8], and that the ω-phase is reversible upon heating and cooling [30]. Furthermore, ω-phase formation is closely related to the e/a ratio of the alloy [1, 2, 3] as is the martensite start composition. Much evidence suggests that both the martensite transformation and the ω-phase formation are the result of a soft phonon mode in the β-phase [30, 31]. Questions have been raised whether or not such transformations are undisturbed by rapid quenching from the melt.

First, it is shown that the martensite transformation starts at a specific composition ($Ti-Fe_{2.8}$) for a fixed cooling rate. This is illustrated by Fig. 11b, in which co-existing martensite structures and ω-phase are found. A small shift toward Ti-rich composition ($Ti-Fe_{2.5}$) results in a fully martensitic structure (Fig. 11a); in a more Fe rich composition ($Ti-Fe_{3.0}$), the martensitic structure disappears (Fig. 11c). This is a

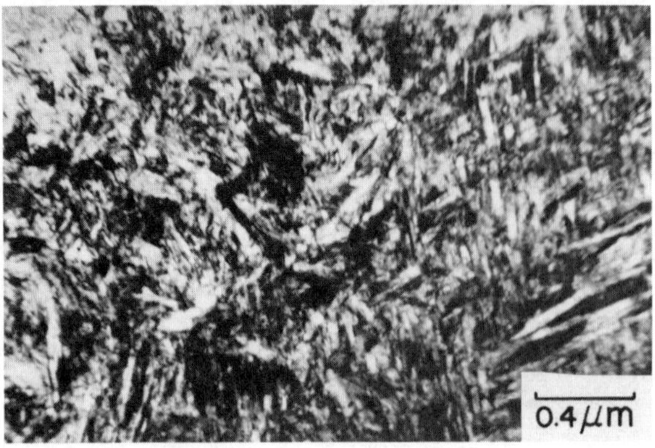

Figure 11

TEM Micrographs of As-Quenched Ti-Fe Alloys

a) $Ti_{97.5}Fe_{2.5}$ (~40μm thick)

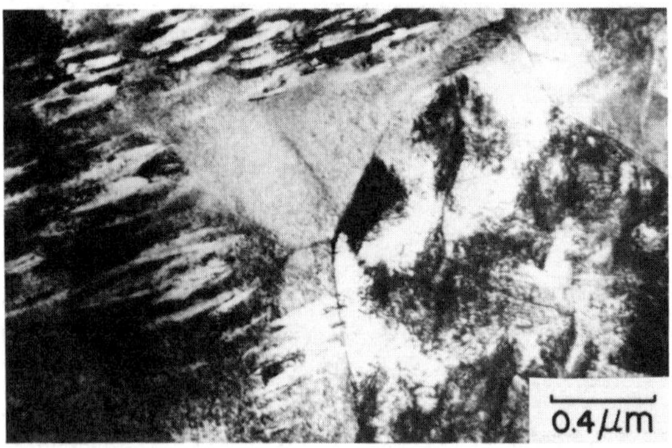

b) $Ti_{97.2}Fe_{2.8}$ (~40μm thick)

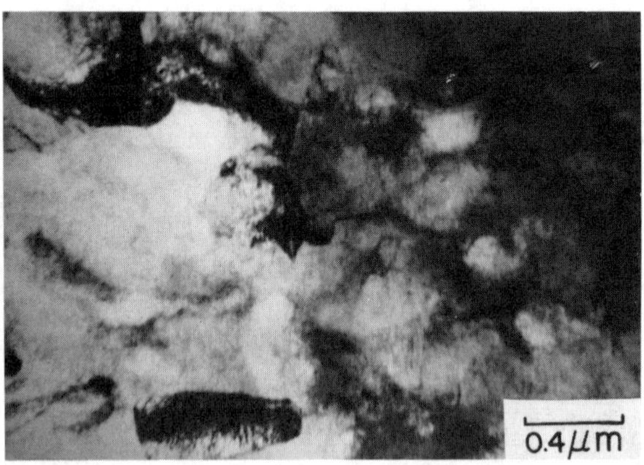

c) $Ti_{97}Fe_{3.0}$ (~40μm thick)

clear indication that a well defined martensite transformation exists.

On the other hand, the cooling rate of the alloy can be increased by reducing the thickness of the foil since, in the Newtonian regime, the cooling rate is inversely proportional to the thickness of the foil produced [32]. TEM micrographs of Ti-$Fe_{2.8}$ with different thicknesses (~20μm, ~40μm and ~65μm) are shown in Fig. 12a, 12b, and 12c, respectively. The 20μm thick alloy (presumably produced at the highest cooling rate) does not show any martensitic structure; the 40μm thick sample a fine martensitic and a dispersed ω-phase; and the 65μm foil exhibits a fully grown martensitic structure and a dense ω-phase in the matrix. This is a demonstration that the formation of martensite and the ω-phase is a sensitive function of the cooling rate, contrary to previous observations [33]. Because of the difference in the cooling rate between the 20μm sample and

Figure 12

TEM Micrographs of As-Quenched $Ti_{97.2}Fe_{2.8}$ Alloys

a) ~20μm Thick Foil

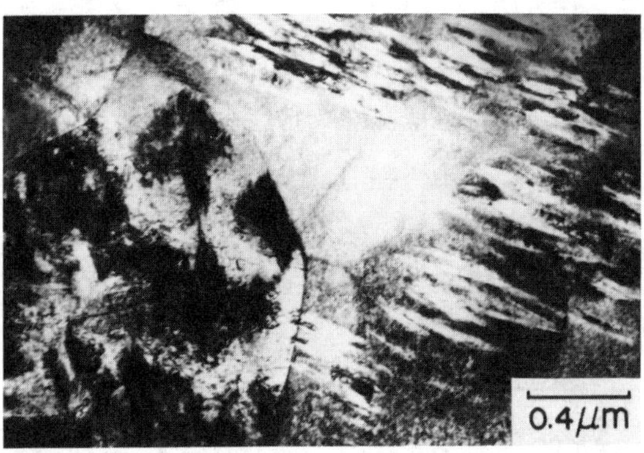

b) ~40μm Thick Foil

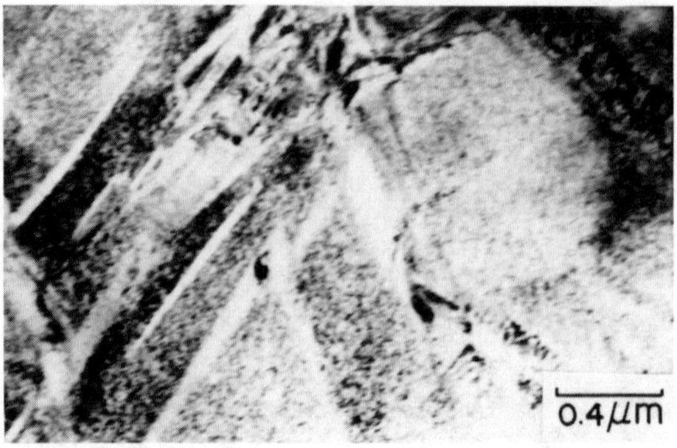

c) ~65μm Thick Foil

the 65μm sample, the time delay for the 65μm sample to reach the same temperature would be a few microseconds for each cooling temperature, assuming that the cooling rate for the 20μm sample is 10^6 °C/sec. When the cooling takes place over 1000°C in both samples, the time delay for the 65μm sample accumulates to an order of 10^{-3} sec.

If the martensite transformation is time-dependent as suggested by these micrographs, the martensite-start temperature may be drawn individually for each cooling rate as shown in Fig. 13. This analysis leads to the conclusion that the transformation is not a diffusionless one if titanium is alloyed with β-stabilizing elements.

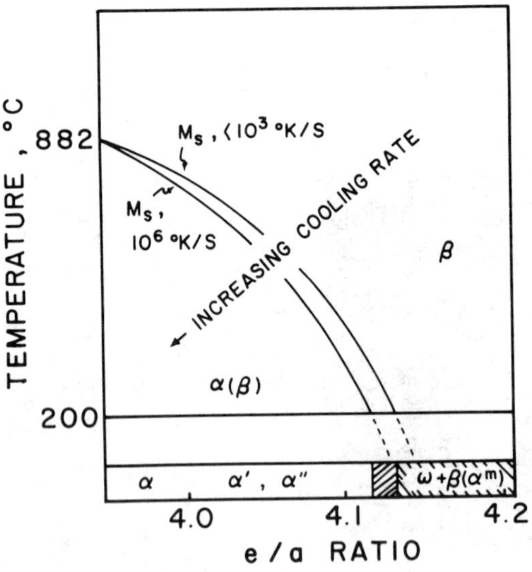

Figure 13

A Schematic Diagram Showing a Relationship Between the Cooling Rate and the Martensite Start Composition

Conclusion

1. The extension of metastable α and β, and glass formation in binary Ti alloys may be explained by the equilibrium phase diagram features considering both thermodynamic and kinetic criteria.

2. The solubility extension and glass forming tendency were predicted with some success by fundamental parameters such as atomic radius, electronegativity, electron density, and heat of formation. In fact, in the present study, no one scheme was found to be superior to the other.

3. The martensite transformation appears to be time-dependent in a strict sense, as deduced from the microstructural observations by varying the cooling rate. Formation of the ω-phase is also closely related to the martensite initiation, which is a confirmation of previous observations.

Acknowledgements

The author gratefully acknowledges the support of the Office of Naval Research for this work (Contract No. N00014-85-K-0787).

I would like to thank Drs. C.S. Chi and C.J. Yang and Mr. Y.D. Hahn for their assistance in the preparation of this manuscript.

References

1. C.A. Luke, R. Taggart and D.H. Polonis: Met. Trans ASM, 57 (1964), pp. 142-148.

2. Yu A. Bagariatskii, I.G. Nosova, and T.V. Tagunova: Sov. Phys. Dokl., 3 (1959), pp. 1014-1018.

3. B.S. Hickman: Trans AIME, 245 (1969), pp. 1329-1336.

4. J.C. Williams: Titanium Science & Technology, Ed. R.I. Jaffee and H.M. Burte (New York, N.Y., Plenum Press, 1973), pp. 1433-1494.

5. J.C. Williams: Ti and Ti Alloys, Science & Technology Aspects, (New York, N.Y., Plenum Press, 1982), pp. 1477-1494.

6. E.W. Collings: The Physical Metallurgy of Titanium Alloys, (Metals Park, OH: American Society for Metals, 1984), pp. 75-110.

7. J.H. Perepezko: Rapid Solidification Processing, Principles and Technologies, II, (Baton Rouge: Claitor's Publishing Division, 1980), pp. 56-67.

8. P. Predecki, A.W. Mullendore, and N.J. Grant: Trans. Metall. Soc., AIME, 233 (1965), p. 1581.

9. S.H. Whang: Materials Science & Engineering, 57 (1983), pp. 87-93.

10. R. Ray, B.C. Giessen, and N.J. Grant: Met. Trans., 3A (1972), pp. 672-679.

11. S.H. Whang: Review on Rapid Solidification Ti Alloy for High Temperature Applications, J. Mat. Sci., July 1986.

12. Y.Z. Lu and S.H. Whang: "Precipitate Hardening of Rapidly Quenched Ti-Zr-B Alloys" (Annual Report to ONR, Barnett Institute of Northeastern University, Boston 1984).

13. R. Ray, M. Segnini, and B.C. Giessen: Solid State Communications, Vol. 10 (1972), pp. 163-167.

14. A.R. Miedema: J. Less-Common Met., 32 (1973), p. 117.

15. A.R. Miedema, F.R. DeBoer and R. Boom: CALPHAD Vol. 1 (1977), pp. 341-359.

16. D.E. Polk and B.C. Giessen: Metallic Glasses, (Metals Park, OH: Am. Soceity for Metals, 1978), pp. 1-35.

17. B.C. Giessen and S.H. Whang: J. Phys. C8 (1980), p. 95.

18. R.W. Cahn: J. Phys. C9 (1982), p. 55.

19. J. Hafner: "Theory of the Structure, Stability, and Dynamics of Simple-Metal Glasses", Glassy Metals 1, Ed. H.J. Guntherodt and H. Beck (New York, N.Y.: Springer-Verlag, 1981), pp. 93-139.

20. F. Sommer: Z. Metallkde, 72 (1981), p. 219.

21. D. Turnbull: See Proceedings of this Symposium.

22. S.H. Whang: J. Non-Cryst. Solid., 70 (1985), pp. 85-92.

23. W. Hume-Rothery, G.W. Mabbott and K.M. Channel-Evans: Phil. Trans. R. Soc. London, Ser. A 233 (1934), pp. 1-97.

24. L. Darken and R. Garry: Physical Chemistry of Metals, (New York, N.Y.: McGraw-Hill Book Pub., 1952), pp. 86-90.

25. J.R. Chelikowsky: Phys. Rev. B, 19 (1979) pp. 686-701.

26. J.A. Alonso and S. Simozar: Phys. Rev. B, 22 (1980), pp. 5583-5589.

27. J.A. Alonso, J.M. Lopez, S. Simozar and L.A. Girifalco: Acta Metall., 30 (1982), pp. 105-107.

28. D.K. Sood: Phys. Lett. A, 68 (1978), pp. 469-472.

29. H. Jones: Proc. Materials Research Society, Vol. 28 (1984), pp. 303-315.

30. D. DeFontaine, N.E. Paton, and J.C. Williams: Acta Metall., 19 (1971), pp. 1153-1162.

31. S.L. Sass: J. Less-Comm. Met, 28 (1972), pp. 157-173.

32. R.C. Ruhl: Met. Sci. Eng., 1 (1967), pp. 312-320.

33. K.S. Jepson, A.R.G. Brown and J.A. Gray: The Science, Technology & Application of Titanium, Ed. R.I. Jaffee and N.E. Promisel (London: Pergamon Press, 1970), pp. 677-690.

PART II: Techniques of Undercooling and Metastable Phase Formation

METASTABLE STATES IN PULSED LASER QUENCHING

Frans Spaepen

Division of Applied Sciences
Harvard University
Cambridge, MA 02138

It is reviewed how irradiation of a metallic surface with a short laser pulse leads to the successive formation of these metastable states: overheated crystal, overheated liquid, undercooled liquid, and metastable crystal or glass. The stability of a liquid overheated above its melting temperature is analyzed. The thermodynamics and kinetics of melting and crystallization of pure metals and alloys are reviewed. The results of quenching experiments on the Nb-Si system are used to illustrate how phases with high isotropy and small unit cell size are kinetically favored.

Introduction

Irradiating the surface of a strongly absorbing solid (metal, semiconductor) with a short laser pulse is a method for depositing thermal energy in a very thin surface layer, which, as a result, is first melted and subsequently quenched at a very high rate. Picosecond laser pulsing can produce quench rates exceeding 10^{12} K/s, making it the fastest liquid quenching method available.

In the first part of this paper the three stages of the process (energy deposition, melt-in, and cooling) are analyzed in some detail, and three metastable states are identified: the overheated crystalline solid (above its melting point), the overheated liquid (above the boiling point), and the undercooled liquid (below the melting point).

The second part of the paper is a detailed analysis of the overheated liquid state. It is shown that the liquid can remain in this metastable state over a considerable temperature range above its boiling point, and that, as a result, the loss of mass or energy by evaporation is negligible in picosecond laser quenching.

In the third part of the paper the thermodynamic restrictions on the type of transformations that can occur on the very short timescale of the laser quenching process are discussed. The selection of the actual transformation from the remaining allowed ones depends on the kinetics of the competing crystallization processes. Depending on the degree of structural rearrangement required to form a crystal from the melt, the process can be collision- or diffusion-controlled. It will be shown that only for the simplest crystal structures or dilute solutions is growth collision-controlled and fast enough to be insuppressible, even by picosecond laser quenching.

In the last part of the paper, two applications are presented. First, it is demonstrated that fast pulsed laser irradiation is one of the few methods available for overheating metals far above their melting temperature. As an example of the solidification products obtained by pulsed laser quenching, a set of recent experiments on the Nb-Si system is reviewed. It is an illustration of the kinetic hierarchy, discussed by Turnbull in the first paper of these proceedings, in which the formation of isotropic systems and systems with small unit cells is kinetically favored.

Laser Processing Mechanism

A number of detailed analyses have been made of the processes of energy deposition, heating and cooling in pulsed laser irradiation of solid surfaces

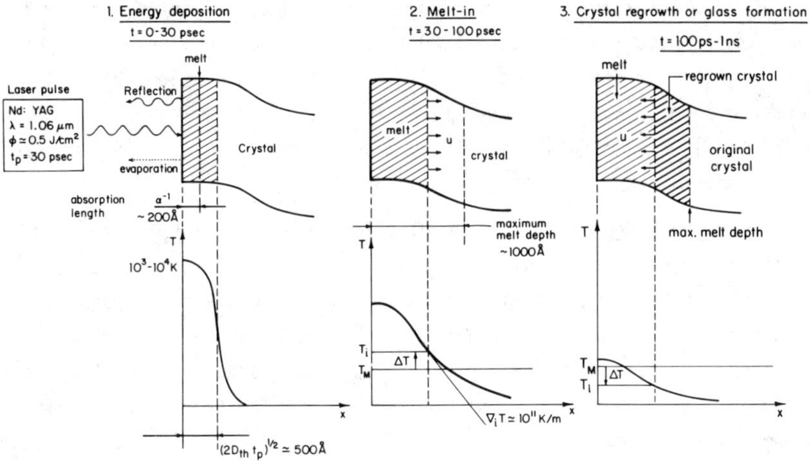

Fig. 1 - Schematic illustration of the mechanism of pulsed laser quenching.

(1-3). These three stages are illustrated in Figure 1, with parameters corresponding to the experiments to be described later.

During the first stage, the laser energy is deposited in a layer of thickness α^{-1}, where α is the absorption coefficient. For optical light on metals, or for UV light on semiconductors, this layer is a few hundred Å thick. Since the transfer of the energy from the electrons, which interact with the light, to the lattice occurs within less than 1 ps (4), the energy deposition process, which lasts for the duration of the pulse, t_p, can be considered a purely thermal one (5). The thermal diffusion length during that time is $\ell_T = (2D_{th}t_p)^{1/2}$, which, for metals is between 250 and 700 Å, depending on the thermal diffusivity ($D_{th} = 10^{-4}$ to 10^{-5} m^2/s). For fluences, ϕ, in the range indicated in Figure 1, a layer of thickness ℓ_T is melted and heated to several thousand degrees.

Note that for a pulse duration of less than 1 ps, ℓ_T is less than α^{-1}. The thickness of the initial molten layer, and hence also the thermal gradients, are then independent of the pulse duration. Femtosecond pulses, therefore, although very useful for time-resolved spectroscopy of the various phenomena associated with the irradiation process, do not produce a higher quench rate. A general discussion of the regimes associated with the relative values of ℓ_T and α^{-1} can be found in Bloembergen's paper (1).

Following the termination of the irradiation pulse, the heat from the very hot liquid continues to diffuse into the substrate, which results in further melting, until the temperature of the crystal-liquid interface drops to the equilibrium melting temperature. For the conditions of Figure 1, the maximum melt depth, d, is about 1000 Å. As will be shown below, the overheating of the crystal during the melt-in process is considerable. The thermal gradients are very high (10^{11} K/m), as is the melt-in velocity ($u \sim 10^3$ m/s). The duration of the melt-in stage of the process is therefore on the order of $d/u \sim 100$ ps.

The third stage is that of resolidification, either by regrowth of the underlying crystal, by growth of a new crystal, or by glass formation if the melt cools to its configurational freezing point before the crystallization front passes. A number of sophisticated numerical analyses have been made of the thermal quantities of this process, but their order of magnitude can easily be estimated from dimensional arguments. The temperature scale is set by the melting temperature, T_m, which is on the order of 10^3 K. The length scale is set by the melt depth $d \sim 10^{-7}$ m. The corresponding temperature gradient is then $\nabla T = T_m/d \sim 10^{10}$ K/m, and the cooling rate $\dot{T} = D_{th}\ \nabla T/d \sim 10^{12}$ K/s. The lifetime of the melt can be estimated as $\tau = T_m/\dot{T} = 10^{-9}$ s, which is in agreement with the experimental values obtained from transient reflectivity measurements (6-9). The velocity with which the isotherms move toward the surface during cooling is an important quantity to be compared to the crystal growth velocity, and can be estimated as $u_T = \dot{T}/\nabla T \sim 100$ m/s. These parameters, together with those for a "conventional" melt quenching process such as melt spinning, are summarized in Table I, using the thermal data for iron.

Table I. Thermal Parameters in Melt Quenching

		Laser Quenching	Melt Spinning
Melt temperature	T_m(K)	10^3	10^3
Melt thickness	d(m)	10^{-7}	5×10^{-5}
Temperature gradient	∇T(K/m)	10^{10}	2×10^7
Cooling rate	T(K/s)	10^{12}	4×10^6
Melt lifetime	t(s)	10^{-9}	not applicable
Isotherm velocity	u_T(m/s)	100	0.2
Heat-flow limited crystal growth velocity	$u_{||}$(m/s)	230	0.5

The Overheated Liquid State

As discussed in the previous section, a molten layer with a thickness ℓ_T and a temperature of several thousand degrees is formed in during the energy deposition stage of the pulsed laser processing. Since the boiling point, T_b, of, for example, liquid iron is 3160 K (at 1 atm), it is important to consider whether the liquid can be overheated and hence be preserved for later quenching, or is lost by fast evaporation above T_b.

1. Thermodynamic Stability Limit

That liquids can be overheated considerably above their boiling temperature is a well-known experimental fact (10,11) that can be explained simply by the van der Waals equation of state:

$$\left(p + \frac{a}{V^2}\right)(V-b) = RT \tag{1}$$

where a and b are constants.

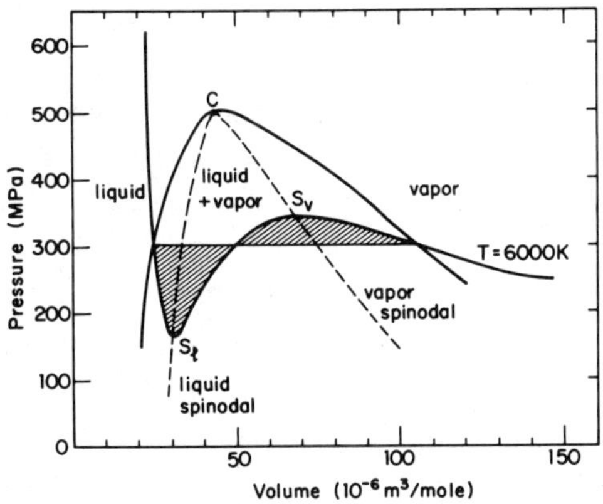

Fig. 2 - P-V diagram for iron, showing one isotherm, the liquid-vapor coexistence domain, and the spinodal lines. The diagram is based on a van der Waals model using the critical point parameters of ref. (12).

Figure 2 shows an isotherm, calculated from the critical point in Fe (12):

$$T_c = 8a/27Rb = 6753 \text{ K} \tag{2}$$

$$p_c = a/27b^2 = 501 \text{ MPa} \tag{3}$$

$$V_c = 3b = 43 \text{ cm}^3/\text{mole} . \tag{4}$$

It should be kept in mind that these values for p_c and T_c are probably lower limiting ones. For example, a critical temperature as high as 10000 K has been quoted for Fe (12).

The equilibrium pressure and volumes for the liquid-vapor transition at a given temperature are found by the Maxwell equal area construction, as shown by the cross-hatched areas on the p-V diagram of Figure 2. The locus of the equilibrium points forms the coexistence curve (solid line), inside which the equilibrium state consists of a mixture of vapor and liquid.

As long as the compressibility of the liquid is positive, however, (i.e., the slope of the isotherm is negative) the liquid can be overexpanded into the coexistence region up to S_ℓ, a point first labeled by van der Waals as the *spinodal point*. Similarly, the vapor can be overcompressed down to the spinodal point S_V. The locus of the spinodal points is easily obtained from the equation of state (1), by requiring $dp/dV = 0$, which gives:

$$p = \frac{a}{V^2}\left(1 - \frac{2b}{V}\right) \tag{5}$$

and is indicated by the dashed line on Figure 2.

In the region between the coexistence line and the spinodal line, the liquid or vapor are *metastable*. This means that in order to form the vapor phase from the liquid it is necessary to nucleate a bubble, which occurs at a finite frequency. Similarly, in order to form a liquid from the metastable vapor requires nucleation of a droplet. Inside the spinodal line the nucleation barrier vanishes: a fluctuation of any size is stable, and vaporization or condensation occurs rapidly by *spinal decomposition* of the liquid or vapor into two phases.

Figure 3 shows the metastable regions for the liquid and vapor of Fe on a p-T diagram. The spinodal lines and the upper part of the equilibrium line (solid line) have been calculated from the van der Waals model and the critical parameters listed above. The dashed part of the equilibrium line has been adjusted to go through the boiling point at 1 atm (0.1 MPa).

In laser heating of a molten surface at atmospheric pressure, rapid vaporization by spinodal decomposition throughout the liquid can only occur if the temperature exceeds that of the liquid spinodal line at 1 atm, $T_{s,\ell}$. This can be calculated precisely by putting $p = 0$ in equation (5), and using the critical point relations (2)-(4):

$$T_{s,\ell} = \frac{27}{32} T_c \ . \tag{6}$$

For iron, using the lower limiting critical parameters (2)-(4), this temperature is estimated to be at least 5700 K, as can be checked on Figure 3. It could actually be as high as 8400 K, if T_c is 10000 K.

If the temperature of the liquid exceeds $T_{s,\ell}$, the decomposition proceeds very rapidly, but an analysis of its time dependence requires a quantitative kinetic theory. The theory of spinodal decomposition is well

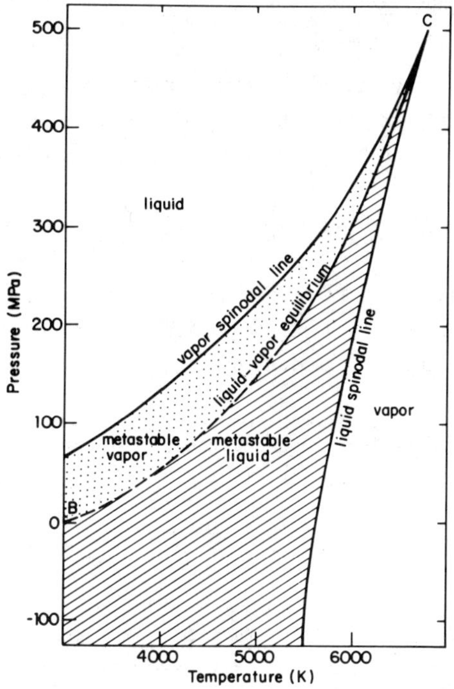

Fig. 3 - P-T diagram showing the liquid-vapor equilibrium line, the spinodal lines and the metastable regions for liquid and vapor. The diagram is based on a van der Waals model using the critical point parameters of ref. (12).

developed (13-15), and explains the minimum length scale for decomposition by taking into account the contribution of very steep gradients (in density or concentration) to the free energy of an inhomogeneous system. Although an estimate of the gradient energy coefficient for liquid metals can in principle be obtained from the same mean field interactions that lead to the van der Waals model, we can simply use the average interatomic distance in the equilibrium vapor phase at $T_{s,\ell}$ (~6 Å) as a lower limit for the length scale of the density fluctuations. An approximate argument similar to the one used below to estimate the time required for assembling a critical vapor nucleus (section 3) can then be invoked here for the formation of a fluctuation. It is therefore possible that, because of the very short time scale of picosecond pulsed laser irradiation, the liquid can even sustain some overheating above $T_{s,\ell}$.

2. Steady State Nucleation

If the temperature of the molten layer does not reach $T_{s,\ell}$, formation of the vapor inside the liquid requires nucleation of a bubble. Classical nucleation theory (16-18) give the following expression for the *steady state* nucleation frequency (number of nuclei per unit volume and time):

$$I = \frac{NkT}{h} \exp\left(- \frac{16\pi\sigma^3}{3p^2kT}\right) \tag{7}$$

where N is the number of atoms per unit volume, k is Boltzmann's constant, h is Planck's constant, p is the equilibrium pressure at the temperature T (p is much larger than the ambient pressure here), and σ is the liquid-vapor surface tension.

The minimum nucleation frequency required to cause vaporization throughout the liquid layer in laser quenching can be estimated as:

$$I_{min} = \frac{\alpha}{a^2 t_p} \tag{8}$$

where a is the maximum distance traveled in a time t_p by the vapor-liquid interface of a nucleated bubble; a = 1 μm is taken as an absolute upper limit. This gives a value for I_{min} of 2×10^{30} $m^{-3}s^{-1}$. The maximum temperature to be considered here is $T_{s,\ell}$, so that an estimated T = 5500 K is used in the analysis below; it should be kept in mind that the results are very insensitive to the choice of the values for T and I_{min}. The equilibrium vapor pressure at this temperature can be read from Figure 3: p = 200 MPa.

Equation (1) can now be used to calculate a maximum value for the liquid-vapor surface tension, σ, for the nucleation frequency to exceed I_{min}. This gives a value σ_{max} = 0.17 N m^{-1}. The surface tension of liquid iron at the melting point (1812 K) is 1.788 N m^{-1} (19). Since the surface tension decreases roughly linearly with increasing temperature and vanishes at the spinodal $T_{s,\ell}$ (14), the surface tension drops below σ_{max} only if the temperature comes within a few hundred degrees from $T_{s,\ell}$.

3. Transient Nucleation

Equation (7) describes the isothermal *steady state* nucleation frequency, i.e., the rate, per unit volume, at which bubbles form if a steady state population of vapor embryos has been established. Establishing such a population after a sudden change in temperature, however, requires a certain amount of time, during which the nucleation rate increases to its steady state value (20). Usually, this transient effect is only important in transformations in condensed systems, where the atomic mobility is relatively low. In vapor-liquid transitions, the transient time is much shorter and can usually be ignored. On the very short time scale of pulsed laser irradiation, however, transient effects may be important, and may be a stabilizing factor against copious nucleation in the overheated liquid.

Although an exact analysis of the transient problem is very time-consuming, a lower limit of the transient time, τ, can be estimated by considering the time it takes to form a critical vapor nucleus by evaporation of the liquid.

For large equilibrium vapor pressures p, the critical nucleus radius is given in the classical theory by:

$$r^* = \frac{2\sigma}{p} \tag{9}$$

which, for the nucleation example above is 17 Å.

The rate of evaporation can be estimated from the kinetic theory of gases. The maximum number of atoms evaporating from a unit surface per second is, in equilibrium, equal to the number impinging on that surface from the vapor, which is:

$$\Gamma = \frac{p}{\sqrt{2\pi mkT}} \tag{10}$$

where m is the atomic mass. For the above conditions, this is about $10^{30}\ m^{-2}s^{-1}$. If the sticking coefficient, q, of the vapor atoms is less than unity, Γ is reduced to (17):

$$\Gamma' = \Gamma \frac{q}{2-q} \quad . \tag{11}$$

The time to form a spherical nucleus of radius r^* can then be estimated by considering that at any time during the growth of the vapor bubble the total volume of the atoms evaporating from the surface of the bubble (of instantaneous area A) is equal to the instantaneous increase in the bubble volume, or:

$$A\Omega\Gamma\, dt = A\, dr \tag{12}$$

where Ω is the atomic volume of the atoms in the vapor phase. Figure 2 shows that for the conditions considered here the molar volume of the vapor is around 150 cm^3, which corresponds to $\Omega = 250$ Å^3. Equation (12) then integrates simply to give the minimum time to form the critical nucleus:

$$\tau = \frac{r^*}{\Omega\Gamma} \quad . \tag{13}$$

Combining equations (9), (10) and (12) gives a value for τ of 6.8 ps. Given that this is a lower limit, it seems possible that the transient effects enhance the stability of the overheated liquid against nucleation in picosecond pulsed laser experiments.

4. Surface Evaporation

The arguments made in the previous two sections show that rapid vaporization throughout the molten layers, either by spinodal decomposition or by nucleation of bubbles, is only possible if the temperature is very close to the liquid spinodal temperature. In pulsed laser experiments designed for studying liquid quenching, the fluence can usually be kept low enough so that this temperature is not reached; the heat content of a liquid layer of thickness ℓ_T at a temperature a little below $T_{s,\ell}$ is large enough for further melt-in to a thickness, d, on the order of 1000 Å.

The only remaining evaporation mechanism is then atoms leaving the liquid free surface. The extent of the loss of energy and mass due to this process can easily be estimated from the evaporation rate, Γ, given by equation (10). At the boiling point, the equilibrium vapor pressure is 1 atm, and $\Gamma = 10^{27}$ $m^{-2}s^{-1}$, or 10^8 monolayers per second. Over the lifetime of the melt ($\sim$ 1 ns), this corresponds to a loss of less than a monolayer. Even if the temperature reaches the spinodal, $T_{s,\ell}$, where $\Gamma = 10^{30}$ $m^{-2}s^{-1}$ or 10^{11} monolayers per second, the duration of this stage is so short ($< t_p$ = 30 ps), that the evaporation loss is at the very most a few tens of monolayers.

The energy loss rate due to evaporation can be estimated as:

$$J_{ev} = \frac{\Gamma \Delta h}{N_{Av}} \tag{14}$$

where Δh is the molar heat of evaporation (around 10^5 J/mole for metals) and N_{Av} is Avogadro's number. Under the present conditions J_{ev} is at most 10^8 W cm^{-2}, which is still considerably below the absorbed laser intensity of 10^9 W cm^{-2}, and therefore negligible as an energy loss.

5. Appearance of the Irradiated Surface

After irradiation with a laser pulse in an experiment designed to study quenching of the liquid, often holes are observed at the spots that received the highest intensity. As discussed above, the fluence in these experiments is not high enough to cause extensive loss of material by vaporization or evaporation.

The holes, therefore, are not a result of evaporation *per se*, but of *mechanical* displacement of the liquid layer by the large recoil pressure of the evaporation ("splashing"). The morphology of the displaced material around the hole is often direct evidence of this splashing phenomenon, as is the finding of substrate atoms near the hole on the top side of supported thin films (21).

Direct evidence of the recoil pressure can be obtained from the following experiment (22). A thin ribbon, clamped at one end, is irradiated by a laser pulse at the free end, which results in visible vibration. The impulse (force × time) of the recoil pressure calculated from the vibration amplitude is in general agreement with the order of magnitude of the temperatures and times discussed above.

Thermodynamics and Kinetics

The short time scale of the picosecond melt quenching process puts restrictions on the transformations that can occur. For the entire liquid layer, of thickness, $d = 1000$ Å, to crystallize during the cooling time, τ, of 1 ns, the velocity must be at least, $u = d/\tau \sim 100$ m/s. The time, t_1, to crystallize a monolayer at this speed is about 3 ps, and the distance an atom can diffuse in that time, $(D_\ell t_1)^{1/2}$, is less than an interatomic distance. (D_ℓ is the diffusivity in the liquid phase). This means that under these short pulsed laser quenching conditions no long-range diffusion can occur, and that the only possible transformations are *partitionless* ones, either growth of a crystal of the same composition of the liquid, or glass formation.

The thermodynamic restrictions imposed by the partitionless conditions are illustrated by the phase diagram of Figure 4, which contains two primary solid solutions (α,δ) and two intermetallic compounds (β,γ). The free energy diagram in Figure 4 shows how the T_0-lines for the transformations of these four phases from the liquid are constructed. At a sufficiently low temperature, T_g, the atomic transport rate in the liquid state becomes negligibly small, so that no more transformations can occur on a reasonable time scale. The liquid is then configurationally frozen and becomes a glass. For a composition in the range $x' - x''$ in Figure 4, no crystallization is possible under picosecond laser quenching conditions, and therefore only glass formation is thermodynamically possible. This criterion for glass formation is only a *sufficient* one, and not a necessary one, as has been claimed (23,24). In fact, more often than not, picosecond quenching experiments yield glasses over a wide range of compositions below the T_0-lines for both intermetallic compounds and primary solutions. To explain this, it is necessary to consider the kinetics of partitionless crystallization in some detail.

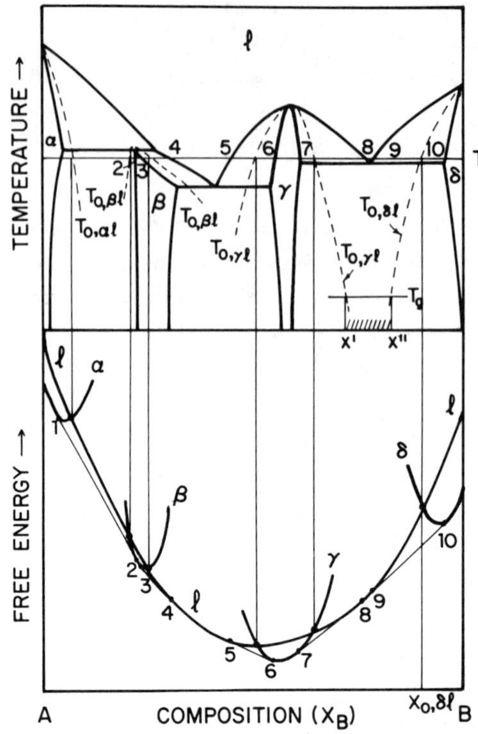

Fig. 4 - Schematic phase diagram (top), and corresponding free energy diagram at temperature T_1 (bottom), illustrating the construction of the T_0-lines for the primary solid solutions and the intermetallic compounds. In the composition range x' - x" (dashed), only glass formation is possible under conditions of partitionless solidification. The numbers correspond to the same compositions in the top and bottom diagrams.

The basic theory for crystal growth far from equilibrium has been formulated by Turnbull, and can be found in a number of publications (25,26), including the first paper in the present volume. The crystal growth velocity can be written as:

$$u = fk\lambda\left[1 - \exp\left(- \frac{\Delta G_c}{RT_i}\right)\right] \tag{15}$$

where k is the atom jump frequency across the crystal-melt interface, f is the fraction of interface sites that can incorporate a new atom, λ is the interatomic distance, ΔG_c is the difference in molar free energy between crystal and melt, and T_i is the interface temperature.

If the undercooling of the interface, $\Delta T_i = T_m - T_i$, is small, the driving free energy can be approximated by:

$$\Delta G_c = \Delta S_c \Delta T_i \tag{16}$$

where ΔS_c is the molar entropy of crystallization. The exponential in equation (15) can be linearized to:

$$u = fk\lambda \frac{\Delta S_c}{R} \frac{\Delta T_i}{T_i} \tag{17}$$

where, for metals, $f \sim 1$ and $\Delta S_c \sim R$.

The crystal growth rate also depends on the rate at which the heat of crystallization, $\Delta h_{c,v}$ per unit volume, can be removed. Since this must occur by heat flow down the temperature gradient at the interface, $\nabla_i T$, the heat flux, J, must obey the relations:

$$J = u\Delta h_{c,v} = \kappa \nabla_i T \tag{18}$$

where κ is the thermal conductivity. Conversely, equation (18) can be turned around to define a heat flow-limited velocity:

$$u_h = \frac{\kappa \nabla_i T}{\Delta h_{c,v}} \quad . \tag{19}$$

Table I gives values for u_h for the laser quenching conditions as well as for melt spinning.

Combining equations (17) and (19) then gives for metals:

$$\frac{u_h}{k\lambda} = \frac{T_m - T_i}{T_i} \quad . \tag{20}$$

In conventional solidification, the interface kinetics are much faster than the rate of heat removal, so that $u_h << k\lambda$, and ΔT_i is small. This type of growth is called *heat flow-limited*. In laser quenching, depending on the nature of the interface rearrangements (see below), the rate of heat removal can be much faster than the interface kinetics, so that $u_h >> k\lambda$, and the undercooling, ΔT_i, is large, so that metastable phases can be formed. This corresponds to *interface-limited* growth.

The maximum growth velocity of a crystal, $u_{max} = k\lambda$, is determined by the nature of the atomic rearrangements that control the growth process. In pure metals, dilute alloys or compounds with simple crystal structures, k can be taken as the frequency of thermal vibration. (~ Debye frequency), so that $u_{max} \sim u_s$, the speed of sound in the liquid, which is on the order of several thousand m/s. This is called the *collision-controlled* regime, because the growth occurs by simple impingement of the liquid atoms onto the crystal.

In more concentrated alloys or in compounds with a more complicated crystal structure, where a change of the nearest neighbor configuration, i.e. diffusion, is necessary, the jump frequency can be taken as D_ℓ/λ^2, and $u_{max,D} = D_\ell/\lambda$, which is on the order of 10 m/s. This is the *diffusion-controlled* regime. Table I shows that for picosecond laser quenching $u_{max,D} << u_T, u_h$, so that this type of growth can easily be suppressed. Equation (20) shows that under these conditions T_i falls far below T_m, and glass formation becomes possible.

Although in partitionless growth of crystals there is no *long-range* diffusional transport, the growth can still be diffusion-controlled if a drastic change of the nearest neighbor environment is necessary to transform the liquid structure into the crystalline one. This is the case, for example, in metal-metalloid alloys such as Fe-B, where the chemical interaction leads to a very pronounced short-range order in the liquid (9 Fe atoms around each B in a capped trigonal prism arrangement (27)), which is very different from that in the crystal (B interstitial in bcc Fe (28)). As a result, it is possible to produce Fe-B glasses containing as little as 5 at.% B (28), which is far below the T_0-line. Similarly, if the unit cell of an intermetallic compound is large, or if long-range chemical order must be established, crystal growth is diffusion-limited, and the growth of these compounds is easily suppressed in picosecond pulsed laser quenching. Numerous experiments on a variety of binary metallic systems (Ni-Nb, Ni-Mo, Co-Mo, Nb-Co (8, 30-34)) have demonstrated that glasses can be formed at the intermetallic compositions. On the other hand, in pure metals with simple crystal structures, and by extension, disordered alloys with the same simple structures (e.g., disordered fcc in Cu-Co or Ag-Cu), growth is collision-limited and cannot be suppressed even by picosecond pulsed laser quenching. This is confirmed by direct measurements of the growth velocity of pure metals and disordered alloys (Au, Cu, Cu-Au) by transient reflectivity (8,9) which yield large values (> 100 m/s) that can only be accounted for by a collision-controlled mechanism.

Applications

1. Overheating of Crystals

It is well known that the interior of large crystals can be heated well above the melting temperature if the kinetics of melting are much slower than the rate of thermal diffusion, as is the case for oxides (35). In metals the kinetics of melting are much faster, so that overheating is much more difficult to achieve. One method, in which conventional heating is used, consists of surrounding a small single crystal with a higher melting coherent layer to prevent surface nucleation of the interior melt. Overheating of silver crystals, coated by a gold layer, by 7.5 K has been reported (36). "Conventional" rapid heating methods, for example by sending a high current pulse through a thin metallic wire, also give overheating of no more than a few degrees (37).

Rapid heating with pulsed lasers provides a unique method for overheating crystals considerably above their melting temperature, as can be seen from the following argument. During the initial stages of the melt-in phase the temperature gradients are an order of magnitude greater than during regrowth. This means that the heat-flow limited crystal growth velocity, given by equation (19), is also an order of magnitude greater than the value listed in Table I, or about 2000 m/s. For pure metals, where $u_{max} = u_s \sim 3000$ m/s, equation (20) gives for the overheating: $\Delta T_i = 0.66\ T_i$, which, for a melting point of 1000 K, corresponds to an overheating of 2000 K.

The obvious drawback of the method is the short duration of the overheating (< 100 ps), which requires fast diagnostic techniques for studying the crystal or possible homogeneous nucleation of the melt in this state.

2. Quenching of Nb-Si Alloys

The Nb-Si system is of interest because of the prediction that the Nb_3Si compound with the A-15 structure should have a very high superconducting transition temperature. Since at 25 at.% Si there are no stable compounds in this system (see the phase diagram of Figure 5), a variety of methods have been used to produce metastable phases. As illustrated in Figure 5, conventional melt quenching methods produce an amorphous phase in the vicinity of the eutectics (38). Non-stoichiometric A-15 phases have also been found (39,40). At the 25 at.% Si composition, however, conventional melt quenching yields a compound with the tetragonal Ti_3P structure (41).

We have recently completed a study of this system using pulsed laser quenching of thin alloy films (~ 1000 Å) supported on various substrates (42-44). Although we did not succeed in obtaining an A-15 phase (the quench rates were too high in all attempts for this), it is still useful to review the various phases that were obtained at different cooling rates. Figure 5 illustrates the results.

At the highest quench rates (> 10^{12} K/s), obtained in the thinnest sections of the spots irradiated with a 30 ps pulse on a copper substrate the amorphous phase was formed over a wide composition range. Such broadening of the range of amorphous phase formation is consistent with our earlier observations on other metal-metalloid and metal-metal systems.

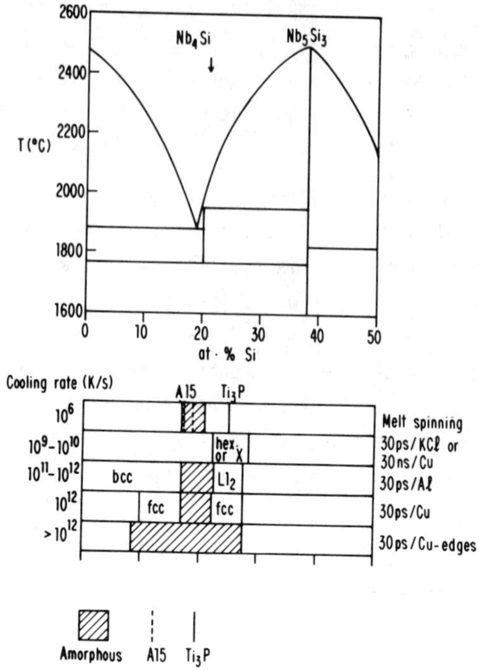

Fig. 5 - Phase diagram of the Nb-Si system (top), and schematic illustration of the phases obtained by different methods of liquid quenching. The melt spinning results are from refs. (38-41); the laser quenching results are from refs. (42-44). The pulse length and the substrate for the laser experiments are indicated.

At slightly lower quench rates (10^{12} K/s), obtained in the remaining area of those spots, a new phase was found: a disordered f.c.c. solid solution (43). Although the formation of disordered f.c.c. solid solutions in metal-metal alloys cannot be suppressed by picosecond laser quenching, the growth rate of the f.c.c. phase in this system, although still the highest one observed, is probably slightly lower due to the short-range order around the Si.

By lowering the quench rate to 10^{11} - 10^{12} K/s by supporting the films on an aluminum substrate, a supersaturated b.c.c. solid solution was found at low Si contents, and an ordered f.c.c. phase ($L1_2$, Cu_3Au-type) around 25 at.% Si.

A cooling rate of 10^9 - 10^{10} K/s was obtained, either by supporting the ps-irradiated films on a salt substrate, or by irradiating the films on a copper substrate with a 30 ns pulse. Under these conditions two entirely new phases were found around 25 at.% Si: a hexagonal phase with a 250 Å^3 unit cell and an as yet unidentified phase (44).

It is interesting to note the degree of isotropy and the unit cell size of the phases formed at 25 at.% Si in order of decreasing quench rate: amorphous (isotropic, no unit cell), disordered f.c.c. (1 atom unit cell, cubic point group), ordered f.c.c. (4 atom unit cell, cubic point group), hexagonal (~ 16 atom unit cell, uniaxial point group), Ti_3P-type (32 atom unit cell, uniaxial point group). This is a good illustration of the principle, stated in Turnbull's paper in this volume, that phases are increasingly kinetically favored (i.e., are more difficult to suppress by quenching) with increasing isotropy and decreasing unit cell size.

Acknowledgements

The author's work in this area is supported by the Office of Naval Research under contract number N00014-85-K-0684.

References

1. N. Bloembergen, in *Laser-Solid Interactions and Laser Processing*, ed. by S.D. Ferris, H.J. Leamy and J.M. Poate, AIP, New York (1979), p. 1.

2. F. Spaepen and C.J. Lin, in *Amorphous Metals and Non-Equilibrium Processing*, ed. by M. von Allmen, Les Editions de Physique, Les Ulis, Frace (1984), p. 65.

3. F. Spaepen, in *Laser Surface Treatments of Metals*, ed. by C.W. Draper and P. Mazzoldi, NATO ASI Series, Martinus Nijhoff, Dordrecht, The Netherlands (1986), in press.

4. For a review of these relaxation processes, see W.L. Brown, Mat. Res. Soc. Symp. Proc., 23, 9 (1984).

5. If the laser power is high enough to ignite a plasma in front of the surface (see e.g., J.A. McKay and J.T. Schriempf, in *Laser-Solid Interactions and Laser Processing*, ed. by S.D. Ferris, H.J. Leamy, and J.M. Poate, AIP, New York (1979, p. 55), the energy deposition time can be longer than the pulse duration, depending on the relaxation time of the plasma. In the quenching experiments described here, however, the fluence is carefully selected to be large enough to melt the surface, and yet small enough to avoid visible surface damage, which would be caused also by the shock wave emitted by the plasma expansion. In practice this was done by producing a small damage hole at the center of the indicated spot; the remaining area has thus simply undergone melting and resolidification.

6. J.M. Liu, R. Yen, H. Kurz, and N. Bloembergen, Appl. Phys. Lett., 39, 755 (1981).

7. C.A. MacDonald, Ph.D. Thesis, Harvard University (1986).

8. C.A. MacDonald and F. Spaepen, in *Laser Surface Treatments of Metals*, ed. by C.W. Draper and P. Mazzoldi, NATO ASI Series, Martinus Nijhoff, Dordrecht, The Netherlands (1986), in press.

9. C.A. MacDonald, A.M. Malvezzi and F. Spaepen, Mat. Res. Soc. Symp. Proc., 51 (1986), in press.

10. H. Wakeshima and K. Takata, J. Phys. Soc. Japan, 13, 1398 (1958).

11. R.E. Apfel, J. Acoust. Soc. Am., 49, 145 (1971).

12. Landolt-Börnstein Tables, Volume II, Part 1, Mechanisch-Thermische Zustandgrössen, ed. by H. Borchers et al., Springer, Berlin, (1971), p. 332.

13. J.W. Cahn, Acta Metall., 9, 795 (1961).

14. J.W. Cahn and J.E. Hilliard, J. Chem. Phys., 28, 258 (1958). It should be noted that close to the spinodal temperature σ deviates from the rough linearity, and approaches zero as $(T-T_{s,\ell})^{3/2}$. (D.R. Nelson, private communication).

15. J.E. Hilliard, in *Phase Transformations*, American Society for Metals, Metals Park, Ohio, (1970), 497.

16. M. Volmer, *Kinetik der Phasenbildung*, Steinkopff, Dresden and Leipzig, (1939), p. 156.

17. R.E. Apfel, Ph.D. Thesis, Harvard University (1970).

18. J.C. Fisher, J. Appl. Phys., 19, 1062 (1948).

19. A.V. Grosse, J. Inorg. Nucl. Chem., 26, 1349 (1964).

20. K.F. Kelton, A.L. Greer and C.V. Thompson, J. Chem. Phys., 79, 6261 (1983). This paper also contains an extensive review of earlier work.

21. C.J. Lin and F. Spaepen, Scripta Metall., 17, 1259 (1983).

22. C.J. Lin, Ph.D. Thesis, Harvard University (1983).

23. T.B. Massalski, Proc. 4th Int. Conf. on *Rapidly Quenched Metals*, ed. by T. Masumoto and K. Suzuki, Jap. Inst. Metals, Sendai (1982), p. 203.

24. W.J. Boettinger, Proc. 4th Int. Conf. on *Rapidly Quenched Metals*, ed. by T. Masumoto and K. Suzuki, Jap. Inst. Metals, Sendai (1982), p. 99.

25. D. Turnbull, in *Physical Processes in Laser-Materials Interactions*, NATO-ASI Proc., ed. by M. Bertollotti, Plenum, NY (1983), p. 117.

26. F. Spaepen and D. Turnbull, in *Laser Processing of Semiconductors*, ed. by J.M. Poate and J.W. Mayer, Academic, NY (1982), p. 15.

27. For a review of the short-range order in metal-metalloid glasses, see J.M. Dubois and G. Le Caer, Acta Metall., 32, 2101 (1984).

28. R. Ray and R. Hasegawa, Sol. St. Comm., 27, 471 (1978).

29. C.J. Lin and F. Spaepen, Appl. Phys. Lett., 41, 721 (1982).

30. C.J. Lin, F. Spaepen and D. Turnbull, J. Non-Cryst. Solids, 61/62, 767 (1984).

31. C.J. Lin and F. Spaepen, Mat. Res. Soc. Symp. Proc., 28, 75 (1984).

32. C.J. Lin and F. Spaepen, in *Chemistry and Physics of Rapidly Solidified Materials*, ed. by B.J. Berkowitz and R.O. Scattergood, TMS-AIME, NY (1983), p. 273.

33. C.J. Lin and F. Spaepen, "Nickel-Niobium Alloys Obtained by Picosecond Pulsed Laser Quenching," Acta Metallurgica, 34, 1367 (1986).

34. F. Spaepen, Proc. Int. Conf. on *Cryogenic Materials*, ed. by T. Orlando and S. Foner, Plenum, NY (1986), in press.

35. N.G. Ainslie, J.D. MacKenzie and D. Turnbull, J. Phys. Chem., 65, 1718 (1961).

36. J. Däges, H. Gleiter and J.H. Perepezko, Mat. Res. Symp. Proc., 57, (1986), in press.

37. S.E. Khaikin and H.P. Benet, Compt. rend. acad. sci. U.R.S.S., 23, 31 (1939).

38. A. Inoue and T. Masumoto, Sci. Rep. Res. Inst. Tohoku University, A28, 165 (1980).

39. R.M. Waterstrat, F. Haenssler and J. Müller, J. Appl. Phys., 50, 4763 (1979).

40. K. Togano, H. Kamakura and K. Tachikawa, Phys. Lett. 76A, 83 (1980).

41. W. Rossteutscher and K. Schubert, Z. Metallk., 56, 813 (1965).

42. W.K. Wang, C.J. Lobb and F. Spaepen, in preparation.

43. W.K. Wang and F. Spaepen, J. Appl. Phys., 58, 4477 (1985).

44. W.K. Wang and F. Spaepen, J. Appl. Phys. (1986), in press.

REVIEW ON DROP TOWERS AND LONG DROP TUBES

R. J. Bayuzick*, W. H. Hofmeister*, and M. B. Robinson+
*Department of Mechanical and Materials Engineering
Vanderbilt University
Nashville, Tennessee 37235

Abstract

At present, there are only three drop towers in existence: the 145 meter and the 30 meter drop towers at the NASA Lewis Research Center, and the 100 meter drop tower at the NASA Marshall Space Flight Center. All of these can be used for experiments in materials science under gravity levels down to 10^{-5}g. However, to date, only one experiment on immiscible alloys has been reported on, and none at all have been reported in other areas such as deep undercooling.

There are likewise three drop tubes currently in existence. These are the cryogenic drop tube and the force-free drop tube at the Jet Propulsion Laboratory, and the 100 meter drop tube at the NASA Marshall Space FLight Center. A fourth drop tube, a 30 meter tube, has been used at the NASA Marshall Space Flight Center in the past, but it is not now in operation.

All four of the tubes have been used to some degree to conduct experiments on materials. Those at the Jet Propulsion Laboratory have been used for the production of glass microshells and of metallic microshells. Those at the Marshall Space Flight Center have been used largely in the study of deep undercooling of metals and alloys. Undercoolings as large as $0.25T_m$, where T_m is the equilibrium liquidus, have been observed in millimeter size samples of Nb alloys. Interesting microstructural effects have accompanied the large undercoolings.

Further applications of drop tubes can be seen by calculations using the 100 meter tube as an example. Providing significant partial pressures of inert gas in the drop tube can be tolerated, aluminum alloys and other non-refractory alloys, as well as the refractory alloys, can be significantly undercooled during the fall time available in the 100 meter drop tube.

+NASA Marshall Space Flight Center, Marshall Space Flight Center, Alabama 35812

Introduction

A drop tube is an enclosure in which a molten sample can be solidified while falling. Usually the tube can be evacuated and backfilled with a gas of choice; at the very least there is a capability of purging the tube with a gas of choice. This is distinct from a drop tower in which the entire experiment package, possibly including both a furnace and instrumentation as well as the specimen, is dropped within an enclosure. The package is housed within a falling canister and there is no need for vacuum or gas control within the enclosure. The distinction is worth noting because the terms are sometimes erroneously interchanged both in the vernacular and in the literature, and this leads to confusion.

Like many modern techniques, experiments, and notions, drop tubes find their roots in antiquity (1). Small lead-shot formation was described in some detail in 1665 by Robert Hooke (2). The technique is said to have been invented by Prince Rupert in 1650 and consisted of dropping molten lead (alloyed with arsenic trisulfide) through a pan from a small height into a tub of water. Out of this, of course, grew the "shot towers" tens of meters high for producing lead shot.

A Description of Drop Towers

Three drop towers currently exist. The 145 meter and the 30 meter drop towers are located at the NASA Lewis Research Center, and the 100 meter drop tower is located at the NASA Marshall Space Flight Center. In the latter two facilities, packages containing experiments are dropped in an environment open to air. The 30 meter drop tower relies solely on a drag shield to isolate the experiment from aerodynamic drag and obtain approximately 10^{-5} g during unguided free-fall. An auxiliary thrust is provided to overcome aerodynamic drag during free-fall in the 100 meter drop tower; this allows experimentation between 4×10^{-2} g and 1×10^{-5} g. The 145 meter drop tower is evacuated to 10^{-2} torr to reduce aerodynamic drag down to 10^{-5} g. The free-fall time for the 30 meter tower is 2.2 seconds, and for the 100 meter tower it is 4.3 seconds. In the case of the 145 meter tower, the free-fall time from the top of its vacuum chamber is 5.2 seconds. However, the experiment capsule can be accelerated from the bottom of the vacuum chamber and then allowed to free-fall for a total low g time of 10 seconds.

Figure 1 shows the configuration for the 100 meter drop tower. The figure is, of course, specific to that particular facility, but, in the generic sense, represents the characteristics of all the drop towers. The size of the package represented in Figure 1 is about 1 m on each side with a maximum test package weight of 204 kg. Similar volumes and weights would apply in the other cases.

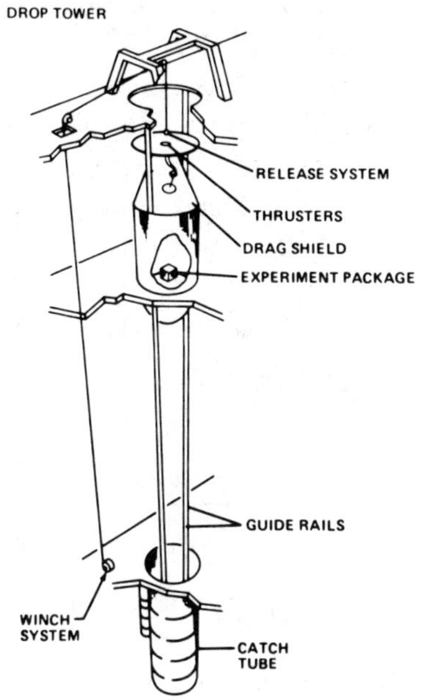

Figure 1 - The 100 meter drop tower. Experiment packages typically measure about one meter on a side.

At present, the 145 meter and 30 meter drop towers have been used to conduct experiments in combustion and fluid physics. They have not been used for experiments in materials, but they would obviously have applicability in such work. Some experiments on materials are now being conducted in the 100 meter tower and one experiment has been reported on in the past. Lacy and Otto processed GaBi and found a finer dispersion of Ga in a Bi matrix relative to the 1 g processed samples (3). Accompanying this was the observation of an anomalous negative slope in the resistivity curve over the temperature range from about 100 K to 300 K.

The Cryogenic and Force-Free Drop Tubes

The Jet Propulsion Laboratory currently has a 13.7 meter and a 13.2 meter drop tube facility (4, 5). The first, shown in Figure 2, is a force-free drop tube in which the velocity of the air flowing through the tube matches the velocity of the falling drop. While this facility offers the advantage of small drag forces on the sample, it has the disadvantage of allowing samples to be processed in air environments only. The second facility is a cryogenic drop tube, shown in Figure 3, where a large section of the tube is cooled to liquid nitrogen temperatures. The internal pressure of the cryogenic drop tube can be controlled between 10^{-5} torr to 1 atmosphere. Both facilities at JPL have free-fall times of 1.7 seconds and were designed principally for the production of glass and metallic microshells for applications such as nuclear fusion targets and light weight structures. Formation of metallic shells of tin in the 750 to 2000 μm range, aluminum in the 460 μm to 3000 μm range, and amorphous shells of $Au_{55}Pb_{22.5}Sb_{22.5}$ of 500 to 1500 μm diameter have been obtained.

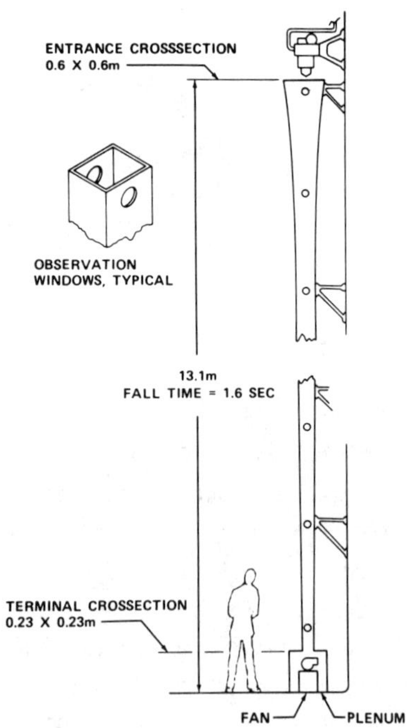

Figure 2 - The force-free drop tube. The velocity of the air flowing through the tube matches the velocity of the falling drop.

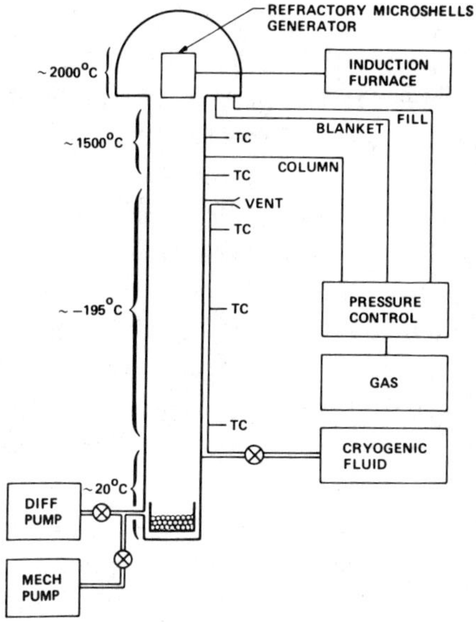

Figure 3 - The cryogenic drop tube. A large section of the tube is cooled to the temperature of liquid nitrogen.

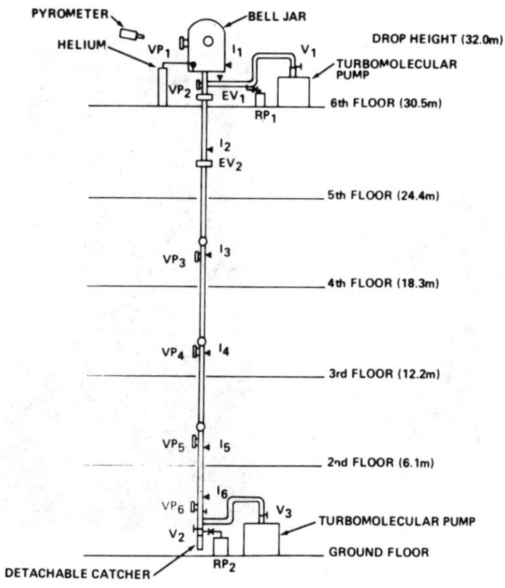

Figure 4 - The thirty meter drop tube. The free-fall time is 2.6 seconds.

The 30 Meter Drop Tube

Initial drop tube work at the NASA Marshall Space Flight Center was performed in the 30 meter facility described in Figure 4 (6). This facility could be maintained under high vacuum (10^{-5} torr) or backfilled with an inert gas and provided 2.6 seconds of free-fall time. Silicon detectors were used to monitor drop brightness and record recalescence if it occurred. Samples of pure niobium of 3 - 5 mm diameter were undercooled during free-fall in vacuum after electron beam melting (7). It was found that the niobium samples undercooled consistently 525 K or 19% of the melting point of niobium, 2741 K. Metallurgical analysis revealed interesting differences between undercooled and non-undercooled samples. Undercooled samples were single crystalline and contained no internal voids or shrinkage cavities. Non-undercooled samples were polycrystalline with very large shrinkage cavities. A 5 mm undercooled niobium sample is shown in Figure 5 along with a Laue diffraction pattern showing the single crystal BCC structure. The surfaces of the samples also showed differences as seen in Figure 6. The surfaces of non-undercooled samples were extremely smooth (Figure 6A), whereas undercooled samples showed interdendritic shrinkage channels on the surface (Figure 6B).

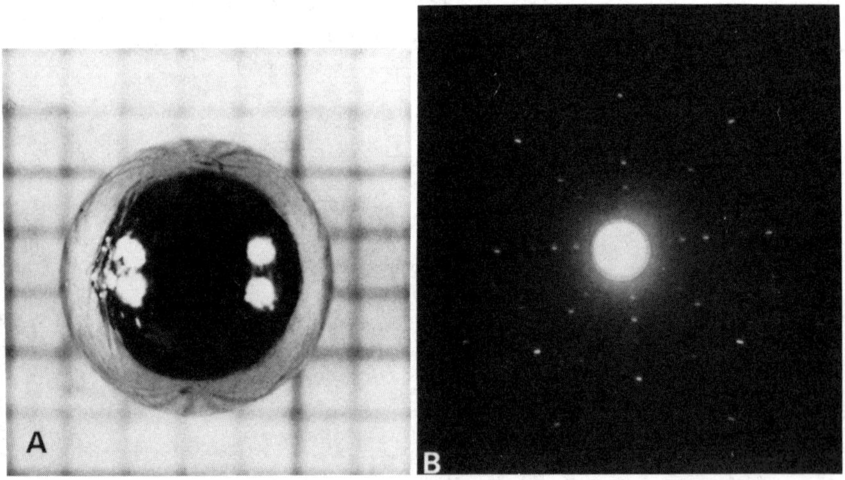

Figure 5 - Undercooled niobium drop. The Laue x-ray diffraction pattern shows a BCC single crystal oriented in a <110> direction.

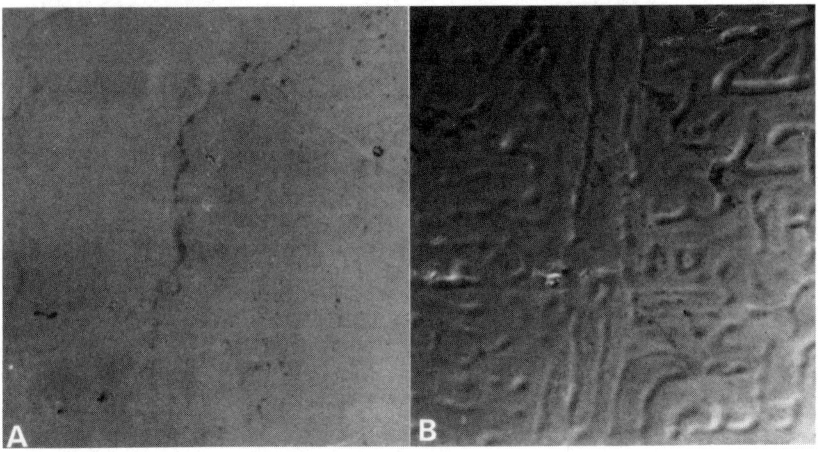

Figure 6 - Scanning electron micrographs of niobium drops at a magnification of 300X. The surface of a non-undercooled drop is given in (A). The surface of a drop undercooled 525K is given in (B).

The 30 meter facility was also used to undercool 3mm diameter Nb-Ge alloys of germanium compositions in the 18 to 27 atomic percent range (8, 9). Samples undercooled the complete length of the tube during free-fall in vacuum and recalesced only after impacting the quenching oil in the catcher section. Undercoolings achieved were approximately $0.13T_m$ to $0.22T_m$ where T_m is the equilibrium liquidus.

In addition, bulk amorphous $Pd_{77.5}Si_{16.5}Cu_6$ droplets have been produced in the 30 meter facility (10). Samples up to 1.5 mm diameter were obtained by dropping through 400 torr helium gas to enhance cooling. An estimated cooling rate of 500 Ks^{-1} was achieved.

The 100 Meter Drop Tube

The only other long drop tube currently in existence is the 100 meter drop tube at the NASA Marshall Space Flight Center. This tube is constructed of stainless steel and has an inner diameter of 0.254 m (see Figure 7). Either an electron beam furnace or an electromagnetic levitation furnace can be mounted at the top; a catch bucket is mounted at the bottom. There are several viewports installed down the length of the tube to provide access for infrared detectors. The system can be evacuated to 8×10^{-6} torr and backfilled with inert gas. The length of the tube from sample to catch system is actually 105 m and gives a free-fall time of 4.6 seconds.

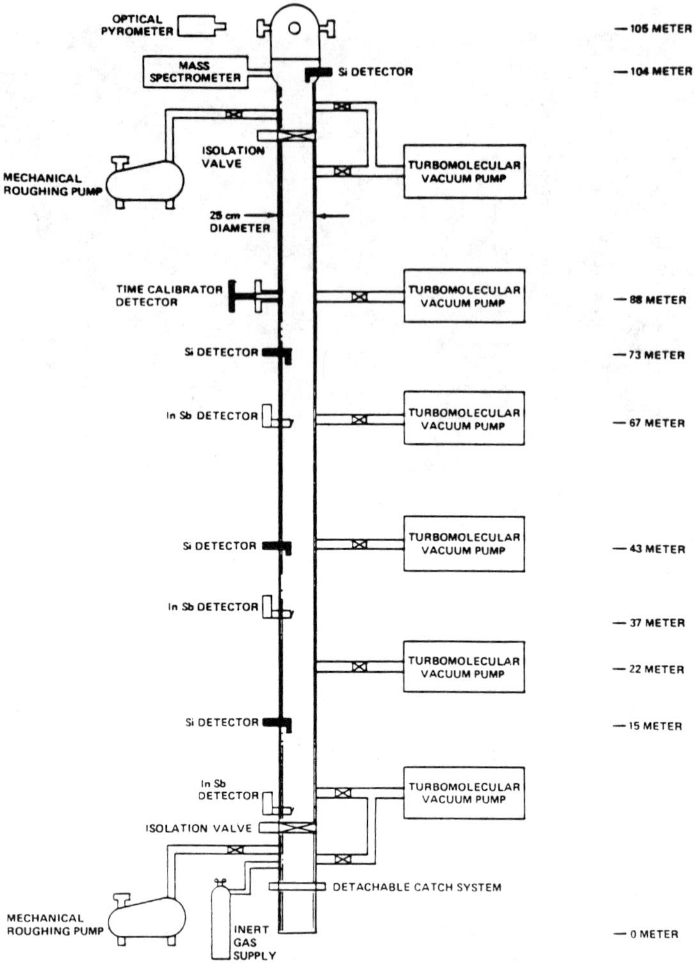

Figure 7 - The 100 meter drop tube. The free-fall time is 4.6 seconds.

Data collected for each experiment includes the drop release temperature measured by optical pyrometry and the voltage output of the infrared detectors. The infrared detectors do not give a direct indication of temperature. They can, however, record the changes in brightness of the sample as a function of fall time. Figure 8 shows such a brightness curve for a Nb-25 atm% Ge alloy. The sharp increase in brightness at around 1.2 seconds fall time is due to recalescence at an undercooling of about 240 K (about 11% below the liquidus). The temperature is calculated through the solution of equation 1 as discussed below.

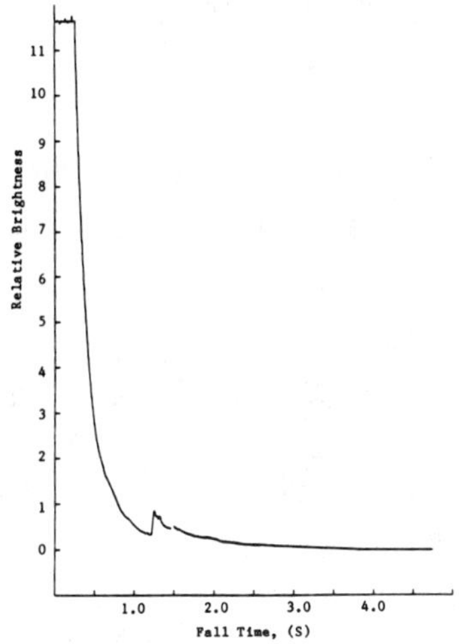

Figure 8 - A typical brightness curve. Solidification is marked by the sudden increase in brightness during recalescence.

Potential for Undercooling in the 100 Meter Drop Tube

In order to gain insight into the cooling capabilities of the drop tube, calculations have been made on several pure materials to estimate the maximum possible undercooling (11). The release temperature of the drop is taken to be the melting temperature and undercooling for the entire 105 m free fall is assumed. The technique follows those of Robinson (12) and Wallace (13). For a spherical shape, the heat loss can be expressed as: (14)

$$dQ/dt = -\varepsilon_T A\sigma (T^4 - T_0^4) - hA(T - T_0) \qquad (1)$$

where dQ/dt is the time rate of heat loss, ε_T is the emissivity of the sample, A is the area of the spherical drop, σ is the Stefan-Boltzmann constant, h is the convective heat transfer coefficient, T is the temperature of the drop, and T_0 is the ambient temperature.

To calculate the temperature of the drop at any given point during free-fall, an iterative computer technique involving the following steps is used:

1. The velocity of the drop is calculated.
2. Knowing the average velocity for a given time increment, the corresponding Reynolds number is calculated. A time increment of 0.001 seconds has been found to be appropriate.

3. The convective heat transfer coefficient is obtained from the Reynolds number using the empirical formulations of Yuge (15).

4. The emissivity is determined from a curve fit to known values of emissivity as a function of temperature.

5. The heat loss during the time interval is calculated.

6. The heat loss during the time intervals is summed to determine the total heat lost. The temperature can then be determined from the intial release temperature and the specific heat.

7. Iterate for the next time interval

In both the pragmatic and fundamental sense, it is useful to normalize the maximum undercooling obtainable under given conditions to the hypercooling temperature, below which recalescence to the equilibrium melting temperature will not occur. The results of such calculations for a variety of metals in the 100 meter drop tube are given in Figure 9 for cooling in vacuum and in Figure 10 for cooling in 200 torr He.

Note that in vacuum only metals and alloys whose equilibrium freezing points are above 2000 K can be solidified during free fall and even then only if the drop mass is about 70 milligrams. At 200 torr He, solidification in the hypercooling regime can be achieved with metals whose melting point is about 900 K as long as the drop mass is less than 200 milligrams.

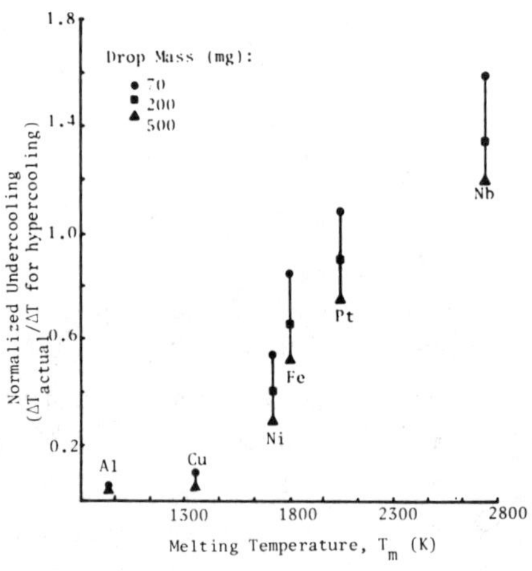

Figure 9 - Maximum possible undercooling in the 100 meter tube under vacuum. Only the more refractory metals are significantly undercooled.

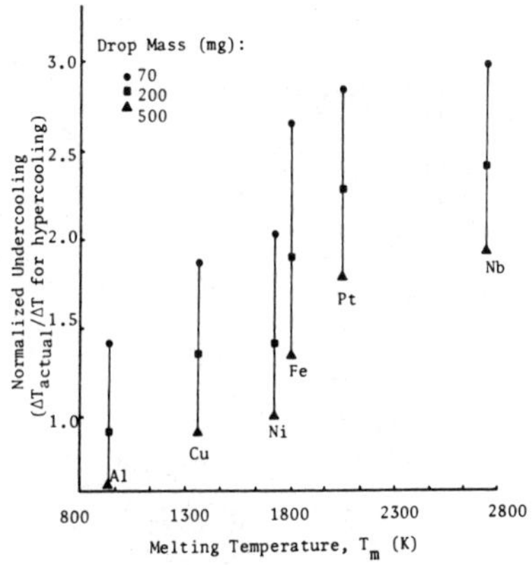

Figure 10 - Maximum possible undercooling with 200 torr He in the 100 meter drop tube. Significant undercooling can be obtained in a wide variety of metals.

The results indicate that it would be necessary in some cases to use significant gas pressure in the drop tube in order to obtain significant undercooling before impact at the bottom. Hence, it could be useful to know the drag force on the drops in such experiments. To indicate the magnitudes involved, the drag force has been calculated for Nb-27 atm% Ge drops of three different masses at He pressures from 100 to 800 torr and expressed as a fraction of earth gravity. Spherical liquid drops were assumed to be released at the melting temperature and allowed to undercool for 100 meters. At the end of free fall the velocity, the Reynolds number and the gas properties were noted. The drag coefficient was determined for White's empirical curve fit for spheres, (16)

$$C_D = 24/Re + 6(1+/Re) + 0.4 \qquad (2)$$

where C_D = drag coefficient and Re is the Reynolds number. The drag coefficient is then used according to the method of Incropera and DeWitt (17) to determine the drag force as:

$$F_D = C_d \; A \; \rho \; V^2/4 \qquad (3)$$

where F_D is the drag force, A is the area normal to the free stream, ρ is the density of the gas at the thin film temperature, and V is the velocity of the drop. The drag forces as a function of size and pressure are shown in Figure 11. These calculations represent the maximum drag force experienced at the maximum drop velocity.

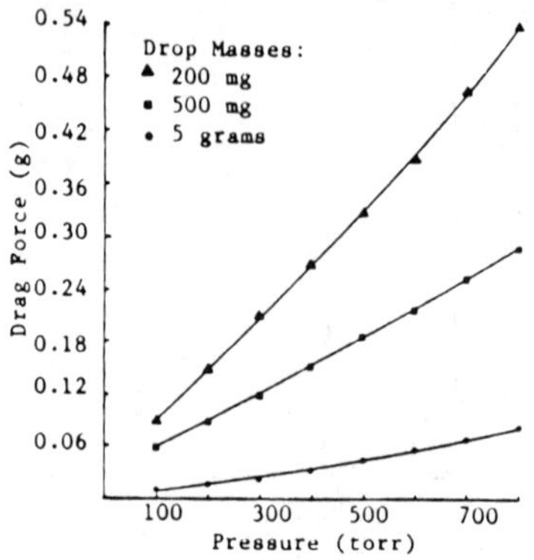

Figure 11 - Drag force at the 100 meter level on Nb-27 atm% Ge drops. Appreciable drag forces can be developed at 200 torr He and above for typical drop masses.

Observed Undercooling

Deep undercooling has been observed in our experiments on pure metals and binary alloys, notably Nb-Ge alloys. The drop sizes in all these experiments were in the millimeter range (1 - 5 mm). Table 1 gives the amount of undercooling observed in pure metals thus far. Each of these were electron beam melted and dropped in vacuum.

Table 1. UNDERCOOLING OF PURE METALS IN THE 100 METER DROP TUBE

Metal	Melting Point (K)	ΔT	$\Delta T/T_m$
Niobium	2741	525	0.19
Hafnium	2500	530	0.21
Iridium	2727	430	0.16
Rhodium	2239	342	0.15

A summary of undercooling in Nb-Ge alloys is given in Figure 12 where the temperatures at which recalescence occurred are superimposed on the Nb-rich portion of the Nb-Ge phase diagram. Undercoolings ranged from $\Delta T/T_m$, where ΔT is the amount of undercooling and T_m is the equilibrium liquidus, equal to 0.03 (about 40 K below the liquidus) in a Nb-25 atm% Ge specimen to $\Delta T/T_m$ equal to 0.25 (about 530 K below the liquidus) in still another Nb-25 atm% Ge specimen. Also, undercooling of $\Delta T/T_m$ equal to 0.23 was seen in both a Nb-30 atm% Ge specimen and a Nb-26 atm% Ge specimen. However, the highest frequency of undercooling was at $\Delta T/T_m$ approximately equal to 0.17

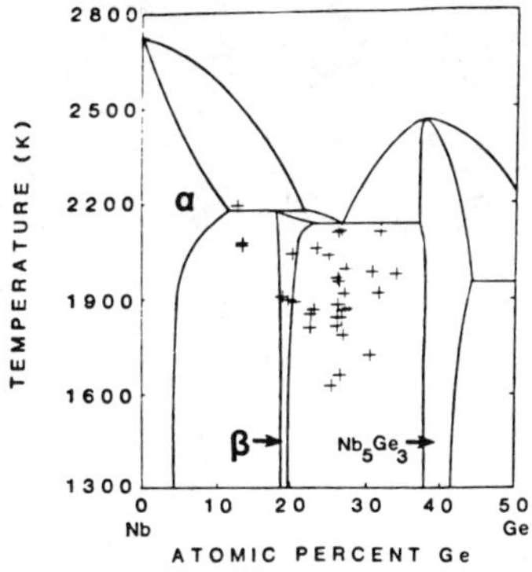

Figure 12 - Niobium rich side of the niobium-germanium phase diagram. Crosses indicate the observed nucleation temperature of alloy drops.

Some Undercooled Microstructures

Interesting microstructures resulting from deep undercooling have been observed in the Nb-Ge alloy system (8, 9, 11, 18). In compositions near 20 atm% Ge, a metastable $\alpha + Nb_5Ge_3$ eutectic has been seen. A transmission electron micrograph of a metastable eutectic region in a Nb-20 atm% Ge specimen undercooled 350 K is given in Figure 13. Convergent beam electron diffraction was used to positively identify the metastable eutectic; analytical work in transmission gave a composition of 3 atm% Ge for α and 33 atm% Ge for the solid solution based on Nb_5Ge_3. The "peppery" appearance in the α phase in Figure 13 is due to ion beam damage from specimen preparation.

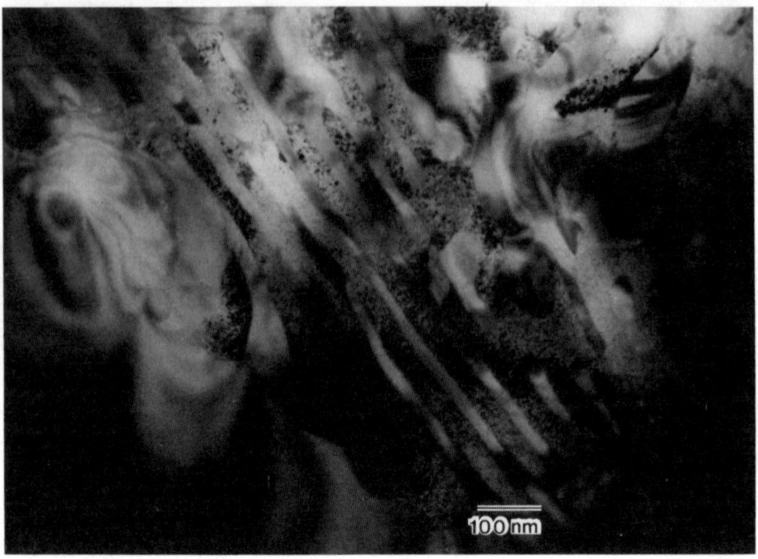

Figure 13 - Transmission electron micrograph of a Nb-20 atm% Ge drop undercooled 350 K. The metastable α + Nb_5Ge_3 eutectic is clearly shown.

The distribution of the metastable eutectic between β cells in a Nb-20 atm% Ge specimen undercooled 220 K can be seen in Figure 14. The metastable eutectic is the sole occupant of the intercellular regions. On the other hand, the intercellular regions of the Nb-23 atm% Ge specimen undercooled 300 K contained both a solid solution based on Nb_5Ge_3 (34 to 37 atm% Ge) and the metastable eutectic as shown in Figure 15. Some follow-on experiments involved splat quenching undercooled liquid on a copper block at the bottom of the drop tube following free-fall (19). The undercooling prior to the splat quench was about 300 K. The resulting microstructures were examined by transmission electron microscopy using conventional imaging, convergent beam electron diffraction, and energy dispersive analysis. In portions of the specimen, but less than 10% by volume, the convergent beam electron diffraction pattern shown in Figure 16 was obtained; such a pattern is typical of amorphous material. The chemical composition of these same portions was equal to the bulk alloy composition of 17 atm% Ge. Both observations lead to the conclusion that part of the specimen was glassy. The remainder of the specimen was comprised of β phase containing sparsely distributed Nb_5Ge_3 precipitates.

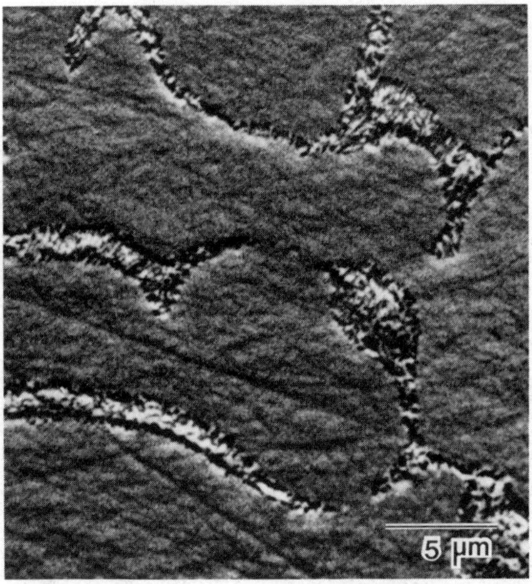

Figure 14 - Nb-20 atm% Ge undercooled 300 K. The metastable eutectic is observed between β cells.

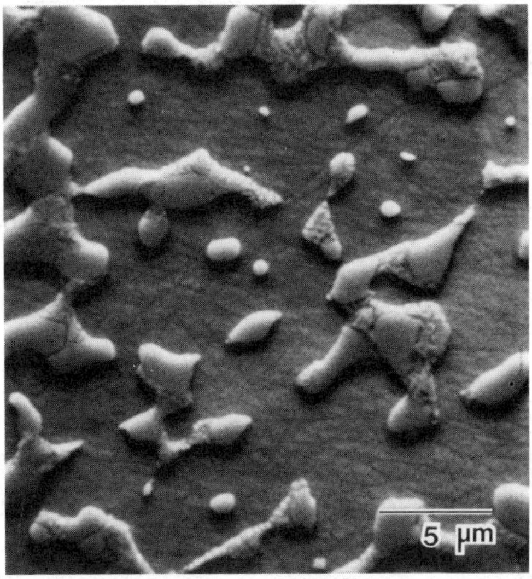

Figure 15 - Nb-23 atm% Ge undercooled 300 K. Both Nb_5Ge_3 and the metastable eutectic appear between β cells.

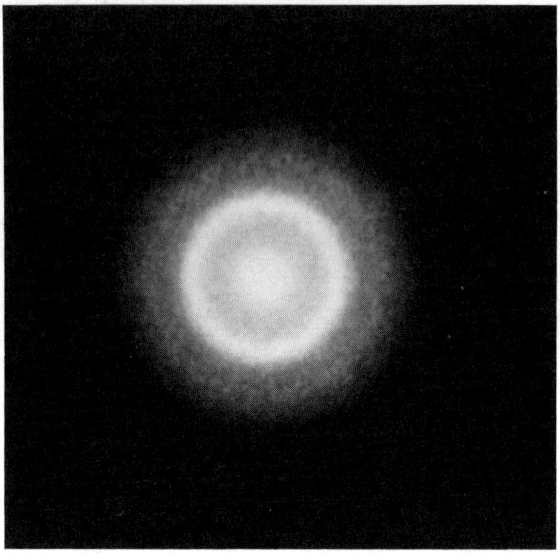

Figure 16 - Convergent beam electron diffraction pattern of a Nb-Ge alloy undercooled 300 K before splatting. The pattern is typical for that of an amorphous phase.

Results on specimens with a nominal composition of 27 atm% Ge serve as good examples of the nature of recalescence effects in highly undercooled specimens in long drop tubes. Figure 17 shows the distribution of fine microstructures (dark regions) and coarse microstructures (light regions) in a drop undercooled 270 K. The fine microstructural regions in a specimen undercooled 200 K can be seen in Figure 18 to be comprised of an interlocking structure of β and solid solution based on Nb_5Ge_3 and are due to rapid solidification at large undercooling. Such a microstructure makes an abrupt transition to a coupled eutectic (the coarse regions in Figure 17) as represented by Figure 19 and is due to recalescence and solidification controlled by heat transfer.

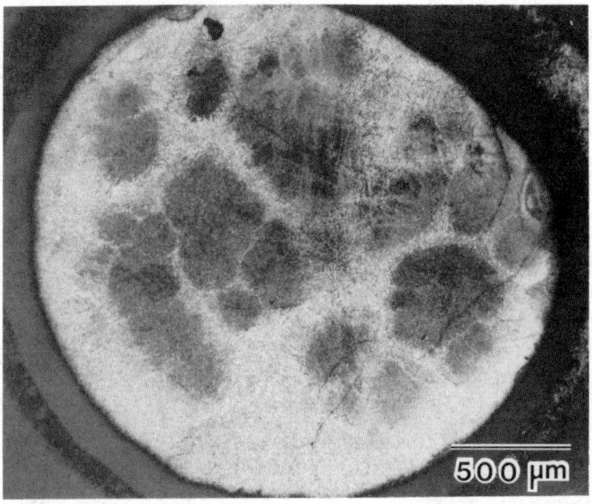

Figure 17 - Microstructural inhomogeneities in a Nb-28 atm% Ge alloy undercooled 270 K. Dark regions are cells of fine microstructure and light regions are coarse microstructure.

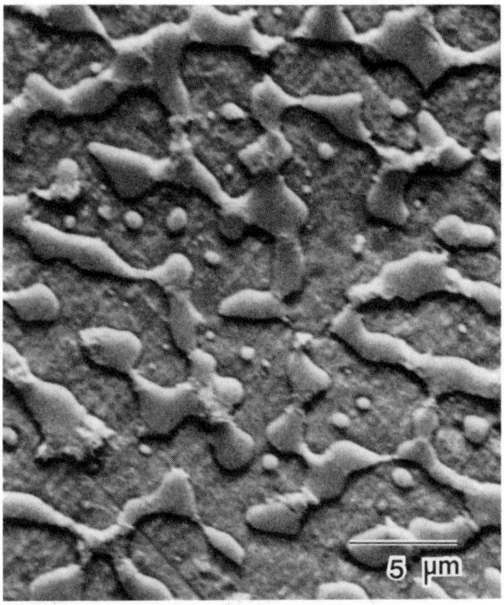

Figure 18 - Representative fine microstructure region in undercooled drops. This eutectic alloy has β cells with Nb_5Ge_3 and metastable α + Nb_5Ge_3 eutectic between cells.

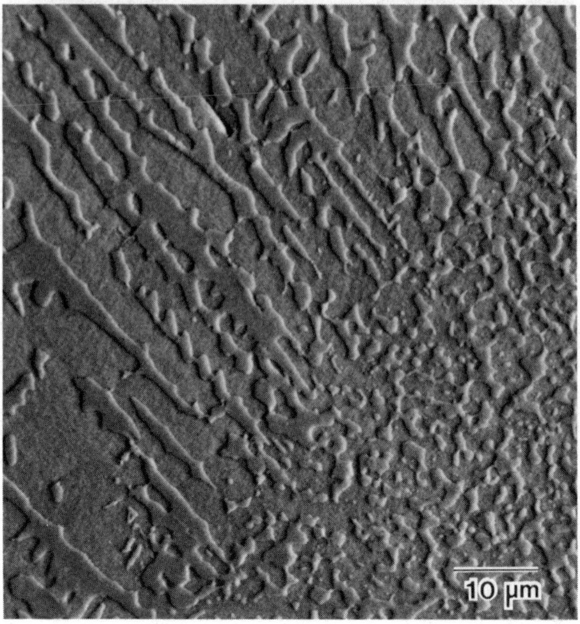

Figure 19 - Transition in morphology in a Nb-25 atm% Ge drop undercooled 190 K. The microstructure changes abruptly as heat transfer becomes the dominant factor in solidification.

The composition of the β phase in highly undercooled alloys with nominally 27 atm% Ge was commonly limited to 23 atm% Ge. This is the equilibrium composition of the beta phase at the equilibrium eutectic temperature and is therefore the equilibrium limit of solid solubility in the beta phase. However, solute trapping occurred during solidification in a specimen that undercooled 530 K ($\Delta T/T_m$ = .25). In the micrograph of a fine microstructural region, Figure 20, submicron Nb_5Ge_3 precipitates can be seen within the beta cells. Energy dispersive analysis of entire cells showed compositions of 25 atm% Ge, equal to the bulk composition of the alloy. Therefore, it is concluded that the initial beta phase had a non-equilibrium composition of 25 atm% Ge. However, recalescence caused the precipitation of Nb_5Ge_3, bringing the resultant beta composition to the equilibrium solubility limit at the eutectic temperature.

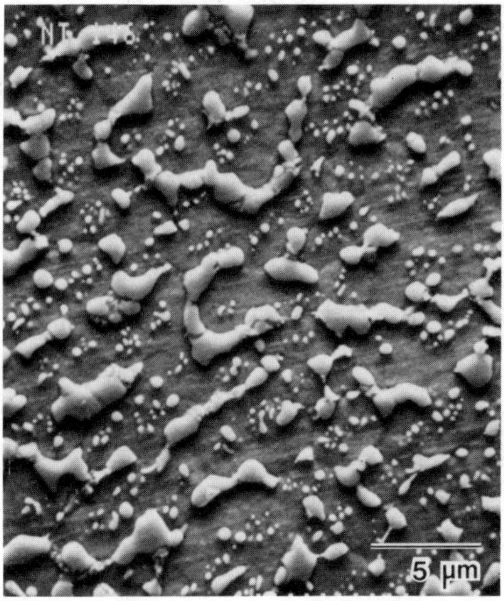

Figure 20 - Microstructure of Nb-25 atm% Ge following undercooling of 530 K. Solute trapping of Ge occurred because of the large degree of undercooling, but recalescence caused precipitation of excess solute in the form of Nb_5Ge_3.

Drop Tube Processing of Immiscible Alloys

Experiments on immiscible alloys in drop tubes are just now beginning. Only one result is presently available, and even it is in the preliminary stage (20). A Au-65 atm% Rh sample was solidified in free-fall in the 100 meter tube. As can be seen in Figure 21, the alloy was in the miscibility gap, but just inside the monotectic composition. Neither the overheating above the liquidus nor the amount of undercooling was known, but nonetheless an interesting observation was made. Arc casting resulted in complete separation of Au and Rh due to sedimentation as shown in Figure 22. However, the drop tube processed sample shows a coarse but definite dispersion of Au in Rh as seen in Figure 23.

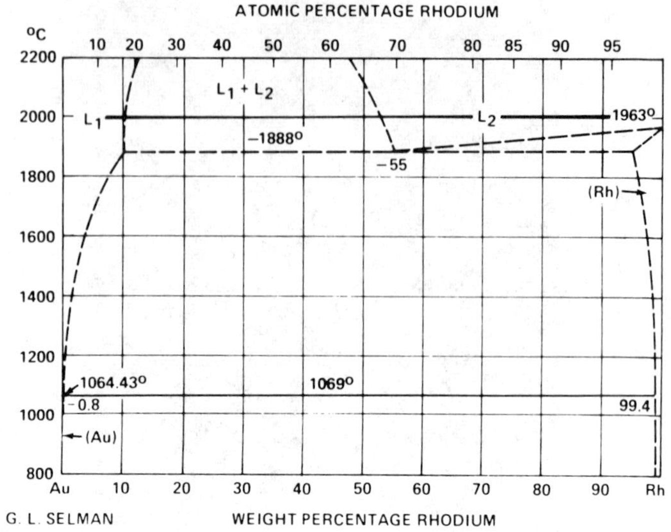

Figure 21 - The Au-Rh phase diagram. The Au-65 atm% Rh lies within the miscibility gap.

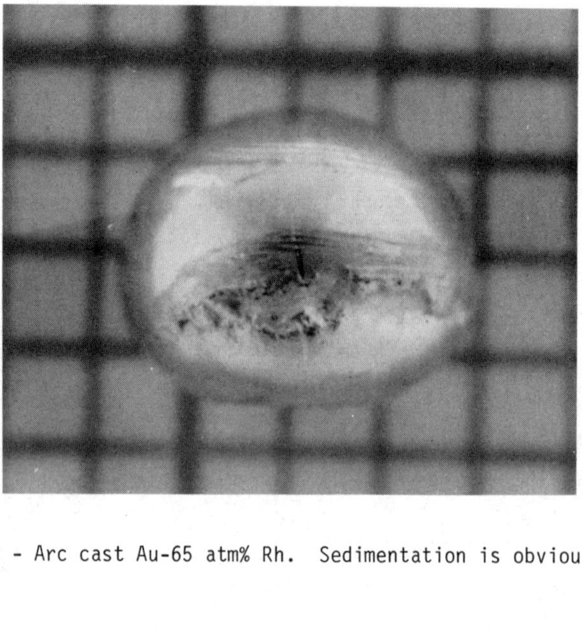

Figure 22 - Arc cast Au-65 atm% Rh. Sedimentation is obvious.

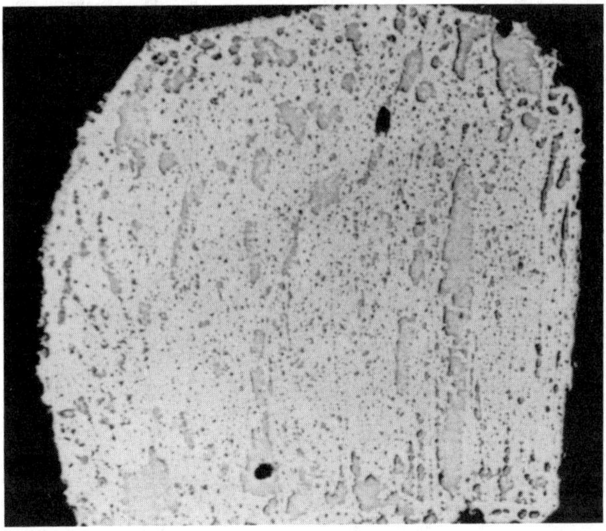

Figure 23 - Drop tube processed sample. Dispersion of Au in Rh has resulted.

A New Drop Tube Facility

Soon to be added to the small family of drop tubes is the 5.5 m drop tube, sketched in Figure 24, at the NASA Lewis Research Center. This particular tube is unique in that four high-speed two-color pyrometers will be spaced approximately equidistant along the column. Hence, measurement of the amount of undercooling will be direct as well as potentially of greater accuracy and to lower temperatures than in previous drop tubes. The rated temperature is 700 to 4000 °C. The tube will have the capability to be evacuated to 10^{-6} torr and backfilled with Ar or He. A disadvantage is the short free-fall time of 1 second, thereby limiting the combination of sample size, gas environment and sample freezing point.

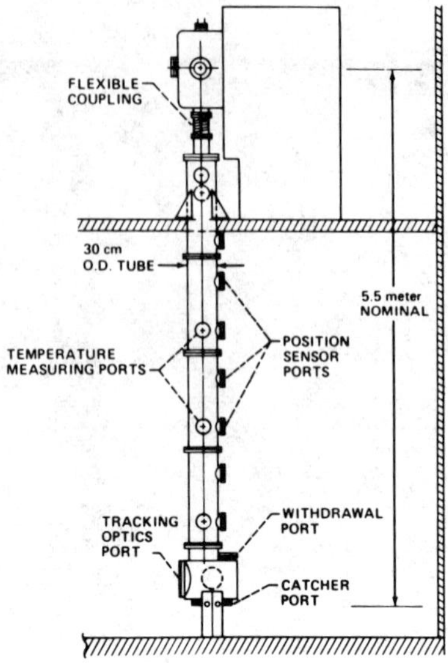

Figure 24 - The 5.5 m drop tube. This drop tube features four two-color pyrometers distributed along the column.

Summary

Drop tubes and drop towers both have the potential for studies on the deep undercooling of bulk materials as well as a variety of low-g solidification studies. These facilities have, however, barely been utilized. The two drop towers (145 meters high and 30 meters respectively) at the NASA Lewis Research Center have yet to be used for experiments on materials. Only one study, that involving immiscibility between Ga and Bi, in the 100 meter tower at the NASA Marshall Space Flight Center has been reported to date.

In comparison, a larger number of experiments have been done in long drop tubes. The force-free drop tube and the cryogenic drop tube at the Jet Propulsion Laboratory have been used to produce glass and metallic microshells for application in nuclear fusion targets and light weight structures. The 30 meter drop tube at the NASA Marshall Space Flight Center has been used in undercooling 3 to 5 mm diameter spheres of pure niobium and niobium-germanium alloys. Niobium samples undercooled to 0.19 T_m; niobium-germanium alloys undercooled in the range 0.13 to 0.22 T_m depending on the particular sample.

Most of the experiments in long drop tubes have been done in the 100 meter drop tube at the NASA Marshall Space Flight Center. Large degrees of undercooling have been obtained in Nb, Hf, Ir, Rh, and in alloys of Nb and Ge. Drop sizes range between 1 to 5 mm in diameter and undercoolings were as large as 0.25 T_m in the case of Nb-Ge alloys. Metastable configurations were formed in the Nb-Ge alloys as a result of deep undercooling. These were a metastable eutectic, an amorphous phase, and a solute enriched β phase. The latter, however, transformed by precipitation during recalescence to a stable configuration. Recalescence in the deep undercooled alloy drops also resulted in a dual morphology beginning with an interlocking β and Nb_5Ge_3.

Lastly, an experiment on free-fall solidification of a Au-Rh alloy with composition in the miscibility gap was done in the 100 meter drop tube. Sedimentation was seen to be prevented.

It is expected that greater use of drop towers and long drop tubes will be made in the future. This should take place for studies on undercooling, homogeneous dispersions in immiscible alloys, and other low-gravity applications. A welcome addition to the family will be the instrumented 5.5 m drop tube at the NASA Lewis Research Center.

Acknowledgments

The authors are grateful for the support of NASA under Grant #NAS8-34676. Additionally, R. J. Bayuzick expresses his appreciation for the support of the Center for the Space Processing of Engineering Materials, Vanderbilt University, established under NASA Grant #NAGW-810.

References

1. W. Johnson, A. G. Mamalis, and H. Hunt, "Small Spherical Lead Shot Forming from the Liquid, Using a Shot Tower," Metallurgia and Metal Forming, (March, 1976), 68-72.

2. R. Hooke, Micrographia, 1665. Cited in W. Johnson, A. G. Mamalis, and R. Hunt, "Small Spherical Lead Shot Forming from the Liquid Using a Shot Tower, " Metallurgia and Metal Forming, (March, 1976), 68.

3. L. L. Lacy and H. G. Otto, "Electrical Resistivity of Gallium-Bismuth Solidified in Free Fall," A.I.A.A. Journal, 13(2)(1975), 219-220.

4. M. C. Lee, "Metal Shell Technology," Sampe Journal, 19(6)(1983), 7-11.

5. J. M. Kendal, M.C. Lee, and T. G. Wang, "Metal Shell Technology Based Upon Hollow Jet Instability," J. Vac. Sci. Technol, 20(4)(1982), 1091-1093.

6. L. L. Lacy, M. B. Robinson, and T. J. Rathz, "Containerless Undercooling and Solidification in Drop Tubes," J. Cryst. Growth, 51(1981), 47-60.

7. L. L. Lacy, T. J. Rathz, and M. B. Robinson, "Containerless Undercooling and Solidification of Bulk Metastable Nb_3Ge Alloys," J. Appl. Phy., 53 (1)(1982), 682-689.

8. L. L. Lacy et al., "Solidification Studies of Nb-Ge Alloys at Large Degrees of Undercooling, in Materials Processing in the Reduced Gravity Environment of Space, ed. G. E. Rindone (New York: Elsevier, 1982), 87-94.

9. N. D. Evans, W. H. Hofmeister, R. J. Bayuzick, and M. B. Robinson, "Solidification of Nb-Ge Alloys in Long Drop Tubes," to be published in Metallurgical Transactions A, (June, 1986).

10. J. Steinberg, et al., "Production of Bulk Amorphous $Pd_{77.5}Si_{16.5}Cu_6$ in a Containerless Low Gravity Environment," Appl. Phys. Lett., 38(3) (1981), 135-137.

11. R. J. Bayuzick, et al., "A Review of Long Drop Tubes as a Supplement/ Alternative to Space Experiments," Adv. Space Res., 4(5)(1984), 85-90.

12. M. B. Robinson, "Undercooling Measurement in a Low-Gravity Containerless Environment" (M.S. thesis, University of Alabama in Huntsville, 1981).

13. D. B. Wallace, "Heat Loss and Drag of Spherical Drop Tube Samples," (NASA Report Ce-161855, NASA George C. Marshall Space Flight Center, 1981).

14. M. W. Zemansky, Heat and Thermodynamics, 5th edition, (New York, NY: McGraw-Hill, 1968).

15. T. Yuge, "Experiments in Heat Transfer for Spheres Including Combined Natural and Forced Convection," Journal of Heat Transfer, (Trans. ASME, 1960), 214-220.

16. F. M. White, Viscous Fluid Flow (New York, NY: McGraw-Hill, 1974).

17. F. P. Incropera and D. P Dewitt, Fundamentals of Heat Transfer (New York, NY: John Wiley & Sons, 1981).

18. W. H. Hofmeister, et al., "Microstructures of Niobium-Germanium Alloys Processed in Inert Gas in the 100 Meter Drop Tube," accepted for publication in Metallurgical Transactions.

19. N. D. Evans, R. J. Bayuzick, and E. A. Kenik, "Metastable Phases Formed in Rapidly Solidified Nb-Ge Alloys," Proceedings of the 44th Annual Meeting of the Electron Microscopy Society of America, ed. G. W. Bailey (San Francisco, CA: SanFrancisco Press, 1986), in press.

20. J. B Andrews and M. B. Robinson, "Containerless Low Gravity Processing of Hypermonotectic Gold-Rhodium Alloys" (Paper presented at the 115th TMS Annual Meeting, New Orleans, Louisianna, 4 March 1986), 68.

SURFACE TENSION MEASUREMENTS ON OSCILLATING DROPLETS OF UNDERCOOLED LIQUID METALS AND ALLOYS

J. Schade, A. McLean, and W.A. Miller

Department of Metallurgy and Materials Science
University of Toronto, Toronto, Ontario, Canada

The levitation technique has been used to measure the surface tensions of undercooled liquid metal and alloy droplets. The natural frequency of droplet oscillation is determined using a Fourier wave analyzer; surface tension is then calculated from Rayleigh's equation. Studies have been carried out to date on pure Fe, Ni, Cr, and Co as well as on various metal-silicon alloys. It has been found that quite large undercoolings can be obtained with this technique in certain high melting point metals; liquid Fe has been undercooled by 538°C ($\Delta T/T_m = 0.30$) and liquid Ni by 452°C ($\Delta T/T_m = 0.26$), considerably exceeding the value of $0.20T_m$ which is often taken as the undercooling at which homogeneous nucleation occurs in these metals. In all cases, the linear variation of surface tension with temperature which was observed in the superheated range was found to extrapolate back throughout the undercooled region. Surface tension measurements on binary Fe-Si and Co-Si alloys indicate that an ordered liquid may persist at compositions corresponding to the ε intermetallic phases in the solid state. Such ordering, if it exists, did not cause a minimum in the degree of undercooling observed at the corresponding compositions.

Introduction

The levitation method of measuring the surface tension of oscillating liquid metal droplets was first developed by Lu and his coworkers (1-3). The surface tension is determined by measuring the natural oscillatory frequency of a droplet suspended in an electromgnetic field. The principal advantages of this method are:

(a) levitation provides a container less environment
(b) contact angles need not be determined, and
(c) there is no requirement to evaluate the density of the melt as the surface tension is a function of droplet weight and oscillatory frequency only.

To date, the levitation melting technique has been used to determine the surface tension values for pure Cu, Ag, Ni, Au, Co, Cr, Sn, and Fe, as well as for several binary alloys (4-8). Lacy and coworkers (9) have pointed out that undercooling is a natural extension in a levitation environment due in part to the elimination of container-induced nucleation. The use of a reducing atmosphere, such as hydrogen gas, further facilitates undercooling as heterogeneous nucleation catalysts (such as oxide or sulphide inclusions) are reduced and removed. The levitation method has proven to be valuable for determining the surface tension of pure metals over large temperature ranges. The results have generally been found to lie in the upper quadrant of data available in the literature (7), with the relatively high results being attributed to the lack of contamination from a container or substrate such as would be expected from static measurements using, for example, the sessile drop method.

Surface tension is calculated in the levitation technique from the measured value of the natural frequency of oscillation of the droplet. Following Keene's innovation (5), recent work (6-8, 10) has made use of Fourier analysis of the droplet's motion to determine the frequency. Surface tension is then calculated using Rayleigh's equation:

$$\gamma = (3/8)\pi m\omega^2$$

where γ is the surface tension (in dyne cm^{-1}, equivalent to milliNm^{-1}), m is the droplet mass (in g), and ω is the frequency (in Hz). The main assumptions involved in deriving this relationship are:

(a) the amplitude of deviations from the spherical shape is small,
(b) the liquid viscosity is low in order that damping effects on the

natural oscillations of the liquid can be neglected, and
(c) the liquid approaches incompressibility.

The temperature dependence of surface tension, $d\gamma/dT$, which is directly related to the surface entropy and the surface excesses, is often expressed in the form:

$$\gamma_T = \gamma_m - (d\gamma/dT)\ (T - T_m)$$

where γ_T is the surface tension at temperature T and γ_m is the surface tension at the melting point. The units of $d\gamma/dT$ are milli$Nm^{-1}K^{-1}$. Because of the relationship to surface entropy, changes in the structure of the liquid with temperature may be reflected in the temperature coefficient of surface tension (11).

This study is concerned with surface tension phenomena in pure Fe, Co, and Ni in both the stable and metastable liquid ranges, as well as in the binary systems Fe-Si and Co-Si.

Experimental

Samples of Fe, Ni, and Co (Johnson Matthey, Specpure grades, (10 ppm t.m.i.) and of Si(Alpha Ventron, Puratronic grade) were prepared using a procedure described elsewhere (5-7, 10). The sample mass was held essentially constant at 0.5 ± 0.5 g. The samples were introduced into the levitation chamber (Fig. 1) on a pushrod and levitated in the electromagnetic field generated by a 450 kHz R.F. generator having an output at 2.5 kW. The levitated samples were inductively melted, with temperature being recorded with a two-colour pyrometer.

Temperature control was achieved mainly by varying the gas flowrate (up to a maximum of 5.5 L/min); all gases were first passed through a gas train to trap water vapour and oxygen. The use of hydrogen as the ambient medium maintains a reducing atmosphere around the droplet, permitting the removal of any dissolved oxygen or sulphur prior to carrying out the surface tension measurements. Tests were carried out to verify that hydrogen has no effect on the surface tension of the metals studied in this work (10).

Following the surface tension measurements, each droplet was solidified in the levitated state (by increasing the gas flowrate to a maximum), after which it was removed for chemical analysis. It was found that the concentration of the surfactant impurities O and S could be reduced to ac-

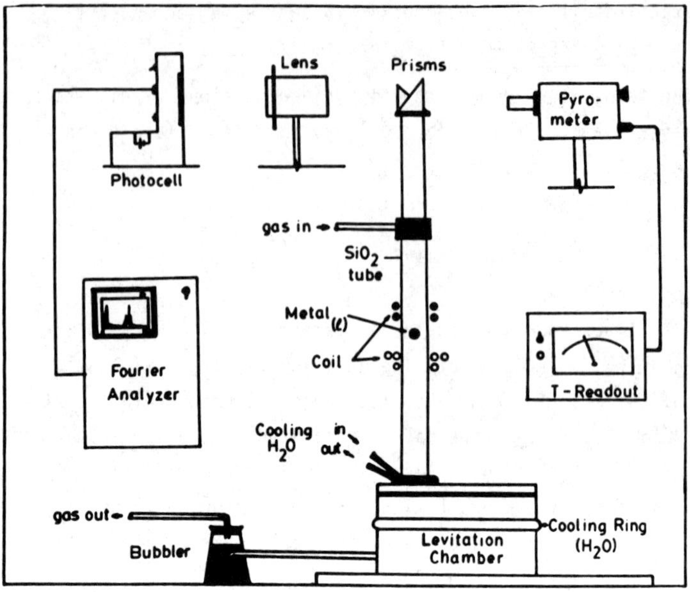

Figure 1 - Experimental levitation apparatus

ceptable levels in the pure metals (< 10 ppm) by levitation in hydrogen. Table 1 summarizes the oxygen analysis results obtained using a LECO RO 16 analyzer.

Alloying was achieved by co-levitating the appropriate amount of either Fe, Ni, or Co with Si to take advantage of the benefits of container less alloying. Even though the induction generator used was not powerful enough to levitate pure Si, alloys with Si contents up to ~ 60 wt % were successfully levitated by 1) melting the metal, 2) reducing its oxygen content by holding it in a hydrogen atmosphere, and 3) bringing the solid Si into contact with the liquid metal. Once the liquid wets the Si, it absorbs it rapidly to form an alloy. For small additions of Si(<12 wt %), an instantaneous temperature increase was noted at this point. This is due to the exothermicity of the M-Si reaction. This behaviour was not observed at higher Si levels, presumably because the larger thermal mass of cold Si tends to lower the temperature of the system considerably, masking the heat generated by the reaction. However, a rapid return to temperature occurred after alloying was complete.

A levitated droplet undergoes three distinct modes of motion in the field, viz. rotation, translation, and oscillation. This motion is detected by projecting the image of the droplet onto a polycrystalline photo-

Table 1 Oxygen content of the metal samples

Metal or Alloy	Oxygen Content (ppm) As Received	After Levitation
Metal		
Fe	27-38	< 10
Ni	22-29	< 10
Co	19-27	< 10
Alloy		
Fe- 4Si	> 12	12
Fe-33Si	> 15	15
Co-15Si	> 9	9

cell, as shown schematically in Fig. 1. The relay signal is sent to a NSC 4468 Fourier analyzer, which processes the input and displays it as either an oscilloscopic trace or as an amplitude-versus-frequency spectrum. The spectra can be used to differentiate unambiguously between solid and liquid, since oscillations occur in liquids only, while translation and rotation occur in both solids and liquids.

At low temperatures where the droplets are solid, only low-frequency rotational and translational components of motion are observed. These will typically lie in the 5 to 10 Hz range. As the specimen temperature is increased, the amplitude of each of these modes increases, reflecting increasing droplet instability as it approaches the melting point. As the droplet melts, a set of peaks in the 40 to 70 Hz region appear (the exact values of frequency depending on the magnitude of γ). These peaks characterize the oscillations of the molten droplet and correspondingly disappear and reappear if solidification or remelting occur. The method is therefore an accurate means of continuously monitoring the presence or absence of the liquid state.

Results and Discussion

a) Surface Tension of Pure Fe, Co, and Ni

The results of surface tension determination for pure Fe, Co and Ni are illustrated in Figs. 2 to 4 respectively. Data obtained by the levitated droplet method are represented by circles, while non-circular symbols denote literature data determined by the more commonly used static techniques such as the sessile drop and pendant drop methods. It may be seen that in all cases our results extend to far lower temperatures than those obtained from the static techniques, and that the linear variation of surface tension with temperature which is observed for superheated liquids extends into the undercooled range.

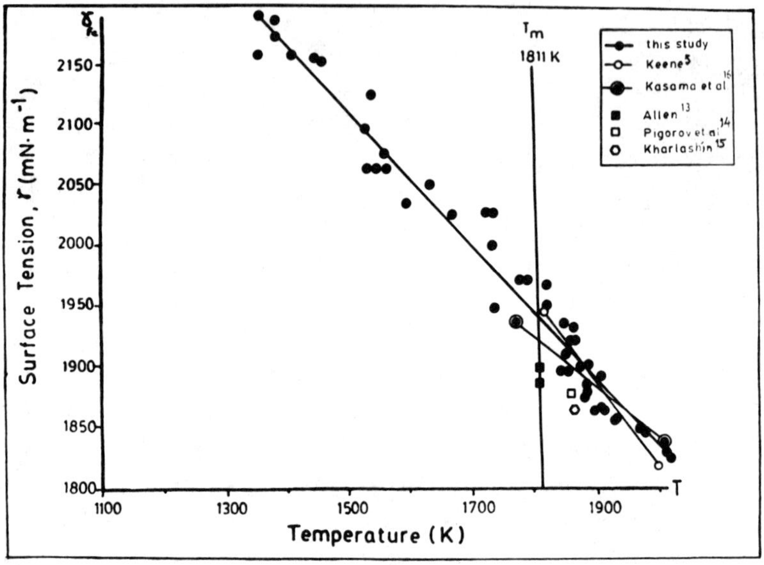

Figure 2 - Surface tension of pure liquid iron

The literature dealing with the liquid surface tension of Fe and its binary alloys has been surveyed by Keene (12). As Fig. 2 indicates, the results of Allen (13), Pigorov et al. (14) and Kharlashin and Molonov (15) show quite good agreement with our results. Our data are also in agreement with previous levitation studies by both Keene (5) and Kasama et al. (16). The surface tension of Fe is represented by the equation

$$\gamma_{Fe} = 1938 - 0.54(T - 1811)\text{mNm}^{-1} \quad 1300\text{K} < T < 2000\text{K}$$

where T is in K. The temperature coefficient of surface tension of -0.54 $mNm^{-1}K^{-1}$ is in relatively good agreement with values contained in the literature.

The values obtained for Co and Ni (Figs. 3 and 4, respectively) also show excellent agreement with previous measurements. The following equations describe this temperature dependence of the surface tension for these metals:

$$\gamma_{Co} = 1931 - 0.56(T - 1763)\ \text{mNm}^{-1}\quad 1610\text{K} < T < 1925\text{K}$$

and

$$\gamma_{Ni} = 1946 - 0.25(T - 1725)\ \text{mNm}^{-1}\quad 1300\text{K} < T < 1900\text{K}$$

Very few property measurements have been carried out on metals in the undercooled condition. A levitation study by Shiraishi and Ward (17) found that the density of undercooled Ni varied linearly with temperature. Viscosity rises far more rapidly with falling temperature (18), but most of the data for undercooled metals involve extrapolations of measurements made in the stable range. Assumption (b) referred to earlier in connection with Rayleigh's equation suggests that a large change in viscosity might influence the accuracy of surface tension data obtained from measurements of droplet oscillation frequency. However, our results indicate that γ varies linearly with temperature over wide temperature ranges, so that any viscosity effect on the surface tension data at low temperature must be quite small.

b) Undercooling Measurements

The maximum values of undercooling observed in this study for Fe, Ni, and Co are listed in Table II along with values obtained from the literature.

Iron and Ni were undercooled even further than is indicated by our values shown in the table. The pyrometer used in this study had a lower reading limit of 1000°C, and accurate determinations of lower temperatures were not possible. Other available pyrometers required emissivity corrections and these are not accessible for highly undercooled liquid Fe and Ni. Nonetheless, these results exceed values recorded in the literature by a wide margin. Note that for Co an undercooling of 350 K was observed, although surface tensions could be measured to an undercooling of only 250 K due to an unexplained deterioration of the oscillatory spectrum for Co at undercoolings of more than 250 K. The same type of spectrum deterioration has also been observed for Cr (6).

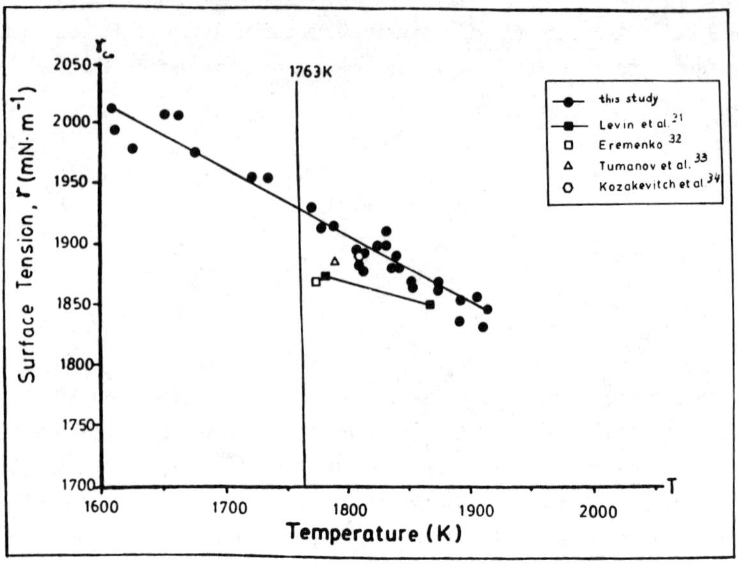

Figure 3 - Surface tension of pure liquid cobalt

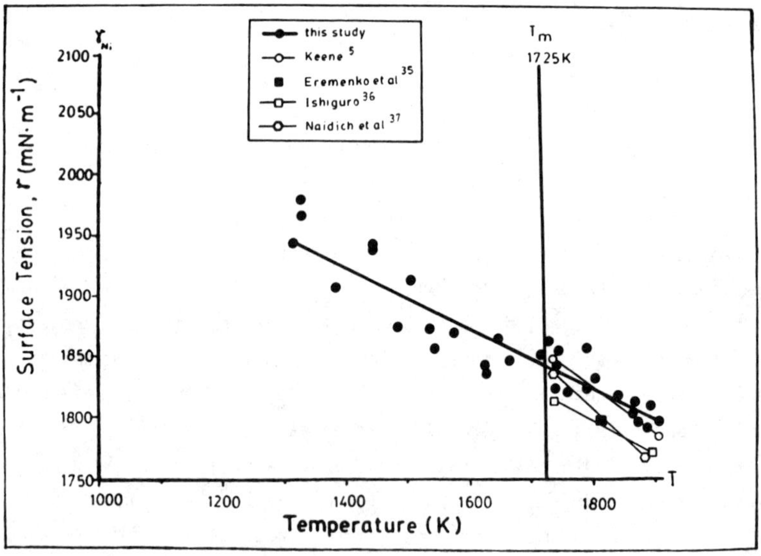

Figure 4 - Surface tension of pure liquid nickel

Table II - Experimental values of maximum undercooling for pure liquid Fe, Ni, and Co

Metal (T_m)	Maximum Undercooling: $\Delta T(K)$ and $\Delta T/T_m$ Fe(1811K)	Ni(1725K)	Co(1763K)
This study	538+ / 0.297	452+/ 0.262	350 / 0.199
Reference (19)	295 / 0.164	219 / 0.185	330 / 0.187
Reference (17)	not invest.	302 / 0.175	not invest.
Reference (20)	350 / 0.193	410 / 0.238	not invest.
Reference (21)	345 / 0.191	not invest.	not invest.
Reference (22)	293 / 0.162	not invest.	not invest.
Reference (23)	360 / 0.199	410 / 0.238	330 / 0.187

There is no generally accepted value for the maximum undercooling which may be achieved for liquid metals, although this limit is expected to be fixed by the onset of volume dependent homogeneous nucleation, unless vitrification below the glass transition temperature occurs first (24). While it has often been stated since the results of Turnbull and Cech (19) that the undercooling limit for homogeneous nucleation is 0.2 T_m, more recent results indicate that values in excess of 0.3 - 0.4 T_m may be reached. Low melting point metals such as Hg, Sn, Bi, and Ga have been undercooled by as much as 0.38 to 0.58 T_m using an emulsification method (24). No comparable results have been obtained for higher melting point metals, for which previously recorded maximum undercoolings were about 0.2 T_m for both pure metals (19-23) and alloys (9, 25).

The large undercoolings observed in this work are attributed to the absence of heterogeneous nucleation sites in the levitated droplets. It would appear that under ideal conditions of droplet purity, droplet shape and mass, and gas atmosphere control, undercoolings of at least 0.25 - 0.3 T_m can occur before nucleation takes place. It is interesting to note that this undercooling is quite close to a limit proposed some time ago by Shiraishi and Ward (17). They measured the density of levitated droplets of undercooled liquid Ni and suggested that the maximum undercooling might correspond to the temperature below which the density of the undercooled liquid would exceed the density of the solid. For Ni, that point is reached at a temperature of ~960°C (17) corresponding to an undercooling of about 0.29 T_m. This value is only slightly greater than the undercooling of 0.26 T_m observed in this study.

c) Metal-Silicon Alloys

Fe-Si and Co-Si alloys were studied in an attempt to clarify some unusual surface tension effects which have been reported in the literature. Several workers (26, 27) have noted a change in slope of the surface tension versus composition curve in the vicinity of the ε (FeSi) phase (the experimental data from the literature are shown in Fig. 5). It has also been reported that the temperature coefficient of the surface tension exhibits a maximum at the intermetallic composition (12). A large value of $d\gamma/dT$ may be associated with a low value of surface entropy, and it has been suggested that these results reflect the existence of an ordered liquid structure which is not completely lost when the intermetallic phase melts (10). Since an ordered liquid structure might also be expected to exhibit a relatively small undercooling, experiments were conducted to see what if any correlation exists between γ, $d\gamma/dT$, and ΔT measurements in these alloys.

The effect of Si additions on the surface tension of Fe-Si and Co-Si alloys was determined isothermally at 1550°C, in the stable liquid range. The results for Fe-Si are shown, superimposed on the phase diagram, in Fig. 6. No results beyond a Si content of 57.1 wt % were obtained, since the output of the R.F. generator used was insufficient to levitate specimens containing more than this amount of Si. It is clear that no abnormal change in the slope of the surface tension curve occurs at compositions close to the intermetallic phase. Previous observations of such an effect may have been due to incomplete fusion above the relatively high melting point of the intermetallic compound, or may reflect an interaction between the sessile droplet and the substrate (27). Our experimental results are compared in Fig. 6 to theoretical curves for binary alloys based upon Guggenheim's (28) and Utigard and Toguri's (29) models. These

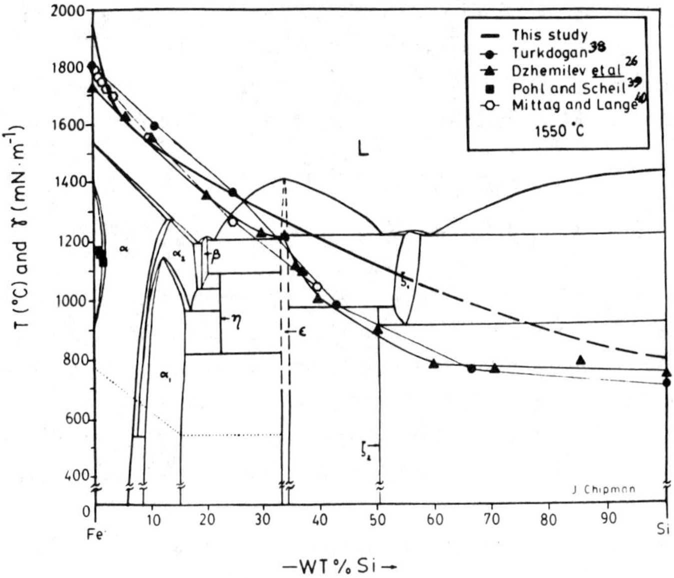

Figure 5 Compositional dependence of the surface tension in the Fe-Si system at 1550^{0}C (Comparison of present results with literature data)

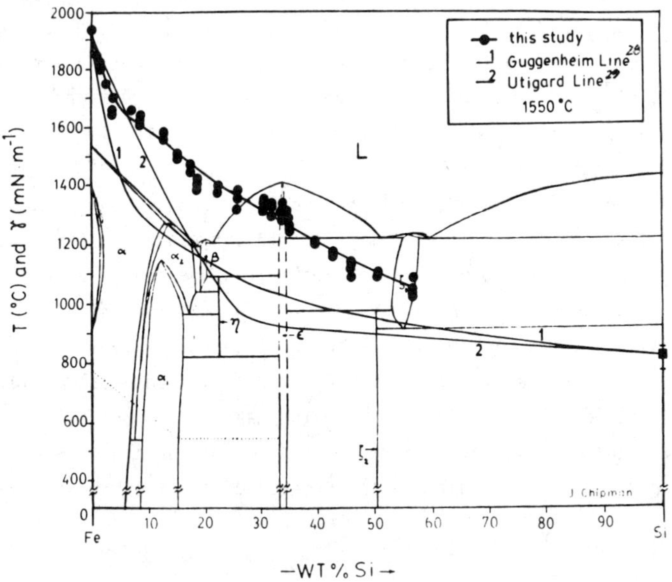

Figure 6 Compositional dependence of the surface tension in the Fe-Si system at 1550^{0}C (Comparison of present results with theoretical models)

models permit the calculation of alloy surface tension as a function of composition based upon experimental values of the surface tensions of each pure metal. The value used for the surface tension of pure Si (=820 mNm^{-1}) was that of Hardy (30). It is clear that the calculated surface tension curves do not agree with the experimental data at Si contents exceeding 5 - 10 wt %, but whether this disagreement can be associated with the presence of intermetallic compounds in this system remains in doubt.

While no pronounced effect of the ε-phase was noted in the isothermal surface tension measurements, a clear effect exists in the $d\gamma/dT$ data. Table III shows the measured values of $d\gamma/dT$ for a series of Fe-Si alloys ranging from pure Fe to Fe - 57.1 wt % Si, and it may be seen that a definite maximum (least negative value) occurs near the intermetallic composition (33.4 wt % Si). These data are therefore not inconsistent with the concept of some ordering occurring in the liquid structure as the intermetallic composition is approached.

The undercooling data for Fe-Si in Table III show that alloy droplets can also be undercooled significantly (maximum undercoolings of at least 0.15 - 0.25 T_L were observed for all compositions, where T_L is the liquidus temperature). Because the lower limit of detection of the pyrometer used in this study was surpassed during cooling by most of the alloy droplets (see Table III), it is not possible to correlate the undercooling behaviour with the $d\gamma/dT$ data. In spite of this technical difficulty, however, the table seems to indicate that an undercooling maximum may exist near the intermetallic composition. This is just the opposite of what would be expected as a result of ordering in the liquid state near the intermetallic composition.

The experimental results obtained for the Co-Si alloys are analogous to those described above for Fe-Si. A smooth curve of γ versus composition is obtained with no observable breaks in slope near the intermetallic compositions. A pronounced maximum in $d\gamma/dT$ occurred at a composition corresponding to the ε-phase in the solid state. All alloys showed undercoolings in the range of 0.17 - 0.20 T_L, but again the problem caused by the lower detection limit of the pyrometer means that most alloys were actually undercooled to even lower temperatures. The data also suggest that an undercooling maximum may be associated with the ε-phase composition.

Table III - Temperature coefficients of surface tension and undercooling of Fe-Si alloys

Si-Content (wt %)	$d\gamma/dT$ ($mNm^{-1}K^{-1}$)	Maximum Undercooling ΔT(K)	$\Delta T/T_L$
0	-0.54	538 +	0.297
4.2	-0.50	462	0.266
9.7	-0.42	340*	0.208
12.8	-0.37	300	0.191
14.8	-0.38	275	0.178
19.5	-0.29	210	0.142
26.3	-0.25	312	0.198
31.0	-0.17	276*	0.167
33.4	-0.052	410*	0.244
33.8	-0.044	395*	0.235
40.2	-0.18	345*	0.210
46.2	-0.18	316	0.200
50.0	-0.24	250	0.163
57.1	-0.30	230	0.152

*The specimens denoted by an asterisk solidified at or above 1000°C, the lower limit of the pyrometer used in this work. All other specimens solidified at an undetermined temperature below 1000°C, so that the actual undercoolings are larger than those indicated.

Summary

The levitation technique is well suited for the study of undercooled metallic phases. Surface tensions of undercooled droplets can be readily measured by determining the natural oscillation frequency of liquid droplets by using a Fourier wave analyzer. The frequency spectrum produced by the analyser serves to differentiate unequivocally between solid and liquid states of the specimen. A temperature-time cycle for a typical experiment is shown in Fig. 7.

The results of the undercooling experiments have shown that both pure Fe, Ni, and Co and their binary alloys with Si can be maintained in the liquid state to undercoolings which approach 0.3 T_m. Surface tension measurements on the metal-silicon alloys indicate that an ordered liquid may persist at compositions corresponding to the ε intermetallic compound, but this effect was not corroborated by measurements of the detree of undercooling as a function of Si content in the vicinity of the intermetallic compounds.

The levitation technique may prove in future to be valuable for studying surface segregation in liquids at low temperatures, where even weakly surface-active species may show extensive segregation effects.

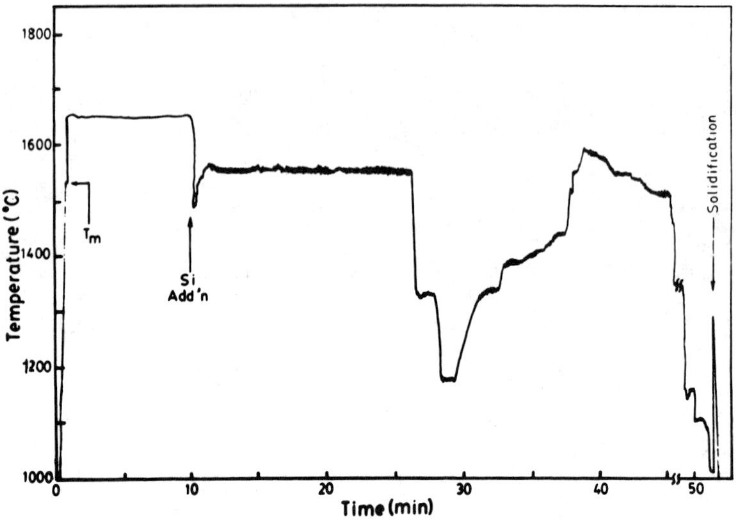

Figure 7 Temperature-time variation for a levitated droplet of Fe - 33.8 wt % Si

The Si was added after 10 minutes, allowing for prior deoxidation of the Fe. Surface tension measurments were conducted at 1550°C, after which the gas flowrate was increased in several stages until the sample was cooled to 1180°C, 225 K of undercooling. Surface tension measurements were repeated at this temperature. The gas flowrate was then decreased to allow reheating into the stable liquid range prior to solidification at 1010°C (ΔT = 394 K).

References

1. M.E. Fraser, W.K. Lu, A.E. Hamielec and R. Murarka, "Surface Tension Measurements on Pure Liquid Iron and Nickel by an Oscillating Drop Technique", Met. Trans., 2(1971), 817-823.

2. R.N. Murarka, W.K. Lu, and A.E. Hamielec, "Surface Tension of Pure Liquid and Supercooled Iron", Met. Trans., 2(1971), 2949-2950.

3. R.N. Murarka, W.K. Lu, and A.E. Hamielec, "Effect of Dissolved Oxygen on the Surface Tension of Liquid Iron", Can. Met. Quart, 14(1975), 111-115.

4. H. Soda, A. McLean, and W.A. Miller, "Surface Tension Measurements of Liquid Copper Droplets in the Temperature Range 1000 to 1330°C", Trans. Japan Inst. Metals, 18(1977), 445-454.

5. B.J. Keene, "The Use of a Fourier Analyzer for the Determination of the Surface Tension of Liquid Metals by the Levitating Drop Technique" (Report DMA A(56), National Physical Laboratory, U.K., 1982).

6. J. Schade, "Surface Tension Measurements of Liquid Chromium Using the Levitation Melting Technique (B.A.Sc. Thesis, Department of Metallurgy and Materials Science, University of Toronto, 1984).

7. B.J. Keene, K.C. Mills, A. Kasama, A. McLean and W.A. Miller, "Comparison of Surface Tension Measurements Using the Levitated Drop Method", Met. Trans. B, 17(1986), 159-162.

8. K. Nogi, W.B. Chung, A. McLean and W.A. Miller, "Comparison of Surface Tension of Some Binary Iron and Nickel Alloys", to be published.

9. L.L. Lacy, M.B. Robinson, and T.J. Rathz, "Containerless Undercooling and Solidification in Drop Tubes", J. Cryst. Growth, 51(1981), 47-60.

10. J. Schade, "Surface Tension Measurements of Liquid Fe-Si, Ni-Si, and Co-Si Alloys Using the Levitation Melting Technique" (M.A.Sc. Thesis, Department of Metallurgy and Materials Science, University of Toronto, 1986).

11. K. Nogi, K. Ogino, A. McLean and W.A. Miller, "The Temperature Coefficient of the Surface Tension of Pure Liquid Metals", Met. Trans. B, 17(1986), 163-170.

12. B.J. Keene, "A Survey of Extant Data for the Surface Tension of Iron and its Binary Alloys" (Report DMA A(67), National Physical Laboratory, U.K., 1983).

13. B.C. Allen, "The Surface Tension of Liquid Metals", in Liquid Metals, S.Z. Beer, ed. (New York: Marcel Dekker Inc., 1972), 161-212.

14. N.A. Pigorov, S.A. Bliznyakov, and Yu.V. Kryadovskii, "Effect of Boron and Cerium on the Surface Tension of Liquid Iron and Steel", Fiz. Khim. Poverkl. Yavlenif Vys. Temp., V.N. Eremenko, ed. (Kiev: Naukova Dumka, 1971), 106.

15. P.S. Kharlashin and G.D. Monolov, "Density and Free Surface Energy of Fe-As Melts", Russ. Metall., 5(1977), 78.

16. A. Kasama, A. McLean, W.A. Miller and M.J. Ward, "Surface Tension of Liquid Iron and Iron-Oxygen Alloys", Can. Met. Quart, 22(1983), 9-17.

17. S.Y. Shiraishi and R.G. Ward, "The Density of Nickel in the Superheated and Supercooled Liquid States", Can. Met. Quart., 3(1964), 117-122.

18. P.H. Shingu et al., "Solidification Rate of Rapidly Quenched Metals" (Paper presented at the Second International Conference on Rapidly Quenched Metals, M.I.T., November 17-19, 1975), 45-50.

19. D. Turnbull and R.E. Cech, "Microscopic Observation of the Solidification of Small Metal Droplets", J. Appl. Phys., 21(1950), 804-810.

20. D.W. Gomersall, S.Y. Shiraishi, and R.G. Ward, "Undercooling Phenomena in Liquid Metal Droplets", J. Aust. Inst. Metals, 10(1965), 220-222.

21. D.E. Ovsienko and V.P. Kostyuchenko, as reported by T. Ohashi et al., ref. (41).

22. W. Patterson, S. Eengler, and R. Moser, as reported by T. Ohashi et al., ibid.

23. A.N. Krestovnikov, K.M. Rozin, and V.N. Vigdorovich, "Some Physicochemical Features of Surface Phenomena in High-Temperature Melts", Surface Phenomena in Metallurgical Processes, A.I. Belyaev, ed. (New York: CBE Inc. 1965), 45-49.

24. J.H. Perepezko and J.S. Paik, "Undercooling Behaviour of Liquid Metals", Rapidly Solidified Amorphous and Crystalline Alloys, B.H. Kear, B.C. Giessen, and M. Cohen, eds. (New York: Elsevier, 1982), 49-63.

25. Y. Shiohara et al., "Solidification Mechanisms of Highly Undercooled Metal Alloys", Rapidly Solidified Amorphous and Crystalline Alloys, B.H. Kear, B.C. Giessen, and M. Cohen, eds. (New York: Elsevier, 1982), 65-69.

26. N.K. Dzhemilev, S.I. Popel, and B.V. Tsarevskii, "Isotherm of the Density and Surface Energy of Iron-Silicon Melts", Fizika. Metal., 18(1964), 77-81.

27. K. Monma and H. Suto, "Experimental Studies on the Surface Tension of Molten Metals and Alloys", J. Japan Inst. Metals, 24(1960), 163-167.

28. E.A. Guggenheim, "Statistical Thermodynamics of the Surface of a Regular Solution", Trans. Farad. Soc., 41(1945), 150-156.

29. T.A. Utigard and J.M. Toguri, Met. Trans. B, 16(1985), 333-338.

30. S.C. Hardy, "The Surface Tension of Liquid Silicon", J. Cryst. Growth, 69(1984), 456-460.

31. E.S. Levin, P.V. Gel'd, and L.V. Shinpanova, as reported by B.J. Keene, ref. (5).

32. V.N. Eremenko, as reported by B.J. Keene, ibid.

33. V.I. Tumanov et al., as reported by B.J. Keene, ibid.

34. P. Kozakevitch and G. Urbain, as reported by B.J. Keene, ibid.

35. V.N. Eremenko and V.I. Nizhenko, as reported by B.J. Keene, ibid.

36. K. Ishiguro, "The Surface Tension of Liquid Nickel by the Sessile Drop Technique" (B.A.Sc. Thesis, Department of Metallurgy, University of Osaka, 1980).

37. Yu.V. Naidich, V.M. Perevertaylo, and G.M. Nevodnik, as reported by B.J. Keene, ref. (5).

38. E.T. Turkdogan, Physical Chemistry of High Temperature Technology (Toronto: Academic Press, 1980), 92-129.

39. D. Pohl and E. Scheil, "Ueber die Oberflaechenspannung von Gusseisen", Giesserei, 42(1956), 833-839.

40. U. Mittag and K.W. Lange, "Experimentelle Ergebnisse fuer die Grenzflaechenspannung von Eisen-Kohlenstoff-, Eisen-Silicum-, Eisen-Chrom- und Eisen-Stickstoff Legierungen", Arch. Eisenhuettenwesen, 47(1976), 65-99.

41. T. Ohashi et al., "Effect of Oxides on the Nucleation Behaviour in Supercooled Iron", Trans. ISIJ., 17(1977), 262-270.

UNDERCOOLING AND MICROSTRUCTURAL EVOLUTION IN GLASS FORMING ALLOYS

M.J. Kaufman
Metallurgy Division
National Bureau of Standards
Gaithersburg, MD 20899

and

H.L. Fraser
Dept. of Materials Science
University of Illinois
1304 W. Green St.
Urbana, IL 61801

The microstructural evolution of highly undercooled submicron powders of a glass forming alloy (Al-30Ge) is considered. A simple calculation is developed for predicting the number of α-Al crystals that form during solidification using classical nucleation and growth equations appropriate for glass forming systems. In addition, a reasonable agreement between theory and experiment is achieved only when a solid-liquid temperature gradient, generated by recalescence effects, is considered.

Introduction

Recently, the synthesis of non-equilibrium structures by rapid solidification has been shown to correlate directly with the ability to both undercool a material from its liquid state by substantial amounts and limit the extent of recalescence during the solidification process. This realization has led to an increasing interest in homogeneous and heterogeneous nucleation phenomena in materials (1-4). As a result, numerous investigators have attempted to extend the pioneering droplet studies of Turnbull and coworkers (5-10) as well as later studies by other authors (11,12) using similar and modified experimental approaches. These investigations have addressed both conventional determinations of undercooling from the liquid state as well as undercooling from the semi-solid state where the solid phase may catalyze nucleation of other solid phases from the liquid. Many of these studies have been performed on small droplets (1-100 μm in diameter) in order to minimize the number of heterogeneous nucleation sites and, thereby, approach homogeneous nucleation conditions. Recently, however, techniques have been developed for producing even smaller droplets (< 1 μm), and when these are melted and solidified, substantial increases in attainable undercoolings are achieved. For example, in a recent study of submicron Al-Ge alloys powders (13), it was shown that the undercoolings obtained experimentally were significantly greater than those obtained in previous studies of larger samples (Table 1). Based on these observations, it was concluded that the suggestions advanced in the previous studies, that homogeneous nucleation had been achieved, were incorrect, whereas it was concluded that homogeneous nucleation had been achieved in the submicron powders. This conclusion was based on experimental observations where highly undercooled droplets solidified into structures comprised of many small crystals (< 10 nm in diameter) of α-Al and β-Ge solid solutions, consistent with homogeneous (volume-dependent) nucleation.

An interesting observation made in the studies on Al-Ge (13) involved the microstructure that formed at the largest undercoolings, in these cases, the structure that developed most frequently consisted of a single crystal of α-Al surrounded by an amorphous shell (e.g. Fig. 1). A model was proposed to explain the evolution of this structure based on the nucleation of an α-Al nucleus and its subsequent growth during which solute rejection and recalescence occurred until the remaining liquid reached a temperature and composition below the glass transition temperature, T_g (see Fig. 13 of Ref. 13).

TABLE 1

Comparison of undercoolings obtained in Al-Ge alloys in previous studies

Material	Transformation	Technique (Droplet size, μm)	T_l (°K)	ΔT (°K)	$\Delta T/T_l$	Ref.
pure Al	liquid→solid	droplets (10-20 μm)	933	130	0.139	7
pure Ge	liquid→solid	droplets (10-20 μm)	1232	227	0.184	7
Al-20Ge	liquid→solid	EHD powders (<1 μm)	783	225	0.287	13
Al-30Ge	liquid→solid	EHD powders (<1 μm)	696	173	0.249	13
Al-36Ge	liquid→solid	EHD powders (<1 μm)	773	266	0.344	13
solid Ge/ liquid Al	semisolid→solid	droplets (10-30 μm)	726	88	0.121	14
solid Ge/ liquid Al	semisolid→solid	droplets (10-30 μm)	?	120	?	15
solid Ge/ liquid Al	semisolid→solid	EHD powders (<1 μm)	754	166	0.220	13
solid Al/ liquid Ge	semisolid→solid	droplets[a] (10-30 μm)	873	291	0.333	
solid Al/ liquid Ge	semisolid→solid	EHD powders (<1 μm)	908	345	0.380	13

[a] Southin and Chadwick (14) used the technique of melting and undercooling droplets within a solid matrix.

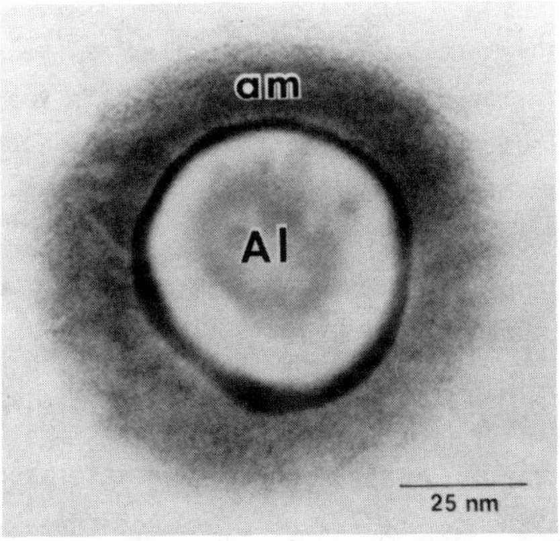

Figure 1. Bright field electron micrograph of a small Al-Ge eutectic powder which was quenched rapidly in the microscope(13) to form the single crystal/amorphous structure.

Although these ideas seemed to account for the formation of the single crystal/amorphous microstructures, no attempt was made to explain the observed transition, with increasing undercooling, from many crystals (homogeneous or volume-dependent nucleation) to one. The purpose of the present study is to address this transition directly. The approach taken is to calculate the number of Al grains which form in highly undercooled Al-30Ge submicron powders using the nucleation and growth equations believed to be applicable to glass forming systems.

Approach

Before discussing the calculation in any detail, it is useful to describe initially the nucleation and growth equations which will be used and to develop the ideas concerning the solid-liquid interfacial temperature gradient.

Homogeneous Nucleation

Because of the observed microstructures, only homogeneous nucleation will be considered. Following (16,17), the nucleation equation can be written as:

$$I_v = \frac{10^{36}}{\eta} \exp \left\{\frac{-16\pi}{3RT} \frac{\sigma^3}{\Delta G_v{}^2}\right\} \qquad (1)$$

where I_v = steady state nucleation rate ($m^{-3}s^{-1}$),

η = viscosity of the undercooled liquid (poise)

σ = nucleus: liquid interfacial energy

ΔG_v = molar Gibbs free energy difference between the melt and crystal

R = gas constant

T = absolute temperature

Analysis of Eqn. 1 reveals the well-known critical dependence of the nucleation rate on ΔG_v and σ since these quantities appear in the exponential term raised to the power of two and three, respectively. Even though there have been many attempts to estimate these quantities theoretically, the lack of experimental data, especially for highly undercooled liquids, has made it difficult, if not impossible, to utilize this equation for making quantitative predictions. Even so, the behavior can be estimated at least qualitatively using this equation and, therefore, it is useful for making approximate predictions.

In addition to these unknowns, the value of the viscosity, η, is not known for highly undercooled liquids. For the present analysis, η will be estimated from the Vogel (18) and Fulcher (19) relationship for glass forming alloys which was later modified by Turnbull (17) as:

$$\eta = 10^{-3.3} \exp \{3.34\ T_l (T - T_b)\} \tag{2}$$

where T_l is the liquidus temperature and T_b is a temperature slightly below T_g. When T_b equals absolute zero, Eqn. 2 reduces to the simple Arrhenius relationship commonly accepted for liquid behavior. However, when the value of T_b is greater than 0°K and is approached by an undercooled liquid, the viscosity behaves as depicted schematically in Fig. 2. It can be seen that, as the liquid temperature approaches T_g, the liquid viscosity is predicted to increase rapidly.

In the present analysis the value of ΔG_v is estimated using the equation derived by Thompson and Spaepen (20) who approximated ΔG_v in a binary alloy as a function of composition by assuming regular solution behavior for the liquid and ideal or dilute solution behavior for the solid. In the limiting case of nucleation of essentially pure A in a binary A-B alloy, the equation for ΔG_v can be written:

$$\Delta G_v = T_l - T)(\Delta S_f^A - R \ln X_L^A) \tag{3}$$

where ΔS_f is the molar entropy of fusion of element A and X_L^A is the mole fraction of A in the liquid. In support of their analysis, the values of ΔG_v determined using Eqn. 3 were shown to be in reasonable agreement with those estimated from the results of undercooling studies by both Perepezko, et al. (2) and Cech and Turnbull (21).

The value of σ will be estimated from the equation put forward by Spaepen and coworkers (22,23) who considered σ to be primarily of entropic origin and, therefore, a function of temperature. The relationship developed can be expressed as:

$$\sigma = \frac{\Delta S_f \alpha_m}{(NV^2)^{1/3}} T \tag{4}$$

where N is Avogadro's number, V is the molar volume and α_m is a constant.

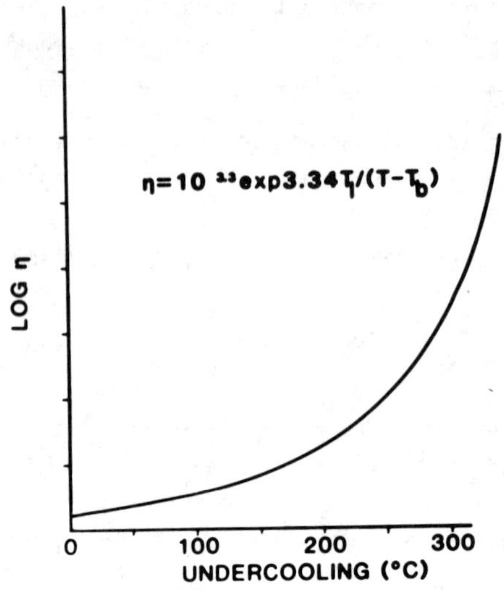

Figure 2. Schematic representation of the viscosity-undercooling relationship calculated using a modified version of the Vogel-Fulcher equation.

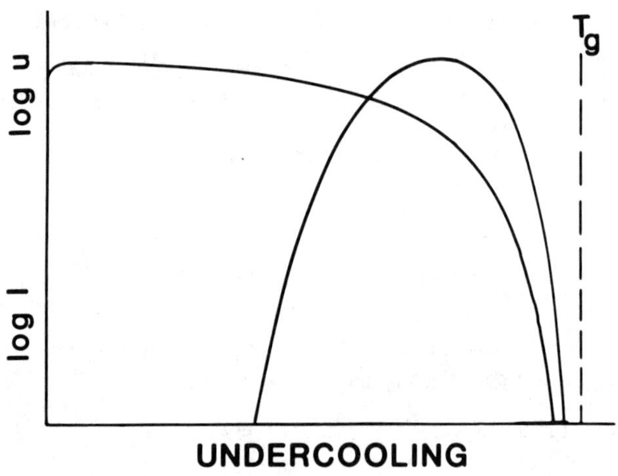

Figure 3. Nucleation and growth functions vs. undercooling calculated using Eqns. 1 and 5 with T_b = 300°C and α_m = 0.75.

In their analysis, Spaepen and Meyer (23) modeled the solid-liquid interface to estimate α_m, and concluded that for close-packed surfaces, α_m = 0.86 for fcc and hcp structures and 0.71 for bcc structures. In addition, other authors (e.g. 17) proposed that α_m = 0.6 at the onset of measurable nucleation. From Eqn. 4 it is evident that, for alloys, the value of σ is dependant upon composition, since both V and ΔS_f are functions of alloy composition, i.e. for linear dependence in a binary alloy, $V = X_A V_A + X_B V_B$ and $\Delta S_f = X_A\ \Delta S_f^A + X_B\ \Delta S_f^B$.

Some researchers (e.g. 8) have disagreed with the temperature dependence of σ, based on their own experimental results, and concluded that this term is essentially constant and equal to the value determined at the equilibrium melting temperature. Consequently, the actual value of σ is still a source of both uncertainty and controversy in classical nucleation theory. It should be mentioned that, for very small nuclei (< 10 nm), σ may also decrease with nucleus size (24,25) leading to further uncertainties in the nucleation equation.

Crystal Growth

There have been numerous theories (e.g. 26,27) describing crystal growth rate as a function of driving forces, whereas experimental measurements have been extremely limited (28). For the case of rapid solidification experiments, it is generally believed that the solid-liquid interface kinetics occur in the continuous growth regime (29), as opposed to the linear or transitional regimes and, as a result, the present study will deal solely with continuous growth kinetics. The equation for growth velocity, u, can be written (30):

$$u = \frac{\beta k T}{3\pi a^2 \eta} \{1 - \exp(\frac{\Delta G_v}{RT})\} \qquad (5)$$

where β is the correction factor proposed by Cahn et al. (26) and a is the step height which is typically of the order of an interatomic distance. For metals β is usually between 1 and 100. Again, the value of η is estimated from Eqn. 2.

The use of Eqn. 5, which only considers interface attachment kinetics, in the present consideration of dendritic solidification of an alloy may not be valid because no account is taken for the effect of solute build-up at the advancing solid-liquid interface. Unfortunately however, the available treatments of dendritic growth into undercooled alloy melts are limited to very small undercoolings (31,32), where interface attachment kinetics are not important and, therefore, do not account for the behavior of alloys at the large undercoolings characteristic of rapidly solidified alloys near their glass transition temperatures. As a result, the present qualitative model will use Eqn. 5 to estimate the growth velocity as a function of undercooling, ignoring the contribution of solute build-up. The possible effects of solute build-up on the calculations will be addressed in the discussion of the limitations of the present approach.

Using the above equations, it is useful to plot the nucleation and growth functions as a function of undercooling (Fig. 3). From this figure, it is apparent that, for Al-30Ge and other glass forming systems, the growth velocity reaches a maximum at relatively low undercoolings and decreases rapidly as T_g is approached. Similarly, the nucleation rate rises rapidly at the homogeneous nucleation temperature and falls again as T_g is approached.

Interface Temperature Gradient

In order to develop the ideas concerning the interfacial temperature gradient, it is useful to contrast the behavior expected for solidification of highly undercooled glass forming droplets with that expected for pure (non-glass forming) metal droplets. To do this, we adopt the ideas proposed by Field et al. (33) to account for the microstructural development of liquid metals undergoing atomization. For example, the thermal history of a liquid droplet undergoing atomization and rapid solidification can be described effectively on a temperature-enthalpy diagram (29). For large initial undercoolings the solidification history may be described as follows. The liquid droplet undercools below the equilibrium melting temperature, T_m, until a nucleus forms at T_n (Fig. 4a). This nucleus begins to grow, releasing latent heat in the process. However, the rate of recalescence is initially low because, not only is the undercooled liquid relatively cold causing sluggish diffusional processes at the crystal/liquid interface, but also, at these levels of undercooling, the critical nucleus size is expected to be very small resulting in relatively slow overall crystallization rates. Consequently, the droplet continues to undercool and, if the nucleation rate is high, as expected for homogeneous nucleation conditions, many nuclei should form. As the overall solidification rate increases the droplet will recalesce because the latent heat of fusion being released cannot be compensated for by the heat

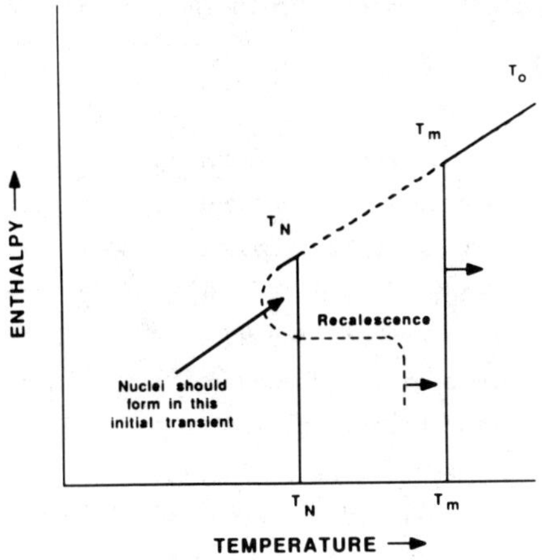

a

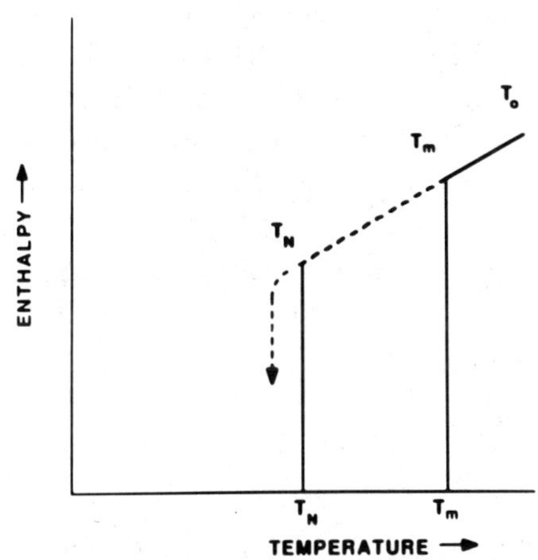

b

Figure 4. Schematic enthalpy diagrams displaying expected behavior for (a) non-glass forming alloys and (b) glass forming alloys.

extraction to the surroundings. If and when the recalescence temperature increases to values where the nucleation frequency becomes negligible, i.e. above T_n (Fig. 4a), then the number of crystals should remain constant (or decrease due to coarsening) throughout the duration of the solidification process.

The other possibility, especially relevant to glass forming alloys, is that once nucleation has occurred at some temperature below T_n, the temperature of the droplet remains constant or continues to decrease because the crystal growth velocity remains low due to the viscous nature of such undercooled liquids (Fig. 4b). Thus, if the droplet continues to undercool then the temperature will eventually fall below T_g causing the remaining liquid to form a glass. However, if the latent heat released by the growing crystal during solidification is considered, it seems likely that the local temperature near the solid/liquid interface will be greater than that of the remaining liquid, especially if the rate of external heat extraction is fairly low due to a small temperature difference between the droplet and the environment. This is depicted schematically in Fig. 5. Similar temperature gradients were predicted to develop during the solidification of undercooled liquid droplets during atomization (29), even for Biot numbers as low as 10^{-4} where Newtonian conditions (no temperature gradients) normally would apply. If such temperature gradients do exist, it is conceivable that significant crystal growth could occur even after the remaining liquid reaches a temperature where the nucleation rate becomes negligible. As a result, the temperature which will be used in the growth equation should be higher than that used in the nucleation equation. This temperature difference, ΔT, will be used as a variable in the calculations below. The effects of various values of ΔT on the growth function are illustrated in Fig. 6.

Calculations

In the present treatment, the following simplifying assumptions are made. Firstly, homogeneous nucleation of essentially pure α-Al in a liquid of eutectic composition (Al-30Ge) is considered. Thus, the value of ΔG_v required in both the nucleation and growth equations can be approximated using values of ΔS_f for pure Al. Secondly, the interfacial energy term is calculated from Eqn. 3 for pure Al using values of α_m between 0.6 and 0.86. Finally, the melt viscosity, η, is estimated from Eqn. 2.

The calculations were made using the qualitative ideas described

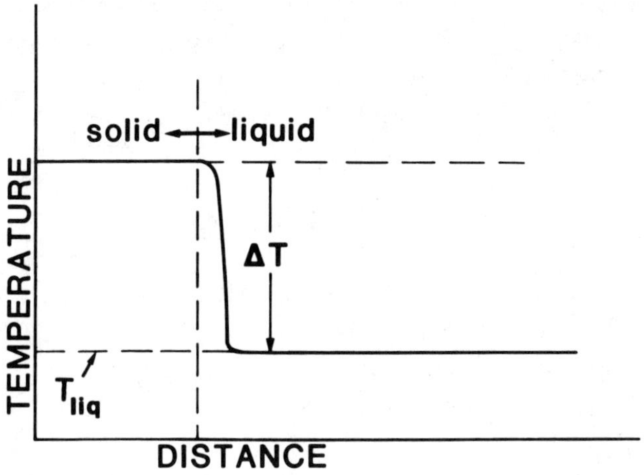

Figure 5. Schematic illustration of the temperature gradient, ΔT, developed during the solidification of highly undercooled eutectic droplets.

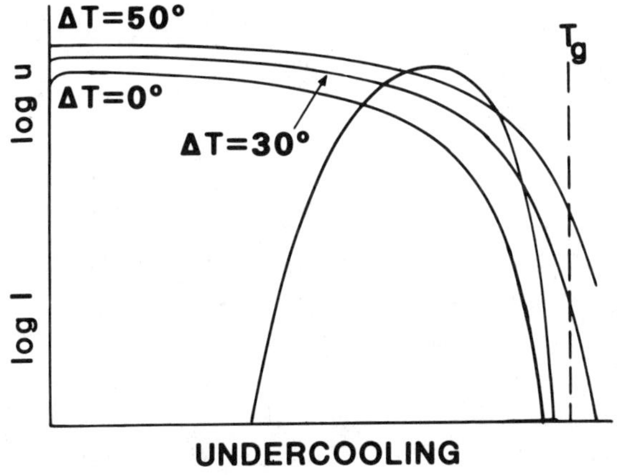

Figure 6. Calculated nucleation and growth curves showing the effect of increasing the temperture in the growth equation by 30°C and 50°C.

above. Specifically, the number of α-Al crystals which form in eutectic powders as a function of the important processing variables is calculated. In the calculation, the nucleation equation is integrated at different time intervals as a function of undercooling as follows. In every time interval, Δt, nuclei form according to:

$$J_{k+1} = J_k + I_{k+1}\ \Delta t \qquad (6)$$

where J_{k+1} = the integrated number of nuclei from time, t = 0, to the current time step k+1

J_k = same at time step k

J_{k+1} = nucleation rate at time step k.

Whenever J_{k+1} reaches a value greater than one then the number of growing nuclei increases by one.

The growth equation used is a modified version of Eq. 5 and includes a term for the effect of interfacial energy. In addition, the temperature used in this equation is modified when ΔT > 0. This equation may be written:

$$u = \frac{\beta RT}{3\pi a_2 \eta} \{1 - \exp \frac{\Delta G_v + \Delta G_{surf}}{RT}\} \qquad (7)$$

where ΔG_v has the same meaning as before and ΔG_{surf} is the surface energy contribution to the driving free energy. This latter term may be expressed as:

$$\Delta G_{surf} = 2\sigma/r \qquad (8)$$

where r is the radius of the growing crystal. It is evident that ΔG_{surf} is positive and opposes the growth of spherical nuclei (it is assumed that the α-Al crystals are spherical, an assumption which should be valid for small sizes based on previous in-situ solidification (13) and amorphous film crystallization observations (34)). It can be seen from the latter equation that crystals of different sizes will have correspondingly different growth rates.

As in the nucleation equation, the growth of the nuclei are calculated at time intervals, Δt, as follows. For a given particle i, the radius changes according to:

$$r_{i,k+1} = r_{i,k} + u_i\ \Delta t \qquad (9)$$

where $r_{i,k+1}$ = radius of the ith crystal at time interval k+1

$r_{i,k}$ = same at time interval k

u_i = growth velocity of ith crystal.

As a result, the growth of each crystal is considered throughout the solidification process.

The solidification histories of submicron eutectic droplets are modeled in an attempt to simulate the undercooling experiments described in Ref 13. Specifically, the model assumes that the liquid droplets are quenched very rapidly to some preselected substrate temperature (Fig. 7) and the solidification histories of such droplets are calculated at this temperature using the equations described above. It is assumed that the solidification process is essentially isothermal (negligible overall recalescence of the droplet) and that the temperature at the solid-liquid interface deviates from the pre-selected temperature by a constant amount during solidification (Fig. 5). For the various temperatures under consideration, the time is incremented, and the nucleation and growth functions integrated, until either solidification is complete (i.e. when the fraction of Al is exhausted) or glass formation occurs (i.e. when the nucleation and growth rates become negligible prior to completion of the process).

Results and Discussion

A typical plot of the number of crystals produced as a function of undercooling is shown in Fig. 8. From this figure it is evident that the present model predicts that the number of crystals should increase as the undercooling is raised until a temperature is reached where a rather sudden decrease in this number occurs. This prediction is consistent with the present set of experimental observations of the transition from a multi-crystalline microstructure to one containing only one or two crystals. The location and magnitude of this curve was found to be very sensitive to the various experimental parameters, especially σ (Fig. 9) and T_b (Fig. 10). Also, the value chosen for the temperature increment, ΔT, at the solid-liquid interface has considerable effects on the predictions (Fig. 11); from this figure, it appears that the temperature increment at the solid-liquid interface would not be required to obtain agreement for the observed transition from the multicrystalline structure to one containing only one grain. However, if no interface gradient is imposed then the volume

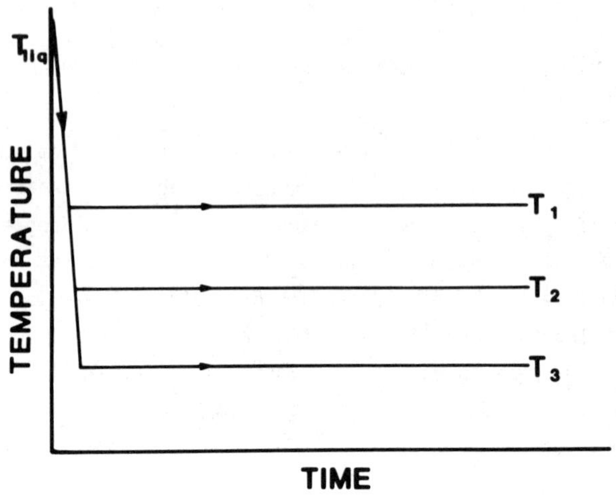

Figure 7. Schematic illustration of the model used to simulate the in-situ undercooling experiments.

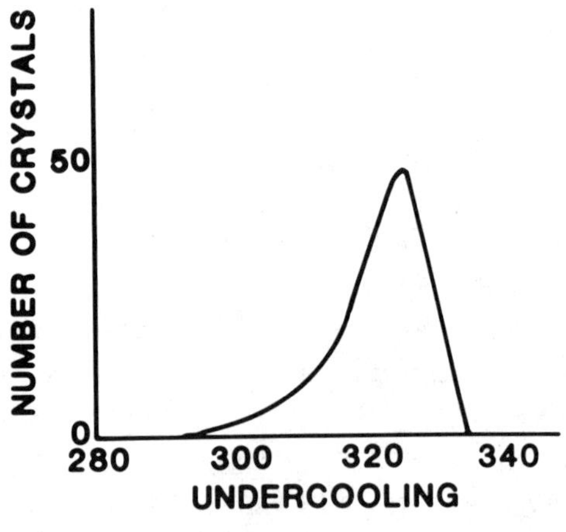

Figure 8. Typical results of model showing how the number of crystals is expected to vary with undercooling. (α_m = 0.75, T_b = 300°C, T = 30°C).

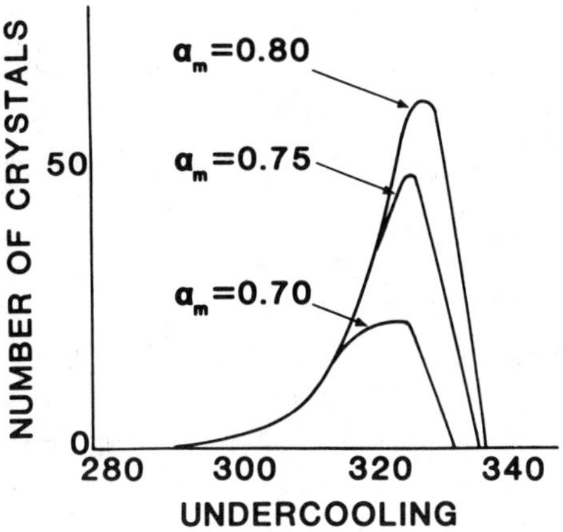

Figure 9. Calculated curves obtained from the present model demonstrating the effects of changing the value of α_m. (T_b = 300°C, ΔT = 30°C).

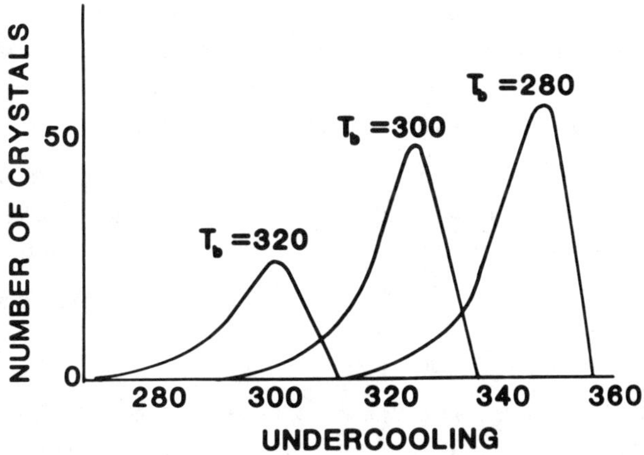

Figure 10. Calculated curves obtained from the present model demonstrating the effects of changing the value of T_b. (α_m = 0.75, ΔT = 30°C).

fraction of solid also approaches zero at this transition temperature, which is not consistent with experiment. On the other hand, if this gradient is chosen to be too large, the model predicts that the multicrystalline structures would never be produced, i.e. the peak of the curve in Fig. 11 would never reach a level commensurate with the observations. Thus, this gradient is required to predict the observed behavior if, indeed, the nucleation and growth equations used are applicable to this system.

When the results in Figs. 8-11 are compared with the experimentally determined undercoolings (13), it becomes clear that there exists substantial discrepancies in the model predictions. This disagreement is not unexpected in view of the uncertainties in the values of the input parameters, the nature of the simplifying assumptions (addressed below) and the very approximate nature of the nucleation and growth equations. However, the qualitative agreement is encouraging and, since the predictions are very sensitive to only small changes in the input parameters, it seems plausible that only minor alterations in the model would be required to obtain a more reasonable fit with the data.

Various factors which were not taken into consideration but are expected to play a role in the predictions are considered below. Firstly, as solidification of the Al-rich crystal(s) progresses, there will be a substantial amount of solute rejection which will lead to an increase in the Ge concentration in the liquid at the solid-liquid interface. As a result, the value of ΔG_v, as calculated using Eqn. 3, will not be constant as assumed, but rather will decrease leading to a lowering of both the nucleation and growth rates of the α-Al crystals.

Secondly, because there is a substantial amount of Al in the amorphous phase that forms in the single crystal/amorphous powders (13), the total quantity of Al in the crystalline portion will be lower than that in the multicrystalline structures where $X_{Al} = 0.7$. This factor will cause the calculations to reach completion earlier giving rise to fewer predicted crystals at the higher undercoolings.

A third factor which was not considered is the effect of recalescence. As has been pointed out previously (35), the initial solidification of highly undercooled alloy powders occurs adiabatically followed by solidification dominated by external heat extraction. However, as pointed out above, the overall amount of recalescence which occurs during solidification of highly undercooled droplets of glass forming alloys is expected to be limited, except possibly at the solid-liquid interface.

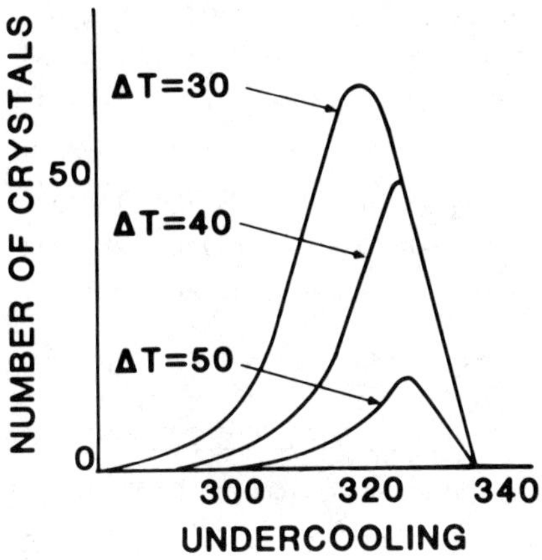

Figure 11. Calculated curves obtained from the present model demonstrating the effects of changing the value of ΔT. (α_m = 0.75, T_b = 300°C).

Finally, it is useful to discuss the relationship of the present results to other glass forming alloys. For example, it is fairly well-established that the characteristic features of glass forming alloys are their high resistance to nucleation and/or growth such that, if the rate of heat extraction from the liquid or vapor prior to reaching T_g is high enough, a glass should form. However, a major difference between Al-Ge alloys and many other glass forming alloys is the relatively low T_g of the former ($T_g \approx$ room temperature). Since the rate of heat extraction depends on the temperature difference between the liquid and the substrate (T_o) it is clear that as these powders approach T_g their cooling rates will decrease accordingly. Therefore, it seems likely that an interfacial temperature gradient in this situation will be much more stable than that which might develop in alloys where there is a greater difference between T_g and T_o and correspondingly higher heat extraction rates as the liquid temperature approaches T_g.

Summary

It has been shown that during growth of solid particles in highly undercooled liquid a solid-liquid interfacial temperature gradient may exist. Such a temperature gradient implies that growth of the solid phase occurs at a slightly higher temperature than the nucleation of new solid in the undercooled liquid. The effect of this temperature gradient, especially in the case of an eutectic alloy taken from the Al-Ge system, is to cause a change from many nuclei to very few when the degree of undercooling prior to nucleation is increased.

Acknowledgements

The authors are grateful to Drs. R.D. Field and C.G. Levi for stimulating discussions. This work was supported by the Division of Materials Science, U.S. Department of Energy under contract #DE-AC02-76ER01198.

References

1. J.H. Perepezko, D.H. Rasmussen, I.E. Anderson and C.R. Loper, Jr., Sheffield International Conf. on "Solidification and Casting of Metals", (Metals Society, London, 1979) p. 169-174.

2. J.H. Perepezko, Proc. 2nd Int. Conf. on "Rapid Solidification Processing, Principles and Technologies", R. Mehrabian, B.H. Kear and M. Cohen, eds., (Claitor's, Baton Rouge, 1980) p. 56-67.

3. J.H. Perepezko, Y. Shiohara, J.S. Paik and M.C. Flemings, Proc. 3rd Int. Conf. on "Rapid Solidification Processing, Principles and and Technologies", R. Mehrabian, ed., (National Bureau of Standards, Gaithersburg, MD, USA, 1982) p. 28-44.

4. J.H. Perepezko and J.S. Paik, "Rapidly Solidified Amorphous and Crystalline Alloys", B.H. Kear, B.C. Giessen and M. Cohen, eds., (Elsevier, Amsterdam, 1982), p. 49-63.

5. D. Turnbull and J.C. Fisher, J. Chem. Phys., 17 (1949) 71.

6. D. Turnbull, J. Chem. Phys., 18 (1950) 198.

7. D. Turnbull and R.E. Cech, J. Appl. Phys. 21 (1950) 804.

8. D. Turnbull, J. Appl. Phys., 21 (1950) 1022-1028.

9. D. Turnbull, J. Chem. Phys., 18 (1950) 18.

10. D. Turnbull, J. Chem. Phys., 20 (1952) 411.

11. G.M. Pound and V.K. LaMer, J. Am. Chem. Soc., 74 (1942) 2323.

12. Y. Miyazawa and G.M. Pound, J. Crystal Growth, 23 (1974) 45.

13. M.J. Kaufman and H.L. Fraser, Int. J. Rapid Solidification, 1 (1984-5) 27-50.

14. R.T. Southin and G.A. Chadwick, Acta Met., 26 (1978) 223-231.

15. R.S. Wagner, in Metallurgy of Advanced Electronic Materials, AIME Press, 19 (1963) 239-248.

16. D. Turnbull, Sol. St. Phys., 3, F. Seitz and D. Turnbull, eds., (Academic Press, New York, 1976) p. 205.

17. F. Spaepen and D. Turnbull, "Proc. 2nd Int. Conf. on Rapidly Quenched Metals", N.J. Grant and B.C. Giessen, eds., (MIT Press, Cambridge, MA, 1976) p. 205.

18. H. Vogel, Phys. A, 22 (1921) 645.

19. G.S. Fulcher, J. Am. Ceram. Soc., 6 (1925) 339.

20. C.V. Thompson and F. Spaepen, Acta Met., 31 (1983) 2021.

21. R.E. Cech and D. Turnbull, J. Metals, 191 (1951) 242.

22. F. Spaepen, Acta Met., 23, (1975) 729.

23. F. Spaepen and R.B. Meyer, Scripta Met., 10 (1976) 257.

24. R.C. Tolman, J. Chem. Phys., 17 (1949) 333.

25. G.C. Benson and R. Shuttleworth, J. Chem. Phys., 19 (1951) 130.

26. J.W. Cahn, W.B. Hillig and G.W. Sears, Acta Met., 12 (1964) 1421.

27. K.A. Jackson, D.R. Uhlmann and J.D. Hunt, J. Crystal Growth, 1 (1967) 1.

28. G.J. Abbaschian and S.F. Ravits, J. Crystal Growth, 28 (1975) 16.

29. C.G. Levi and R. Mehrabian, Metall. Trans., 13A (1982) 221.

30. M.J. Kaufman, PhD Thesis, University of Illinois, 1984.

31. J. Lipton, M.E. Glicksman and W. Kurz, Mat. Sci. Eng., 65 (1984) 45.

32. M.E. Glicksman, Mat. Sci. Eng., 65 (1984) 45.

33. R.D. Field, C.G. Levi and H.L. Fraser, unpublished results.

34. M.J. Kaufman and H.L. Fraser, Acta Met., 33 (1985) 191-203.

35. W.J. Boettinger and J.H. Perepzko, Proc. Symp. on "Rapidly Solidified Crystalline Alloys", N.E Regional meeting of TMS-AIME, Morristown, 1985.

UNDERCOOLING EFFECTS IN GAS ATOMIZED POWDERS

I. E. Anderson and M. P. Kemppainen*

Physical Metallurgy Branch
Naval Research Laboratory
Washington, DC 20375-5000
USA

Abstract

The microstructure of gas-atomized Cu alloy powders has been found to be a strong function of particle size, especially in ultrafine (<10μm) particles, where undercooling rather than cooling rate appears to control the solidification morphology. Microstructural analysis using SEM of sectioned particle size fractions of both Ar-atomized and He-atomized Cu-8.3 wt.% Al powders revealed a transition at about 10μm in the predominant morphology from cellular-like in larger particles to apparently featureless in smaller. X-ray diffraction measurements of the relative product phases correspond well to the microscopy results and reveal a β'_1 martensite as the preferred product in <10μm powders. An Ar-atomized Cu-6.0 wt.% Si alloy powder was also studied as a function of particle size with SEM and X-ray diffraction techniques and a similar microstructure transition was observed, with β(hcp) phase as the preferred product in <10μm powders. Differential thermal analysis permitted undercooling measurements of particle size fractions in the range of 10μm and detected the trend for increased undercooling of particles <10μm in both of the Cu alloy powders. The solidification reactions of these two alloys at high undercooling are discussed in terms of metastable phase relationships and of the mote isolation effect in highly dispersed melts.

* Now Research Assistant, Michigan Technological University, Houghton, MI 49931.

Introduction

The effective utilization of rapid solidification processing (RSP) for advanced powder metallurgy applications requires, in part, a supply of powder with uniform, optimal, rapidly-solidified particulate microstructures. Such a powder would be likely to exhibit uniform deformation characteristics, albeit at high flow stresses, desirable for effective consolidation processing. Gas atomization is already a well established commercial technique for metal and alloy powder production and shows promise also for optimized RSP powder production.

In studies of gas-atomized particle microstructures, the tendency for particle size dependent changes in solidification morphology has been observed often (1,2,3,4). A reduced dendrite arm spacing (DAS) in gas-atomized particles of a reduced diameter (1,2,3) has been explained as a particle size dependent cooling rate effect, in a direct extension of the chill rate dependence of cast microstructures (2). However, the appearance of non-dendritic microstructures (1,3) and the insensitivity of DAS results to gas thermal conductivity changes, especially in powders smaller than 75μm (3), could not be explained by cooling rate effects alone. In fact, some recent work (4) has shown that more dramatic microstructure modifications occur in atomized particles smaller than about 10μm, attributed to undercooling effects. The ultrafine powders, less than 10μm in diameter, contain a highly refined "microcellular" morphology with intercellular spacings of about 0.05μm (4). Results of metal droplet emulsion experiments demonstrate that high liquid undercoolings are achieved most readily in ultrafine metal droplet samples (5). Indeed, many examples of rapidly solidified particle microstructures have been generated at high (.2T_m to .4T_m) undercoolings and at very low (10-20°C/min.) cooling rates (5).

In general, a mean droplet size of 10 μm or less is required to realize the maximum reported undercoolings (5). Since typical commercial gas atomized powders have a mean size of about 40-70 μm (6), a project was initiated to develop a modified gas atomization process to produce high yields of ultrafine (<10 μm) powders to attempt to promote high undercooling in most of the powder yield.

Atomization of Powders

Studies of gas atomization have revealed the increased particle size refinement capacity of atomizers operated at moderately elevated gas pressures (7). To probe the ultimate extent of this effect the gas inlet pressure design limit for our atomizer was set as high as possible. While the system is capable of a 28 MPa inlet gas pressue, the maximum operating pressure thus far has been 17 MPa (8). These values compare to the highest commercial gas atomization pressures of about 5-8 MPa (9) which is also the normal operating range of the gas nozzle used in our system, an "ultrasonic" type (10). The initial studies of high pressure gas atomization (HPGA), chiefly with Ar gas, demonstrated a significant increase in particle size refinement with increased gas pressure and a previously unreported reversal of the size refinement trend above a critical pressure level (8). For Ar gas, the critical inlet pressure level for maximum size refinement was established at about 12 MPa (8).

The Cu-8.3 wt.% Al and Cu-6.0 wt.% Si alloys used in this study were atomized with Ar at 12 MPa inlet pressure. An additional sample of the Cu-8.3 wt.% Al alloy was atomized with He at 14 MPa. The resulting powder samples contained a large fraction of ultrafine particles, as shown in Figure 1, which shows the particle size distribution of the Ar-atomized Cu-8.3 Al powder, for example. The mean particle diameter of the Cu alloy powder samples ranged from 10-15 μm. The HPGA powders are much finer than several other Cu alloy powders prepared with a similar gas atomization nozzle at lower pressures which had mean diameters from 60-100 μm (8). As Figure 1 illustrates, a powder sample contains a range of particle sizes. Thus, microsieving and other powder sizing techniques permitted each sample to be separated into size fractions to facilitate characterization as a function of particle size.

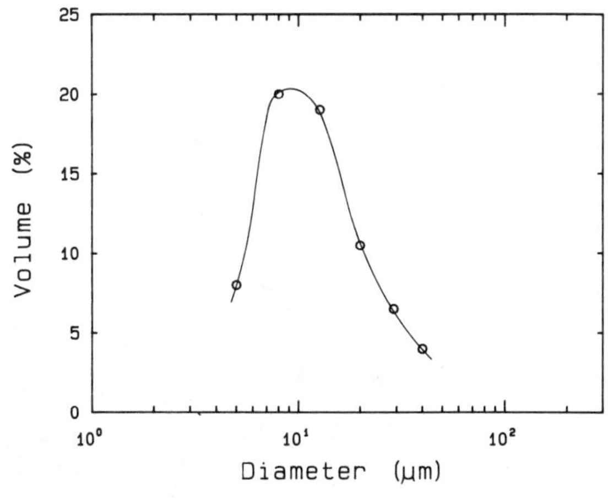

Figure 1 - Particle size distribution of Ar-atomized Cu-8.3 wt.% Al powder. Data from electrozone counter analysis of powders, <62μm corrected with data from sieve analysis of the entire yield.

Particulate Characterization

Cu-Al Results

X-ray diffraction measurements of the powder samples were performed on particle size fractions from the as-atomized powders to identify the product phases. Indexing of the diffraction patterns revealed that the Cu-8.3 wt.% Al alloy powders contained a mixture of two phases; the α-Cu (fcc) primary solid solution and a martensite phase designated β'_1 (11). Since β'_1 martensite forms on cooling of a β(bcc) parent phase (11), β'_1 x-ray reflections are considered evidence of initial β-phase solidification. Equivalent supersaturation of α was detected in both Ar and He atomized powders from all size fractions. The error bar in Figure 2 encompasses the a_0 lattice parameter values obtained from linear least squares analysis of the X-ray data from Ar-atomized powders sieved to 62-30 μm, 30-25 μm, 25-20 μm, 20-15 μm, 15-10 μm, and <10 μm size fractions. When the initial alloy composition is chosen to establish the coordinates of the point on the lattice parameter vs. composition dependence (12) in Figure 2, a good fit is obtained. A nearly constant minor fraction of β'_1 is observed in both Ar-atomized and He-atomized powders larger than about 10 μm, as given in Figure 3. However, the peak intensity ratio results in Figure 3 reveal a significant increase in β'_1 in the 10-5 μm and the <5 μm size fractions of powders atomized with either Ar or He.

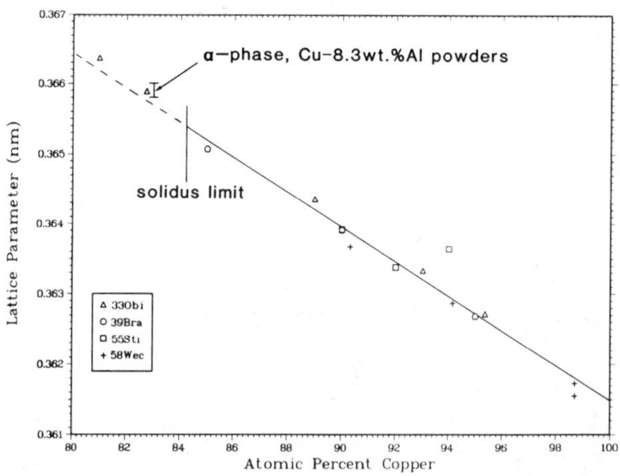

Figure 2 - Data bar represents summary of "a" lattice parameter results from all powder size fractions of the Ar-atomized Cu-8.3 wt.% Al powder plotted with previous data (12). Values obtained from linear least squares analysis using the Nelson-Riley function for α(fcc) diffractometer reflections of Cu $K\alpha_1$ radiation. Average standard error of each calculated lattice parameter is 0.0006A.

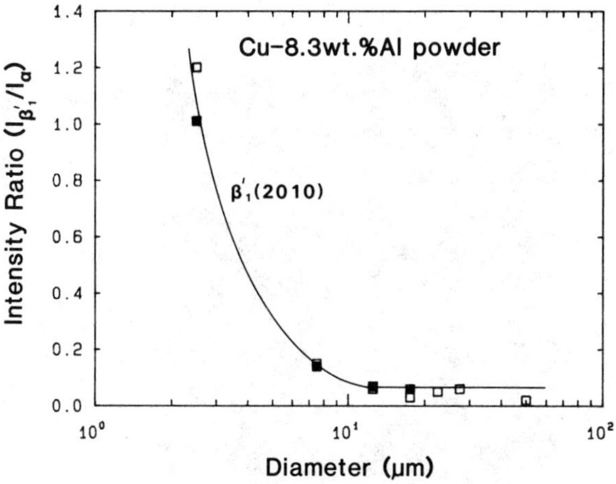

Figure 3 - Measurements of Bragg peak ratio of β'_1 (2010)/α(200) for various size fractions of Cu-8.3 wt.% Al powders. Open squares are Ar-atomized data and solid squares are He-atomized data. Each particle diameter value represents the midpoint of a size fraction.

Microstructural analysis of size-fractioned samples of the Cu-Al powders was performed on sectioned and etched particles with SEM to observe typical particle solidification morphologies. Microscopy revealed a cellular-like morphology in most particles larger than about 10 μm, as given in Figure 4a showing 20-15 μm Ar-atomized powders and in Figure 5a showing 20-15 μm He-atomized powders. The observation of uniformly etched particles apparently of a single phase, was much more common in powders smaller than 10 μm, as shown in Figures 4b and 5b.

Differential thermal analysis (DTA) permitted assessment of the undercooling potential of gas atomized particles that are remelted and slowly cooled at about 20°/min to complete solidification. The powder samples are prepared for DTA by suspending dispersed alloy particles, of a given size fraction, in a thick ethanol slurry loaded with .05 μm Al_2O_3 powder. A dried slurry sample contains about 10% by volume of Cu alloy powder particles in an alumina powder matrix to minimize alloy droplet coalescence or interaction and thereby, to permit observation of independent particle melting and freezing events throughout the sample. A summary of the DTA results for the indicated size fractions of Ar-atomized Cu-Al powder is given in Figure 6. The He-atomized powder was not studied with DTA because no difference in undercooling behavior would be expected for the same size powders with the same composition, assuming the same particle surface condition. The top thermogram in Figure 6 represents the typical heating and melting endotherm

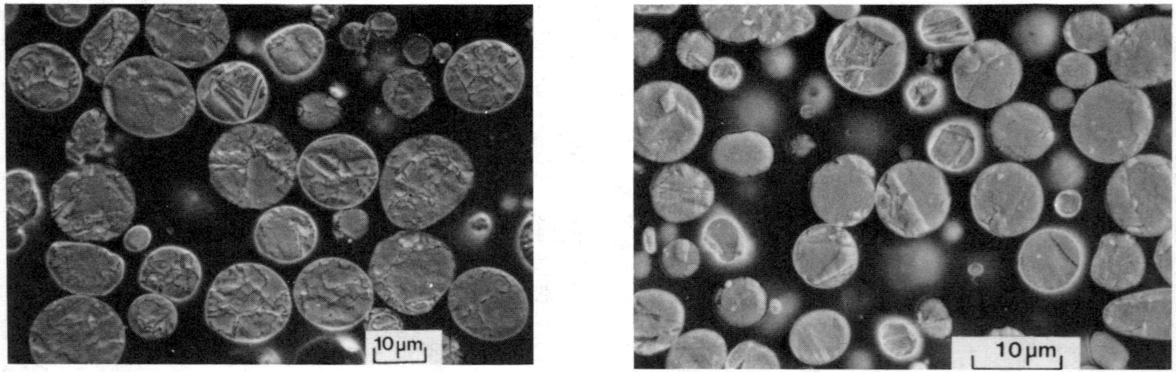

a) 20-15 μm

b) <10 μm

Figure 4 - SEM micrographs of size fractioned Ar-atomized Cu-8.3 wt.% Al powder. Particles, embedded in epoxy, were metallographically polished and etched with aqueous solutions of potassium dichromate/sulfuric acid/hydrochloric acid and of ferric chloride/hydrochloric acid, as recommended in ref. (19).

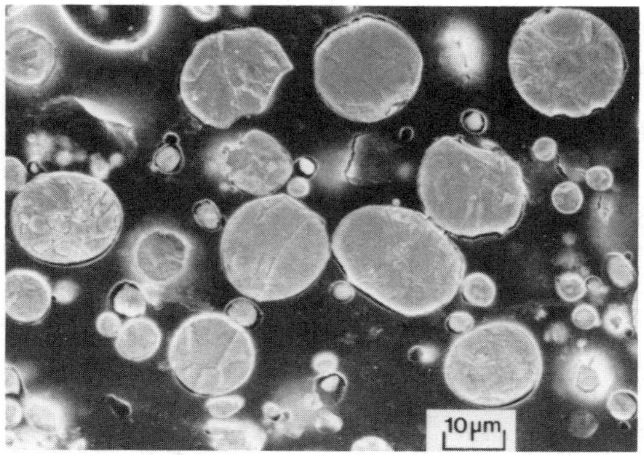

a) 20-15 μm

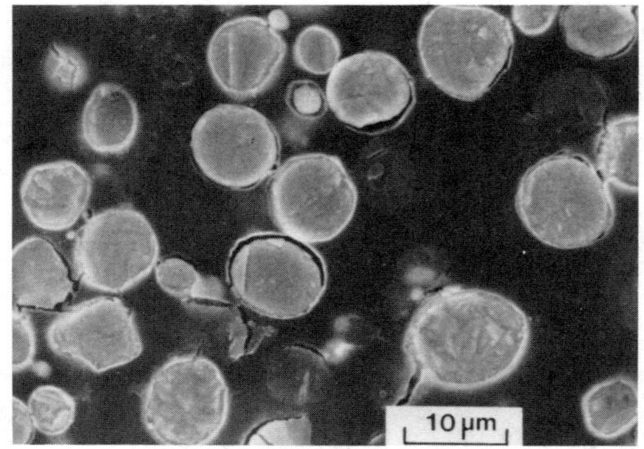

b) <10 μm

Figure 5 - SEM micrographs of size fractioned, He-atomized Cu-8.3 wt.% Al powder. Sectioned particle microstructures prepared as in Figure 4.

of a dried slurry sample of this alloy powder. The other three traces in Figure 6 show sample cooling and particle solidification exotherms along with dashed lines to indicate the approximate thermogram baseline of each. A major exotherm and a rather minor exotherm appear during cooling of the 15-20 μm sample in Figure 6, which indicates that the majority of the particles in the sample experience only a minimal undercooling prior to solidification. However, the samples of smaller size fractions exhibit an increased exothermic peak area at lower temperatures indicating that a greater population of these ultrafine alloy particles achieved high undercoolings before solidification.

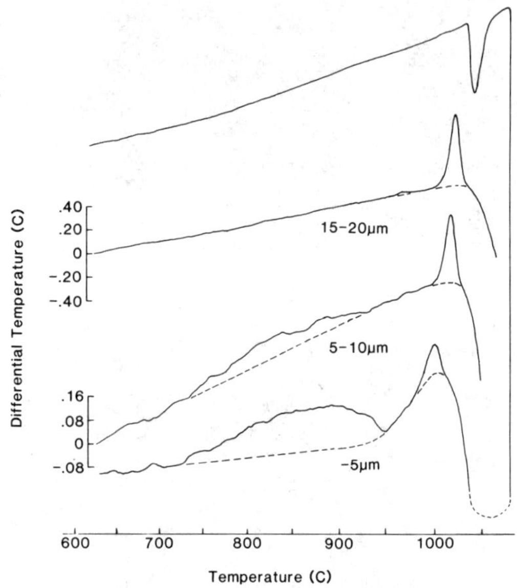

Figure 6 - Summary of DTA results for size fractioned, Ar-atomized Cu-8.3 wt.% Al powders, dispersed in alumina powders, with heating and cooling rates of 20°C·sec^{-1}. Differential temperature sensitivity of bottom two cooling thermograms is higher than the other, as indicated. Eutectic temperature measured at endotherm onset in top heating thermogram.

Cu-Si Results

X-ray diffractometer measurements of size-fractioned Cu-6.0 wt.% Si alloy powders revealed three product phases; α-Cu (fcc) primary solid solution, γ (bcc), and β (hcp). No supersaturation of α was found by comparison of lattice parameter measurements to published data (13), using the above mentioned analysis procedure. The intensity of the γ (110) reflection relative to the α (200) peak remained minor and approximately constant over the range of particle size fractions that were studied, as Figure 7 indicates. A strong neighboring peak obscured the γ (110) peak in the diffractometer trace of the <5 μm powders and this peak ratio is not given in Figure 7. On the other hand, the intensity of the β (10.0) peak relative to the α (200) peak, as shown in Figure 7, exhibits a significant particle size dependence. At particle sizes smaller than about 10 μm the β (10.0) intensity ratio increases strongly, while particles larger than about 20 μm show a more moderate β (10.0) ratio increase. In addition, the X-ray diffraction results for the <5 μm size fraction indicate the development of two significant peaks which could not be indexed, adjacent to the β (10.1) reflection (46.28° 2Θ) at 45.55° and 47.13° 2Θ. These adjacent peaks may represent "streaking" of the reflection, as previously reported (14).

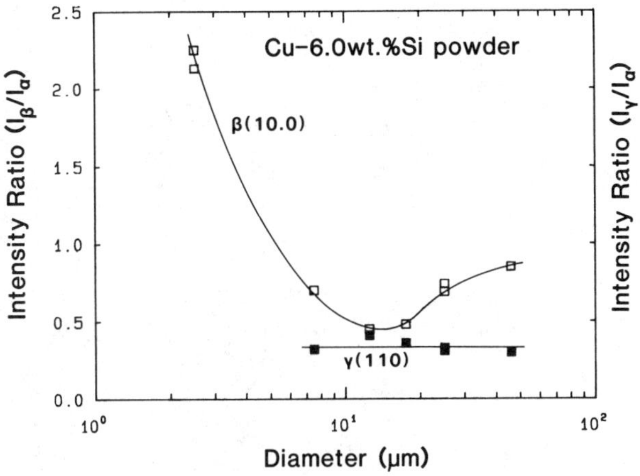

Figure 7 - Measurements of Bragg peak ratios for various size fractions of Ar-atomized Cu-6.0 wt.% Si powders. Open squares are β(10.0)/α(200) ratio. Each particle diameter value represents the midpoint of a size fraction.

Scanning electron microscopy of the Ar-atomized Cu-Si particles, prepared by sectioning and etching the epoxy-mounted and sieved powder samples revealed three distinct solidification morphologies. In general, cellular-like or dendritic microstructures (or a mixed cellular/dendritic morphology) appeared in particles larger than about 10 μm, as shown in Figure 8a. Numerous particles without any obvious microsegregation pattern were observed in the <10 μm powder sample, as Figure 8b reveals.

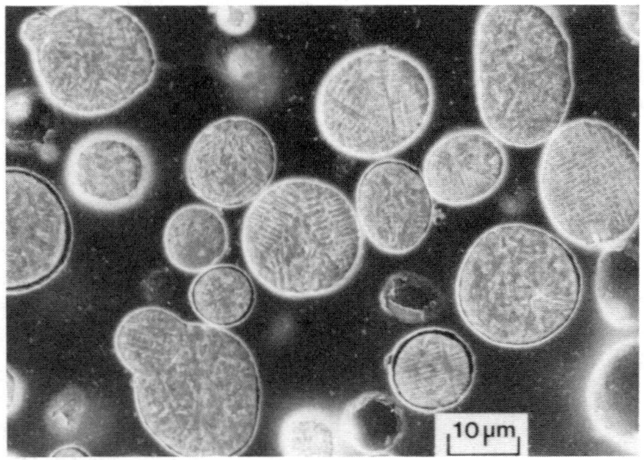

a 20-15 μm

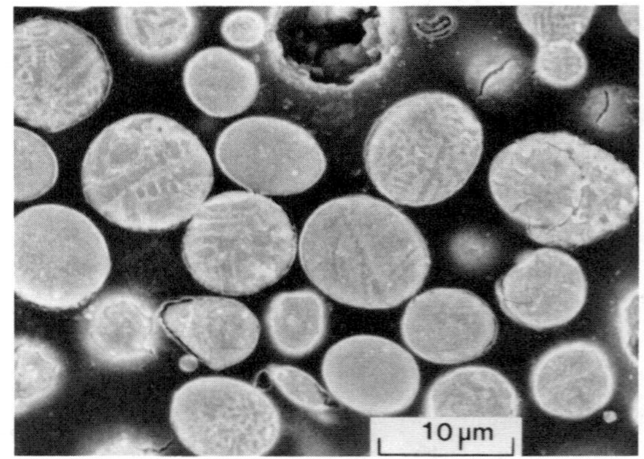

b) <10 μm

Figure 8 - SEM micrographs of size fractioned, Ar-atomized Cu-6.0 wt.% Si powder. Sectioned particle microstructures prepared as in Figure 4.

The undercooling behavior of the Cu-Si powders is summarized in Figure 9 for three particle size fractions. Each DTA powder sample was prepared as described above, heated above the liquidus temperature as the top thermogram in Figure 9 shows, and cooled at 20°C min.$^{-1}$ to observe the exothermic pattern representing particulate solidification. The thermogram which shows cooling and solidification of Cu-Si particles between 15 and 20 μm in diameter contains two exotherms of similar peak areas, suggesting similar populations of high and low undercooling particles. The other two thermograms of powders smaller than 10 μm indicate that a significantly greater population of these ultrafine particles achieves a high undercooling before solidification.

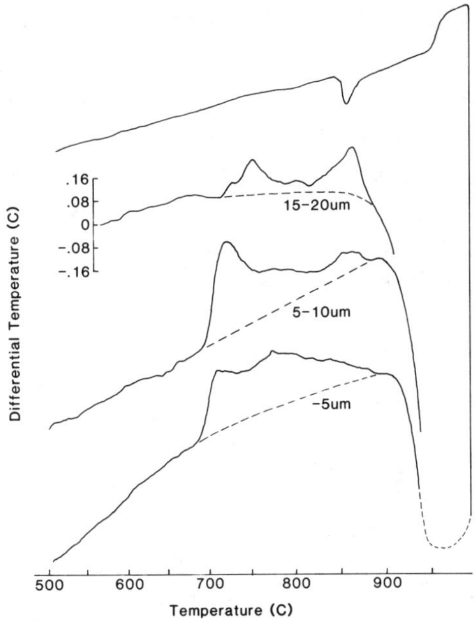

Figure 9 - Summary of DTA results for size fractioned, Ar-atomized Cu-6.0 wt.% Si powders, dispersed in alumina powders, with heating and cooling rates of 20°C·sec^{-1}. Differential temperature sensitivity of all three cooling thermograms is equivalent. Liquidus temperature measured at onset of return to baseline of melting endotherm in top heating thermogram.

Discussion

Ultrafine Particle Undercooling

The particulate characterization results indicate that the ultrafine (<10 μm) powder fractions of these gas atomized samples achieved high undercoolings prior to solidification. The high undercoolings permitted production of solidification morphologies without apparent solute partitioning. Support for the hypothesis of undercooling control of the solidification morphology of gas atomized powders was found in the observation of a constant maximum particle size threshold of about 10 μm required for generation of the "partitionless," single phase particles, especially in the Cu-Al powders. In other words, although the calculated convective cooling rates of Ar and He differ by a factor of 10 for gas atomization of any specific size particle (7), the microstructure of equivalent particle sizes revealed no significant differences between the powders produced with either gas. Strong evidence for this observation is found also in the overlap of the sharp onset of extensive β'_1 formation in Ar-atomized and He atomized Cu-Al powders, as given in Figure 3.

Turnbull's hypothesis that a "mote" isolation effect would occur in a finely subdivided melt (15) is offered as an explanation for the high undercooling behavior of ultrafine gas atomized particles (5). In terms of probability, if a certain residual population of potent crystal catalysts, or motes, is distributed randomly in a molten mass, as represented in Figure 10a, the subdivision of that melt reduces the probability of finding a mote in any specific melt volume element, i.e.:

$$X = \exp(-Mv) \qquad [1]$$

where X is the fraction of mote-free volume elements, e.g. atomized particles, M is the mote concentration in the mass, and v is the volume of each independent volume element (5). As Figures 10b and 10c indicate, continued melt subdivision can result in a large mote-free fraction, approaching unity, of the total particle population. On cooling, the large mote-free fraction of the particle population could solidify at high undercoolings not limited by the potent catalysis effect of the motes.

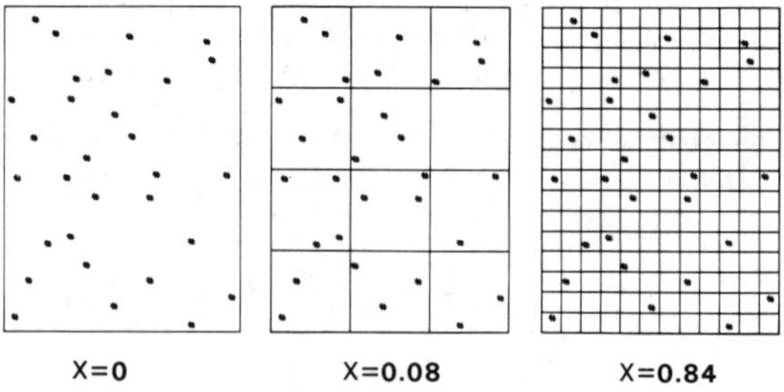

Figure 10 - Schematic illustration of mote isolation effect with X values as defined in eq. [1].

Undercooling Limits

Further evidence to corroborate the microstructural indications of mote isolation was obtained by differential thermal analysis, which permitted measurement of the increased undercooling of ultrafine powder size fractions of the atomized powders. A significant increase in the apparent number of particles that achieved high undercooling was observed as a function of decreased particle size, consistent with earlier DTA results reported for gas atomized Al powders (16). The lowest nucleation peak temperature, T_N^{MIN}, of the thermogram corresponding to the <5 μm particle size fractions was used to derive the maximum undercooling, ΔT^{MAX}, listed in Table I for each alloy, i.e.:

$$T_M - T_N^{MIN} = \Delta T^{MAX} \qquad [2]$$

where T_M is the alloy melting temperature, listed as the eutectic temperature and the liquidus temperature for the Cu-Al and Cu-Si alloys, respectively. An estimated hypercooling limit (17), ΔT^{HYP}, for each alloy was derived from:

$$\frac{\Delta H_M^A X_A + \Delta H_M^B X_B}{C_p^{L(A)} X_A + C_p^{L(B)} X_B} = \Delta T^{HYP} \qquad [3]$$

where $\Delta H^A{}_M$ and $\Delta H^B{}_M$ are the enthalpies of fusion in cal·mole^{-1} for the two pure alloy components (18), $C_p{}^{L(A)}$ and $C_p{}^{L(B)}$ are the heat capacities at constant pressure in cal·mole^{-1}·K^{-1} for the pure liquid components (18), and X_A and X_B are the mole fractions of the alloy components. Comparison of the measured values for ΔT^{MAX} with the calculated ΔT^{HYP} values suggests that substantial undercoolings were achieved in the slow-cooled DTA samples, especially in the Cu-Al powders. The observation of apparently partitionless solidification microstructures in the as-atomized ultrafine powders, which is consistent with hypercooled conditions (17), suggests either that the particles experienced a greater undercooling during the actual gas atomization process than during the DTA cycle or that the estimated hypercoolings were too large. While the assumptions of a simple rule of mixtures and of equilibrium pure metal values for the alloy thermodynamic quantities are simplified (5), the possibility is also reasonable that the measured ΔT^{MAX} may represent a lower bound on undercooling potential during gas atomization where particle cooling rates are much higher and particle surface conditions are much more nascent (17).

Table I

Alloy	T_m (C)	ΔTmax (C)	ΔThyp (C)
Cu-8.3 wt. % Al	1036	291	390
Cu-6.0 wt. % Si	943	218	560

Solidification Reactions at High Undercooling

Rapid solidification of a highly undercooled alloy liquid often produces product microstructures which are difficult to predict a priori. Increased undercooling effectively broadens the range of product phase choices (5) and creates the initial thermal gradient required to advance the solidification growth front rapidly (17). Microstructural analysis and x-ray diffraction techniques allow characterization of the results of the nucleation kinetics competition and the rapid crystal growth processes that occur during rapid solidification of gas-atomized powder particles. Results suggest that ultrafine powders of both alloys studied solidify as single-phase, single-crystal particles. Of course, high resolution TEM and microanalytical techniques are essential to establish this effect. In fact, Boettinger, Bendersky and Early (4) have demonstrated that the apparently partitionless microstructure in ultrafine Al-8 wt.% Fe powders is actually a microcellular morphology with cell spacings of 25-100 nm.

If at least the product phase nucleation event is assumed to be a composition invariant reaction, the thermodynamic driving forces of competing phases can be compared qualitatively using metastable extensions of equilibrium phase relationships (17). Simple extensions, following the slight natural curvature, were constructed of the equilibrium liquidus and solidus phase boundaries for α-Cu and β phases on the portion of the Cu-Al phase diagram (19) that is shown in Figure 11. The initial composition of the alloy powders is the eutectic composition of 8.3 wt.% Al. Since eutectic solidification apparently was bypassed in favor of a single phase β product, in ultrafine powders, an interesting comparison can be made between the thermodynamic driving force for composition invariant nucleation of either β or α. Inspection of Figure 11 reveals that the T_0 temperatures of both phases (17), which must lie between the metastable liquidus and solidus temperatures (17), are approximately equal at the eutectic composition. Therefore the driving force for nucleation of either α or β at a high undercooling also would be similar (5), assuming similar enthalpy values for the phases. This suggests that the nucleation kinetics of β is more rapid than α, allowing its selection over α, in this case.

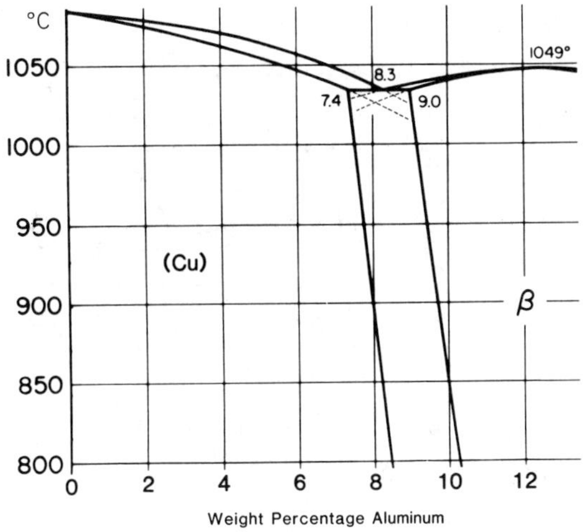

Figure 11 - Portion of Cu-Al phase diagram (19) with extensions of liquidus and solidus curves for the α and β phases.

A similar qualitative driving force assessment of competing phases in the Cu-6.0 wt.% Si alloy also was conducted with similar results but with more complex extrapolations of phase boundaries as shown in Figure 12. Because the results suggest that the peritectic reaction shown in Figure 12 is bypassed in highly undercooled Cu-Si particles in favor of composition invariant solidification of β, the competition between β and α-Cu nucleation, both of the same composition, appears reasonable for this analysis. Although an estimate of T_0 for α of 6.0 wt.% Si can be produced by simple bisection of a tieline connecting the equilibrium liquidus and solidus (17), a metastable β liquidus and solidus must be constructed without direct extensions of equilibrium β+L phase boundaries in order to estimate a T_0 for β nucleation (20). As Figure 12 indicates, the α+L and β+L regions overlap and reasonable estimates of the T_0 temperatures for α and β of the 6.0 wt.% Si alloy are also nearly equivalent. Thus, the driving force for nucleation of either α or β at high undercooling is nearly the same, assuming a similar enthalpy for each phase. As in the Cu-Al alloy case, the nucleation kinetics of β appears to favor its formation instead of α, because of the lack of any obvious difference in the thermodynamic driving force.

Conclusions

1. Ultrafine gas atomized powders can sustain high undercoolings prior to solidification.

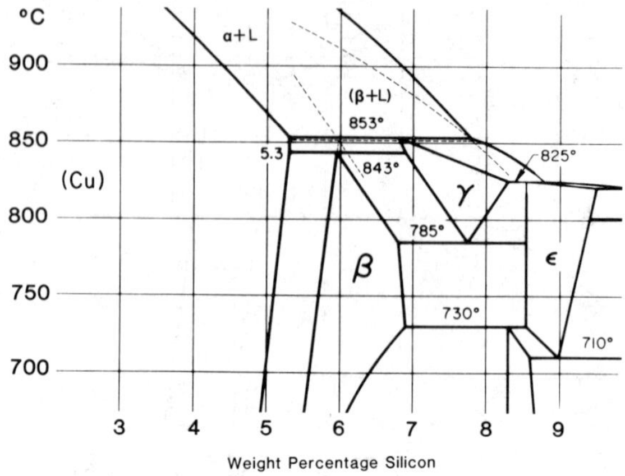

Figure 12 - Portion of Cu-Si phase diagram (19) with metastable β+L phase field shown within the equilibrium α+L region.

2. The solidification morphology of fine gas-atomized powders appears to be influenced more by undercooling than by cooling rate.

3. The enhanced undercooling levels may be related to a mote isolation effect in the highly dispersed melt.

4. Measured undercooling levels may actually represent lower bounds on undercooling behavior during atomization.

5. High undercooling can shift a particle solidification reaction to a metastable path whose driving force may be assessed using extensions of equilibrium phase relationships.

Acknowledgments

The authors wish to thank Dr. J. D. Ayers and Mr. W. P. Robey for their analytical and technical assistance with this project.

References

1. B.H. Kear, P.R. Holiday and A.R. Cox, "On the Microstructure of Rapidly Solidified IN-100 Powders," Met. Trans. A, 10A (1979), 191-197.

2. P.A. Joly and R. Mehrabian, "Complex Alloy Powders Produced by Different Atomization Techniques: Relationship Between Heat Flow and Structure," J. of Mat'ls. Sci. 9(1974), 1446-1455.

3. R.H. VanStone, F.J. Rizzo and J.F. Radavich, Rapid Solidification Processing: Principles and Technologies, II (Baton Rouge, LA: Claitors Publishing Division, 1980), 260-272.

4. W.J. Boettinger, L. Bendersky and J.G. Early, "An Analysis of the Microstructure of Rapidly Solidified Al-8 wt.% Fe Powder," submitted to Met. Trans. A (1985).

5. J.H. Perepezko and I.E. Anderson, Synthesis and Properties of Metastable Phases (Warrendale, PA: The Metallurgical Society of AIME, 1980), 31-63.

6. N.J. Grant, "Rapid Solidification of Metallic Particulates," Journal of Metals 35(1)(1983), 20-27.

7. A. Lawley, "An Overview of Powder Atomization Processes and Fundamentals," Int. J. of Pow. Met. and Pow. Technology 13(3)(1977), 169-188.

8. I.E. Anderson and B.B. Rath, Rapidly Solidified Crystalline Alloys (Warrendale, PA: The Metallurgical Society of AIME, 1985), 219-244.

9. G. DelCorso, Carpenter Technologies, private communication with author, National Bureau of Standards, 31 July 1985.

10. V. Anand, A.J. Kaufman and N.J. Grant, Rapid Solidification Processing: Principles and Technologies, II (Baton Rouge, LA: Claitors Publishing Division, 1980), 273-286.

11. H. Warlimont and M. Wilkens, "Die Struktur der Martensitphase β'_1 in System Kupfer-Aluminium," Z. Metallkde. 55(1964), 382-387.

12. J.L. Murray, "The Aluminum-Copper System," Int. Metals Reviews 30(5) (1985), 211-233.

13. W.B. Pearson, A Handbook of Lattice Spacings and Structures of Metals and Alloys (New York, NY: Pergamon Press, 1958), 610-611.

14. S.D. Dahlgren, W.F. Flanagan and D.H. Polonis, "An X-ray Study of the FccHcp Transformation in a Cu-Si Alloy," Trans. of the Met. Soc. of AIME 236(1966), 1071-1077.

15. D. Turnbull, "Kinetics of Solidification of Supercooled Liquid Mercury Droplets," J. Chem. Phys. 20(3)(1952), 411-424.

16. S.E. LeBeau, J.H. Perepezko, B.A. Mueller and G.J. Hildeman, Rapidly Solidified Powder Aluminum Alloys (Philadelphia, PA: ASTM, in press).

17. W.J. Boettinger and J.H. Perepezko, Rapidly Solidified Crystalline Alloys (Warrendale, PA: The Metallurgical Society of AIME, 1985), 21-58.

18. R. Hultgren, et al., eds., Selected Values of the Thermodynamic Properties of the Elements (Metals Park, OH: American Society for Metals, 1973) 25-31, 149-159, 465-471.

19. J.H. Perepezko and W.J. Boettinger, Alloy Phase Diagrams (New York, NY: Elsevier Science Pub. Co., 1983), 223-240.

20. T. Lyman et al., eds., Metals Handbook, vol. 8 (Metals Park, OH: American Society for Metals, 1973), 259, 298.

21. W.J. Boettinger, National Bureau of Standards, private communication with author, U. S. Naval Research Laboratory, 21 February 1986.

PART III: Undercooled Solidification - Theory and Experiment

SOLIDIFICATION OF UNDERCOOLED ALLOYS

J. H. Perepezko, B. A. Mueller and K. Ohsaka

Department of Metallurgical and Mineral Engineering
University of Wisconsin
1509 University Avenue
Madison, WI 53706

Abstract

During solidification the amount of undercooling is an important factor in determining microstructural development by controlling phase selection during nucleation. Usually crystallization is initiated at a heterogeneous catalytic site at a relatively small undercooling. The use of fine droplet samples is an effective method of obtaining a substantial undercooling into the range of $(0.3\text{-}0.4)T_m$. A number of factors have been identified which influence the undercooling in droplets, including: droplet coating, droplet size, cooling rate and melt superheat. Kinetics analysis of the nucleation undercooling response during cooling can provide a basis to evaluate transformation diagrams for the operative solidification reaction. The catalytic effect of heterogeneous nucleation sites can be examined under well-defined conditions in two phase (L+S) alloys droplets. With such samples metastable phase equilibria may be determined and specific reactions or product structures selected to allow for controlled microstructural modification.

Introduction

The importance of undercooling in the liquid to solid transformation has been recognized in the literature since the early thermometry work of Fahrenheit more than 200 years ago. However, until the major contributions by Turnbull (1-5), observations of undercooling appeared mainly as scattered reports. Largely based on Turnbull's early experimental and theoretical studies on metals, undercooling effects have been identified as central to solidification processing and as a major influence in determining the nucleation behavior. Indeed, an important guideline in grain refining and modification is the amount of attainable undercooling. The recent development of novel microstructures based on non-equilibrium phases during rapid solidification processing (RSP) has emphasized further the fundamental role of undercooling in structural development and control.

The observation of a large liquid undercooling prior to the onset of crystallization during cooling at any rate does not necessarily signify the absence of heterogeneous nucleation. Within a large continuous sample of liquid a spectrum of possible nucleant sites or motes can be considered to be present and to display a range of catalytic activity. Only the effect of the most potent nucleant site will be observed to establish the amount of undercooling which under usual conditions in bulk samples is typically 10-20°C or less. However, several experimental approaches have been developed which are effective in extending the range of observable undercooling (6-8). In particular, the droplet sample configuration which has been employed extensively by Turnbull and coworkers (1-3), has been demonstrated to be a successful and experimentally useful approach. With droplet samples a progressive removal or isolation of nucleants is effected so that an undercooling level will be observed which is characteristic of the most potent nucleant still remaining or which represents the limit set by homogeneous nucleation or a glass transition. For a liquid containing a nucleant concentration, m, the dispersal of the sample into a large number of droplet volumes will yield a significant number of nucleant-free droplets provided that the droplet concentration exceeds the nucleant concentration by a large amount. In order to realize this objective the droplet surface coating which is necessary to maintain independence of the dispersed droplets should have a low catalytic potency for crystal nucleation. When a fine droplet sample is undercooled, the drops containing the most potent nucleants will freeze at low undercooling, but the majority of the drops will not freeze until reaching the maximum undercooling nucleation temperature. The capability to attain large undercooling during slow cooling permits the measurement of the nucleation temperature, solidification kinetics and the properties of the metastable liquid state. With the advent of improved sample preparation methods and more extensive structural determinations, the droplet method has provided also a basis for the evaluation of the phase selection nucleation kinetics that govern structure development during solidification processing (9).

The study of undercooling behavior in liquids has become more widespread since Turnbull's early work so that a complete overview is beyond the scope of the present discussion. In order to highlight some of the recent advances that have developed, several features of undercooling behavior will be examined and related to the processing conditions that control the amount of undercooling and the solidification reaction kinetics. One processing condition of interest is the effect of imposed rapid cooling on the competitive nucleation kinetics of undercooled alloy phases

Undercooling Behavior and Processing Conditions

In large continuous liquid samples a single exothermic crystallization temperature is expected at an undercooling determined by the most active catalytic site. For polydisperse droplets, a single crystallization temperature is not always observed as indicated by the thermograms in Figure 1 for a high purity (99.999%) Sn sample (10). For each of the droplet samples represented in Figure 1, the processing conditions were identical except for the degree of liquid dispersion. As a result, the crystallization behavior is expected to represent the effect of the degree of dispersal on undercooling.

Overall, the undercooling behavior and the trends in crystallization temperature provide support for the basis of the droplet method. The coarse dispersion with a most frequently occurring size, $\overline{D}$, of 275μ displays a major crystallization exotherm centered at 184°C corresponding to an undercooling of 48°C. There are also several subsidiary exotherms evident extending to about 50°C below the main crystallization peak. As $\overline{D}$ decreases with increasing refinement, there is a clear trend for the crystallization exotherm at 184°C to decrease in magnitude. A similar behavior is exhibited by other crystallization exotherms in Figure 1. In addition, accompanying each level of droplet refinement there is a development of further crystallization exotherms at decreasing temperatures.

In effect the sequence of crystallization characteristics developed during the cooling of dispersions of progressively finer droplets reveals

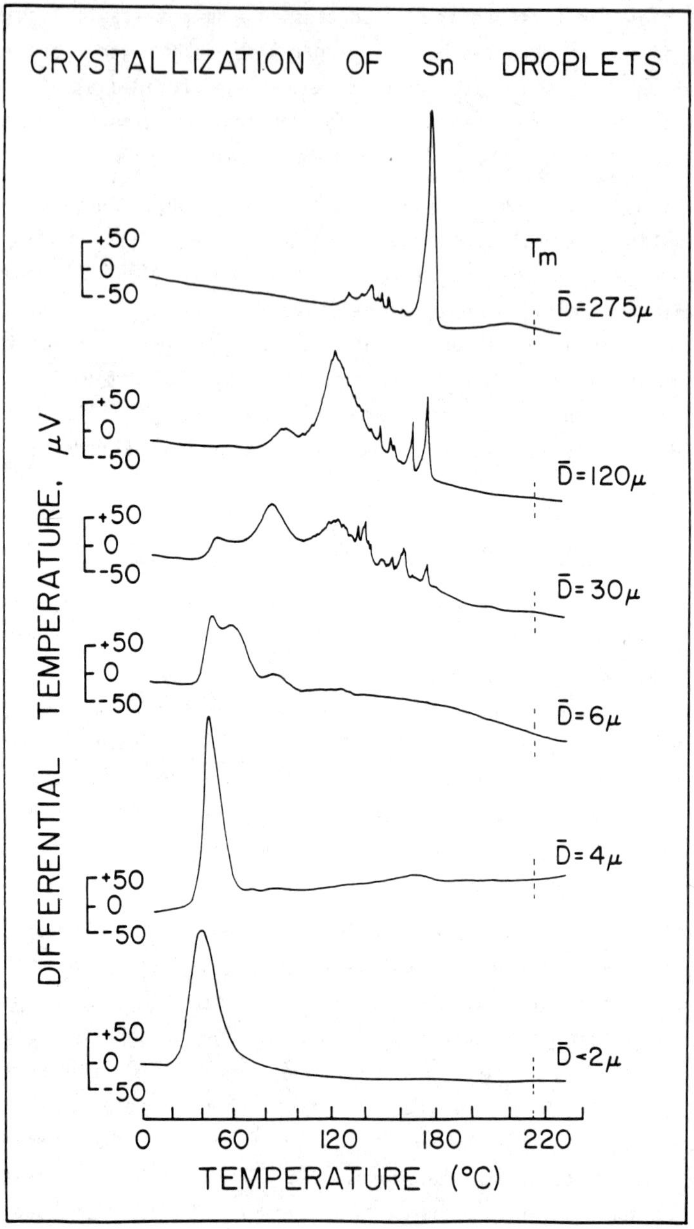

Figure 1 Crystallization of pure Sn droplets at different levels of size refinement.

the spectrum of nucleants with varying catalytic potency within Sn. It is clear that the active nucleants are not all equally effective and are not distributed in equal abundance within the sample. The occurrence of crystallization exotherms at about the same temperature in several dispersions with large size differences suggests that in each case crystallization is initiated by a common catalytic agent. At the same time, the shift of crystallization exotherms to lower temperature with increasing refinement indicates that even for dispersions that may be considered to be relatively fine (i.e. $\overline{D}$ of about 30μ) several nucleants are likely to be present and effect many of the droplets. Actually, the average size is not the only distribution characteristic relating to the level of refinement, since the range of sizes within each emulsion represented in Figure 1 varied considerably. The sample with a $\overline{D}$ value of 275μ contained a spread of droplet sizes from 160 to 400μ, while at $\overline{D}$ = 30μ the size spread ranged from 10 to 60μ and at $\overline{D}$ = 2μ, droplets were detected in the range between less than 1 and 5μ. For example, the undercooling for a Sn droplet sample with a broad crystallization and multiple peaks is shown in Figure 2 together with the measured droplet size distribution giving the mean, μ_v, and standard deviation, σ_v, on a volume basis. The fit given in Figure 2 indicates that the droplet distribution produced by mechanical shearing is log-normal with a unimodal character even though the crystallization shows multiple peak behavior. The greatest level of undercooling and the most uniform crystallization behavior in terms of a narrow temperature range for a given droplet surface coating, are obtained for droplet dispersions with a fine and narrowly distributed size range. In fact, the crystallization temperature for the dispersion in Figure 1 with $\overline{D}$ = 2μ is 41°C which represents the largest reported undercooling of 191°C for Sn.

The observations relating to the influence of droplet dispersal highlight the influence of particle size on the undercooling, nucleation and crystallization behavior of liquids. If the nucleating sites contained within the original liquid volume are distributed randomly, the arrangement of nucleants among the droplets for a high degree of dispersal may be described by a Poisson distribution. For this case the nucleant-free droplet fraction X among droplets of volume v is represented by X = exp (-mv) where mv is the average number of nucleants per droplet. With such a nucleant distribution it is necessary for a sample to contain an active catalyst concentration of $3x10^{17}$ m^{-3} to yield crystallization for all droplets larger than 6 μm in size. However, a size refinement to 2 μm will yield X = 0.9. While the example indicates the importance of particle size refinement in achieving a large undercooling, it reveals also that very

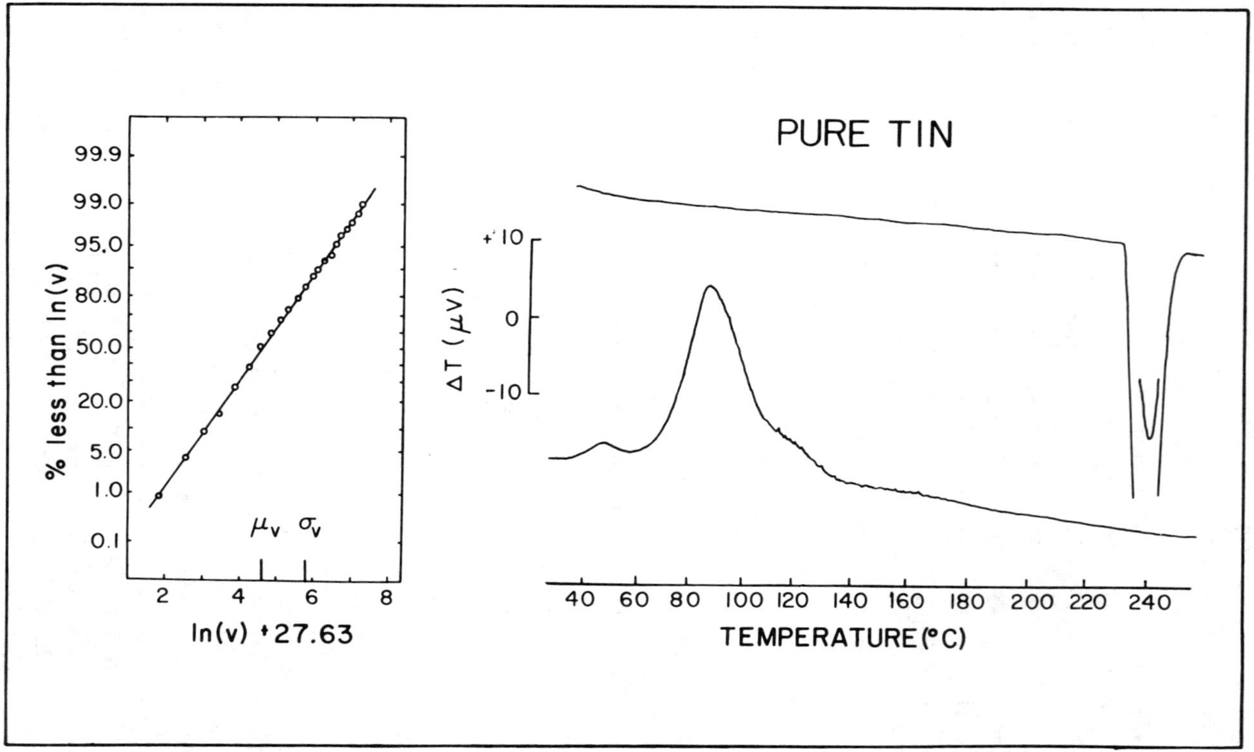

Figure 2 Log-normal size distribution of a Sn droplet sample and the corresponding undercooling behavior.

small active catalyst concentrations can influence the undercooling strongly. Furthermore, since mv is proportional to the total number of active nucleants contained in the starting liquid and the size of the droplets, the fraction of the nucleant-free droplets in some cases may be related to the purity of the starting materials as well as to the degree of size refinement. For example, at the 99.9999% purity level a material has an impurity concentration of about 10^{22} m^{-3}. An examination of this feature is presented in Figure 3 for the undercooling behavior of aluminum droplets. In this case for the two sizes indicated there is a minor effect on the undercooling of purity changes from 99.999% to 99.83%. At the same time the change in droplet size indicated has an effect on the undercooling level. While for aluminum specific impurity levels do not seem to be crucial, other examples have been noted where the impurity content can be an important factor. For example in Figure 4 two thermograms are given to illustrate the effect of purity on the undercooling of Bi droplets of the same size range and treated with the same surface coating procedure. A change of purity from (99.9999%) to (99.98%) yields a measurable decrease in undercooling, with samples prepared from (99.98%) pure material crystallizing over a broader temperature range. Usually, the largest undercooling is observed with high purity liquids but, even with melts of nominal purity, an undercooling may be realized which is substantial.

When nucleants are fractionated effectively by dispersing a liquid metal into a large number of droplets, the undercooling limit of the emulsion sample can be determined by the catalytic nature of the droplet surface coating (10). Since surface coatings of a non-crystalline type will tend to wet the liquid phase and to discourage a surface catalysis, a droplet surface treatment with surfactants which have a complex and bulky structure is expected to yield a larger undercooling than with surfactants of a simple structure. This condition has been demonstrated for a number of metals and is demonstrated in Figure 5 for pure Sn droplets processed under identical conditions using two different coating treatments. In sample A where an inorganic acid surfactant was used the undercooling to nucleation is 31°C lower than that for sample B where an organic surfactant treatment was applied during dispersal.

In the presence of a large nucleant population which is comparable with the total number of droplets, a significant fraction of droplets may still contain nucleants. In addition to internal nucleants, metal oxide surface patches which may be formed during the emulsification process may also become active catalytic sites during cooling. Part of the high temperature

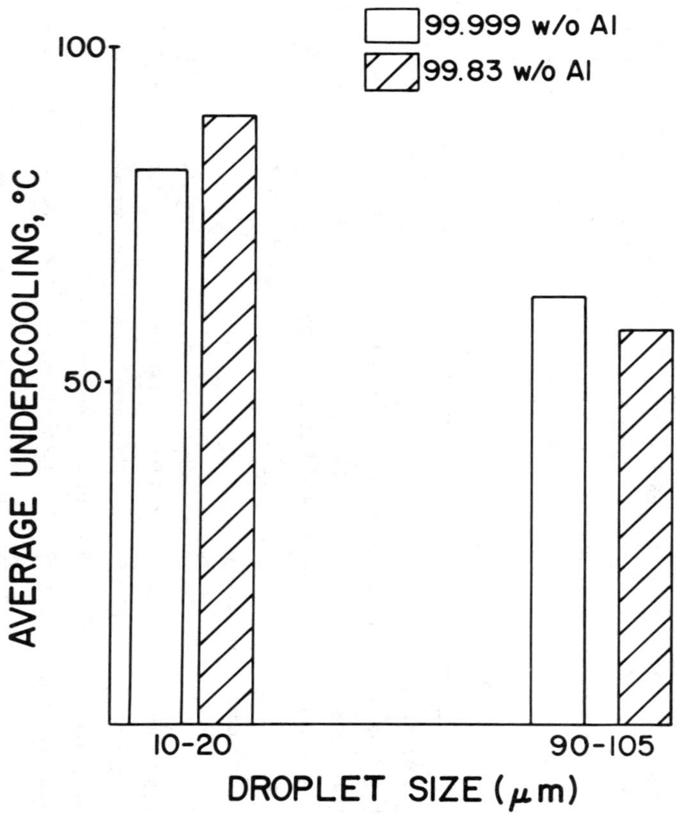

Figure 3 Undercooling behavior of Al powders at different levels of purity and size refinement.

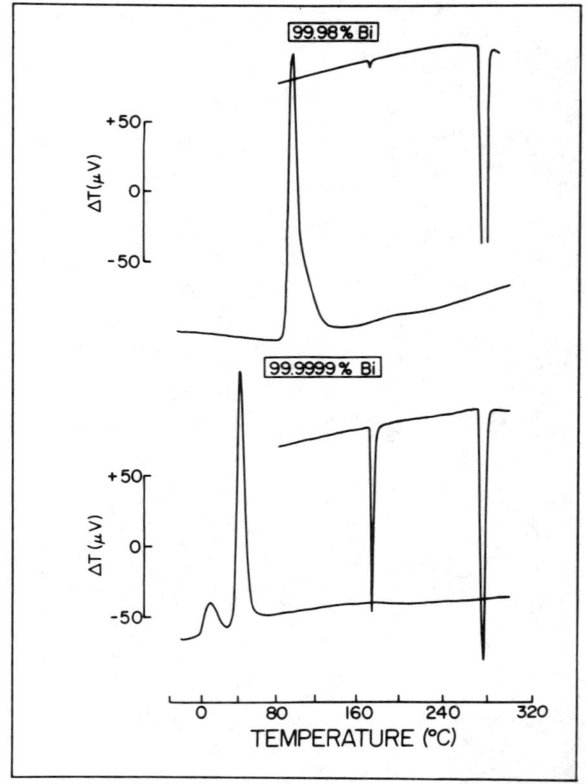

Figure 4 The influence of changes in total impurity content on undercooling of Bi droplets prepared with similar size distribution ($\overline{D}$ ~ 10μ) and coating treatment.

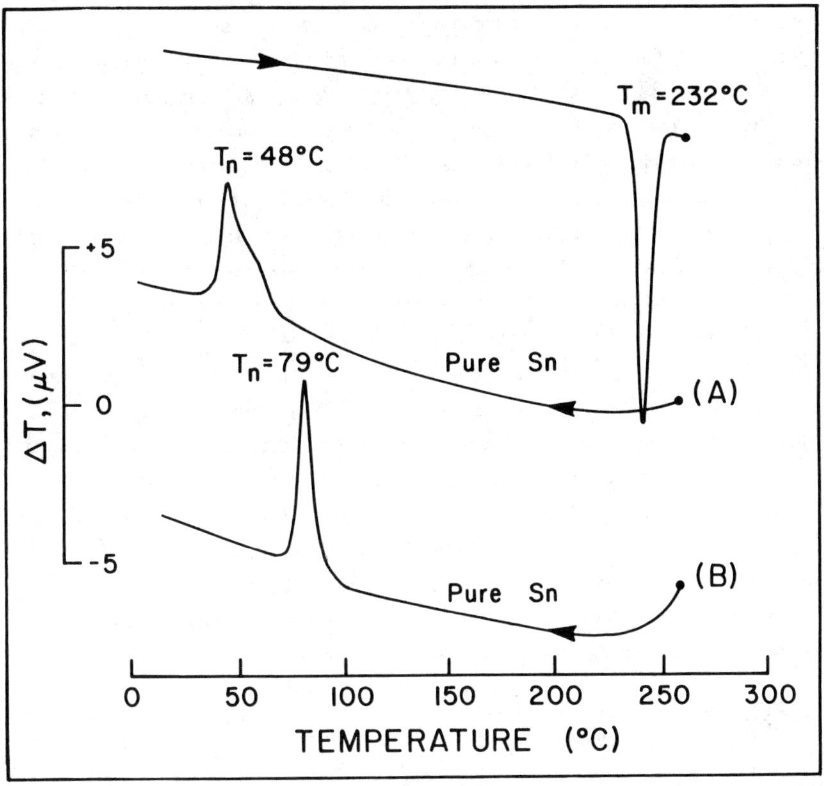

Figure 5 The undercooling behavior of Sn droplet dispersions with the same purity (99.999%) and similar size distribution ($\overline{D} \sim 10\mu$) but with different surface coating treatments.

crystallization of liquid indium emulsions has been found to be related to the catalytic action of indium oxide (11). Under these conditions, subdivision of liquid metals is not always effective in isolating nucleants into a small fraction of a droplet population. For surface sites the distribution for X becomes $X = \exp(-m_a a)$ where m_a is the concentration/cm^2 of surface sites in a droplet of area, a. With surface sites it appears that the catalytic potency may also be a function of the site size so that smaller sizes become active at larger undercooling. In such cases, specific surface coating treatments are necessary to deactivate the nucleants in order to increase the undercooling.

When all of the processing parameters that have been identified to control undercooling are considered, a considrable effort is necessary in order to attain the optimum undercooling level for a given metal. One method to approach the optimization of undercooling has been demonstrated recently for Al(12). A convenient and powerful statistical method for performing the analysis of processing conditions on a response such as undercooling is factorial design analysis. Factorial designs are a class of proven statistical methods which analyze a large number of variables both accurately and precisely in relatively few runs per factor studied (13). To perform a factorial design, a fixed number of levels are selected for each variable and experiments are conducted at all possible variable combinations. As an illustration of the analysis four variables were examined at two levels, a 2^4 factorial. The undercoolings, defined as the difference between the melting point of Al and the peak of the largest nucleation exotherm, and the levels at which the variables were analyzed, are shown on a factorial cube in Figure 6. Each undercooling is the average of two or three runs at that condition. Replicate runs were conducted to provide an estimate of the standard error. The Yates algorithm was used to analyze the factorial data. The significant effects were determined from normal probability plots. The estimates of the effects are given in Table 1 where the carrier fluid salt represents droplet coating, DS is droplet size, $\dot{T}$ is cooling rate and T_S is melt superheat. Additional parameters can be examined, but the number of trials would increase.

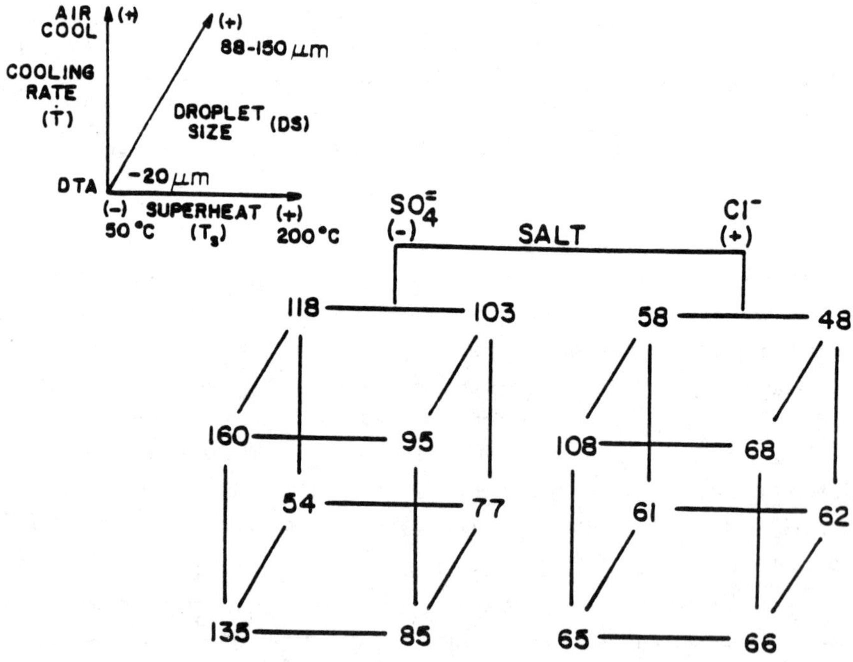

Figure 6 Factorial cube with the measured undercoolings of Al powders indicated at cube vertices for each variable combination.

TABLE 1: Estimates of the factorial analysis effects

	Effect	Estimate (°C)
	Average	85*
Main Effects	Salt	-36*
	DS	-25*
	$\dot{T}$	19*
	T_S	-19*
2-Factor Interactions	Salt x DS	6
	Salt x $\dot{T}$	-12*
	Salt x T_S	7
	DS x $\dot{T}$	1
	DS x T_S	19*
	$\dot{T}$ x T_S	-13*
3-Factor Interactions	Salt x DS x $\dot{T}$	-15*
	Salt x DS x T_S	-12*
	Salt x $\dot{T}$ x T_S	0
	DS x $\dot{T}$ x T_S	1
4-Factor Interaction	Salt x DS x $\dot{T}$ x T_S	7

*Significant effect

While the factorial analysis does not explore fully a wide range of the variables, it does indicate major trends. The analysis suggests that salt is the most important variable with powders produced in the sulfate salt averaging 36°C more undercooling than those produced in the chloride salt. Droplet size, cooling rate, and melt superheat also influence undercooling; however, not as significantly as salt. Additionally, synergism was detected between variables. This is indicated by significant two factor (Salt x $\dot{T}$, DS x T_S, $\dot{T}$ x T_S) and three factor (Salt x DS x $\dot{T}$, Salt x DS x T_S) interactions. The factorial analysis implies that the finest powders produced in the sulfate salt, cooled at the highest rates from low superheats, will achieve the lowest nucleation temperature. As a means of testing this hypothesis, -20 μm droplets produced in the sulfate salt were quenched into water ($\dot{T}$=500°C/sec) from low superheats. Under these conditions the nucleation temperature decreased to 485°C which corresponds to an undercooling of 175°C or $0.19T_m$. This undercooling exceeds the previously reported maximum of 160°C for powders prepared in the chloride salt (14).

Each metal or alloy system often requires specific handling and treatment during the preparation of droplet dispersions. However, on the basis of the experience just discussed, the undercooling of droplets may be optimized through the use of high purity samples with a fine and narrowly distributed droplet size and involving surfactant treatments that tend to promote the formation of thin non-crystalline surface films. The maximum undercooling limits for several metals in droplet form that have been determined by thermal analysis measurement are summarized in Table 2. All current undercooling values have been measured at a cooling rate of 20°C/min except for Hg where 5°C/min and Al where 500°C/sec were used. On the basis of the comparison presented in Table 2, the current droplet studies have demonstrated that the previous undercooling values actually represent a heterogeneous process. At present, the increases in observable undercooling are not uniform in each case. Further studies of nucleation rates are necessary before a complete kinetic characterization of the revised undercooling limits is possible. These studies are in progress.

TABLE 2 - Maximum undercooling limits

Element	Previous Studies			Current Studies	
	ΔT (°C)	$\Delta T/T_m$	Reference	ΔT (°C)	$\Delta T/T_m$
Al	130	0.14	2	175	0.19
Sb	135	0.15	2	210	0.23
Bi	90	0.16	2	227	0.41
Cd	-	-	-	110	0.19
Ga	150	0.50	15	174	0.58
In	-	-	-	110	0.26
Pb	80	0.13	2	153	0.26
Hg	80	0.34	3	88	0.38
Te	-	-	-	236	0.32
Sn	117	0.23	16	191	0.38

Competitive Nucleation Kinetics in Undercooled Melts

In the present discussion the main interest in nucleation kinetics will focus on the relationships between undercooling, melt cooling rate and particle size during the solidification of a distribution of liquid droplets. These relationships can be examined to reveal the range of processing conditions that expose different phase selection options or other transitions in the nature of the operating nucleation kinetics. The reaction kinetics may be evaluated by numerical nucleation rate calculations, but the computed results are highly dependent on the specific values assigned to the kinetic parameters. Reliable information is available for some parameters; however, in other cases the kinetic parameters are assigned values based upon models which require experimental verification. In order to minimize uncertainty it is of interest to examine the range of nucleation kinetics in a general manner. With this approach it is possible to develop some useful guiding relationships.

Droplet Nucleation Kinetics

The classical nucleation model is based on the formation of critical size clusters by the addition and dissolution of individual atoms. At steady state, the distribution of cluster sizes is stationary and the nucleation rate is constant. Nucleation of a crystalline phase during solidification has been classified into two mechanistic types. When nucleation occurs sporadically throughout a liquid volume without an association with any catalytic sites, it represents homogeneous nucleation. For the majority of processing conditions, nucleation is initiated at some catalytic site as heterogeneous nucleation. These heterogeneous sites may be distributed on the surface or within the bulk volume of the liquid. Based upon classical theory (3) a general expression for the steady state nucleation rate, J_i, can be represented as

$$J_i = \Omega_i \exp\left[-\frac{\Delta G^* f(\theta)}{kT}\right] \qquad (1)$$

where J_i may represent heterogeneous surface nucleation, J_a; homogeneous nucleation, J_v; or heterogeneous nucleation on catalytic sites distributed within the volume, J_s. Appropriate values for the prefactor, Ω_i, activation barrier, ΔG^*, and contact angle function, $f(\theta)$, are used in Equation (1) and kT has the usual meaning. The terms for Ω_i involve a product of a nucleation site density on a sample surface or volume basis, the number of atoms on a nucleus surface and a liquid jump frequency. For many cases $\Omega_v = 10^{30}\ \eta^{-1}\ cm^{-3}\ S^{-1}$ and $\Omega_a = \varphi\ 10^{22}\ \eta^{-1}\ cm^{-2}\ S^{-1}$ with η, the liquid shear viscosity in poise, given by (17) $\eta = 10^{-3.3} \exp[3.34\ T_L/(T-Tg)]$ in terms of the liquidus temperature, T_L, glass transition, T_g and where φ is the fraction of sample surface sites that are active. For J_v and J_a, ΔG^* is given by $\Delta G^* = b\ \sigma^3/\Delta G_v^2$ where σ is the liquid-solid interfacial energy, ΔG_v is the driving free energy for nucleation of unit volume of product phase and $b = 16\pi/3$ for spherical nuclei with $f(\theta) = 0.25[2-3\cos\theta + \cos^3\theta]$. For heterogeneous volume nucleation the value of Ω_s depends on the specific catalyst concentration, size and the form of ΔG^* which is related to the details of the catalyst-nucleus interaction (18).

For pure metals or single phase alloy compounds ΔG_v can be calculated directly from

$$\Delta G_v = \frac{\Delta H \, \Delta T}{T_m} - \int_{T_N}^{T_m} \Delta C_p \, dT + T \int_{T_N}^{T_m} \frac{\Delta C_p}{T} \, dT \qquad (2)$$

where ΔH is the latent heat and ΔC_p is the heat capacity difference between undercooled liquid and crystal. Most often mesaurements are not available and approximations for ΔC_p are used in Equation (2). For pure metals, measurements (19) indicate that ΔG_v is represented to within a few percent by the relation proposed by Turnbull (3) based upon $\Delta C_p = 0$ yielding $\Delta G_v = \Delta H \, \Delta T/T_m = \Delta H \, (1-T_r)$ where $T_r = T_N/T_m$. In the case of alloys and especially for glass forming ranges, ΔC_p corrections can contribute significantly to ΔG_v.

In alloys, nucleation of a solid of composition C_S from an undercooled liquid involves a free energy change given by

$$\Delta G_v = [(\mu_S^A - \mu_L^A)(1 - C_S) + (\mu_S^B - \mu_L^B) C_S]/V_m \qquad (3)$$

where μ_S^A and μ_S^B are the chemical potentials for species A and B, μ_L^A and μ_L^B are the component chemical potentials for the liquid and V_m is the molar volume. For alloys the free energy for nucleus formation is a function of cluster size and cluster composition so that ΔG^* represents critical values of nucleus composition and size. While ΔG_v is maximized when $\Delta\mu^A = \Delta\mu^B$ (20), the concurrence of a minimum ΔG^* with the maximum in ΔG_v also requires that σ and $f(\theta)$ have a weak composition dependence which has not been established experimentally as yet.

When a liquid sample is solidified in a highly undercooled condition, the nucleation process becomes the solidification rate determining step. Therefore, the nucleation rate is equivalent to the solidification rate. In a fine droplet dispersion, the nucleation event in each droplet is an independent occurrence. On this basis a first order reaction law can be applied to describe the rate of solidification in a distribution of droplets (3) as

$$\frac{dN\,(v_i,t)}{dt} = -\,k_i\,N\,(v_i,t) \qquad (4)$$

or

$$N(v_i,t) = N(v_i,0)\exp(-k_i t) \qquad (5)$$

where $N(v_i,t)$ is the number of liquid droplets belonging to the ith size group at time t with volume v_i and k_i is the operative nucleation frequency. For homogeneous nucleation, $k_i = J_v v_i$ and for the more commonly observed surface area dependent heterogeneous nucleation, $k_i = J_a a_i$ where a_i is the surface area of a droplet with volume v_i.

The time dependence of the liquid droplet distribution or the freezing rate is illustrated schematically in Figure 7 for isothermal conditions. Initially at τ=o the liquid droplet size distribution is log-normal. As time passes, larger droplets freeze before small sizes at a rate given by Equation 4. As a result the initial distribution characteristics of liquid droplets changes as shown in Figure 7. The rate of this size distribution change will be different for volume dependent and surface area dependent kinetics and can be measured experimentally to determine the operative nucleation mechanism (3,21).

Most often a droplet sample contains a distribution of nucleants either internal or on the surface which can become active during cooling. As noted previously the droplet fraction which is free of such nucleants can be evaluated from the Poisson distribution function. Under these conditions the number of liquid droplets of volume v_i that remain in a distribution after cooling from T_M to T_N can be obtained as

$$N(v_i,0) = N_T\, P(v_i)\exp(-m_i\, a_i) \qquad (6a)$$

where N_T is the total number of droplets, $P(v_i)$ is the log-normal distribution function (13), and m_i is the number of surface catalysts per unit area. The isothermal solidification of a droplet dispersion is then described by the total number of unfrozen droplets at time t, $N_T(t)$ as the summation of nucleation events over all size classes

$$N_T(t) = N_T \sum_i P(v_i)\exp(-m_i\, a_i)\exp(-J_a\, a_i\, t) \qquad (6b)$$

In practice, $N_T(t)$ can be monitored by calorimetric (21), dilatometric (2) or direct observation (22) measurements.

Continuous Cooling

In general, the critical condition to observe nucleation experimentally during isothermal holding over a time t is $J_v vt = 1$, $J_a at = 1$ or $J_s st = 1$ depending on the nucleation kinetics controlling solidification. For continuous cooling this condition requires some modification (23) to account for the total number of nuclei formed during cooling from T_m to T_N. For cooling at a constant rate T and surface nucleation kinetics the condition for the observation of nucleation is given by

$$- \frac{a}{\dot{T}} \int_{T_M}^{T_N} J_a(T)\, dT = 1 \qquad (7)$$

where $J_a(T)$ is the nucleation rate at T. The integration involved in equation 7 is the area under the curve given in Figure 8. The area can be approximated by the shadowed region as

$$-\int_{T_M}^{T_N} J_a(T)\, dT \simeq (T' - T_N)\, J_a\,(T_N) \qquad (8)$$

where T' is defined as the intersection between the tangent to $J_a(T)$ at T_N and the T axis or

$$T' = T_N - \frac{\Delta G_v^3\, T_N^2}{A(2T_N\, d\Delta G_v/dT + \Delta G_v} \qquad (9)$$

where $A = 16\pi\sigma^3 f(\theta)/3k$.

Equation (7) can be written as

$$- \frac{a\, T_N^2\, \Delta G_v^3\, J_a(T_N)}{\dot{T}\, A\, (2T_N\, d\Delta G_v/dT + \Delta G_v)} = 1 \qquad (10)$$

or

$$a\, J_a\,(T_N)\, t = \frac{-\, A\, \Delta T\, (2T_N\, d\Delta G_v/dT + \Delta G_v)}{\Delta G_v^3\, T_N^2} = K\,(T_N) \qquad (11)$$

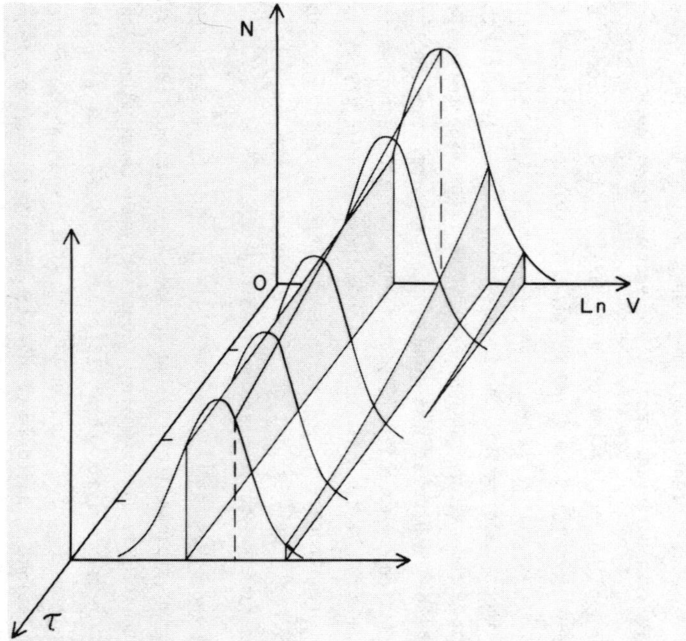

Figure 7 Illustration of the decay of the number of unfrozen droplets during isothermal nucleation at different intervals of $\tau = J_v t$ for a volume dependent nucleation process.

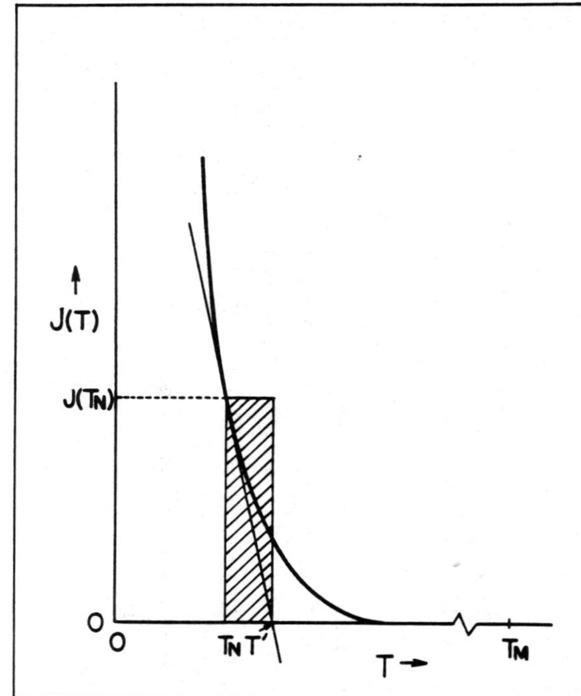

Figure 8 Graphical representation of the approximate evaluation of droplet nucleation during continuous cooling.

where $t = \Delta T/\dot{T}$. Typically, $K(T_N)$ ranges from 100-200. Recalling Equation (1), Equation (11) can be written as

$$\ell n(t) = -\ell n\left[\frac{a\,\Omega_a}{K(T_N)}\right] + \frac{A}{T_N\,\Delta G_v^2} \qquad (12)$$

for which the describes heterogeneous nucleation time for the appearance of solid during continuous cooling at rates that allow for the maintenance of steady state conditions and over a range of nucleation temperatures where the approximation of Equation (8) is reasonable. For the complete analysis of the continuous cooling transformation behavior it is necessary to carry out a full numerical evaluation of Equation (7).

An illustration of the application of a continuous cooling analysis to the crystallization kinetics is represented by some recent measurements for eutectic Pb-56a/o Bi alloy droplets (24). In this alloy extensive undercooling to $0.3T_m$ occurs and yields nucleation of a metastable single phase product with a well defined crystallization onset. Over the cooling rate range from 10°C/min. to 320°C/min., DSC measurements indicate that T_N decreases from 265K to 255K. Also other droplet work indicates that these nucleation temperatures do not represent homogeneous nucleation since lower T_n values are possible with different droplet preparation treatments.

The continuous cooling nucleation results are plotted in terms of equation (12) in Figure 9 where t is given by $\Delta T/\dot{T}$. In this case instead of the usual approximation for $\Delta G_v = \Delta H(1-T_r)$, the ΔG_v values used in Figure 9 were derived from direct calorimetric measurements of the latent of the metastable phase, the temperature dependence of ΔC_p and reported molar volumes. Except for the results at the highest cooling rate, the kinetics measurements can be fitted to equation (12) in terms of a single nucleation frequency with a slope of 1.74×10^7 KJ^2 cm^{-6} and a value for $(\Omega_a a/K(T_N))$ of 3.3×10^{22} sec^{-1}. The prefactor term, Ω_a, is in reasonable agreement with the classical theory result if only a portion of the surface area presents an active catalytic site for metastable phase nucleation.

Based upon the analysis in Figure 9 it is useful to develop the complete transformation diagram that is described by Equation (12). If Ω_a, a and A are treated as approximately constant with temperature, the evaluation of these terms from Equation (12) can be used to develop the isothermal transformation diagram based on $J_a at=1$ that is shown in Figure 10. While the calculation is approximate due to estimates of some of the

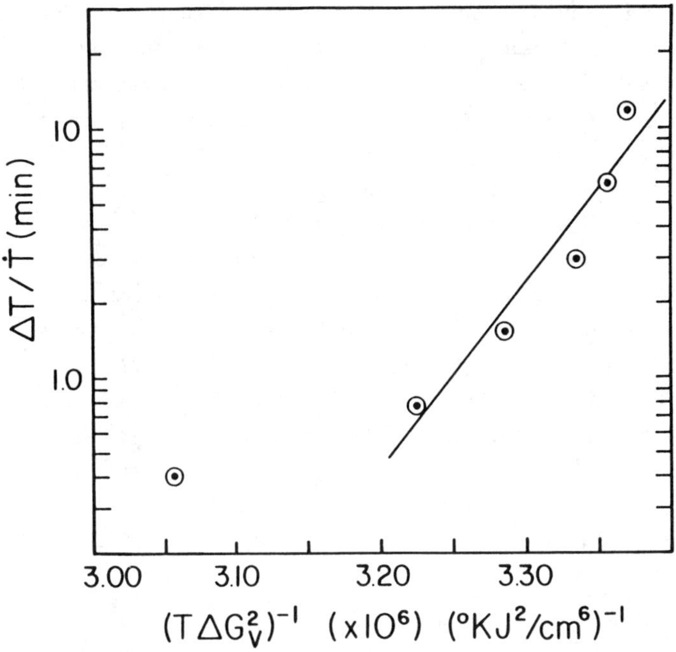

Figure 9 Nucleation kinetics plot based on continuous cooling measurements for Pb-56 a/o Bi droplets crystallizing to a single x phase product.

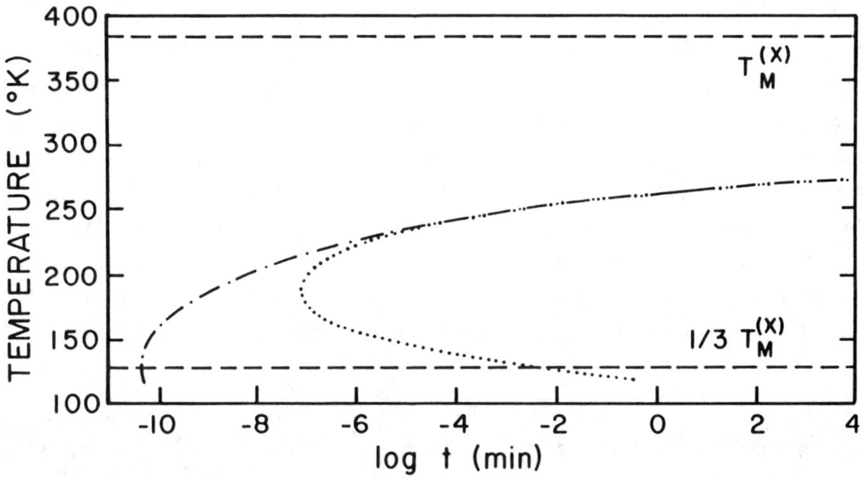

Figure 10 Calculated heterogeneous nucleation kinetics for x phase in Pb-56 a/o Bi droplets. The dotted curve is based on measured ΔG_v and the broken curve is derived from the linear ΔG_v approximation.

parameters, the results do offer an interesting viewpoint on the kinetics. For example, if the linear approximation for ΔG_v had been applied instead of the measured values, the nose of the transformation diagram would be located at $(1/3T_m)$ = 128 K instead of 187 K. Also, the cooling rates to reach the nose differ by a factor of about 10^3 for the different ΔG_v values. This observation reveals the importance of using measured values, even if they represent a limited temperature range, for the nucleation parameters instead of estimates. It is also of interest to compare the droplet kinetics behavior with rapid quenching results. In this regard the calculated transformation diagram is consistent with the observation that suppression of the metastable phase by splat quenching is difficult (25). According to Figure 10, a critical cooling rate in excess of 5×10^7°C/sec is required to bypass metastable phase nucleation. Further, other droplet results indicate that another nucleation site for the metastable phase is active at a greater undercooling than that shown in Figure 10. As a result, at high cooling rate a transition from one type of nucleation site to another for formation of the same phase is possible.

The influence of continuous cooling during the solidification of undercooled liquids can also have important consequences on the development of microstructure. For example in Al-Si alloys the nucleation temperature of hypereutectic and eutectic alloys appears to lie on a plateau indicating that a common catalyst is limiting the undercooling (26). In these alloys the common catalyst appears to be a primary Si phase which forms slightly below the liquidus and catalyzes the remaining liquid at a nucleation temperature of 484°C (Figure 11a). When eutectic composition droplets are quenched at about 500°C/s the nucleation temperature of the sample was depressed to 370°C (Figure 11b). While the solidification structure in eutectic droplets following slow cooling (0.5°C/S) typically contained primary Si particles as shown in Figure 12a, the quenched droplets usually did not reveal evidence for primary Si as shown in Figure 12b. Thus, the increased cooling rate resulting from the quenching treatment suppressed formation of primary Si, yielded a transition to another nucleation kinetics operating at higher undercooling and produced a modified microstructure.

Nucleation Kinetics Transitions

In addition to the effect of size distribution statistics on nucleant isolation which allows for a kinetic transition in product phase selection, it is also important to discern if the size distribution effects can result in a transition in the nature of a nucleation process. In this regard, the

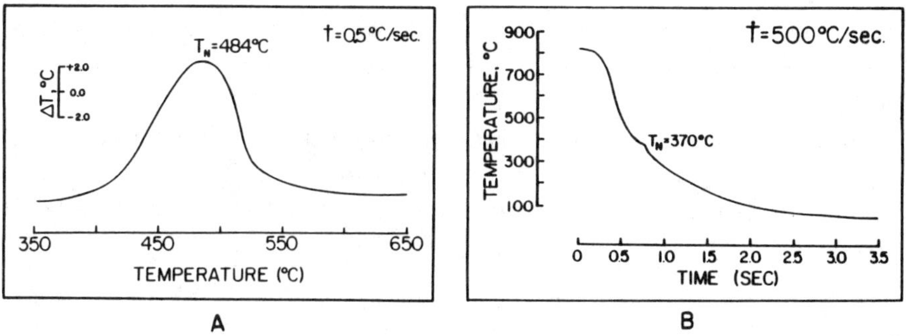

Figure 11 Undercooling behavior in Al-Si eutectic alloy droplets. (a) DTA trace at 0.5°C/S; (b) continuous cooling curve at 500°C/S.

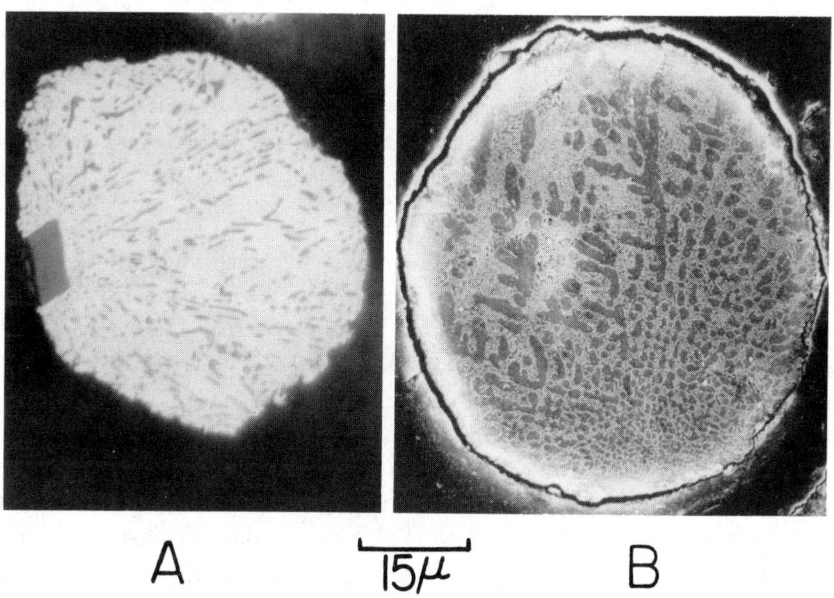

Figure 12 Solidification microstructures developed in undercooled Al-Si eutectic alloy droplets. (a) T_N = 484°C at $\dot{T}$ = 0.5 °C/S; (b) T_N = 370°C at $\dot{T}$ = 500°C/S.

transition from heterogeneous surface to homogeneous volume dependent nucleation is of interest. Of course, with sufficient undercooling, homogeneous nucleation is believed to be possible provided a glass transition does not develop. However, due to the statistical nature of nucleation and sample size distribution effects, an abrupt transition in kinetics is not expected with changing undercooling or droplet size. One way of examining such a case involves considering when J_v and J_a become of comparable magnitude for a given powder size (27). For a monodisperse droplet size this can be expressed in terms of the ratio

$$\frac{J_v}{J_a} = \frac{\Omega_v}{\Omega_a} \exp \left[- \frac{\Delta G^*}{kT} (1-f(\theta))\right] \qquad (13)$$

When J_v and J_a become comparable at one nucleus/drop-sec. for a 20 μm volume element, then $J_v = 2.4 \times 10^8$/sec. cm^3 and $J_a = 7.9 \times 10^4$/sec. cm^2. Further, if sensible nucleation is represented by a limiting value for ΔG^* of 60 kT, then the resulting value of $f(\theta)$ from equation (13) implies a contact angle of about 118°C when $\Omega_v = 10^{35}$ cm^{-3} s^{-1} and $\Omega_a = 10^{27}$ cm^{-2} s^{-1} for a constant catalytic site density. For nucleants described by contact angles in excess of 118°, homogeneous nucleation is expected to dominate against a background of heterogeneous nucleation as demonstrated in Figure 13. This behavior is mainly related to the larger value of the prefactor for homogeneous nucleation.

Another feature to consider when examining the conditions for a kinetic transition is the influence of the possible range of catalytic site densities represented by variation in Ω_a in Equation (13) as well as changing catalytic potencies as reflected by $f(\theta)$. The result of these factors on the observation of a kinetic transition is illustrated schematically in Figure 14 by transformation diagrams. In both cases the broken curve represents homogeneous nucleation. In Figure 14a a set of transformation curves is shown for conditions of relatively low catalytic site density where $(\Omega_a/\Omega_v) \simeq 10^{-20}$. A kinetic transition from heterogeneous surface to homogeneous nucleation is possible in this case depending on the cooling rate and catalytic potency. On the other hand, another set of transformation curves is given in Figure 14b for relatively high catalytic site density $(\Omega_a/\Omega_v) \simeq 10^{-10}$. For this case heterogeneous nucleation dominates and a heterogeneous to homogeneous kinetics transition is not expected by changing cooling rate or catalytic potency.

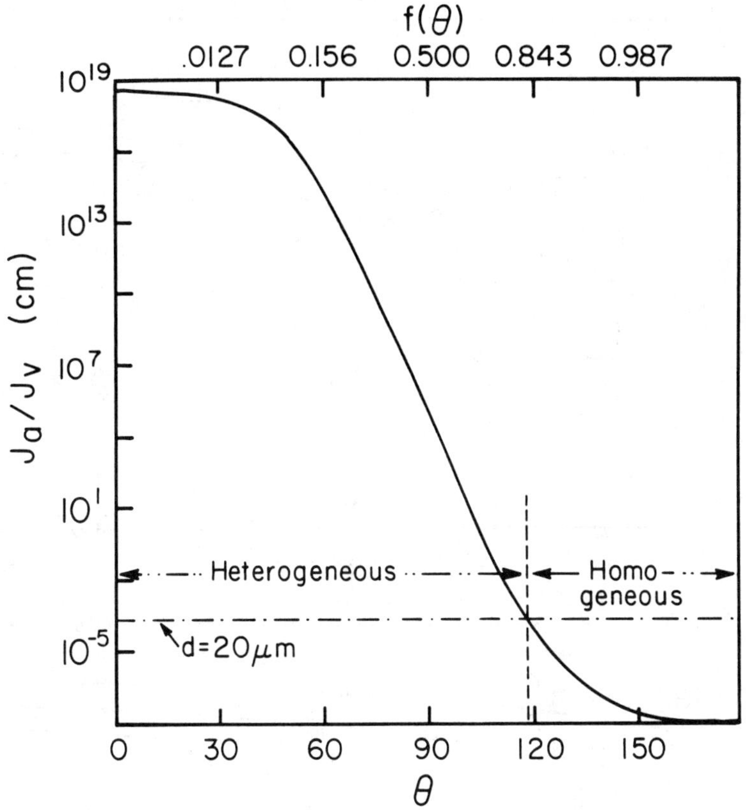

Figure 13 Relative magnitudes of heterogeneous surface to homogeneous nucleation in a 20 μm droplet as a function of contact angle.

The conditions examined in Figure 13 and Figure 14 demonstrate that it is not necessary for a given undercooled volume to be free of all nucleants; it is only necessary for the isolation induced by subdivision or other processes to mitigate the influence of the most potent nucleants in order to observe the onset of homogeneous nucleation. This observation can have a number of important implications extending for example to the understanding of grain size effects during rapid solidification processing.

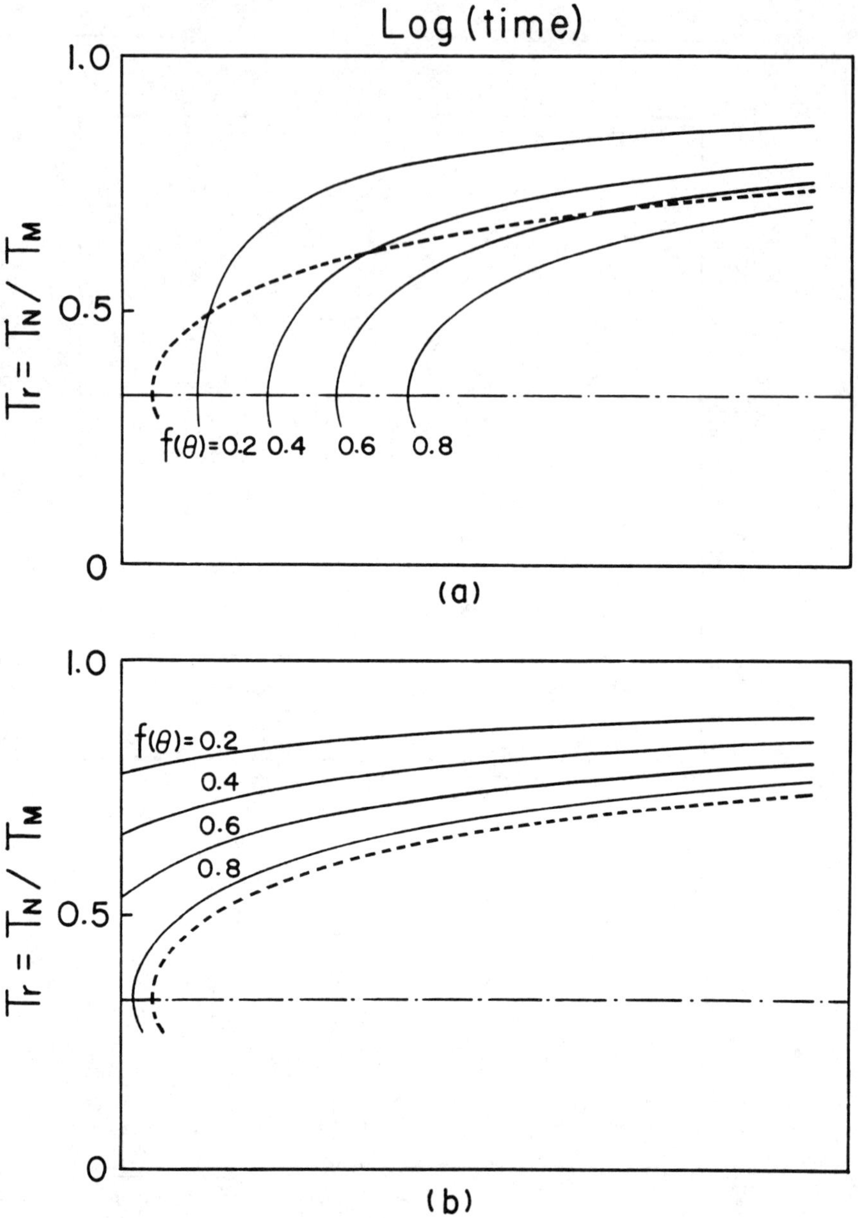

Figure 14 Time-temperature-transformation diagram representation of the transitions in nucleation kinetics that may occur during continuous cooling of undercooled liquids.

Multiple Nucleation

The thermal conditions immediately following the initial nucleation event in an undercooled liquid volume have an important influence on the likelihood of additional nucleation events in the liquid volume. The thermal history reflects the net result of two competing effects which are the imposed external cooling and the recalescence arising from the latent heat emitted by a rapidly advancing solidification front. The solidification of undercooled liquid droplets during slow external cooling invariably yields single grain structures, but during rapid quenching multi-grain structures have been reported. This behavior suggests that rapid quenching may promote multiple nucleation.

Several detailed numerical calculations (28,29) have been developed to treat the thermal history during solidification following an initial nucleation event. However, for small departures from the initial nucleation temperature, T_N, a useful approximate model has been developed recently (27) which accounts for the basic features of the thermal history. The result of the model is illustrated in Figure 15a where the increment in undercooling $(T_N - T_N^*)$ from initial nucleation is

$$(T_N - T_N^*) = \left[\frac{C_p}{9\pi}\frac{\nu_s}{\Delta H}\left(\frac{\dot{T}}{V}\right)^3\right]^{1/2} \tag{14}$$

and occurs after a time

$$t^* = \left[\frac{\nu_s\, C_p\, \dot{T}}{4\pi\, V^3\, \Delta H}\right]^{1/2} \tag{15}$$

where V is the radial crystal growth rate which is assumed constant near T_N and ν_S is the reciprocal of the nucleant density. An undercooling increment may develop if it is considered that immediately following a single nucleation event, V may be large, but the crystal size is small so that recalescence may not overcome the external cooling.

The corresponding change in nucleation rate associated with the increment in undercooling is shown in Figure 15b. The effective number of nucleation events during the period of extra undercooling can be estimated by the product of the shadowed area in Figure 15b and the nucleation site area as

$$\delta I = a\, J_a\, (T_N^*)\, t^* \qquad (16)$$

When δI exceeds unity, multiple grain nucleation is possible. A schematic illustration of the envelope of processing conditions related to equations 14-16 is presented in Figure 16.

For multiple nucleation, processing conditions that favor large values of $(T_N - T_N^*)$ and t^* are of interest. As indicated in Figure 16 rapid cooling rate and low nucleant densities are important, but low solidification velocities are most important in maximizing t^*. In this regard it is known that the growth of multiphase solid with solute partitioning is limited to about 10 cm/sec (30) while solidification of pure metals or superaturated solid solutions can be faster by a factor of 10^2 or more. Thus, multiple nucleation should be favored in eutectic alloys near glass forming conditions as reported recently for Nb-Si alloys where grain sizes on the order of 15 nm were observed (31). The opportunity for multiple nucleation during partitionless solidification is limited although a high value of Ω_a (i.e. high site density) may allow for some extra grain formation as noted for Al-Si alloys (28).

Metastable Phase Selection and Formation

A large number of examples of metastable phase formation have been demonstrated during rapid quenching experiments. Difficulties in nucleation and growth of stable phases that can occur during RSP lead to an expanded utilization of phase diagram concepts; in particular these constraints require a close examination of the metastable equilibrium features of phase diagrams. At the onset it should be emphasized that a metastable equilibrium is a true reversible equilibrium. It is also useful to note that with stable equilibrium, there is a certainty of phase prediction. With metastable equilibrium this certainty of prediction becomes a choice of product phases depending on the particular constraints imposed on the system and the controlling phase selection kinetics. There is a single stable equilibrium phase diagram, but there are several possible metastable phase diagrams for a given alloy system. While the possible reaction paths leading to metastable phase formation in undercooled alloy systems are numerous, a consideration of thermodynamic constraints can be useful in providing guidelines and limiting bounds on the reaction sequences that are possible.

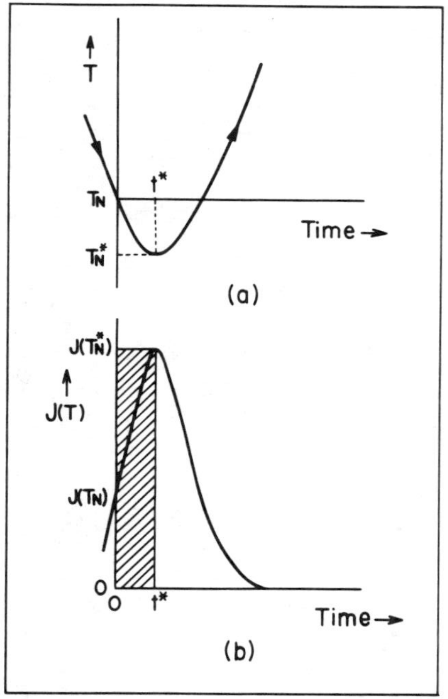

Figure 15 Schematic profiles following nucleation during continuous cooling of an undercooled liquid; (a) temperature and (b) corresponding nucleation rate.

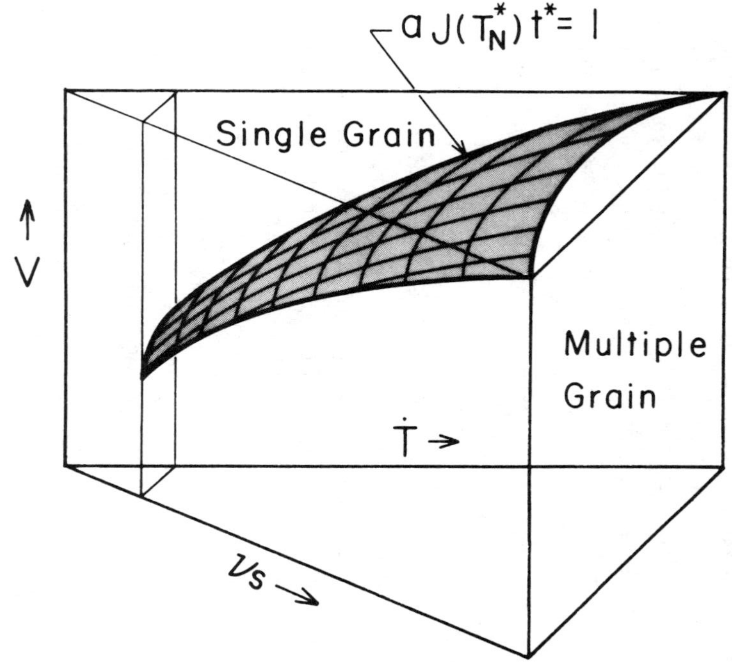

Figure 16 Diagrammatic representation of the envelop of solidification variables that can contribute to multiple nucleation in droplets during continuous cooling.

A thermodynamic analysis is of most value when it can also provide an estimate of the potential metastable intermediate phases that can enter in the kinetic competition. Since the driving force for the formation of a metastable phase is directly related to the amount of undercooling below the melting temperature of the phase, a high level of undercooling can expand the nucleation product selection to an increased variety of types of structures. This is illustrated by the nucleation transition from FCC to BCC in Fe-Ni alloys, stainless steels and tool steels. Other examples have been observed for Pb-Bi, Bi-Cd and Pb-Sn alloys as well as pure Ga, Bi and Sb. For Pb-Bi alloys a multiple phase selection transition has been observed in droplets. As the level of undercoolng increases in a Pb-Bi eutectic alloy the equilibrium eutectic mixture is replaced by a supersaturated (HCP) phase which in turn is superseded by a metastable intermediate phase at the highest undercooling. In addition, for Pb-Sn, Sn-Ge, Ge-Sb and Cd-Sb alloys, the transition in product phases has been related to the catalytic affect of specific heterogeneous sites (32). One type of effective nucleation site is a primary solid solution (33). When the α phase is a potent catalyst, the continued growth of α is possible during cooling in two-phase (α+L) samples to form a supersaturated α product. In eutectic alloys the usual $L \rightarrow \alpha + \beta$ reaction may be suppressed when α is strongly catalytic in promoting the nucleation of a metastable intermediate γ phase. The formation of an equilibrium β phase may not occur if a two phase (α+L) sample is undercooled below the metastable γ phase melting point to allow γ to nucleate on α. One way of summarizing the potential for structure modification in alloys is to construct metastable phase diagrams for the various combinations of phases (34).

The metastable phase diagram approach can be illustrated by reference to the schematic phase diagram shown with metastable phase equilibria in Figure 17. Under conditions of local interfacial equilibrium an alloy of composition C_A can follow any one of several pathways depending on the reaction sequence and the operating nucleation and growth kinetics at the temperature noted in Figure 17. For example, if the β phase forms, the β/L interface should give compositions of $C_{L/\beta}$ and $C_{\beta/L}$. If the other crystalline phases (i.e. α, A_3B and AB) do not form, then the liquid at $C_{L/\beta}$ will reach its glass transition and vitrify even in the presence of crystalline β. If the compound A_3B forms, the compound/liquid interface will involve compositions C_{A_3B} and C_{L/A_3B}. The liquid at C_{L/A_3B} can form β phase, glass or other metastable structures. A number of other possibilities exist including those involving metastable intermediate phases, the devitrification of a glass at C_A during heating and solid state

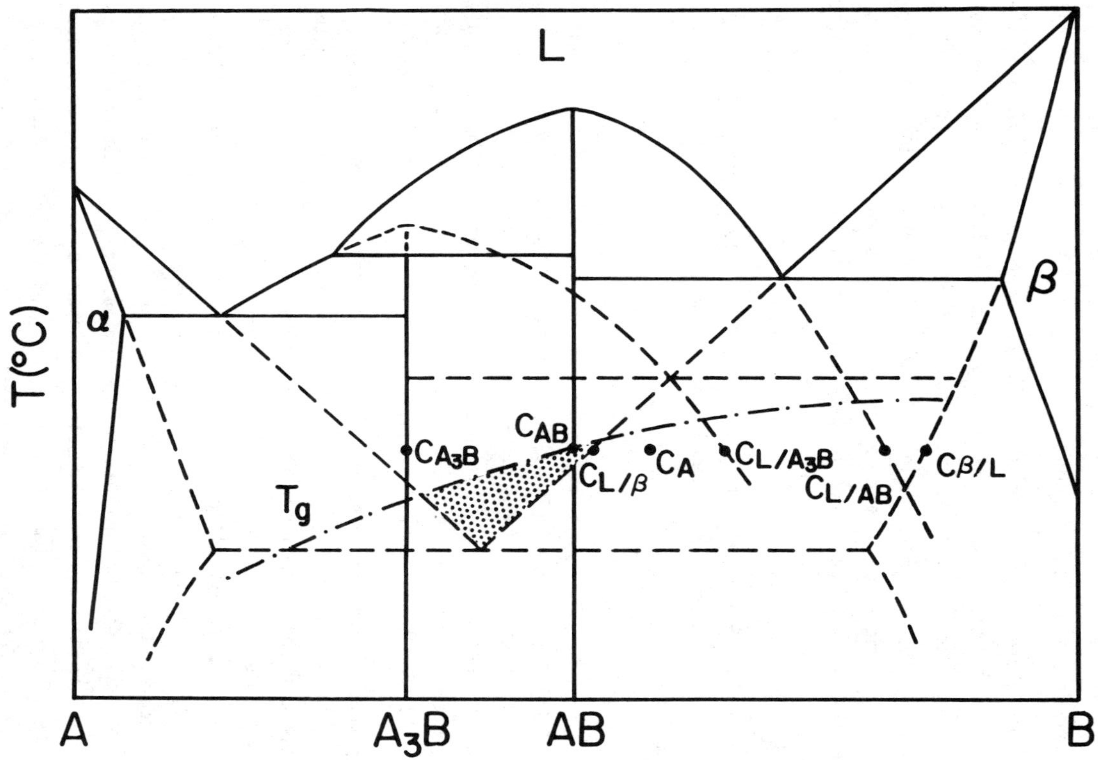

Figure 17 Schematic phase diagram indicating some of the possible metastable phase equilibria that may be selected during solidification of undercooled liquid alloys.

interdiffusion reactions. The metastable equilibria given in Figure 17 provide a clear explanation for the observation of glass formation during some solid state interdiffusion reactions (35). When the compounds A_3B and AB have unfavorable formation kinetics, a diffusion couple of pure A and B will form α and β solutions which will effectively "melt" to form a glass when the interdiffusion treatment is carried out below T_g, but above the metastable eutectic involving α and β.

Metastable phase diagrams are often useful in providing guidance on the range of available product structure options when equilibrium phases do not exhibit favorable formation kinetics. Thus, for an alloy that may exhibit metastable products, it is possible to select a particular phase by controlling the degree of undercooling prior to solidification and by providing a high density of favorable catalytic sites. The precise kinetic prescription for the most potent site for a given product phase structure is uncertain at this time although it appears that similarity of structural arrangement and low lattice misfit disregistry along close packed planes and directions may be important in some cases; especially at relatively low undercooling levels.

Summary

Rapid solidification has stimulated a renewed interest in understanding and controlling undercooling in liquids. As a result, more effort has been directed to the examination of nucleation in undercooled liquids. This increase of attention to nucleation kinetics has been yielding new insights on structure control from several viewpoints.

From a thermodynamic perspective the consideration of metastable phase diagrams has been useful in defining the range of operating kinetics and the reaction pathways for some nucleation processes. A number of important processing variables have been identified in terms of their influence on the optimization of the undercooling level. While control of these variables has allowed for the attainment of new maximum undercooling limits for a number of metals, the observed crystallization kinetics emphasizes the importance of heterogeneous nucleation in dominating the initial product selection in most cases. In addressing the examination of heterogeneous nucleation, good progress has been made in experimentally defining the proper conditions for the valid measurement of the catalytic potency of known heterogeneous nucleation sites. Based upon these experimental advances and new observations on nucleation in undercooled liquids, the operation of competitive kinetics to promote the formation of different product structures has been established and is now receiving a quantitative evaluation.

Overall, it appears that the experimental experience and methods of nucleation kinetics analysis are evolving to a level that should allow for progress beyond the rationalization of post solidification structural observations, to a realization of some predictive capability for phase selection structural control and alloy design in the near future.

Acknowledgement

The support of the ARO (DAAG-29-80-K-0068) and the NSF (DMR-79-15802) is gratefully acknowledged. It is a pleasure to acknowledge many stimulating discussions with Dr. W. J. Boettinger of NBS.

References

1. D. Turnbull and R. E. Cech, J. Appl. Phys., 21 (1950) 804.
2. D. Turnbull, J. Appl. Phys., 21 (1950) 1022.
3. D. Turnbull, J. Chem. Phys., 20 (1952) 411.
4. D. Turnbull and B. Vonnegut, Ind. Eng. Chem., 44 (1952) 1292.
5. D. Turnbull, Trans. AIME, 221 (1961) 422.
6. W. C. Wang and C. S. Smith, Trans. AIME, 188 (1950) 136.
7. T. Z. Kattamis and M. C. Flemings, Trans. AIME, 236 (1966) 1523.
8. S. Y. Shirashi and R. G. Ward, Canadian Met. Quart., 3 (1964) 117.
9. J. H. Perepezko, Mat. Sci. Eng., 65 (1984) 125.
10. J. H. Perepezko, Rapid Solidification Processing: Principles and Technologies II, R. Mehrabian, B. H. Kear and M. Cohen, eds. (Claitors Pub., Baton Rouge, LA, 1980) 56.
11. J. H. Perepezko and J. S. Paik, Rapidly Solidified Amorphous and Crystalline Alloys, B. H. Kear, B. C. Giessen and M. Cohen, eds., (North Holland, Amsterdam, 1982) 49.
12. B. A. Mueller and J. H. Perepezko, to be published.
13. G. E. P. Box, W. G. Hunter and J. S. Hunter, Statistics for Experimenters (J. Wiley, N.Y. 1978).
14. J. H. Perepezko, S. E. LeBeau, B. A. Mueller and G. J. Hildeman, Rapidly Solidified Powder Aluminum Alloys ASTM STP 890, M. E. Fine and E. A. Strake, Jr., eds., (ASTM, Philadelphia, 1986) 118.
15. L. Bosio, A. Defrain and A. Epelboin, J. Phys. (Paris), 27 (1966) 61.
16. G. M. Pound and V. K. La Mer, J. Amer. Chem. Soc., 74 (1952) 2323.
17. D. Turnbull, Contemp. Phys., 10 (1969) 473.

18. D. Turnbull, Prog. Mat. Sci. - Chalmers Ann. Vol., J. W. Christian, P. Haasen and T. B. Massalski, eds., (Pergamon Press, England 1981) 269.
19. J. H. Perepezko and J. S. Paik, J. Non-Cryst. Solids, 61 (1984) 113.
20. M. Hillert, Acta Met., 1 (1953) 764.
21. J. S. Paik and J. H. Perepezko, to be published.
22. G. R. Wood and A. G. Walton, J. Appl. Phys., 41 (1970) 3027.
23. J. P. Hirth, Met. Trans., 9A (1978) 401.
24. J. H. Perepezko, B. A. Mueller, J. J. Richmond and K. P. Cooper, Rapidly Quenched Metals V, S. Steeb and H. Warlimont, eds. (Elsevier, Amsterdam 1985) 43.
25. C. Barromee - Gautier, B. C. Giessen and N. J. Grant, J. Chem. Phys., 48 (1968) 1905.
26. B. A. Mueller, J. J. Richmond and J. H. Perepezko, Rapidly Quenched Metals V, S. Steeb and H. Warlimont, eds. (Elsevier, Amsterdam 1985) 47.
27. W. J. Boettinger and J. H. Perepezko, Rapidly Solidified Crystalline Alloys, S. K. Das, B. H. Kear and C. M. Adam, eds. (TMS-AIME, Warrendale, PA 1985) 21.
28. C. G. Levi and R. Mehrabian, Met. Trans., 13A (1982), 221.
29. T. W. Clyne, Met. Trans., 15B (1984) 369.
30. W. J. Boettinger, Rapidly Solidified Amorphous and Crystalline Alloys, B. H. Kear, B. C. Giessen and M. Cohen, eds. (North Holland, Amsterdam, 1982) 15.
31. L. Bendersky, W. J. Boettinger and J. H. Perepezko submitted to Jnl. Mat. Sci.
32. J. H. Perepezko, Y. Shiohara, J. S. Paik and M. C. Flemings, Proc. 3rd Int. Conf. on Rapid Solidification Processing: Principles and Technologies, R. Mehrabian, ed. (Nat. Bur. Stds., Washington DC 1983) 28.
33. J. H. Perepezko and I. E. Anderson, Synthesis and Properties of Metastable Phases, T. J. Rowland and E. S. Machlin, eds. (TMS-AIME, Warrendale, PA 1980) 31.
34. J. H. Perepezko and W. J. Boettinger, Mat. Res. Soc. Symp. Proc., 19 (1983) 223.
35. R. Schwarz and W. L. Johnson, Phys. Rev. Lett., 51 (1983) 415.

DENDRITIC GROWTH OF AN UNDERCOOLED NICKEL-TIN ALLOY

Merton C. Flemings,
Yuh Shiohara,
Yanzhong Wu,
and Thomas J. Piccone

Department of Materials Science and Engineering
Massachusetts Institute of Technology
Cambridge, MA 02139

Dendrite growth velocities were measured in undercooled Ni-25wt% Sn Alloy by both cinematography and high-speed optical temperature measurement. Measured velocities ranged from about 0.07 meters per second to 16 meters per second. Measured velocities agree well with the theoretical model of Lipton, Kurz, and Trivedi, and with the similar model of Boettinger and Coriell. The high-speed cinematography revealed large dendrite-like isotherms; it is concluded that they comprise an array of very much finer, solute diffusion controlled dendrites.

Optical measurements of the sample surface temperature showed a period of "rapid recalescence" which comprises the time required for the dendrites to sweep across the sample surface, plus the much smaller time required for the supercooling between the dendrite arms to dissipate. This rapid recalescence was followed by a period of "slow recalescence" which is believed to result from the ripening of the very fine dendritic array. The maximum recalescence temperature which the alloy reached is that predicted assuming equilibrium of liquid and solid phases at the maximum recalescence temperature. Microstructures of fully solidified samples are strongly influenced by the ripening which occurs during and after recalescence. Some general implications of these results are discussed.

Introduction

This paper summarizes results of recent experimental studies on solidification of undercooled melts of Ni-25% Sn alloy; detailed experiments are described elsewhere (1,2,3). The Ni-25% Sn is a hypoeutectic alloy which, at equilibrium, would begin to solidify at the liquidus, 1497 K, and would complete its solidification at the eutectic temperature, 1405 K. The alloy composition intersects the extrapolated solidus at 1279 K. Experiments were conducted with initial undercoolings up to 260 K, corresponding to a dimensionless undercooling ($C_p\Delta T/\Delta H$) of 0.87 below the liquidus or 0.14 below the extrapolated solidus, where ΔT is undercooling, C_p is specific heat, and ΔH is heat of fusion.

These undercoolings did not approach the unit undercooling below the solidus that would be necessary to achieve plane front solidification in the absence of solute trapping. However, the high interface velocity at the largest undercooling (16 m/s) makes solute trapping a possibility. We begin by comparing our experimental results to recent analyses of solidification of undercooled melts which assume dendritic growth and no solute trapping, and will show that our results agree well with one of these analyses (that of Lipton, Kurz and Trivedi (4)) over the entire range of our experiments. The remainder of this paper deals with effects of ripening on recalescence and on final structures observed.

Mathematical Models for Dendritic Growth

The most recent model for dendritic growth in undercooled alloy melts is that of Lipton, Kurz, and Trivedi (4) (LKT) which may be written:

$$\Delta T = T_L - T_\infty = \frac{\Delta H}{C_p} Iv(P_t) + m_L C_0 \left[1 - \frac{1}{1-(1-k_e)\,Iv(P_c)}\right] + \frac{2\gamma}{r_t \Delta S} \quad (1)$$

where ΔT is the bulk undercooling, T_L is the liquidus temperature of the alloy, T_∞ is the bulk undercooled liquid temperature, m_L is the slope of the liquidus, C_0 is the initial alloy composition, k_e is the equilibrium solute partition ratio, γ is the solid-liquid interfacial energy, ΔS is the entropy of melting, P_t and P_c are thermal and solutal Peclet numbers, respectively, which are defined as $P_t=Rr_t/2\alpha$ and $P_c=Rr_t/2D$, in which R is the denrite tip velocity, r_t is the dendrite tip radius, α is thermal diffusivity, D is the solute diffusion coefficient in the liquid alloy, and Iv(P) is the Ivantsov function, which is equal to P exp(P) $E_1(P)$, where $E_1(P)$ is the exponential integral function (5). The three terms in this equation represent the effects of thermal diffusion (ΔT_α), solute diffusion (ΔT_D), and interfacial energy (ΔT_r), on the dendrite tip temperature, respectively.

In order to obtain a second equation for determining R and r_t, the marginal stability criterion is applied with the dendrite tip radius, r_t, being set equal to the shortest wavelength which can form under the local growth conditions at the tip (6,7,8). The dendrite tip radius is then given by (4):

$$r_t = \frac{\gamma / (\Delta S\, \sigma^*)}{\dfrac{P_t \Delta H}{C_p}(1-n) + \dfrac{2 m_L P_c C_0 (k_e - 1)}{1 - (1-k_e)\, Iv(P_c)}(1+g)} \tag{2}$$

with

$$n = \frac{1}{\sqrt{1 + 1/(\sigma^* P_t^2)}} \tag{2a}$$

and

$$g = \frac{2 k_e}{1 - 2k_e - \sqrt{1 + 1/(\sigma^* P_c^2)}} \tag{2b}$$

where σ^* is the stability constant. The parabolic eigenstate model provides $\sigma^* = 1/4\pi^2 = 0.025$ (6,7). The results of calculations based on the foregoing equations and applied to Ni-25% Sn alloy are given in Figure 1 and labeled "LKT".

For small Peclet numbers, both n and g approach zero and equation (2) reduces to that of Lipton, Glicksman, and Kurz (9) (labeled "LGK"). When n can be neglected but g cannot, the above equations reduce to the model of Boettinger and Coriell (10) (labeled "BC"). For the alloy and conditions of this study, the LKT and BC analyses provide equivalent results. Results of calculations are shown in Figures 1 and 9. Thermophysical properties used in the calculations are listed in the Appendix.

As Trivedi has shown, thermal undercooling (at the dendrite tip) in the growth of undercooled melts is small compared with solute undercooling at undercoolings less than about the equilibrium solidification freezing range. Therefore, over this temperature range, growth of dendrites in undercooled melts can be described by the same equations used for constrained growth in non-undercooled melts - e.g., Trivedi and Somboonsuk (11,12). The results of this analysis, also plotted in Figure 1 and labeled "TS", agree well with the other analyses up to about 80 K undercooling. A central assumption of all these analyses is the important one of equilibrium interface kinetics, and so it is expected that, at sufficiently high tip velocities, even the LKT and BC analyses may break down due to solute trapping at the tip. Solute trapping is expected to become important at interface velocities within an order of magnitude of 10 m/s (13,14).

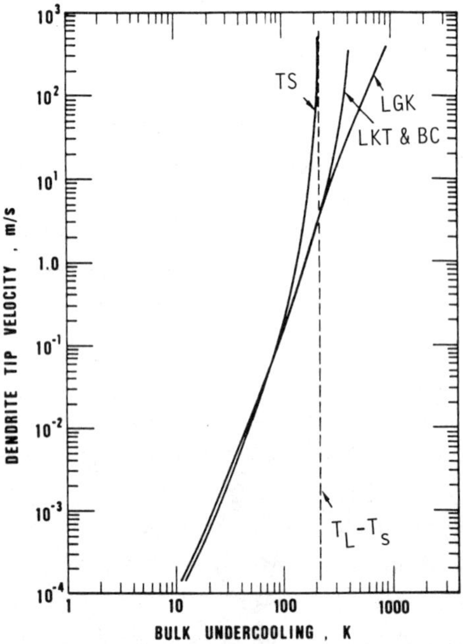

FIGURE 1. Dendrite tip velocity vs. bulk undercooling for Ni-25wt% Sn alloy. Calculations based on the models of Lipton-Glicksman-Kurz (LGK) (9), Lipton-Kurz-Trivedi and of Boettinger-Coriell (LKT and BC) (4,10), and Trivedi-Somboonsuk (TS) (12). From ref. (2).

Measurement of Dendrite Tip Velocity

In our experimental study (1,2,3), the solidification of Ni-25% Sn alloy was examined in small specimens contained within a transparent glass slag, as shown schematically in Figure 2. In some experiments, the samples used were nearly spherical; in others they were rectangular in cross-section to allow better focusing for high-speed cinematography. In the latter, the camera was focused on one flat face and an optical pyrometer on another for temperature measurement.

The sample was undercooled and allowed to nucleate; high-speed photographic measurements (500 frames per second) were made during the period of recalescence. These showed bright "dendrites" moving across the specimen as recalescence occurred. Figure 3 shows four frames from one such experiment, a sample undercooled 78 K; Figure 4 shows results from a similar experiment in which undercooling was 121 K. The Figures are photographic negatives, as the bright "dendrites" appear dark. Note that the "dendrites" at the higher undercooling are smaller and move more rapidly than those at the lower undercooling.

Figure 5 is a composite sketch made of the advancing dendrites of Figure 4, which grew at 121 K undercooling. Note that four dendrite tips advanced along paths marked A, B, C, and D; one location of dendrite thickening was along path E. The location of the advancing solid fronts along each of these lines is plotted in Figure 6. The dendrite tips moved substantially faster than they thickened, and one dendrite, D, slowed abruptly as it impinged on the advancing dendrite B. Except for such slowing due to impingement, the "dendrites" were usually seen to grow at constant velocity. Velocities taken from the slopes of curves such as these are plotted versus bulk undercooling in Figure 7. Dendrite tip velocity increases rapidly with increasing undercooling. These velocities are also plotted in Figure 8 as the large black points and compared with the LKT-BC analysis discussed earlier. The fit is excellent. Other points on this curve were obtained by thermal measurements as discussed below; these also agree well with the theoretical curve. It is interesting to note that the points for the highest undercoolings agree well with this LKT-BC model but not with the other, more approximate models shown in Figure 1. It is also interesting that these points extend up to 16 m/s. Solute trapping, if it is occurring, is insufficient to cause significant deviation of dendrite tip velocity from that predicted by this theoretical model.

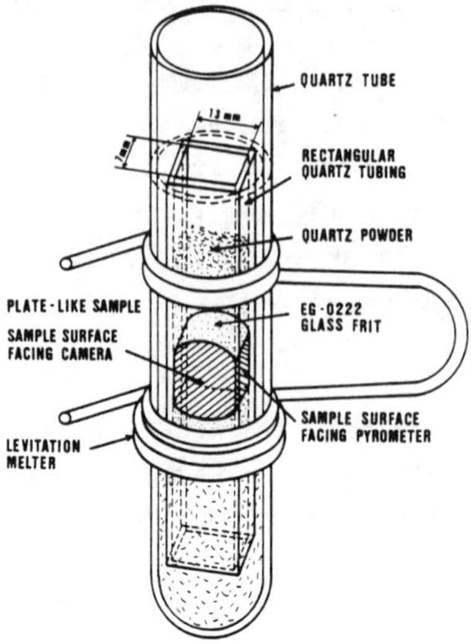

FIGURE 2. Schematic representation of sample assembly.

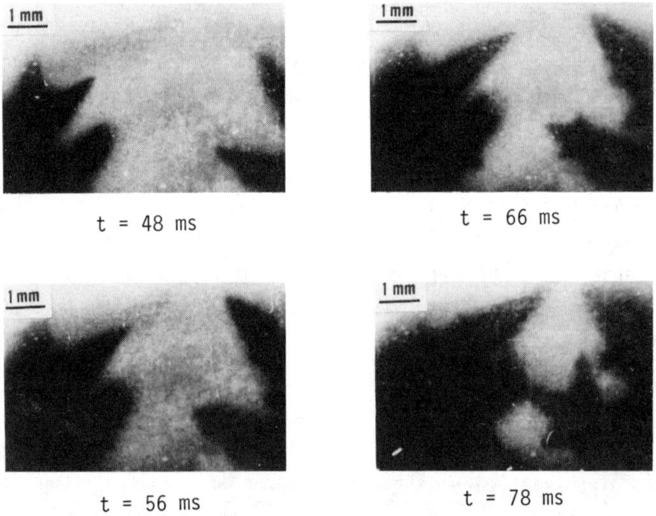

FIGURE 3. Photographs taken at four different times during recalescence of a Ni-25wt% Sn hypoeutectic alloy sample undercooled by 78 K. The dark area is the solidifying portion; the light area is remaining undercooled melt. From ref. (2).

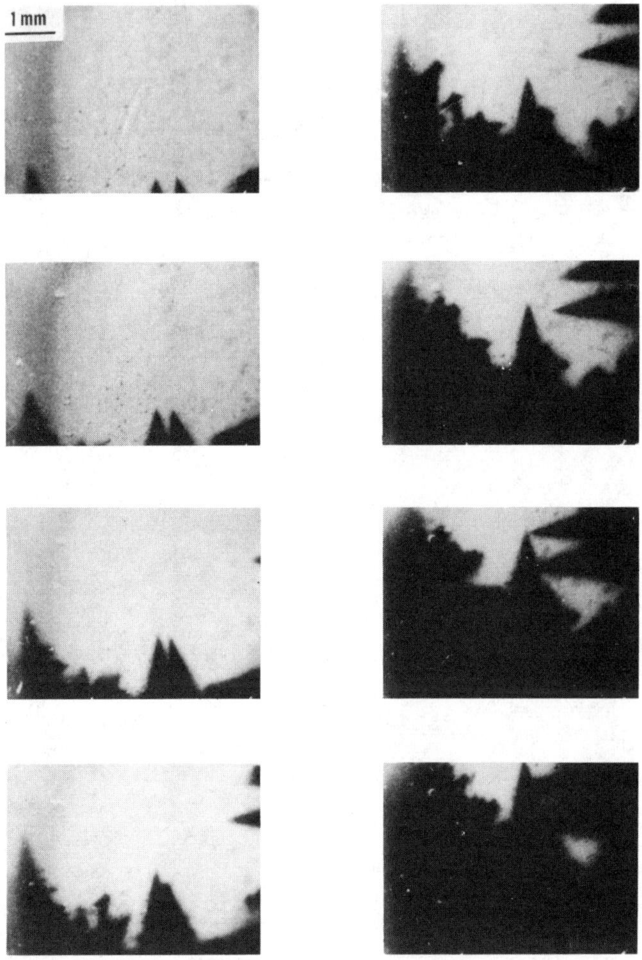

FIGURE 4. Eight photographs, taken at 2 ms intervals during recalescence of a Ni-25wt% Sn hypoeutectic alloy sample undercooled by 121 K. From ref. (2).

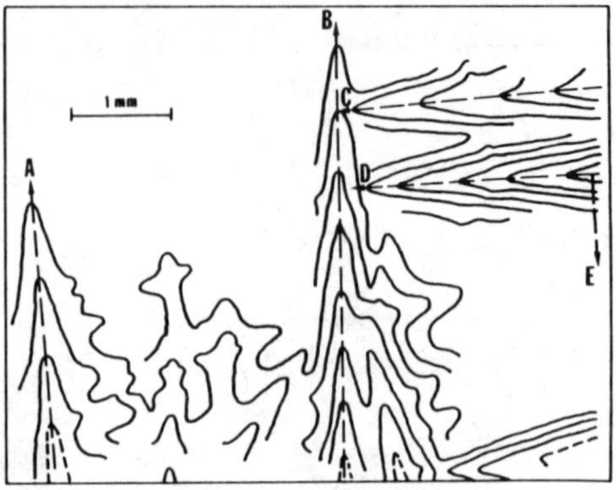

FIGURE 5. Drawing of the advancing solid/liquid interface of a Ni-25wt% Sn hypoeutectic alloy sample during recalescence. Undercooling: 121 K. The time interval between lines is 2 ms. Drawings are traces from photographic negatives. From ref. (2).

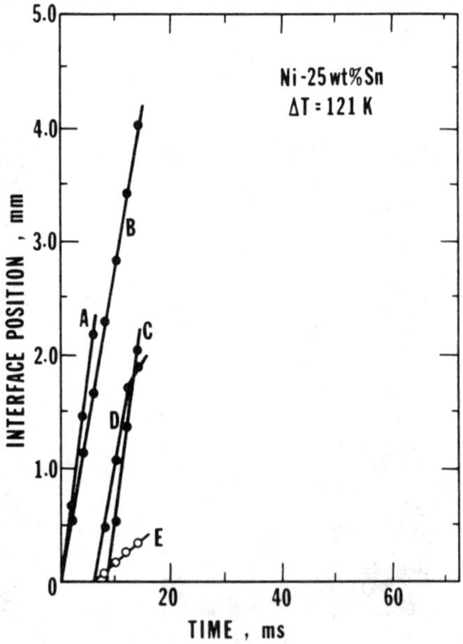

FIGURE 6. Movement of several parts of an advancing solid/liquid interface versus time for a Ni-25wt% Sn hypoeutectic alloy sample undercooled by 121 K. A, B, C, D, and E correspond to dendrites and directions shown in Figure 5. From ref. (2).

We see in Figure 8 excellent agreement of theory with experiment, but there is a major discrepancy. The "dendrites" that we observe by photographic measurements on the surface of the sample are far larger than those predicted by any of the analyses. Figure 9 is a comparison of these measurements with calculations of the three analyses discussed above. The LKT and BC analyses predict dendrite tip radii ranging from about 10^{-6} m (1 μm) at modest undercoolings to about 10^{-8} m (10 nm) at 200 K undercooling. The experimentally measured tip radii of the "dendrites" observed by high-speed cinematography are two or more magnitudes greater than this. It is concluded that the most likely explanation for this large discrepancy is that the "dendritic" structure seen by high-speed cinematography is one whose size scale is determined by the requirement of dissipation of heat, but that this "dendrite" contains an internal structure of very fine solute diffusion controlled dendrites, as sketched in Figure 10. The rate of advance of the large "dendrite" tip is determined by the rate of advance of the fine solute diffusion controlled dendrites (e.g., by the LKT-BC analyses). However, the simultaneous requirement of heat dissipation results in these solute dendrites moving forward in an array, whose morphology is determined by the requirements of dissipation of heat.

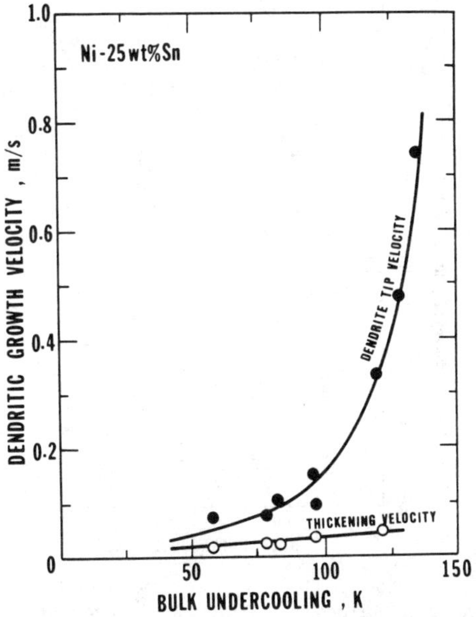

FIGURE 7. Dendrite tip velocity and dendrite thickening velocity versus initial bulk undercooling for Ni-25wt% Sn alloy samples, measured by high-speed cinematography. From ref. (2).

Optical Measurement of Recalescence

The sample temperature was measured optically using a pair of silicon photocells, a pair of matched amplifiers, and a digital storage oscilloscope. The response time of the system was less than 20 μs. Two typical curves of recalescence temperature versus time are shown in Figure 11, for 205 K and 265 K undercoolings, respectively. After nucleation, each of curve shows a period of "rapid recalescence" (between the two arrows marked on the curves), followed by a period of much slower recalescence. The times for these rapid recalescences are plotted in Figure 12 versus undercooling for a wide range of undercoolings. Included are results for samples of both the hypoeutectic alloy discussed herein and of the eutectic alloy (32.5wt% Sn), which is discussed in a separate paper (1) but will not be dealt with here. The points lie on a single curve, from times on the order of 1.0 s at 40 K undercooling to less than 10^{-3} s for undercoolings greater than 200 K. We conclude that

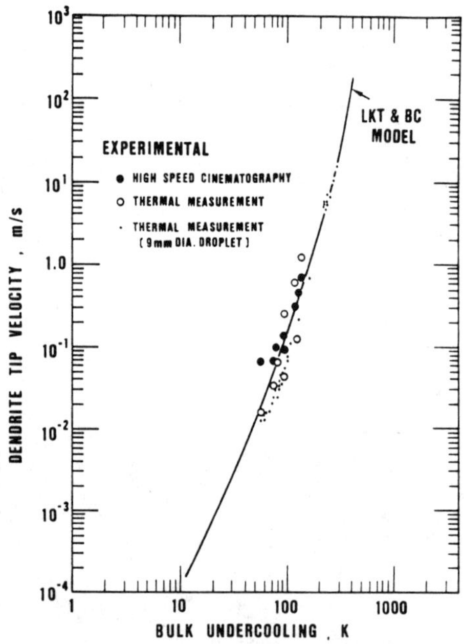

FIGURE 8. Dendrite tip velocity versus bulk undercooling for Ni-25wt% Sn alloy. Experimental results and calculated curves based on the models developed by Lipton, Kurz and Trivedi (4) and Boettinger and Coriell (10). From ref. (2).

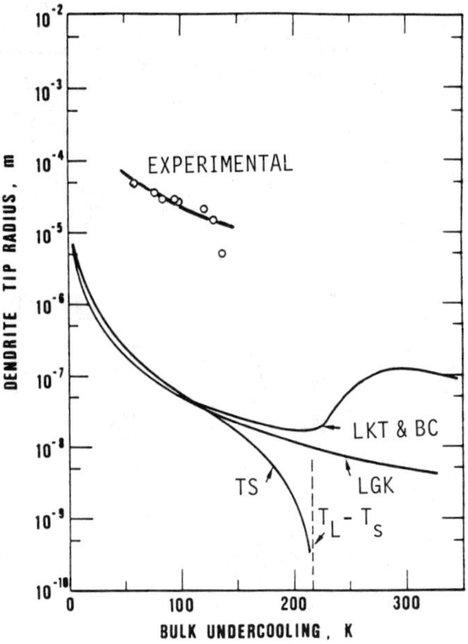

FIGURE 9. Dendrite tip radius vs. bulk undercooling for Ni-25wt% Sn alloy. From ref. (2).

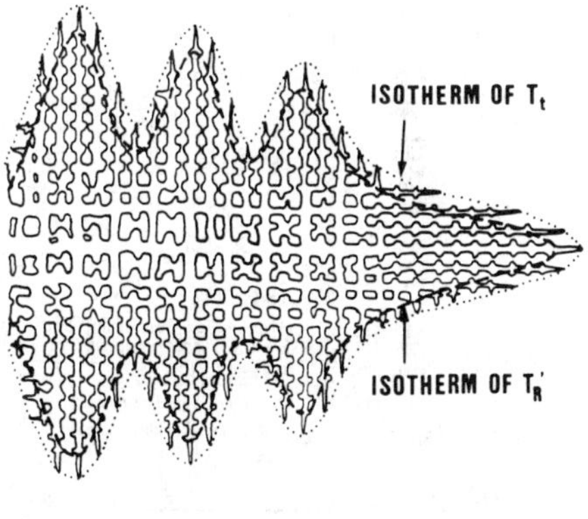

FIGURE 10. Schematic illustration of large "dendritic isotherm" growing into an undercooled melt. From ref. (2).

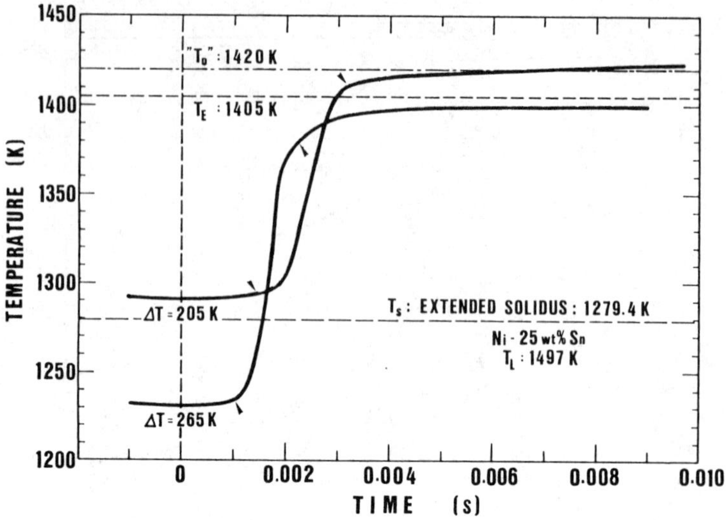

FIGURE 11. Recalescence behavior of two samples of hypoeutectic alloy. From ref. (1).

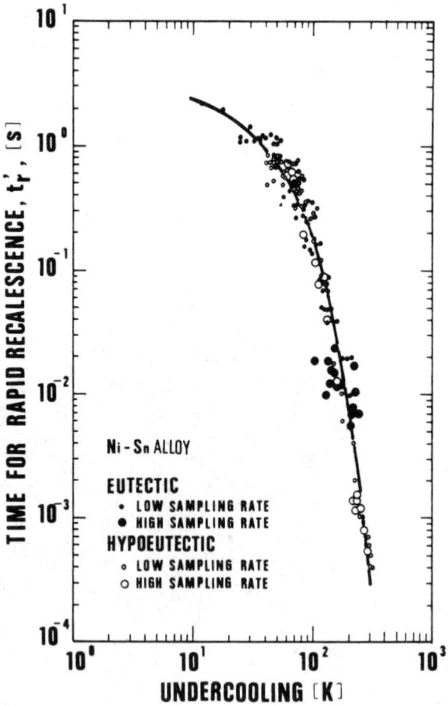

FIGURE 12. Recalescence time versus undercooling. From ref. (1).

this period of rapid recalescence comprises the time during which dendrites sweep across the sample surface plus the (much shorter) time for dissipation of residual supercooling within the dendrite array. Therefore, the dendrite velocity along the sample surface can be estimated by dividing this rapid recalescence time into a length measurement approximately equal to the sample diameter. This calculation was done for all samples known to have recalesced approximately adiabatically, and the results are plotted in Figure 8. Note that these results agree well with those of the photographic measurements of dendrite velocity, and they extend those measurements over a much broader range of undercoolings.

The temperature interval of "slow recalescence" is always much less than that of rapid recalescence, and increases with increasing undercooling, as shown in Figure 13. At least up to about 200 K undercooling, this interval varies in approximately the same way as does ΔT_r from the LKT-BC analysis, and over this range is nearly half the calculated ΔT_r at the dendrite tip. This result strongly suggests that the period of slow recalescence is one in which the temperature rise results directly from the effect of ripening on the equilibrium melting point. Viewing it in the simplest possible way, the undercooling due to the radius of curvature of a cylindrical dendrite arm is one-half of that of the cap of the advancing spherical tip of the dendrite; hence, undercooling after dissipation of interdendritic supercooling should not be more than one-half that due to the curvature at the growing tip. The conclusion that slow recalescence results from ripening is strengthened by plots such as that of Figure 14, in which maximum recalescence temperature minus sample temperature is shown versus time on a log-log scale. The period of slow recalescence has a linear slope and obeys a $t^{-1/3}$ law, as expected from ripening kinetics (15). The pre-exponential factor is also about that to be expected from coarsening theory.

Maximum Recalescence Temperature

The solidification model outlined above is one of an extremely fine dendritic structure growing and rapidly recalescing until undercooling is dissipated, and then continuing to slowly recalesce due to coarsening. The coarsening results in the dissolution of most of the dendrite arms originally present in the structure, with re-precipitation of solid on larger, growing arms. The solid, as it re-precipitates, is not of its original composition, but of a composition that is in equilibrium with the liquid for the given

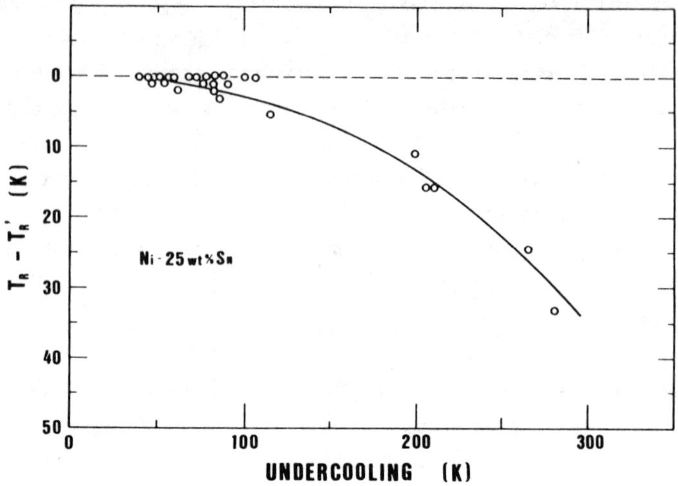

FIGURE 13. Temperature range of slow recalescence, hypoeutectic alloy. From ref. (1).

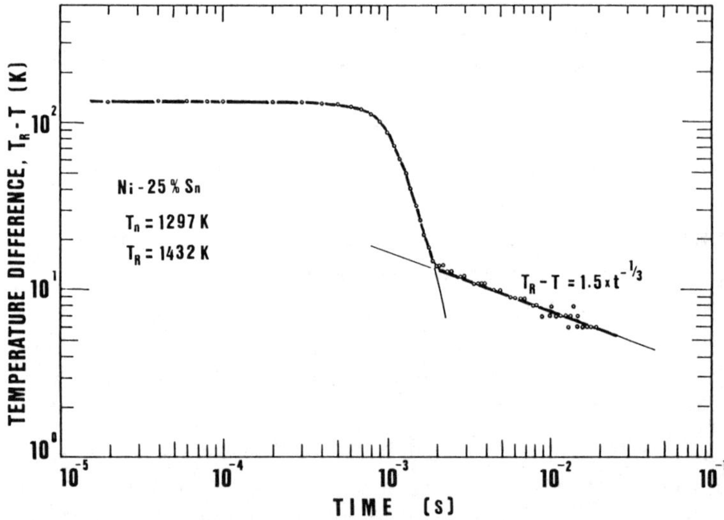

FIGURE 14. Recalescence behaior of hypoeutectic alloy, plotted on a log-log scale to show the linear behavior in the temperature range of slow recalescence. From ref. (1).

temperature and curvature. Thus, the maximum recalescence temperature should approach that temperature where the liquid and solid are in equilibrium. For adiabatic recalescence, this temperature is readily calculated by combining the equilibrium lever rule with the appropriate thermal balance. In the case of non-adiabatic recalescence, the maximum recalescence temperature can also be calculated, but heat losses must be considered in the thermal balance. Figure 15 is a plot of data for maximum recalescence temperature obtained over a wide range of undercoolings, and it is seen that these results agree closely with calculations based on the "equilibrium" model described above.

Microstructures

Following recalescence, undercooled samples proceed to solidify as in the solidification of ordinary castings and ingots. Typical cooling curves for three different undercoolings are shown in Figure 16. For example, a sample undercooled 174 K recalesces to well above the eutectic temperature and then cools to the eutectic temperature, where it exhibits a "hold" while the eutectic solidifies. The microstructures observed after cooling to room temperature must be greatly different from those present during recalescence, due to coarsening during recalescence and during cooling after recalescence. Some examples of structures are shown in Figures 17 and 18.

Figure 17 shows microstructures at different degrees of undercooling and shows the usual change from dendritic to spherical morphology with increasing undercooling, which in this system occurs at about 170 K. A simple illustration of the effect of coarsening after recalescence on structure is shown in Figure 18, in which the size of the spherical "dendrite elements" is shown to increase markedly with distance into the sample. It seems likely that the very large amount of ripening that occurs during and after recalescence is the cause of the spherical morphology we see in highly undercooled samples. The very large amount of ripening occurring in the more highly undercooled samples could fully destroy an originally dendritic structure, resulting in the observed spherical morphology.

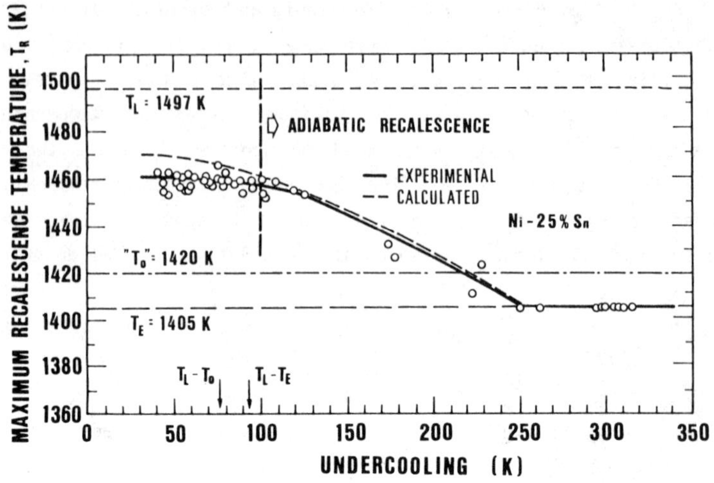

FIGURE 15. Maximum recalescence temperature, T_R, versus undercooling for the hypoeutectic alloy. From ref. (1).

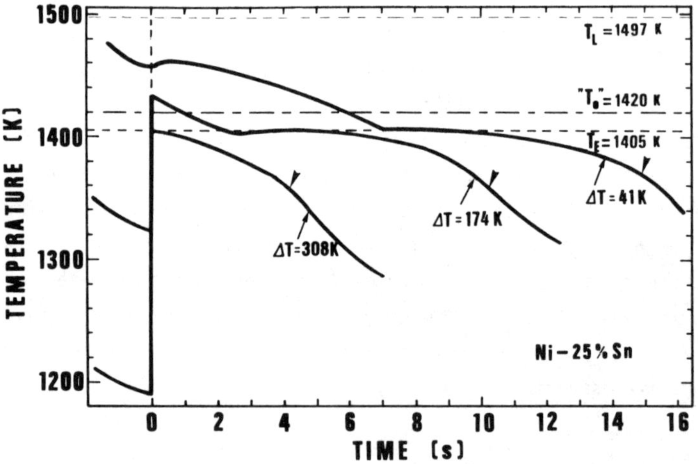

FIGURE 16. Recalescence and cooling curves for three samples of hypoeutectic alloy, low sampling rate. From ref. (1).

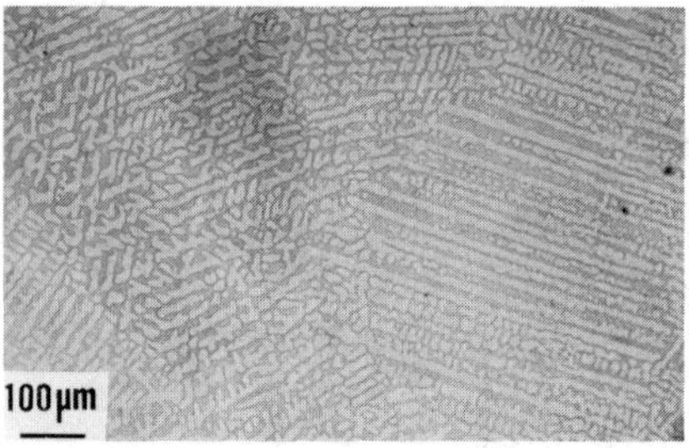

Initial Undercooling ΔT = 56 K

ΔT = 173 K

ΔT = 314 K

FIGURE 17. Cross-sectional microstructures for three Ni-25wt% Sn hypoeutectic alloy samples with different undercoolings. From ref. (3).

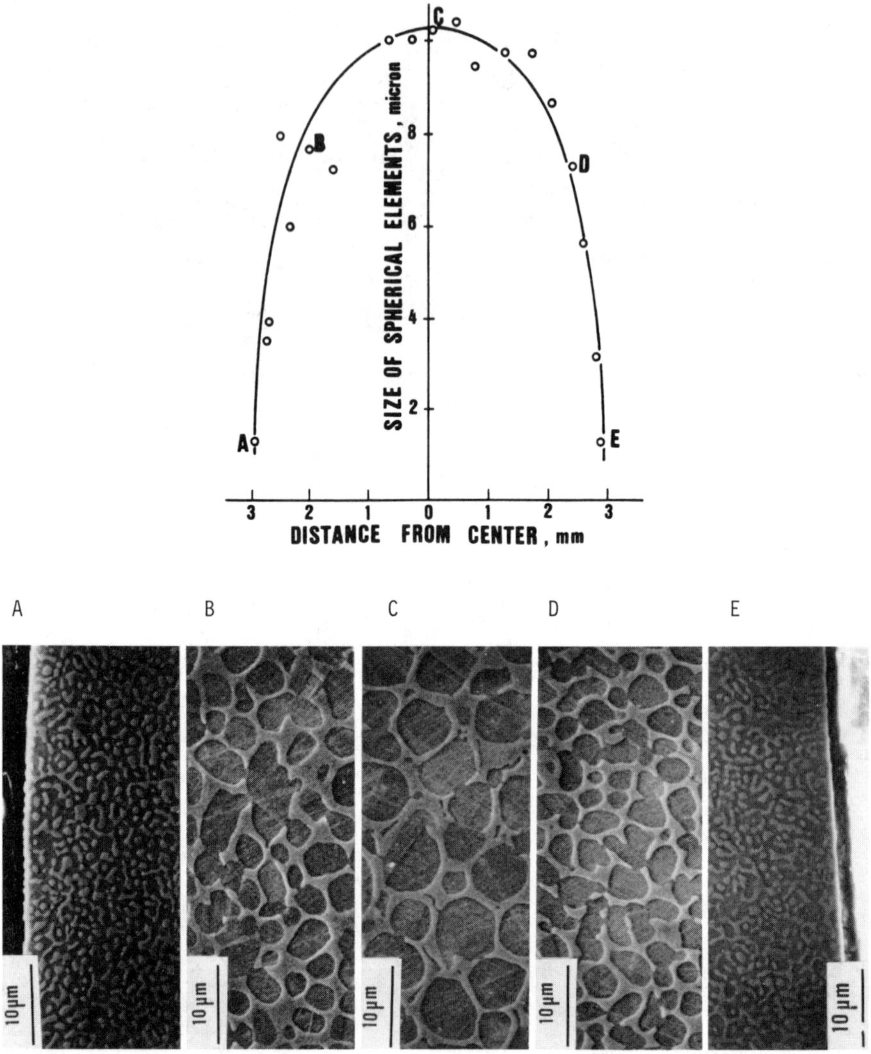

FIGURE 18. Dependence of diameter of spherical elements on the position in a Ni-25wt% Sn hypoeutectic alloy sample undercooled by 314 K. The five photographs correspond to the five data points labeled in the plot. From ref. (3).

Summary and Implications

This paper summarizes results of an ongoing study of solidification of undercooled melts; detailed results are presented elsewhere (1,2,3). Results are reported here for a Ni-Sn alloy solidified with initial undercoolings as high as 260 K and at dendrite tip velocities up to approximately 16 m/s. Tip velocity measurements agree well with theoretical analyses of Lipton, Kurz, and Trivedi, and of Boettinger and Coriell. Both analyses assume equilibrium interface kinetics. In this work no evidence was found of "solute trapping," although some may be occurring at the dendrite tips in the samples solidified at the highest undercoolings.

Solidification of undercooled melts proceeds by dendrites first sweeping across the sample at a rate controlled primarily by solute diffusion. The dendrites are very fine (e.g., 20 nm at 200 K undercooling), and they proceed forward in a "dendritic" array, whose shape is determined by the requirement of heat dissipation. Because of the extremely fine dendrite array, undercooling behind the dendrite tips must be dissipated very rapidly. Thus, the time interval required for the dendrites to sweep across the sample comprises essentially the period of "rapid recalescence." Following this period there is a time of "slow recalescence," during which sample temperature continues to climb as a result of coarsening of the very fine structure present. The melting and re-solidification that occurs during this final period of recalescence results in the solid and liquid phases approaching their equilibrium compositions and fractions at the maximum recalescence temperature.

Figure 19 shows schematically the "solidification path" of undercooled Ni-25% Sn alloy, based on the LKT model and the additional concepts outlined above, and assuming no nucleation of β phase until after undercooling is dissipated. Paths would be similar for all undercoolings studied in this work, since these were significantly less than unit undercooling below the solidus.

ABCD is the "path" of liquid composition at the interface (C_L^*), and A'B'C'D' is the path of solid composition at the interface (C_S^*). Path OA describes the path of the change in solute concentration in front of the advancing dendrite tip. The points A and A' represent the composition and temperature at the dendrite tip. The LKT model gives quantitatively the points A and A', as well as the line OA.

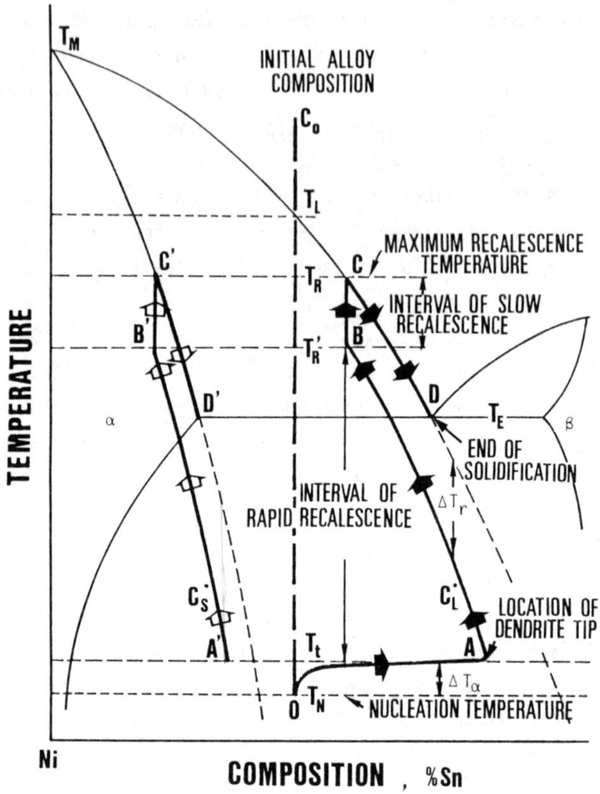

FIGURE 19. Schematic "solidification path" of undercooled Ni-25wt% Sn alloy. ABCD is the path of the liquid composition at the liquid-solid interface. A'B'C'D' is the path of the solid composition at the interface. Path OA describes the change in solute concentration in front of the advancing dendrite tip. The temperature differences between the "paths" and the phase diagram extensions are exaggerated.

The compositions of the liquid and solid at the liquid-solid interface move along the lines AB and A'B', respectively, during the stage of rapid recalescence. Our experiments show that these paths deviate from the equilibrium phase diagram by an amount which decreases from the undercooling at the dendrite tip (ΔT_r) at T_t to about one-half of that amount at the end of rapid recalescence, B and B'.

The "slow recalescence" from coarsening occurs along the lines BC and B'C'. Paths CD and C'D' describe the solidification after recalescence: in this stage, solidification follows a Scheil-type model. The points D and D' indicate the end of solidification, where the sample exhibits a thermal "hold" while the eutectic solidifies.

Coarsening occurs after recalescence (during the final stages of solidification), as well as during solidification. This results in structures that are different, and vastly coarser, than those that originally form. The very large amount of ripening occurring at high undercoolings is probably responsible for the spherical morphology observed in highly undercooled samples.

The results of this study shed some light as well on solidification of undercooled, rapidly solidified, atomized droplets. Here, of course, rate of heat extraction is very much more rapid, but we can expect the early stages of solidification to be similar when undercooling is present. In the later stages of solidification, the rapid rates of heat extraction from droplets can greatly reduce the amount of ripening that occurs. Sufficiently rapid rates of heat extraction from the atomized droplets could also eliminate at least the latter stages of recalescence, and therefore most or all the ripening. The resulting structures, even though equilibrium interface kinetics prevailed, would then have compositional heterogeneities present only on the nanometer scale of the very fine solute dendrites that originally grew. The resulting structures could well be difficult to distinguish from those in which complete "solute trapping" occurs.

Acknowledgments

The authors express their gratitude to NASA for support of this work under Grant No. NSG-7645 and NAG3-597 through Dr. R. E. Halpern and Mr. F. H. Harf. They also acknowledge with thanks a number of students and others who have helped with various phases of the experimental work including Mr. M. D. Dellinger, Mr. M. Yamamoto, Mr. Q. Li and Mrs. M. Wehrle-Due. Appreciation is expressed to Dr. W. J. Boettinger, National Bureau of Standards, and Prof. R. Trivedi, Iowa State University, for copies of their valuable unpublished papers. Discussions during the course of the work with Prof. T. Z. Kattamis, University of Connecticut, and Prof. D. Turnbull, Harvard University, have been particularly helpful.

References

1. T. J. Piccone, Y. Wu, Y. Shiohara, and M. C. Flemings, submitted for publication.

2. Y. Wu, T. J. Piccone, Y. Shiohara, and M. C. Flemings, submitted for publication.

3. Y. Wu, T. J. Piccone, Y. Shiohara, and M. C. Flemings, submitted for publication.

4. J. Lipton, W. Kurz, and R. Trivedi, to be published.

5. G. P. Ivantsov, Dokl. Akad, Nauk S.S.S.R., 58 (1947), 567.

6. M. E. Glicksman, Mat. Sci. and Eng. 65 (1984), 45-55.

7. S. C. Huang and M. E. Glicksman, Acta Met., 29 (1981), 701-716.

8. R. Trivedi, J. Crystal Growth, 48 (1980), 93-99.

9. J. Lipton, M. E. Glicksman, and W. Kurz, Mat. Sci. and Eng. 65 (1984), 57-63.

10. W. J. Boettinger and S. Coriell, Proceedings of NATO Workshop on "Rapid Solidification Technologies", Castle of Theuren, FRG, March, (1985).

11. R. Trivedi, J. Crystal Growth, 49 (1980), 219-232.

12. R. Trivedi and K. Somboonsuk, Mat. Sci. and Eng. 65 (1984), 65-74.

13. D. Turnbull, Proceedings of 2nd Israel Materials Engineering Conference, H. Grill and S. I. Rokhlin (eds.), Ben Gurion University, Beersheva, Israel, (1984), 1.

14. M. C. Flemings and Y. Shiohara, Tetsu-to-Hagane 71, No. 10 (1985), A204-208.

15. P. W. Voorhees and M. E. Glicksman, J. Crystal Growth 72 (1985), 599-615.

Appendix

Table of Thermophysical Properties Used in Calculations

C_p: Specific heat of the alloy = 500 J/kgK
ΔH: Latent heat of fusion = 1.5 x 10^5 J/kg
ΔS: Entropy of melting ($=\rho\ \Delta H/T_L$) = 841,700 J/m^3K
ρ : Density of the alloy = 8400 kg/m^3
T_L: Liquidus temperature (Ni-25wt% Sn) = 1497 K
m_L: Liquidus slope = -12.27 K/wt% Sn
k_e: Equilibrium solute partition ratio = 0.585
C_o: Initial composition of the alloy = 25wt% Sn
α : Thermal diffusivity of the alloy = 5 x 10^{-6} m^2/s
D: Interdiffusion coefficient of tin in the liquid alloy = 5 x 10^{-9} m^2/s
γ : Solid-liquid interfacial energy = 0.25 J/m^2
σ^*: Stability constant = 0.025

FUNDAMENTAL ASPECTS OF RAPID SOLIDIFICATION PROCESSING

Carlos G. Levi
and
Robert Mehrabian

Engineering Materials Program and
Department of Mechanical and Environmental Engineering
University of California
Santa Barbara, CA 93106

Current understanding of rapid liquid to solid transformations in highly supercooled systems is reviewed. Detailed consideration is given to the interplay of supercooling and external heat extraction in controlling the mode and kinetics of rapid solidification processes. Thermal considerations indicate that the achievable liquid-solid interface velocities are much higher in the presence of bulk supercoolings than when solidification is controlled by the external heat extraction rate. On the other hand, thermodynamics and kinetics establish strict requirements of interfacial supercoolings and velocities for metastable phenomena like solute trapping. Solidification of highly supercooled melts results in recalescence as the rapidly moving interface generates heat much faster than the rate at which it can be removed from the system. The degree and extent of metastability is thus dependent on the history of interfacial conditions during solidification. Microstructure evolution in supercooled droplets is discussed in terms of models for the thermal history and the concomitant solute redistribution. Evidence from ultrafine metal powders is presented to support the trends predicted by the models.

Introduction

Ultrafine microstructures and metastable phases produced by rapid liquid-solid transformations have become increasingly important in the development of high-performance materials. A host of experimental programs has provided numerous examples of metastability ranging from dendritic refinement to amorphous metals. Concurrently, we have established a growing effort to advance our theoretical understanding of microstructure evolution and its relationship to process variables. It is now acknowledged that rapid solidification frequently involves substantial departures from interfacial equilibrium which cannot be treated by conventional solidification models. Further, the concept of solidification rate has progressively shifted from cooling rate to interface velocity; it is evident that rapid solidification also occurs under large heating rates during recalescence of highly supercooled systems.

An important component of the effort to establish a science base for rapid solidification has been the development of mathematical models describing the thermal history of liquid-solid interfaces and its relationship to microstructure formation [1,2]. As in many other cases, the contributions of Professor Turnbull to the understanding of supercooling phenomena and solidification kinetics have played a key role in the advancement of this endeavor.

This paper will review the progress in the fundamental understanding of liquid-solid transformations stemming from modelling activities combined with experimental work in highly supercooled droplets. Emphasis is placed on the relative roles of supercooling, kinetics and alloy composition on the thermal history of the interface and the concomitant solute redistribution. Bulk supercooling and its effects on structural refinement are discussed in terms of the competition between nucleation and growth processes in the early stages of the transformation.

Thermodynamic and Kinetic Considerations

Microstructures resulting from liquid-solid transformations are rarely thermodynamically stable, as full equilibrium conditions are only possible if the relaxation times for all nucleation, growth, diffusion and heat transfer processes are much smaller than the time scale of the transformation. Increasing solidification rate will give rise to a hierarchy of non-equilibrium effects manifested as changes in the nature, relative amounts, morphology and chemical homogeneity of the phases in the microstructure [3]. A distinction may thus be made between non-equilibrium solidification regimes in which local equilibrium still occurs at the interface (e.g. segregation in conventional casting), and those in which interfacial equilibrium is lost (e.g. solute trapping by rapid solidification). In the first type the chemical potentials are almost equal at the interface; the loss of equilibrium arises because of limited diffusion, resulting in segregation, or because a stable phase cannot nucleate or grow and is replaced by a metastable one that can form more readily. In either case, the corresponding temperatures and compositions at the interface can be predicted from a stable or metastable phase diagram; in addition, sufficient knowledge exists to predict the relative amounts of phases and, to some extent, the changes in morphology.

Significant loss of interfacial equilibrium is believed to occur when the crystal growth rate exceeds the diffusive speed of solute atoms in the liquid phase [3]. When equilibrium is lost at the liquid-solid interface the chemical potentials are not equal. Although the free energy functions for the phases are still assumed to hold, kinetic considerations must be introduced to predict the interfacial temperature and composition, which do not correspond to the stable (or metastable) phase diagram. Partitionless solidification, a phenomenon of particular interest to this conference, inherently requires interfaces substantially away from equilibrium conditions.

Understanding partitionless solidification first requires a brief description of the thermodynamic and kinetic constraints to solute trapping. Consider the binary phase diagram in Figure 1 which includes metastable extensions of the liquidus and solidus curves. Thermodynamics dictates that solute partitioning may be avoided only if the temperature of the interface remains below T_0, which is the temperature wherein the free energies of the solid and liquid phases for the particular alloy composition are the same. In addition, kinetic considerations require interfacial velocities in excess of the diffusive velocity of the solute in front of the interface; otherwise partitioning will occur even if the system is below the metastable solidus T_S. Existing kinetic models propose a monotonic dependence of partition coefficient on interface velocity [4-6].

The degree of supercooling required to achieve complete partitionless solidification of an alloy must be such that the growth velocity will remain sufficiently high to avoid solute partitioning despite the rising interfacial temperature due to the evolving heat of fusion [7]. During recalescence the growth rate may fall below that needed for complete solute trapping at some temperature, T_C, where $T_C < T_0$. If solidification is nearly adiabatic, as indicated by the forthcoming discussion, nucleation must be suppressed until the enthalpy of the liquid at T_N is less or equal than that of the solid at T_C [7,8].

Fulfillment of the thermodynamic and kinetic conditions for partition-

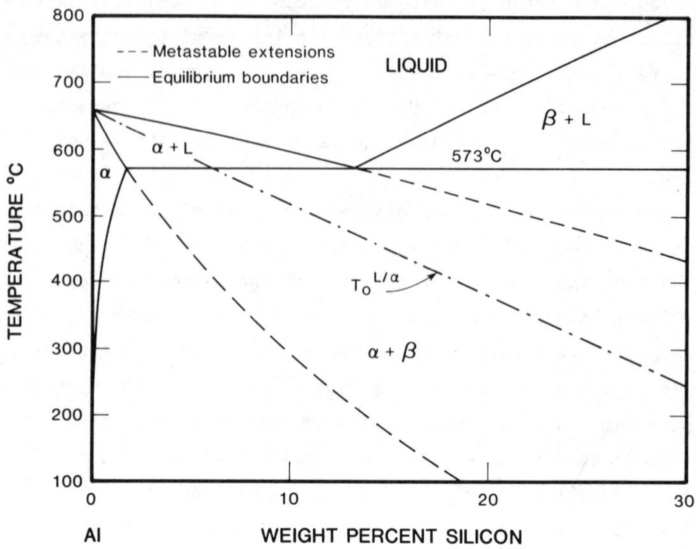

Figure 1 - Al-rich corner of the Al-Si phase diagram, indicating the T_0 curve and metastable extensions for the liquidus and solidus lines.

less solidification thus depends on the evolution of the temperature and velocity of the interface as solidification progresses, i.e. on its thermal history.

Thermal Considerations

The thermal history of liquid-solid interfaces during solidification of supercooled droplets is discussed below. Solidification modes are defined on the enthalpy-temperature diagram and trends are established from numerical analyses of the interface linking heat transfer to thermodynamic and kinetic models. We assume for most of the discussion that the droplet is sufficiently small to minimize internal temperature differences so that its supercooling is identical to that of the interface (Newtonian cooling). The behavior of droplets departing from this model is also examined.

Solidification in Unary Systems

The thermal history of powder solidification can be conveniently described on the enthalpy-temperature diagram for a pure metal A shown in

Figure 2 [9]. The molar enthalpy, H, of a control volume at a uniform temperature T is given by the additive contributions of solid and liquid, whose molar fractions are f and (1-f), respectively.

$$\psi = [1 + \theta]\ (1 - f) + [\kappa_S \theta]\ f \tag{1}$$

where ψ and θ are the dimensionless enthalpy and temperature,

$$\psi = (H - H^A_{SM})\ /\ \Delta H^A_M \tag{2}$$

$$\theta = C^A_L\ (T - T^A_M)\ /\ \Delta H^A_M \tag{3}$$

H^A_{SM} is the enthalpy of the solid metal at its melting temperature T^A_M, C^A_L is the liquid metal heat capacity, ΔH^A_M is the enthalpy of transformation and κ_S is the ratio of solid/liquid heat capacities. A system can conceivably follow any path of decreasing enthalpy in the supercooled region of Figure 2 ($\psi < 1$), starting at the nucleation temperature ($\theta_N < 0$) on the liquid line, and ending at some point on the solid line.

Two extreme cases may be distinguished in this diagram. A vertical path represents an isothermal transformation, of which a limiting case is solidification without significant barriers to nucleation and growth, with the interface at the equilibrium temperature T_M (path one in Figure 2). The solidification rate in this case is completely controlled by the external rate of heat extraction and the interface velocities are typically much lower than those necessary for partitionless solidification.

A horizontal path (e.g. path two in Figure 2) represents isenthalpic (or adiabatic) solidification whereby the system can form an amount of solid proportional to the initial supercooling of the liquid without further heat loss to the environment. The droplet would recalesce as evolving heat of fusion is transformed into sensible heat and solidification would be characterized by a rather large heating rate, instead of the more familiar concept of "cooling rate". Nevertheless, the high supercoolings involved can translate into interface velocities sufficiently high to produce partitionless solidification. A liquid that can completely transform to solid under adiabatic conditions is said to be "hypercooled".

The general case of interest to rapid solidification is one wherein a droplet is highly supercooled prior to nucleation ($\theta_N < 0$) and concurrent heat extraction occurs during solidification (path 3 in Figure 2). The slope of the ψ - θ curve would then reflect the competition between the external rate of heat extraction and the rate at which latent heat is released by the advancing interface. The latter is proportional to the solidification rate (change in fraction solid with time df/dt) which is a function of the liquid-solid interface velocity and its area.

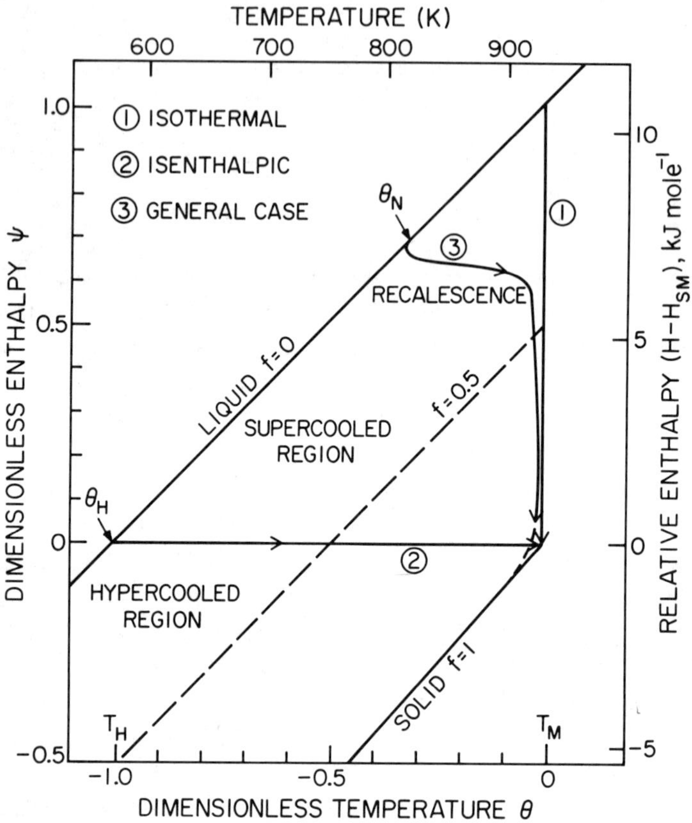

Figure 2 - Enthalpy-temperature diagram for pure Al showing possible solidification paths. θ_N is the dimensionless temperature at the moment of nucleation and f is the molar fraction solid.

Under these conditions, the thermal history typically consists of three distinct solidification regimes. Immediately after nucleation the liquid-solid interface velocity is finite but its area, determined by the size of the critical nucleus, is very small. Consequently, the initial solidification rate and corresponding rate of latent heat release are relatively low. The external heat extraction dominates the thermal history and the droplet temperature drops further below θ_N.

With the progressive formation of solid the interfacial area increases rapidly in size; meanwhile, the interface velocities are high in response to the large supercooling. A second regime develops as heat is released much faster than it can be removed by the external cooling, resulting in recalescence. As the interfacial temperature approaches the melting temperature, its supercooling and velocity decrease, making the external heat removal more competitive. The third regime in the thermal history arises when the rates of release and extraction of latent heat are balanced. This regime is characterized by much lower supercoolings and interface velocities, and the progress of solidification is limited by the external rate of cooling.

Early models for powder solidification assumed that the interface is always close to its equilibrium temperature and the transformation is controlled by heat transfer to the environment [10,11]. Although these models could explain the microstructural refinement with increasing cooling rate, they provided little insight into metastable effects like solute supersaturation. As the role of supercooling and the change in interfacial temperature during solidification were recognized, it was evident that kinetics had to be incorporated into the heat transfer model to predict the interface behavior.

Newtonian models for solidification of supercooled droplets* reveal that the ψ - θ path and interface velocity at a given fraction solid are solely a function of the solidification geometry (including the number of nucleation events), the dimensionless nucleation temperature θ_N and the ratio (K/Bi), where K is a dimensionless kinetic coefficient relating interface velocity and supercooling, and Bi is the Biot number. An important corollary of the Newtonian analysis is that, for a constant heat transfer coefficient, the thermal history is not a function of the particle size. The effect of the latter is reflected only on the time scale of the transformation [9].

Analysis of the thermal history for typical K/Bi ratios encountered in metal atomization (K/Bi $>$ 100) indicates that external heat extraction is practically irrelevant during recalescence, and the droplet approaches the adiabatic solidification case. Under these conditions, the extent of rapid solidification is limited by the ability of the droplet to act as a heat sink, which is mainly a function of the supercooling prior to nucleation.

Figure 3 shows the effect of the initial supercooling on the thermal history for an aluminum droplet solidifying from a single nucleation event at the surface, with a constant K/Bi = 100. It is evident that increasing the achievable supercooling in the liquid translates into a corresponding in-

---

* The basic features of the models are described in the Appendix.

crease in the solidification rate, thus reducing the relative contribution of the external heat extraction. Note however that supercooling (and velocity) at a given fraction solid increase**, hence extending the potential for metastability.

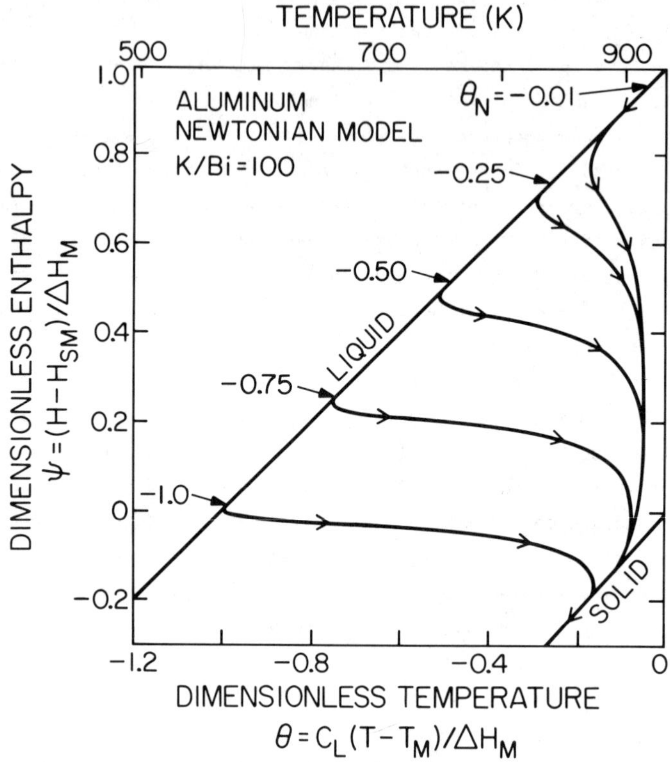

Figure 3 - Calculated enthalpy-temperature curves for the solidification of aluminum droplets after different degrees of supercooling over the range 3 to 360 K. Newtonian heat flow is assumed; the ratio K/Bi = 100 corresponds to extremely rapid cooling conditions for atomized aluminum.

Numerical modelling of solidification when significant temperature differences develop within the droplet (non-Newtonian conditions) reveals that the thermal history is also strongly dependent on the initial supercooling,

---

** Lines of constant fraction solid run between the liquid and solid enthalpy lines and change gradually in slope from 1 at f=0 to κ_S at f=1 (see Figure 2).

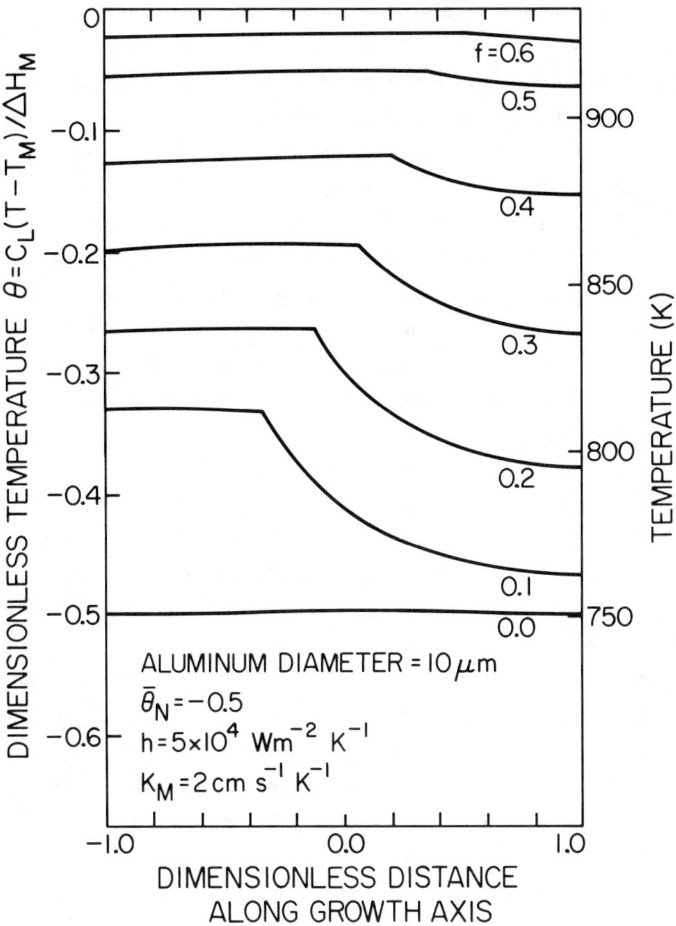

Figure 4 - Thermal profiles across a supercooled aluminum droplet during solidification from a single nucleation event at the droplet surface. Note the temperature differences developed inside the droplet during recalescence.

but instead of being unique for a constant K/Bi ratio, it is affected by the particular values of the droplet radius, kinetic and heat transfer coefficients. In general, the limitations to heat transfer inside the particle become more relevant than the surface conductance in maintaining the interface supercooled. The rapidly moving interface recalesces faster than the rest of the droplet, adjusting its temperature to achieve a proper balance between the rate at which heat is being released (determined by its velocity/supercooling) and the rate at which heat is being withdrawn (determined by the thermal conductivity and resulting gradients in its vicinity).

Calculated temperature profiles as solidification progresses in a recalescing aluminum droplet are depicted in Figure 4. As expected from the Biot number (Bi = 0.0028) the profiles are Newtonian before nucleation and after recalescence. However, substantial temperature differences are developed inside the droplet during the period of rapid growth, i.e the interface recalesces "locally", even under the relatively sluggish kinetics assumed in this figure. Since the interface supercooling is lower than the average droplet supercooling, the concomitant decrease in velocity and rate of latent heat evolution will slow down the overall recalescence of the droplet. Although the droplets would appear to recalesce more slowly as they depart from Newtonian behavior, the extent of metastability in the microstructure would progressively decrease.

In the non-Newtonian analysis the shape of the solidification front is dictated by the relative velocity (supercooling) of different sections of the liquid-solid interface, which in turn depends on the thermal field in its neighboring environment. The interface profile and velocity as a function of particle size are illustrated in Figure 5. As expected, the interface velocity during recalescence decreases with increasing fraction solid and droplet diameter. Note, however, that the initial velocities are much higher than those typical of heat transfer control (e.g. f = 0.75) and within the range necessary for solute trapping, even for the moderate kinetics assumed. Furthermore, the Newtonian results provide an upper limit to the interface velocity for a given supercooling prior to nucleation, but this limit is only approached in droplets below 1 μm in diameter for Al.

Bulk supercooling prior to nucleation thus emerges as the critical parameter in determining the extent of metastable effects associated with substantial departure from interfacial equilibrium during solidification. In general, for a given microstructural modification to be feasible, nucleation should be suppressed until a desired supercooling is achieved. For that to occur in an assembly of volume V, the total number of nuclei, N, given by

$$N = V \int_{T_M}^{T_N} [J(T)/\varepsilon] \, dT \qquad (4)$$

should be less than unity. In this expression, J(T) is the temperature dependent nucleation frequency and ε is the cooling rate. The achievable supercooling is then a function of available sites for nucleation and time; hence, external heat extraction is more important in extending the achievable super-

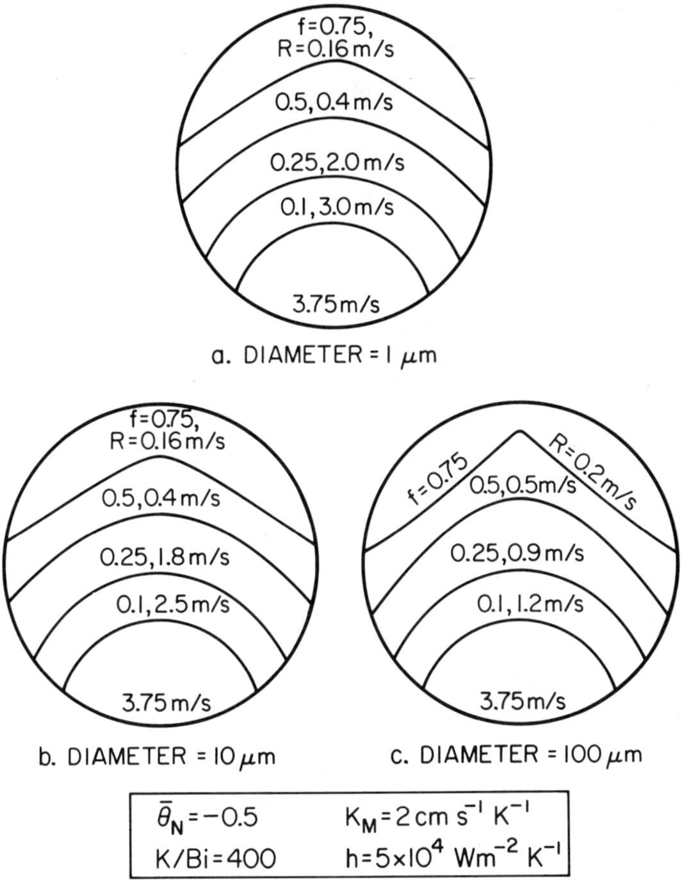

Figure 5 - Interfacial profile and velocity, R, as a function of fraction solidified (f) for three aluminum droplets supercooled 180 K. Calculations are based on non-Newtonian finite difference formulations.

cooling than in slowing down recalescence once solidification starts.

Modelling of nucleation behavior during rapid solidification as a means to predicting achievable supercoolings and grain size refinement has been addressed by a number of publications [8,12-14]. Models developed by the authors are based on a discretized version of the integral in Equation (4). Nucleation and growth kinetics are incorporated into a Newtonian heat transfer formulation to study the competition between these processes during continuous cooling. The nucleation frequency is calculated from the classical

Turnbull-Fisher theory [15] assuming that the distribution of embryos can be relaxed to the corresponding quasi-steady state within the time scale of the process (worst possible case). The achievable bulk supercooling of the liquid is defined as that wherein the number of nuclei in Equation (4) reaches a value of unity.

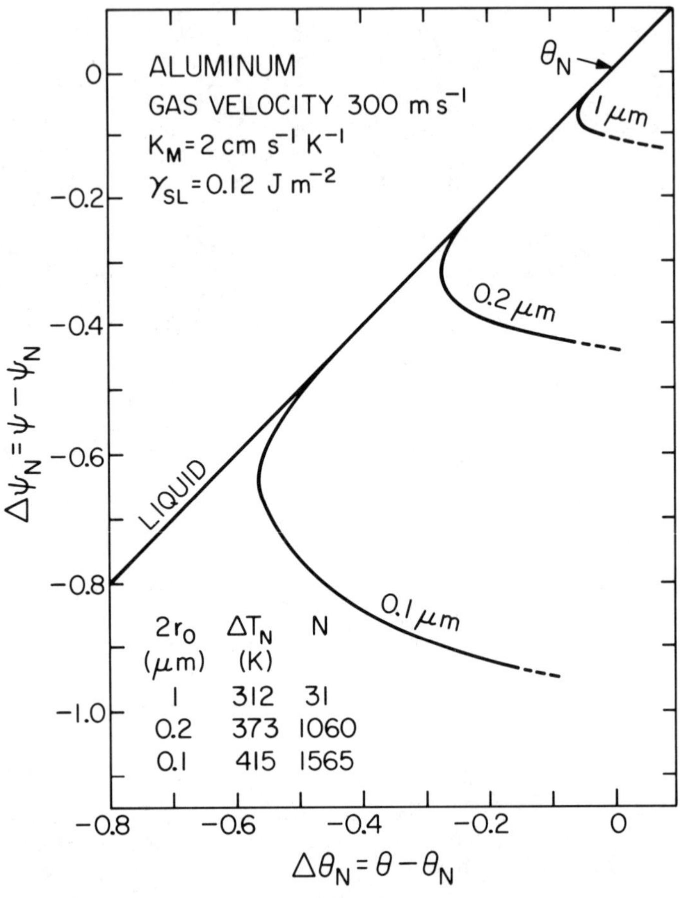

Figure 6 - Effect of droplet size on the initial transient in the thermal history of supercooled Al droplets. The curves have been shifted to make their nucleation points coincide. Note the increase in the number of grains as the thermal excursion beyond θ_N is extended.

Figure 6 shows the calculated thermal history right after nucleation for Al droplets cooled by convection. The period of interest corresponds to the initial transient in the enthalpy-temperature curve depicted in Figure 2.

At the onset of nucleation a competition is established between formation of additional nuclei and growth of the existing one(s). Larger supercoolings reduce the solidification rate in the early stages of growth by decreasing the initial area of the interface and enhancing the hindrance of capillarity and atomic transport on the growth kinetics. On the other hand, nucleation rate increases until it becomes competitive with growth, promoting the formation of multiple grains. The competition is brief since both processes favor recalescence--by increasing the interfacial area--with the consequent reduction in supercooling. The last nucleation event usually occurs shortly after recalescence starts and before the temperature of the droplet surpasses the initial nucleation temperature θ_N.

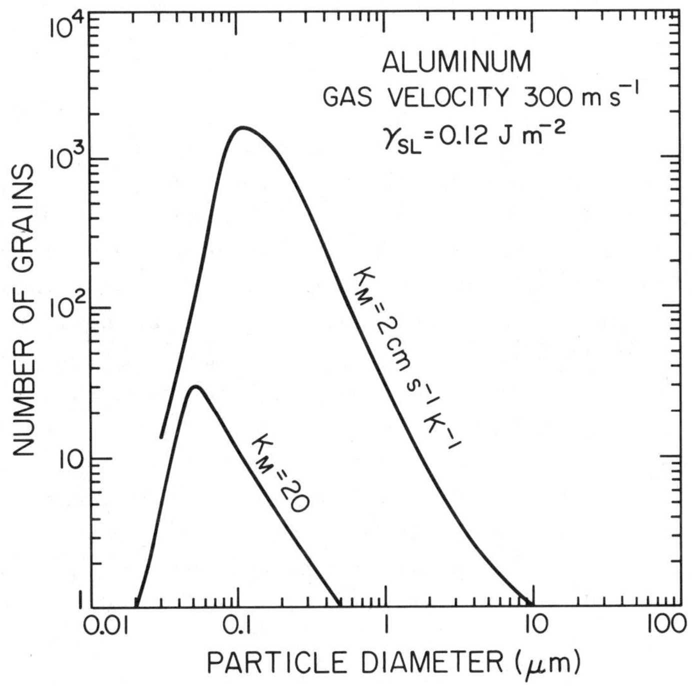

Figure 7 - Massive nucleation produced by increasing supercooling as particle size decreases. Note the reduction in the number of grains as the growth kinetic hindrance decreases. Peaks represent minimum grain sizes of 8 and 14 nm for 0.1 and 0.05 μm droplets.

Figure 7 shows the effect of droplet diameter on the calculated number of grains for pure Al droplets with two levels of kinetic hindrance. Decreasing the droplet size increases the achievable supercooling prior to nucleation by increasing the cooling rate in the liquid and reducing the volume of the assembly, V, in Equation (4), hence decreasing the probability of a fluctuation resulting in a successful nucleation event. Moreover, the thermal excursion beyond the initial supercooling is larger (Figure 6), enhancing the nucleation rate, hindering the growth rate by increasing the influence of capillarity and diffusion, and resulting in higher probability for multiple nucleation. A faster attachment kinetics (higher K_M) accelerates the recalescence process and limits the time available for additional nuclei to form, translating into coarser grains for the same particle size.

It should be noted that the trend in Figure 7 is opposite to that expected when nucleation is controlled by the number of heterogeneous nucleation sites (inversely proportional to the volume of the droplet) and growth is relatively fast.

Solidification in Binary Systems

Although numerical models for pure metals may be used to gain some qualitative insight into the rapid solidification of supercooled alloys, there are several important limitations that must be considered if a more quantitative understanding of the interrelationship between process variables and microstructure is desired. Important features of alloy solidification like solute redistribution, the critical conditions to suppress segregation, the changes in nucleation and growth kinetics as a function of temperature and composition, and the morphological stability of the interface can only be addressed qualitatively. We will now discuss the progress in understanding the rapid solidification of supercooled alloys.

Following the concepts of the enthalpy-temperature diagram for a pure solvent A, the enthalpy of a system AB of composition X_0 consisting of liquid of average composition $\bar{X}_L$ and solid of average composition $\bar{X}_S$ at temperature T and molar fraction solid f may be expressed as [16]:

$$\psi = [1 + \Delta\psi_L + \kappa_L\theta]\ (1 - f) + [\Delta\psi_S + \kappa_S\theta]\ f \tag{5}$$

where ψ and θ are the dimensionless enthalpy and temperature defined by Equations (2) and (3) respectively.

The thermal effects of solute additions may now be described with the help of the schematic isocompositional section of the enthalpy-temperature diagram of Figure 8. Equation (5) contains two excess enthalpy terms, $\Delta\psi_L$ and $\Delta\psi_S$, that reflect the relative effect of alloying on the latent heat of the pure element, and appear as displacements of the intercepts of the liquid and

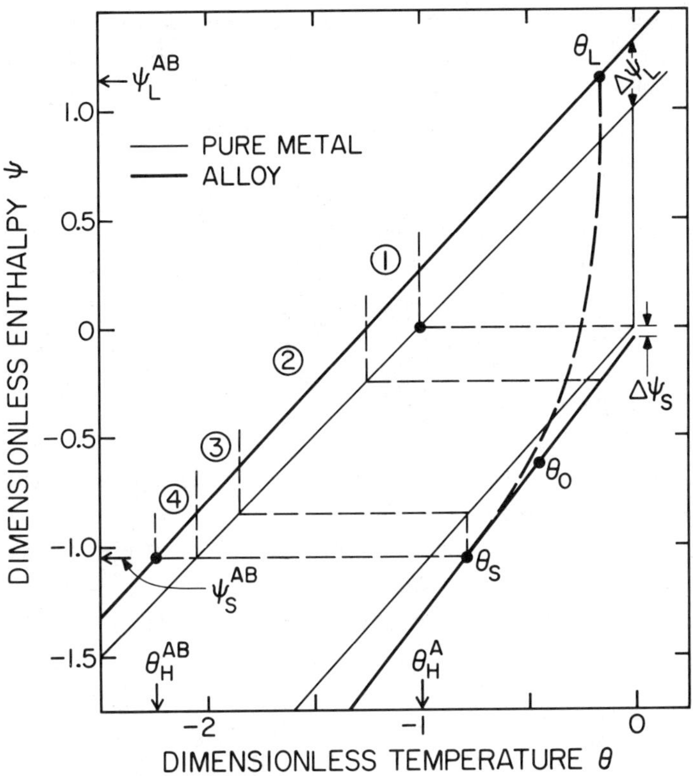

Figure 8 - Modification of the enthalpy-temperature diagram by alloying. The dashed line joining the liquidus and solidus temperatures represents the equilibrium solidification path. Note the depression of the hypercooling temperature and the concomitant increase in the enthalpy to be withdrawn prior to nucleation.

solid enthalpy curves with the isotherm at the melting point of the pure element (T_M^A, $\theta = 0$).

$$\Delta\psi_L = \{[\Delta H_M^B + (C_S^B - C_L^B)(T_M^B - T_M^A) - \Delta H_M^A]\,\bar{X}_L + \Delta H_L^X\,\bar{X}_L(1 - \bar{X}_L)\}/\Delta H_M^A \qquad (6)$$

$$\Delta\psi_S = [\Delta H_S^X\,\bar{X}_S(1 - \bar{X}_S)]/\Delta H_M^A \qquad (7)$$

The term $\Delta\psi_L$ arises because of the differences in latent heat and melting temperature of solute and solvent, and the enthalpy change associated with the non-ideal mixing of the two species. For example, when Si is added to Al the liquid enthalpy curve will be displaced upwards due to the higher heat of fusion of the solute. The effect of the heat of solution would be to reduce this displacement in the case of exothermic mixing or enhance it in the opposite case. Since the enthalpy of pure solid A at T_M^A is the reference point ($\psi = 0$), the term $\Delta\psi_S$ only contains the effect of the non-ideal enthalpy of mixing in the solid.

The coefficients of the dimensionless temperature in Equation (5) are the heat capacities of the liquid and solid alloy normalized by the heat capacity of the pure liquid solvent.

$$\kappa_L = 1 + [(C_L^B - C_L^A)/C_L^A]\ \bar{X}_L \tag{8}$$

$$\kappa_S = (C_S^A/C_L^A) + [(C_S^B - C_S^A)/C_L^A]\ \bar{X}_S \tag{9}$$

The first term on the right hand side of Equations (8) and (9) represents the slopes of the liquid and solid enthalpy lines for the pure solvent, whereas the second term reflects the contribution of the alloying element. This effect is relatively small for Al-Si alloys since solvent and solute have similar heat capacities, but other systems like Al-Cu and Al-Ge show a more pronounced variation [17].

In general, Figure 8 shows that the major thermal effect of alloying is an increase in the enthalpy difference between liquid at θ_L and solid at θ_S resulting mainly from the sensible heat associated with the melting range. This enhanced heat of fusion will be reflected in longer solidification times in the case of solidification near interfacial equilibrium.

In solidification from a supercooled melt, alloying will substantially increase the amount of heat that must be withdrawn from the liquid prior to nucleation, and consequently depress the onset of hypercooling. For example, an Al-6%Si liquid alloy must lose twice as much heat as pure Al in order to prevent recalescence above the [metastable] solidus temperature. This involves an additional supercooling depicted in Figure 8 which arises from: (1) the depression of the liquidus temperature, (2) the opening of the melting range, and the relative displacements and slopes of the solid (3) and liquid (4) enthalpy lines.

Figure 9 shows the calculated supercoolings prior to nucleation to prevent adiabatic recalescence above T_S and T_0 superimposed on the phase diagram for Al-Si alloys. Note that supercoolings below room temperature are required to prevent recalescence above T_S in alloys with more than 4%Si, a situation clearly limited by heat transfer. Furthermore, the required supercooling to avoid recalescence above T_0 also approaches room temperature as Si content

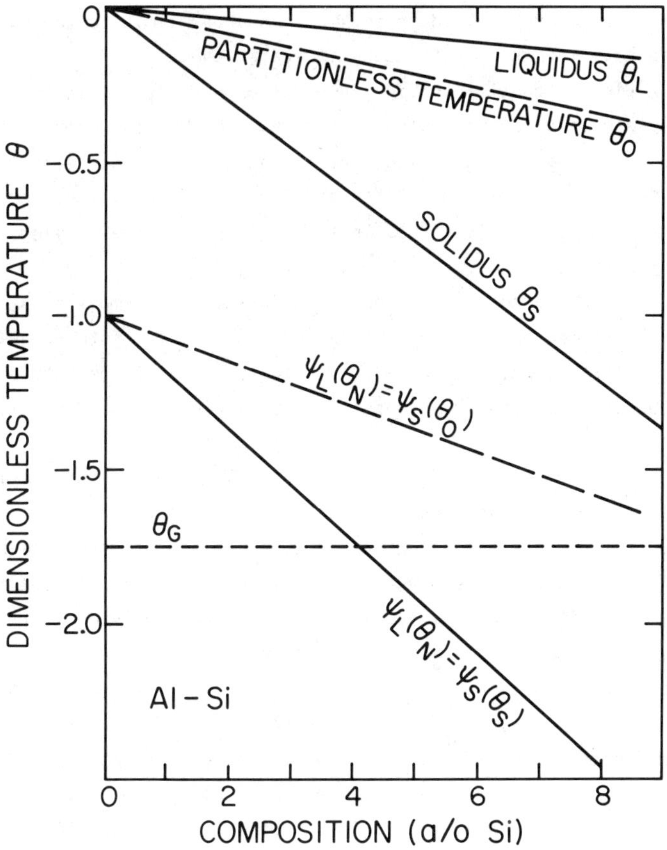

Figure 9 - Effect of solute additions on the critical temperatures to prevent adiabatic recalescence above the solidus θ_S, and the partitionless solidification temperature θ_0.

increases, making the heat extraction from the liquid quite inefficient. For example, an Al-6%Si alloy would have to be supercooled at least 524 K (T= 409 K) to achieve complete partitionless solidification under Newtonian conditions. The cooling rate in the liquid at this temperature would be only 40% of that at the hypercooling point of Al assuming that droplets are cooled with gas at 300 K. Therefore, increasing alloying should progressively decrease the extent of partitionless solidification unless a parallel reduction in kinetics offsets the thermal effects.

The recalescence of binary alloy powders has been studied by the authors using a solidification model which involves the following elements.

(a) A Newtonian heat transfer formulation. The Newtonian model is useful as a first approximation to estimate an upper limit to the extent of rapid solidification effects as indicated above.

(b) A kinetic model involving an equation for the growth of a diffuse interface as a function of supercooling [18,19], and Aziz' continuous model relating the partition coefficient to interface velocity [6].

(c) A dilute solution model for the free energy as a function of temperature and composition [20]. This free energy is the driving force in the growth equation.

(d) A mass balance relating bulk and interfacial compositions. The liquid is assumed of uniform composition since we are interested in conditions close to partitionless solidification. Solute redistribution by diffusion in the solid is considered negligible within the time scale of the process.

(e) An axisymmetric growth geometry starting from one nucleation event in the form of a spherical cap at the droplet surface [9].

Results from this model confirm earlier conclusions from the analysis of pure metals that for the range of interface velocities necessary for partitionless solidification the recalescence is practically adiabatic and insensitive to changes in the external cooling.

Figure 10 shows the solidification path of Al-3%Si droplets on the enthalpy temperature diagram as a function of a kinetic parameter β. This parameter represents the ease of interfacial rearrangement: its value may range from about 1 for a strong kinetic hindrance to about 100 if the attachment kinetics is only limited by the collision of atoms with the interface [16]. Typical interface velocities corresponding to these values would range from 2 to 180 m/s at the nucleation temperature.

The initial supercooling $\Delta\theta_N = 1.0$ (357 K) in Figure 10 would be sufficient to ensure complete solidification of a pure Al droplet if recalescence were adiabatic (K/Bi or $\beta/h \rightarrow \infty$). However, the alloy would only be 53% solid when reaching θ_S (or 84% when reaching θ_0) under the same conditions. Supercoolings of 1.17 (419 K) and 1.53 (549 K) are necessary for this alloy to achieve adiabatic solidification below θ_0 and θ_S respectively, as shown in Figure 9.

The recalescence paths on the ψ - θ diagram of Figure 10 show that the behavior of the alloy is similar to that of the pure solvent. The role of the external cooling is marginal and decreases as the kinetic hindrance diminishes ($\beta/h \rightarrow 0$). Increasing supercooling prior to nucleation would increase the driving force and consequently the rate of the transformation, further reducing the heat transfer to the surrounding environment during recalescence. Note, however, that for the more sluggish kinetics ($\beta < 10$) the driving force decreases sufficiently as the droplet approaches T_0 that the external cooling can prevent the droplet from reaching that limiting temperature. The thermodynamic condition for partitionless solidification is thus satisfied at all times during the solidification process, but more information is needed to determine if the kinetic requirement is fulfilled.

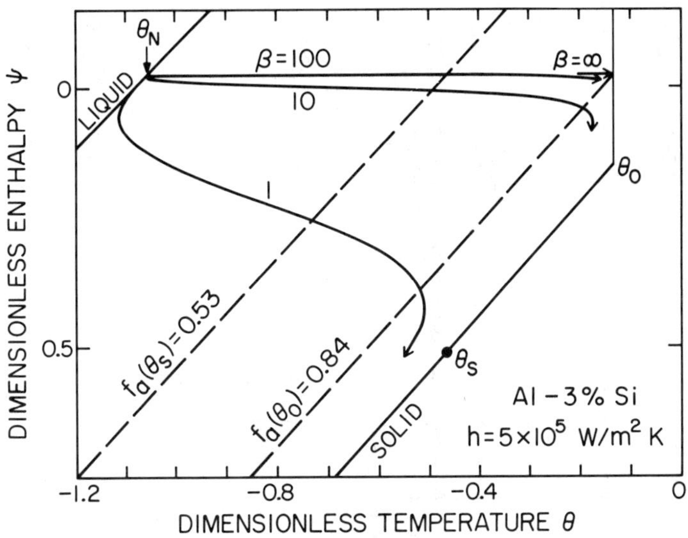

Figure 10 - Effect of interfacial rearrangement kinetics on the thermal history of Al-3%Si droplets with a constant initial supercooling. Lines of constant fraction solid are included for the amount of material that can solidify adiabatically below θ_0 and θ_S.

Microstructure evolution cannot be explained solely on the basis of the ψ - θ path since no information is provided on solute redistribution. Figure 11 shows the calculated interfacial compositions corresponding to the curves in Figure 10, superimposed on the phase diagram. Compositional variations after recalescence have not yet been addressed in the modelling work and are not included in this figure. However, the solid and liquid curves are expected to approach the equilibrium solidus and liquidus and follow them upon cooling until solidification is completed or the second phase nucleates.

It is evident from Figure 11 that partitionless solidification is only approached at the higher β, as expected from Aziz' model. As the kinetic hindrance increases ($\beta \rightarrow 1$) the supercooling at a given fraction solid would

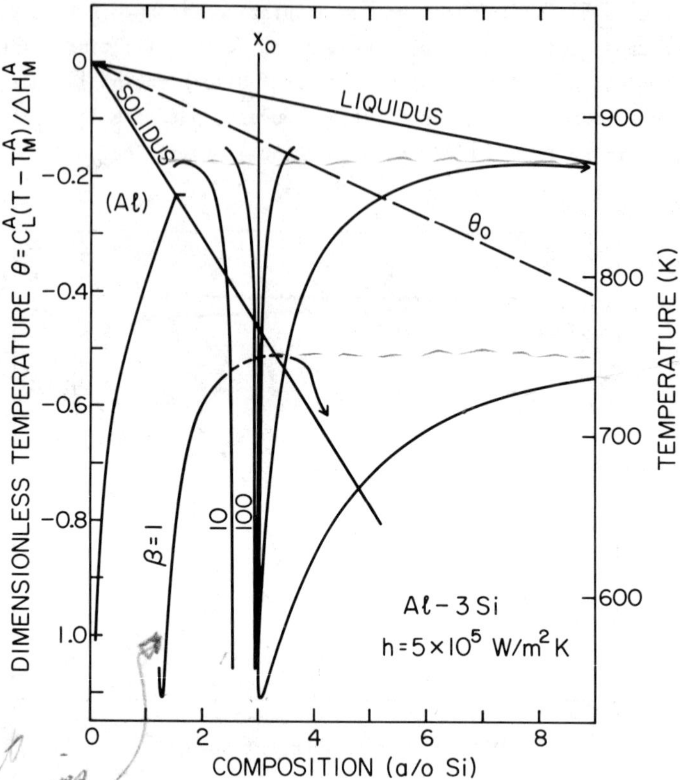

Figure 11 - Solute partitioning during recalescence as a function of interfacial rearrangement kinetics. Curves starting to the left of X_0 depict the solid compositions at the interface X_S^*, whereas those on the right side are the corresponding liquid compositions X_L^*.

be larger but the corresponding interface velocity and partition coefficient would be lower. For example, **k** decreases from 0.98 to 0.42 as β changes from 100 to 1 under the conditions of this figure. Therefore, kinetics limits the extent of supersaturation even when the thermodynamic condition is satisfied.

Finally, let us combine the last two figures to reveal the segregation pattern as a function of fraction solid and alloy composition, Figure 12. These curves were generated with the collision-limited attachment kinetics and a fixed supercooling of 357 K (i.e. equivalent to the onset of hypercooling for pure Al). Note that the segregation patterns consist of a region where solute concentration is very uniform and close to the original liquid composition X_0. As recalescence progresses and the supercooling is exhausted,

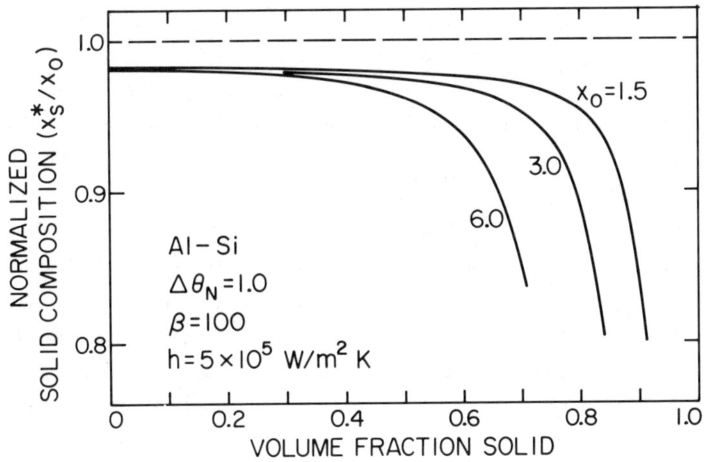

Figure 12 - Effect of alloy composition on the pattern of solute redistribution. Note that the compositions are normalized by the original liquid composition for comparison.

the interface velocity and partition coefficient decrease, the solute supersaturation in the solid rapidly drops producing a solute-depleted region. The onset of partitioning initiates a solute buildup in front of the liquid-solid interface and eventually leads to a cellular or dendritic breakdown.

Figure 12 also shows that increasing the solute content decreases the extent of supersaturation for the same initial supercooling. The partition coefficient at the beginning of solidification is not substantially affected but the onset of segregation is accelerated by increasing solute additions as the supercooling relative to T_0, i.e. the "driving force" for supersaturation, decreases.

Experimental Evidence

Ultrafine powders produced by the electrohydrodynamic (EHD) vacuum atomization technique were first used by the authors to test some of the concepts and theoretical models discussed above [7,21,22]. These powders, while not of commercial importance, are especially suited for fundamental studies of solidification because they are electron-transparent and permit examination of a complete solidification microstructure without recourse to thinning procedures. Although the anticipated cooling rates are moderate at best (radiative heat transfer, $h \leq 10^2$ W m^{-2} K^{-1}) the droplets are likely to supercool substantially prior to nucleation because the active nucleation catalysts will be isolated into a small fraction of the total volume. Furthermore, sub-micron droplets are smaller than the characteristic heat diffusion distance (α/R), which is in the order of a few microns for Al at the interface velocities necessary for extensive solute trapping. Therefore, their thermal history should approach the Newtonian predictions, which optimize the interface supercooling and velocity and lead to the maximum extent of microstructural metastability.

The segregated cellular structure in Figure 13 is representative of the thermal history depicted by path one in Figure 2. It is believed that limited supercooling ($T_N > T_C$) occurred in this Al-4.5%Cu droplet prior to nucleation, so that the interfacial temperature and composition corresponded fairly closely to those dictated by the equilibrium phase diagram. Partitioning starts soon after nucleation, resulting in solute buildup in front of the interface which becomes morphologically unstable due to constitutional supercooling. Solidification rates may be substantial, as evidenced by the fine cellular spacing, but the external cooling cannot produce the conditions necessary for extensive solute trapping in the absence of bulk supercooling.

On the other hand, the Al-6%Si powder in Figure 14 shows no evidence of segregation, even though the Si content is four times the maximum equilibrium solubility. This powder must have achieved a supercooling sufficiently high for the interface temperature and velocity to satisfy both the thermodynamic and kinetic conditions for partitionless solidification at all times during the transformation. Since solidification is completed before the interfacial temperature surpasses the onset of segregation, T_C, the thermal history of this powder represents the extreme case of adiabatic solidification depicted by path two in Figure 2.

The mixed solidification structure of Figure 15 represents the transition in composition and morphology expected from the distinct solidification regimes of path three in Figure 2. The initial supercooling was large enough for $T_N < T_0$ and for the interfacial velocity to be above that necessary for extensive solute trapping ($\mathbf{k} \to 1$). Results of microchemical analysis shown in this figure reveal that the initial region of supersaturation is followed by a solute depleted zone prior to the cellular breakdown, as predicted by the solute redistribution profiles of Figure 12.

Analysis of the microstructure evolution as a function of particle size for Al-6%Si powders has shown that the extent of the supersaturated zone increases with decreasing particle size, indicating a progressive increase in bulk supercooling consistent with our understanding of nucleation kinetics. Experimental evidence supporting the concomitant grain refinement predicted by Figure 7 has also been observed both in pure and alloyed Al [21, 23].

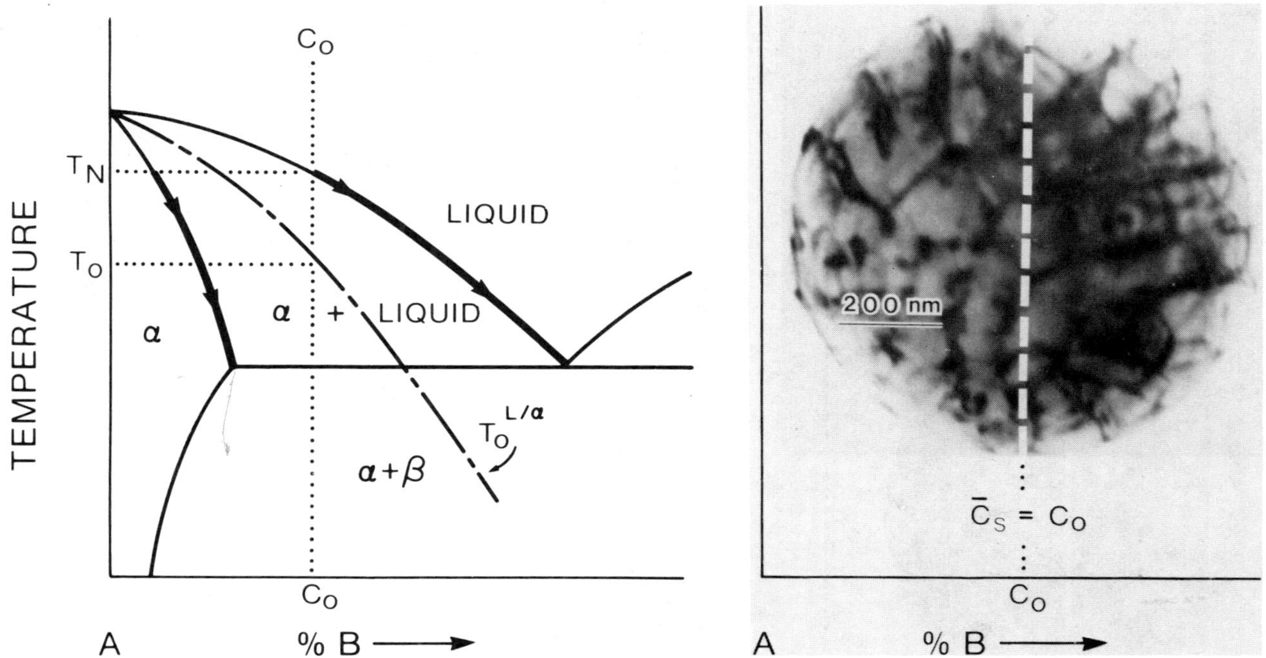

Figure 13 - Schematic phase diagram of Al-Cu system and electron-transparent micrograph of a rapidly solidified Al-4.5%Cu sub-micron droplet. Liquid and solid compositions followed the paths shown on the phase diagram. From [7].

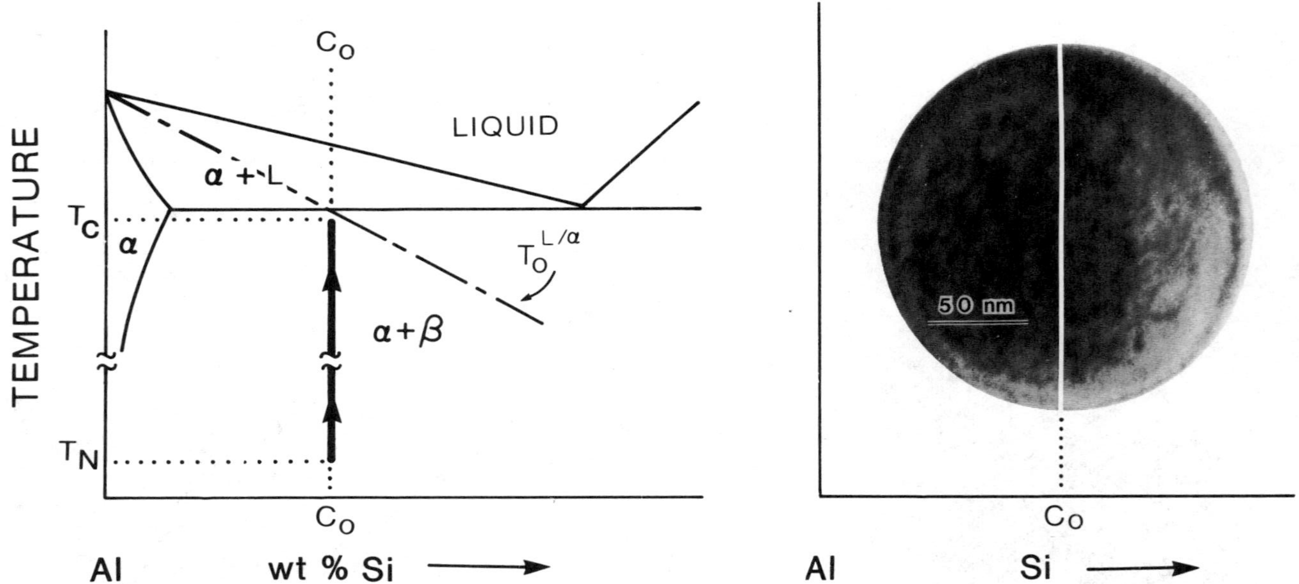

Figure 14 - Schematic phase diagram of the Al-Si system showing the solidification path of an Al-6%Si alloy solidified in a partitionless manner. From [7].

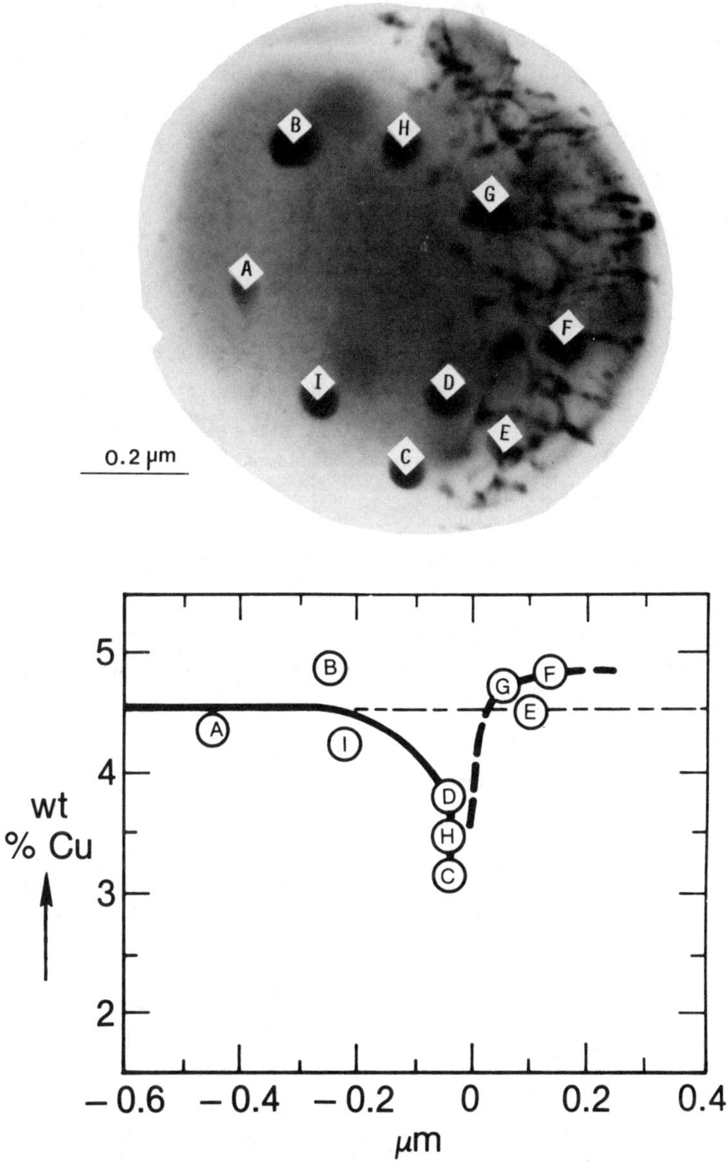

Figure 15 - Copper concentration as a function of distance from the interface between the planar (partitionless) and cellular (segregated) regions in a particle containing an average of 4.5%Cu. From [22].

Concluding Remarks

In conclusion, we believe that substantial progress has been made in understanding the relative roles of heat transfer and kinetics in rapid liquid-solid transformations. Modelling efforts have enhanced our capability to describe microstructure evolution, identify the relevant process parameters and establish relationships between microstructure and processing which are increasingly quantitative in nature.

A continuing activity in this field is necessary to improve our ability to make quantitative predictions which may be used for microstructure optimization and control. Future directions include extension and refinement of the models to deal with solute redistribution and morphological stability in alloys. Better descriptions of thermal and constitutional fields in the vicinity of the liquid-solid interface are necessary to relate supercooling and velocity to microstructure. Phase selection problems shall also be addressed by modelling nucleation and growth processes of competing metastable structures during continuous cooling.

Appendix. Fundamentals of Solidification Models

The models described in this paper are based on a modified version of the enthalpy model for solidification heat transfer originally proposed by Shamsundar and Sparrow [24]. The general equation for the heat balance is

$$(d/dt) \int_V \rho \, H \, dV \quad = \quad \int_A (k \cdot \nabla T) \cdot (\vec{e} \, dA) \qquad \text{(A1)}$$

where A and V are the heat transfer area and volume of a control element, k is the thermal conductivity of the solid or liquid, depending on the phase(s) involved in the direction of heat flow, and $\vec{e}$ is the unit vector perpendicular to the bounding surface of the element. For elements in a sphere of radius r_0 we may rewrite this equation in dimensionless form as

$$(d\psi/dFo) = (4\pi r_0^3 \, / \, V) \, Q \qquad \text{(A2)}$$

where Q is a dimensionless form of the net heat flow in or out of the control volume and Fo is the Fourier number (dimensionless time).

The Newtonian formulation assumes that the droplets are sufficiently small to neglect temperature differences inside the metal. In the case of solidification near equilibrium, the approximation is valid if the surface conductance, h, is much smaller than the metal conductance, k/r_0. For highly supercooled systems, however, it is necessary for the droplet to be smaller than a critical heat transfer length (α_L/R), where α_L is the thermal diffusivity and R is the interface velocity. If Newtonian conditions apply, the appropriate form of the enthalpy balance becomes

$$(d\psi/dFo) = - 3 \phi \text{ Bi } (\theta + \text{Ste}) \tag{A3}$$

where Bi = $(h\ r_0/k_L)$ is the Biot number and Ste = $[C_L^A\ (T_M^A - TG)/\Delta H_M^A]$ is the Stefan number. ϕ is a ratio of molar volumes introduced when dealing with alloys since the dimensionless terms are defined based on the properties of the pure solvent.

Equation (A3) can be solved analytically with the aid of Equation (1) for single phase cooling (constant f) and isothermal solidification ($\theta = 0$). For solidification of supercooled systems both temperature and fraction solid change simultaneously, and the initial condition θ_N (nucleation temperature) is undefined. The solution approach normally involves an implicit finite difference formulation combined with existing models for nucleation and growth kinetics.

Nucleation

The nucleation temperature is typically taken as a parameter when dealing with thermal history and/or solute redistribution during recalescence (Figures 2-5, 10-12). In the analysis of undercoolability and evolution of microcrystalline structures the liquid is initially at or above its melting temperature. As time progresses, temperature changes according to

$$\theta_{k+1} = (\psi_k - 1 - 3\text{BiSte}\Delta\text{Fo})/(1 + 3\text{Bi}\Delta\text{Fo}) \tag{A4}$$

while nuclei formation is predicted by

$$I_{k+1} = I_k + \underline{J}(\theta_{k+1})\ \Delta\text{Fo} \tag{A5}$$

In these equations k and k+1 denote the previous (known) and current (unknown) values, respectively, and $\underline{J}$ is a dimensionless form of the homogeneous or heterogeneous nucleation rate expressions by Turnbull et al [e.g. 15]. A nucleation event occurs every time I reaches an integer value.

Growth

After the first nucleation event the temperature becomes a function of both fraction solidified and heat extraction rate. The governing equation for a binary alloy is

$$\theta_{k+1} = \frac{\psi_k - 3\phi Bi Ste \Delta Fo - (1+\Delta\psi_L)(1-f_{k+1}) - \Delta\psi_S f_{k+1}}{3\phi Bi \Delta Fo + [\kappa_L (1-f_{k+1}) + \kappa_S f_{k+1}]} \qquad (A6)$$

while for a pure metal $\Delta\psi_L = \Delta\psi_S = 0$ and $\phi = \kappa_L = 1$. θ and f are unknown for pure metals and alloys, and in the latter the excess enthalpies and heat capacities are also a function of composition, as indicated by Equations (6) to (9). Other relationships are thus necessary to account for the additional degrees of freedom in order to solve the problem.

The fraction solid is directly related to the interface position by the solidification geometry. On the other hand, the interface position can be determined from its velocity and the time elapsed from nucleation. Further, the interface velocity and temperature are related by kinetic models of the type [18,19]

$$R = \beta\ (D_L/\delta)\ [1 - \exp(\Delta G/R_g T^*)] \qquad (A7)$$

where R is the interface velocity, ΔG is the driving force for solidification, R_g is the gas constant, T^* is the interfacial temperature, D_L is a kinetic constant, typically the self diffusion coefficient of the liquid, and δ is the length of an interface step.

The thermodynamic driving force for pure metals is essentially the entropy of melting multiplied by the supercooling. In the case of alloys, however, one must account for both temperature and composition in ΔG; the following relationship has been proposed for Henrian solutions [20]

$$\frac{\Delta G}{R_g T^*} = (1 - X_S^*)\ \ln \frac{(1 - X_S^*)\ (1 - X_L^e)}{(1 - X_L^*)\ (1 - X_S^e)} + X_S^* \ln \frac{\mathbf{k}}{\mathbf{k}^e} \qquad (A8)$$

where X^* and X^e are respectively the interfacial and equilibrium compositions of solid and liquid at temperature T^*, and $\mathbf{k}^e$ is the equilibrium partition coefficient given by the phase diagram. The non-equilibrium partition coefficient is considered to increase monotonically with velocity, e.g. [6],

$$\mathbf{k} = (X_S^*/X_L^*) = [\mathbf{k}^e + (R\lambda^*/D^*)] \; / \; [1 + (R\lambda^*/D^*)] \qquad (A9)$$

where (D^*/λ^*) is the diffusional velocity of the solute in front of the liquid solid interface.

The applicable equations above are solved simultaneously at every time step in the numerical integration to calculate enthalpy, temperature, fraction solid and interface velocity as a function of time. For the analysis of recalescence the models are constrained to one nucleation event per particle. A slightly more involved algorithm is necessary when studying the competition between nucleation and growth in the early stages of solidification--Figures 6 and 7. In this case one must track the interface of every single nucleus as they form and grow into the liquid. If capillarity considerations are introduced, the interface velocity is also a function of interface curvature, hence position. Furthermore, the nuclei may not only grow but also remelt as recalescence starts.

Non-Newtonian Models

The general approach when significant temperature differences exist within the droplet is similar to the Newtonian case. The formulation is more complicated, however, as the finite difference equations are defined on a bispherical coordinate mesh superimposed on the spherical droplet. The reader is referred to reference [9] for a complete description of this model.

Acknowledgements

This paper contains material developed in programs sponsored by the Defense Advanced Research Projects Agency, the Office of Naval Research, the Aluminum Company of America and the National Science Foundation.

References

1. R. Mehrabian: International Metals Reviews, 27(4)(1982), 185.

2. T.W. Clyne: Metall. Trans. B, 15B(1984), 369.

3. W.J. Boettinger and S.R. Coriell: "Microstructure Formation in Rapidly Solidified Alloys", in Proceedings of the NATO Advanced Workshop on Rapid Solidification Technologies, Castle of Theuern, FRG, March 18-22, 1985.

4. J.C. Baker: Interfacial Partitioning During Solidification, Ph.D. Thesis, MIT 1970, Chapter 5.

5. K.A. Jackson, G.H. Gilmer and H.J. Leamy: in Laser and Electron Beam Processing of Materials, (New York, NY, Academic Press, 1980), 104.

6. M.J. Aziz: J. Appl. Phys., 1982, vol. 53, p. 1158.

7. M.Cohen and R. Mehrabian: in Rapid Solidification Processing, Principles and Technologies III, ed. R. Mehrabian, (Gaithersburg, MD, National Bureau of Standards, 1983), 1-27.

8. J.P. Hirth: Metall. Trans. A, 9A(1978), 264.

9. C.G. Levi and R. Mehrabian: Metall. Trans. A, 13A(2)(1982), 221-234.

10. M.R. Glickstein, R.J. Patterson and N.E. Shockley: in Rapid Solidification Processing, Principles and Technologies, ed. R. Mehrabian, B.H. Kear and M. Cohen, (Baton Rouge, LA: Claitor's Publishing Div., 1978), 46-63.

11. C.G. Levi and R. Mehrabian: Metall. Trans. B, 11B(3)(1980), 21-27.

12. P.G. Boswell and G.A. Chadwick: Scripta Met., 11(1977), 459.

13. H. Reiss and J.L. Katz: in Rapid Solidification Processing, Principles and Technologies I, eds. R. Mehrabian, B.H. Kear and M. Cohen, (Baton Rouge, LA, Claitor Publishing Div., 1978), 64-77.

14. J.H. Perepezko, Y. Shiohara, J.S. Paik and M.C. Flemings: in Rapid Solidification Processing, Principles and Technologies III, ed. R. Mehrabian, (Gaithersburg, MD, National Bureau of Standards, 1983) 28-44.

15. D. Turnbull and J.C. Fisher: J. Chem. Phys. 17(1949), 71.

16. C.G. Levi: in Modelling and Control of Casting and Welding Processes III, (Warrendale, Pa, The Metallurgical Society of AIME, 1986), (in press).

17. S.D. Ridder and C.G. Levi: "Heat Flow and Solidification in Metal Powder Particles" (Paper presented at the The Metallurgical Society of AIME Annual Meeting, New York, NY, 27 February 1985).

18. D. Turnbull: in Thermodynamics in Physical Metallurgy, (Metals Park, OH, American Society for Metals, 1949).

19. J.W. Cahn, W.B. Hillig and G.W. Sears: Acta Met. 12(1964), 1421.

20. J.C. Baker and J.W. Cahn: in Solidification, (Metals Park, OH, American Society for Metals, 1970), 23-58.

21. C.G. Levi and R. Mehrabian: Metall. Trans. A, 13A(1)(1982) 13-23.

22. D. Shechtman, S.D. Ridder and R. Mehrabian: in Rapid Solidification Processing, Principles and Technologies III, ed. R. Mehrabian, (Gaithersburg, MD, National Bureau of Standards, 1983), 96-104.

23. Private discussion with L. Bendersky and S.D. Ridder on microstructural refinement in EHD Al-Mn powders, NBS, Sept. 1985.

24. N. Shamsundar and E.M. Sparrow: Trans. ASME, Series C, 97(1975), 333.

INTERFACE RESPONSE FUNCTIONS FOR RAPID ALLOY SOLIDIFICATION: THEORY AND EXPERIMENT

Michael J. Aziz

Division of Applied Sciences
Harvard University
Cambridge, MA 02138

The deviations from local equilibrium at a rapidly moving solid-liquid interface are well documented. The fraction of solute atoms in the liquid at the interface that joins the crystal during rapid solidification approaches unity and the interface temperature drops. A kinetic theory is presented that predicts the amount of solute incorporation and the amount of interfacial undercooling in terms of the interface speed during rapid solidification. Predictions for systems of interest are presented. Experiments to test the theory are discussed.

Introduction

Rapid solidification experiments have reached a crystal growth regime where deviations from local equilibrium are obvious and interface motion is no longer strictly heat-flow limited. These experiments allow us the opportunity to study the interface reaction kinetics in high-mobility systems for the first time. In the first part of this paper, I review evidence for the departure from local equilibrium. I then describe a simple model, applicable to both metals and semiconductors, for the kinetics of the fundamental atomic processes occurring at the interface during rapid solidification of binary alloys. The result is a pair of "interface response functions" which predict (a) how much solute should be incorporated into the solid, and (b) how fast the interface should move; in terms of the local conditions at the interface, namely undercooling and composition. I then describe some recent experiments that until recently have been possible only in semiconductors; the results support one model and rule out others. Finally, the consequences are discussed for a metallic alloy system of interest.

The analysis of an alloy solidification process typically treats the transport of solute and heat through the bulk of one or both of the phases involved in the transformation. The boundary conditions for the transport equations are the interface response functions, which are often unknown. To facilitate solution of the transport equations, the assumption of local equilibrium boundary conditions has been used extensively. The idea behind such an assumption is that, although under conventional processing conditions, long-range diffusion through a liquid may be too slow to maintain compositional equilibrium throughout the bulk of the material, short-range transport across the solid-liquid interface may still be sufficiently rapid to effect interequilibration of the atoms in the liquid monolayer and the solid monolayer adjacent to the interface. Such is the model of local equilibrium: the compositional relationship between the two phases is the equilibrium relationship across the interface. Local equilibrium implies that, given a composition in the liquid at the interface, the solid will grow at the equilibrium liquidus temperature, with the equilibrium solidus composition (1).

Now in a rapid quenching process such as electrohydrodynamic atomization or splat-cooling of molten alloys, or regrowth from laser-induced melting, the interface temperature is often unknown. Knowledge of the initial and final temperatures and compositions may be insufficient to determine whether local equilibrium existed at the interface. Local equilibrium can hold during such a process only if the interface temperature during the quench corresponds to the equilibrium solidus curve or its metastable extension, e.g., below a eutectic point. Baker and Cahn proved that local equilibrium is <u>not</u> maintained during rapid interface motion produced by splat quenching of Zn-Cd alloys (2). The retrograde solidus of Cd in Zn is characterized by a maximum Cd concentration, at solid compositions beyond which solidification under local equilibrium conditions cannot occur. The formation of solids with Cd concentrations greater than this maximum demonstrated that the solid composition at the interface was far from the equilibrium solidus curve, inspiring efforts to model the deviations from local equilibrium.

Pulsed laser melting of doped semiconductors (3,4) permitted experiments under very well-controlled conditions; for example with one-dimensional experimental geometries, allowing simple but accurate calculations, and even measurements, of important quantities such as impurity depth profiles and interface positions and velocities. Heat-flow calculations and post-irradiation analyses of dopant distributions showed that: (a) maximum equilibrium solubilities could be exceeded by several orders of magnitude, and (b) the partition coefficient k, defined by the ratio of the impurity concentration in the growing solid to that in the liquid at the interface, departs from its equilibrium value and approaches unity. In systems with retrograde solidi it could be shown that the impurity had actually undergone an increase in chemical potential upon solidification, termed "solute trapping" (1,2).

A number of physically plausible models have been proposed (5-13) to explain how the host atoms persuade the impurity atoms to increase their chemical potential and join the crystal. They were in qualitative agreement with the early data, which show the partition coefficient monotonically increasing from its equilibrium value k_e toward unity as the interface speed, v, increases.

The basis for the model described here is based on an idea due to Turnbull viz., that crystallization and interdiffusion at the interface involve processes that are fundamentally different. Conventional modeling of interface kinetics (10) uses the same atomic motions to describe crystallization as those to describe interdiffusion at the solid-liquid interface. In such a case, the growth speed can never exceed the maximum diffusive speed, which is typically on the order of 1-10 m/s, given by $v_D = D_i/\lambda$, where D_i is the coefficient for interdiffusion across (not along) the interface and λ is the interatomic distance. However, the collision-limited growth model developed by Turnbull and coworkers (14-22), which has received recent support from the molecular dynamics work of Broughton et al. (23) and the picosecond laser melting experiments of MacDonald et al. (24) on pure metals, indicate that for metallic and other simple molecular systems such a scaling is invalid and that v_s, the speed of sound, should be the only limit to the crystal growth rate. Consequently, the growth speed can very much exceed the diffusive speed, and when it does so in alloys, we expect suppressed segregation and a diffusionless transformation.

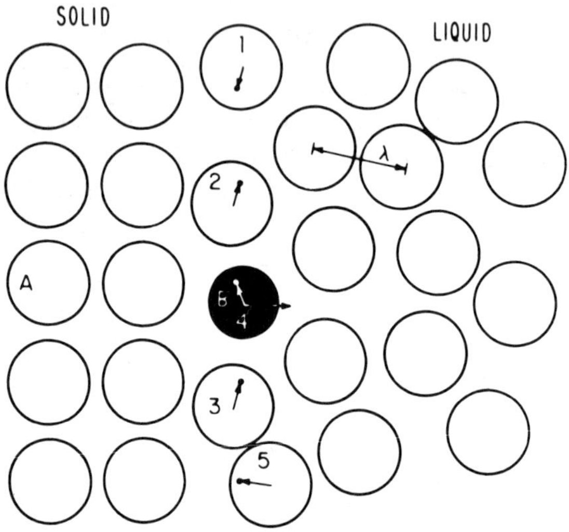

Fig. 1. Choreography of solute trapping. In order to avoid incorporation into the crystal (solid arrow), the "B" atom must: (1) escape from its potential well (dashed arrow) as the well gains energy and moves to the lattice site (solid arrow), (2) push "A" atoms out of the way, and (3) be replaced by an "A" atom on the lattice site.

The reason that crystal growth in metals should not scale with diffusion is illustrated in Fig. 1. A crystallization event consists of an atom shifting its position a small distance to move from a potential well in the liquid structure to a well on the crystal lattice. Since no bonds must be broken, the activation barrier for this reaction ought to be quite small. A diffusive jump, on the other hand, is a different atomic process requiring the cooperation of the neighboring atoms in the system. Position shifts of a fraction of an interatomic distance do not accomplish the solute-solvent redistribution which is necessary to avoid solute incorporation into the crystal. A solute atom must accomplish three things to avoid crystallization: it must escape from its potential well, as shown by the dashed arrow in Fig. 1, as the energy of the well increases and the well moves to the lattice site; it must push solvent atoms out of the way; and it must be replaced by a solvent atom on the lattice site, or else it will be pushed right back to where it came from. Any solute atom that does not accomplish these feats will be incorporated into the crystal.

Atoms in the interface region tend to interdiffuse in order to establish equilibrium compositions on either side of the interface. As the interface passes, solute-solvent redistribution only occurs during a limited time. Before the interface arrives, the driving force for redistribution is zero; afterwards, atomic mobilities are too low to permit significant interdiffusion. When $v \ll v_D$, there is plenty of time for the atoms at the interface to equilibrate. As the growth velocity becomes greater, solute and solvent atoms have less time while they are in the interface region to interdiffuse and maintain local equilibrium. When $v \gg v_D$, the atoms have virtually no time to interdiffuse; the interface passes and locks the liquid composition into the bulk of the solid before the atoms have a chance to react.

If growth occurs by the periodic rapid lateral passage of steps of height λ, then the time between the passage of steps is $t_c = \lambda/v$. The time it takes solvent atoms to assume their lattice positions is negligible compared to t_c, and the shifts can be treated as virtually instantaneous. On the other hand, if continuous solidification occurs, the impurity atom is being dragged toward its lattice site during a time period of order t_c. An analysis of solute diffusion back into the liquid (11) during the interval t_c between the passing steps in the former case yields, for a dilute solution of B in A,

$$k(v,T) = k_e(T) + [1-k_e(T)] \exp(-v_D/v) \tag{1}$$

where the equilibrium partition coefficient $k_e(T)$ is the ratio of the equilibrium solidus composition to the equilibrium liquidus composition at the temperature T of the undercooled interface, and the diffusive speed v_D is the only fitting parameter in the model. This model should apply when the chemical potential of the solute atom is raised from its liquid value to its solid value during a time interval that is very short compared to t_c, alternating periodically with diffusive solute escape between such events. In the latter case of continuous growth, in the limit of strict steady-state fluxes on a microscopic scale, the fluxes and concentrations become time-independent and a similar analysis (12) yields

$$k(v,T) = [(v/v_D) + k_e(T)]/[(v/v_D) + 1] \tag{2}$$

which shows the same qualitative features as in stepwise growth, a transition from equilibrium partitioning ($k = k_e$) to complete solute trapping ($k \to 1$) as the interface speed surpasses the diffusive speed. In the continuous growth model, however, the transition is more gradual than in the stepwise growth model.

Experimental Results

The most significant experimental progress in the field of interface kinetics during rapid solidification has been made with pulsed laser melting of semiconductors. Progress was rapid because defect-free plane-front solidification is routinely attained and because in many cases k_e is so small that when k approaches unity tremendous changes are effected in the final impurity distribution.

The pioneering work of White et al. (3,25) and of Baeri et al. (4,26) showed that k can in fact increase several orders of magnitude above k_e at high interface speeds and demonstrated the velocity-dependence of k by varying the laser and substrate parameters to cover a range of regrowth velocities during pulsed laser melting of ion implanted Si. When compared in detail, the results of the two groups were contradictory in some cases. Certain studies showed an effect termed "saturation", where $dk/dv \rightarrow 0$ at some value of $k < 1$. This effect was reported for Bi (26), In (27), and Sb (28) in Si. In a separate study of Bi in Si, however, the effect was not observed (25). This saturation phenomenon is intriguing because it cannot be explained by any simple model; if correct, it indicates that more complicated solute trapping models (8,13,29) are necessary for understanding the phenomenon.

A study was recently undertaken (30) to examine the saturation phenomenon in detail and to see whether any of the existing models could yield quantitative agreement with the data. To obtain high precision data, we used the transient conductance technique for measuring the interface velocity (31) for Bi-implanted Si, thereby eliminating errors that arose in the previous studies, which could only estimate v by the use of heat-flow calculations. The time-evolution of the melt depth was measured with roughly 1 ns resolution with this technique, which measures changes in the transverse electrical conductance of a thin film specimen as the phase transformation proceeds. The initial and final Bi profiles were determined to high accuracy by Rutherford Backscattering Spectroscopy (RBS) carried out in glancing angle geometry. Analysis of the partition coefficient was performed by numerical simulation with a finite-element solution of the diffusion equation, using the measured as-implanted profile and the measured melt depth as a function of time for each sample. Impurity diffusion was assumed to occur in the liquid phase, but neglected in the solid phase. During solidification, a fraction k of the impurity in the liquid at the interface was incorporated into the crystal. The parameters allowed to vary were the bulk liquid-phase diffusivity D_L and the partition coefficient during solidification k. Comparison with the measured profiles yielded unique values for both D_L and k.

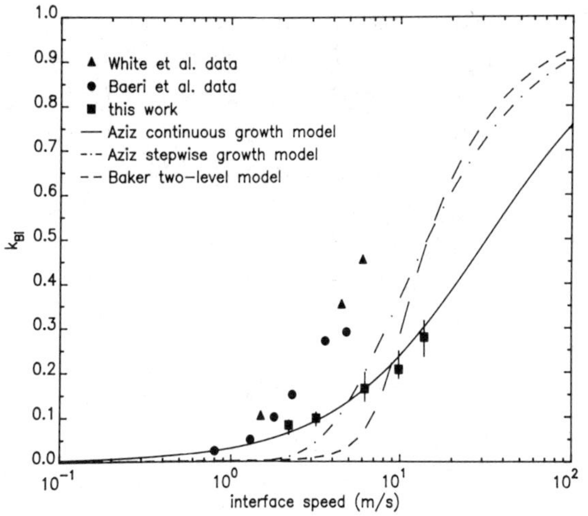

Fig. 2. Dependence of partition coefficient on velocity. Triangles and circles: velocity estimated from heat-flow calculations. Squares: velocity measured by transient conductance. Continuous growth model: v_D = 32 m/s. Stepwise growth model: v_D = 10 m/s. Baker model: v_D = 1 m/s.

In Fig. 2 the results of analysis of several samples are shown. The predictions of the models discussed above for stepwise growth and for continuous growth are shown, as is the two-level Baker model (8,11,20), where the impurity is treated as if diffusing in a continuum down a steep energy gradient at the interface. For this model, k is given by $k = \{\beta + \ln k_e\}/\{\beta+[(1/k_e)(\ln k_e)\exp(-\beta)]\}$, where $\beta \equiv v/v_D$. The diffusive speed v_D is the only free parameter in all three models; it has been chosen in each case to yield the best overall fit to the data. Note that in each of these models v only appears as the ratio v/v_D; thus when plotted as k versus log v, as in Fig. 2, the only effect of varying the free parameter v_D is a rigid horizontal shift of the curve, with no change in the slope. We see that the stepwise growth model and the two-level Baker model produce curves that rise too steeply to account for our data. The continuous growth model, on the other hand, fits our data quite well.

Other solute trapping models were not considered. One model due to Jackson and coworkers (10) produces a k(v) curve similar to that of the continuous growth model discussed here, but has been rejected because it cannot account for solute trapping on both ends of a phase diagram (32). We are unable to compare our results to the models of Hillert and Sundman (33) and of Gilmer (13) because they involve complex computer calculations that we cannot perform.

Data from the original work of Baeri et al. (26) and of White et al. (25) are also shown in Fig. 2. While sufficient to establish that k increases far beyond k_e as v increases, we see here that their data were not accurate enough to allow any conclusions regarding appropriate models or the presence of the saturation effect discussed earlier. Note that there is no evidence in our data for such an effect. It appears that the main source of the disagreement is the use in the original heat flow calculations of an assumed thermal conductivity of amorphous Si that is very much greater than measurements eventually indicated (34). This apparently caused errors in their calculations of the time-evolution of the melt depth, which fed back to produce errors in the apparent k.

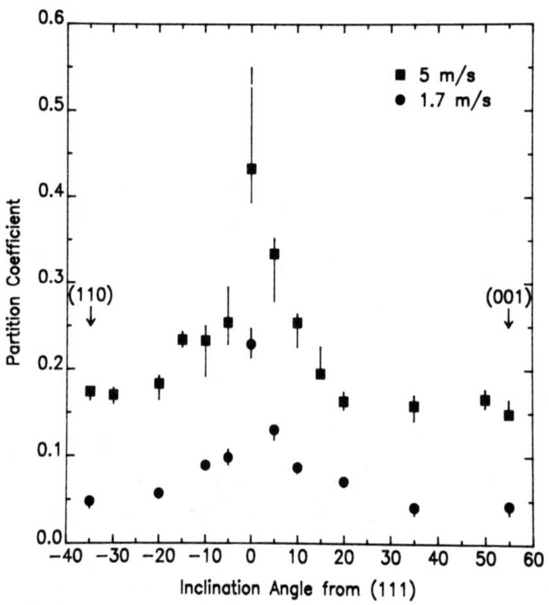

Fig. 3. Orientation dependence of partition coefficient of Bi in Si at two speeds. Squares: 5 m/s. Circles: 1.7 m/s.

Orientation Dependence

Similar agreement with the continuous growth model was obtained for trapping at the (111) interface. The accuracy was limited by the unavailability of (111) Si-on-sapphire, forcing us to rely on heat-flow calculations of the interface position and velocity. It was surprising to find that the stepwise growth model does not apply even at the (111) interface. A full study of the orientation dependence of k was undertaken (35) to resolve this puzzle.

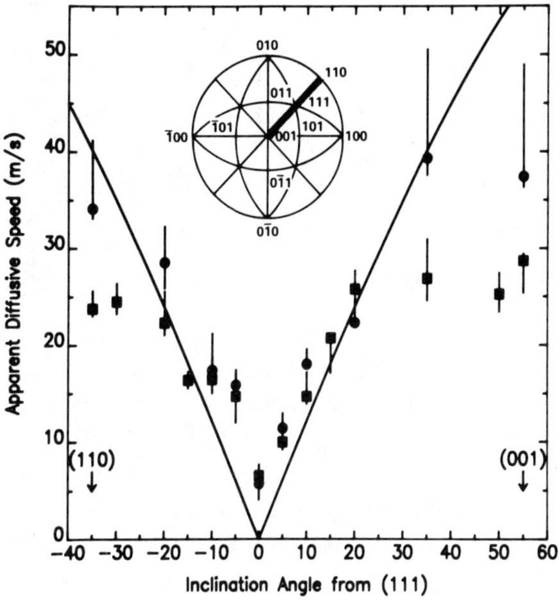

Fig. 4. Orientation dependence of apparent diffusive speed in continuous growth model. Squares: 5 m/s. Circles: 1.7 m/s. Curve is v_D = 70 m/s sin|θ|.

Si wafers were cut at orientations from (110) through (111) to (001) at 5° increments and given identical Bi implants. They were then irradiated with a pulsed excimer laser under two sets of conditions, calculated to yield regrowth speeds of 1.7 and 5 m/s, respectively. As a check on the calculations and the shot-to-shot variation, the laser pulse energy was monitored by using time-resolved reflectivity to measure the melt duration for each shot. Numerical simulations were performed as above to determine k, using the calculated melt depth as a function of time. The results, plotted in Fig. 3, show a sharp peak in k at the (111) orientation. Some insight into these results can be gained by considering the orientation dependence of the apparent diffusive speed, determined by fitting the continuous growth model (eq. (2)) to yield the known partition coefficient at the known velocity, as shown in Fig. 4. The fact that at 5 m/s and 1.7 m/s the values of v_D so obtained are nearly equal for any given orientation is evidence that the continuous growth model describes the velocity dependence of k at all orientations. Fitting, for example, to the stepwise growth model (eq. (1)) at a given orientation requires wildly different values of v_D at these two velocities, which is not a plausible interpretation of the data. It is not clear at this point whether the small differences in v_D at large inclinations from (111) are significant. The curve in Fig. 4, which fits the data reasonably well, is the equation $v_D = 70$ m/s $\sin|\theta|$, where θ is the abscissa.

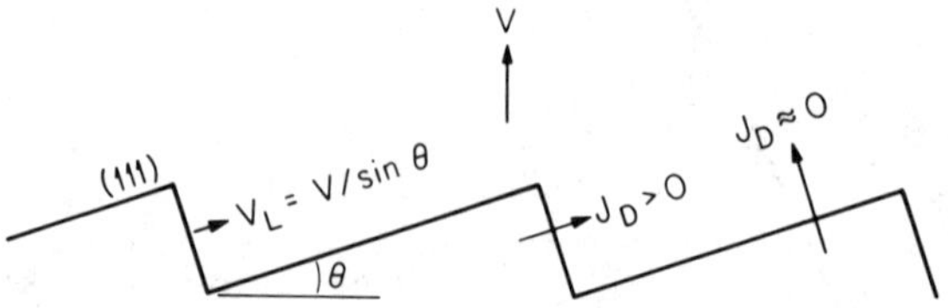

Fig. 5. Solute trapping by lateral passage of ledge, with lateral solute escape flux at moving ledge.

The results are interpreted as crystal growth occurring by the passage of (111) ledges with lateral speed $v_L = v/\sin|\theta|$, as shown in Fig. 5. The absence of cusps at (110) and (001) in Fig. 4 is evidence for the absence of stepped surfaces with these orientations. With this model, the velocity dependence will continue to follow the continuous rather than the stepwise growth model if, contrary to our original assumption in the latter (11), there is negligible solute escape from the terrace once the ledge passes. Rather, lateral escape could occur at the edge of the moving step itself. This interpretation is plausible since we expect the solute atom on the terrace to be triply or quadruply bonded to the crystal once the ledge passes. In addition, the ledges may be several monolayers in height, thus eliminating the possibility for most of the solute to escape through the terrace. Atoms at the edge of the step, on the other hand, will have fewer bonds to the crystal, which would allow more rapid exchange from the solid into the liquid. Since from entropic considerations we expect the steps to be heavily kinked and therefore diffuse laterally, the kinetics of continuous lateral ledge motion vs. lateral solute-solvent redistribution lead to eq. (2), with v replaced by v_L and v_D replaced by a maximum speed v_D^L of redistribution across the step edge:

$$k(v,\theta) = \{[v_L(\theta)/v_D^L] + k_e\}/\{[v_L(\theta)/v_D^L] + 1\}.$$

Although conceptually quite simple, such a model accounts remarkably well for partitioning of Bi in Si over a wide range of orientations and velocities, with a single fitting parameter of $v_D^L = 70$ m/s. Nevertheless, if one looks precisely along (111) or at large inclinations from (111), the possibility for further refinement of the model by, e.g., allowing a small amount of solute escape through the (111) terraces, is evident. Finally, note how close the coefficient for interdiffusion at the step, i.e. the product of v_D^L and the (111) layer spacing of 3.1 Å, is to the measured bulk liquid diffusivity of 2×10^{-4} cm^2/s.

Growth Velocity

Onsager's reciprocity relations (36) imply that if crystal growth affects the amount of solute incorporation, then solute trapping will in turn affect the growth rate. The common assumption of heat-flow-limited growth is in many cases a useful one. It does, however, become unrealistic at high velocities and solute concentrations, where the interface reaction kinetics can come into play. The implications of equation (2) for crystal growth rate are described below.

For pure systems, the crystal growth process is understood as follows (37). In equilibrium, the rates at which atoms join and leave the crystal are equal. When there is a finite driving force for crystallization, the former exceeds the latter by a Boltzmann factor $\exp(-\Delta\mu/RT)$; the growth velocity is given by their difference. The kinetics are manifested by an intrinsic growth-velocity limit, which is the speed of the forward reaction alone, $v_o(T)$. This is the speed with which the interface would propagate at infinite driving force (i.e., when the back reaction disappears). The growth velocity is then written as $v = v_o(T)[1 - \exp(\Delta\mu/RT)]$ (note $\Delta\mu$ is defined to be negative during solidification). This expression becomes the familiar $v = M\Delta G$ at small driving forces. Inverting the above expression, yields $\Delta\mu = RT \ln(1 - v/v_o)$. The left-hand side is the driving free energy; the right-hand side can be thought of as the free energy dissipated by the irreversible crystallization reaction at the interface. Thus, the velocity-driving force relation can be written $\Delta G_{DF} = \Delta G_F$, where ΔG_{DF} is the driving energy, expressed as free energy per mole, and ΔG_F, the "interface friction energy" due to a finite interfacial mobility, is the free energy dissipated per mole of material solidified. The available driving force is dissipated by the irreversible processes at the interface. One can phrase this statement in an alternative way by dividing each side of this equation by the molar volume (38). One would then say that the sum of the forces per unit area on an interface moving at constant velocity is zero.

For alloys, one can write $\Delta G_{DF} = \Sigma\Delta G_{DISS}$, where the ΔG_{DISS} are the losses due to all of the simultaneous dissipative processes. In the steady state, if the system does no work on the surroundings, the driving free energy must be entirely dissipated by irreversible processes during the transformation. If one can calculate the dissipation and the driving force as functions of velocity, one has an equation that predicts the growth velocity. Ignoring heat conduction, the two dissipative processes at the interface are the crystallization and redistribution reactions. The dissipative terms associated with them are, respectively, the "interface friction energy" ΔG_F and the "solute drag energy" ΔG_D (33,38,39). By assuming interface friction to be the same for alloys as for pure systems, these terms can be calculated (12), resulting in $\Delta G_F = RT \ln(1 - v/v_o)$ and, for the dilute solution limit of the continuous growth model (eq. (2)), $\Delta G_D = X_s RT[(1 - k)/k] \ln[k/k_e(T)]$. The driving free energy is given by (33) $\Delta G_{DF} = X_s \Delta\mu_B + (1 - X_s)\Delta\mu_A$; the growth velocity equation becomes

$$\Delta G_{DF}(v) = \Delta G_F(v) + \Delta G_D(v)$$

or, inverted,

$$v = v_o[1 - \exp(-1/RT)(\Delta G_{DF} - \Delta G_D)] \quad (3)$$

The dilute solution approximation is invalid when the mole fraction of solute is no longer negligible or when activity coefficients λ are no longer constants. When the appropriate corrections to the fluxes and driving forces in the continuous growth model are made (12), the results are

$$k = [(v/v_D) + \kappa_o]/[(v/v_D) + 1 - (1 - \kappa_o)]X_{1i}] \quad (4)$$

$$\Delta G_D = X_s[(1 - k)/k]RT \ln(k/k_A \kappa_o) \quad (5)$$

where $\kappa_o(X_s, X_{1i}, T) \equiv \exp[-\Delta(\mu' - \mu_A')/RT]$, $\mu'(X,T) \equiv \mu(X,T) - RT \ln X = \mu^o(T) + RT \ln \gamma(X,T)$, X_{1i} is the mole fraction of B in the liquid at the interface, $k_A \equiv X_s^A/X_{1i}^A$, A = solute, B = solvent, and Δ = "solid minus liquid". The implicit assumption has been made that at the interface the transition state behaves in an ideal manner. In practice, temperature and composition dependence may show up in D_i.

Theoretical Results

The kinetic model requires as input the chemical potential of each species in each phase at all temperatures and compositions of interest. If one hopes to compare the predictions of any kinetic model with experimental results, an accurate thermodynamic model is essential, for if the predictions do not agree with experiment, one must know whether the fault lies in the

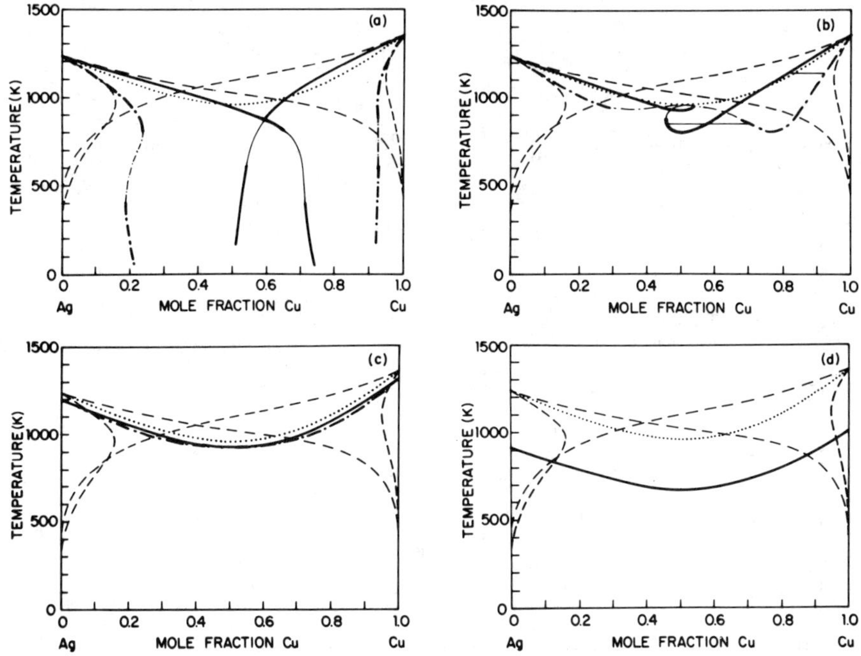

Fig. 6. Kinetic interface condition diagram for Ag-Cu assuming v_o/v_D = 100. The effect of solute drag is included. Dashed lines: solutions for v = 0 corresponding to equilibrium phase diagram. Dotted line: T_0 curve. Solutions for: (a) v/v_D = 0.1; solid line: kinetic liquidus; dot-dash line: kinetic solidus. Thin regions of kinetic solidus and liquidus denote unstable roots. (b) v/v_D = 0.3; (c) v/v_D = 3.0; (d) v/v_D = 30.

thermodynamic or in the kinetic model. The Ag-Cu system has been modeled thermodynamically throughout the undercooled liquid regime by Murray (40); her model is used in the following calculation. The only free kinetic parameters are the speed of interdiffusion at infinite driving force v_D, and the speed of crystal growth at infinite driving force v_o. An estimate of v_D/v_o = 0.01 was used for this system.

The four variables X_S, X_L, T_i, and v are related by the pair of equations (3) and (4). Any pair of variables can thus be expressed in terms of the other pair. The results are displayed in the interface condition diagrams of Fig. 6. They show, for example, the interface temperature and solid composition that result if one imposes a certain velocity and liquid composition at the interface. The solution of equations (3) and (4) for v = 0, displayed as dashed lines in Fig. 6 (a)-(d), are the solidus and

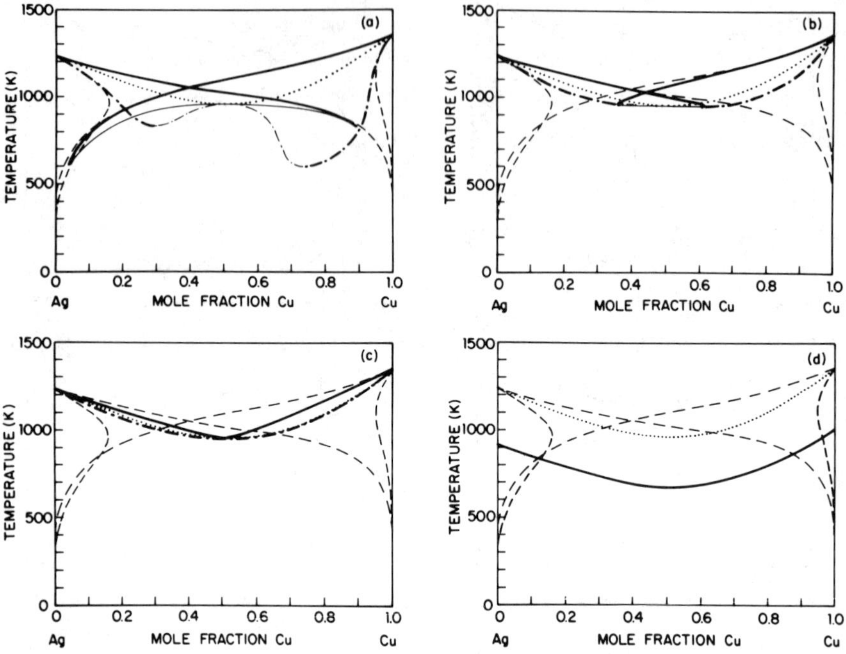

Fig. 7. Kinetic interface condition diagram for Ag-Cu neglecting any effect due to solute drag. Solutions with v_0/v_D = 100 for: (a) v/v_D = 0.01; solid line: kinetic liquidus; dot-dash line: kinetic solidus. (b) v/v_D = 0.1; (c) v/v_D = 0.7; (d) v/v_D = 30.

liquidus of the equilibrium phase diagram. Their metastable extensions below the eutectic point are shown here, and solid-solid equilibrium has not been included in the diagrams. The dotted line is the T_0 curve, which is the locus of points where liquid and solid of identical composition have equal free energies; it represents the maximum temperature at which a solid of a given composition can be formed from a liquid of any composition. The solid and dot-dashed curves in Fig. 6(a) are the "kinetic" liquidus and solidus, solutions to the equations for an imposed interface speed of 10% of v_D. The liquidus and solidus begin to approach each other, reflecting the increasing partition coefficient (which is the ratio of the kinetic solidus and liquidus compositions). The already substantial undercooling of the kinetic liquidus below the equilibrium liquidus is due to solute drag. Note also that the composition at which the two liquidus curves cross each other has shifted substantially. In Fig. 6(b) is plotted the solutions for an imposed interface speed of 30% of v_D. At this speed, the kinetic liquidus is approaching the T_0 curve as the degree of solute trapping and of interfacial undercooling increase. Multiple solutions are found in the middle of the phase diagram. This region of the phase diagram lies under the chemical spinodal for the solid and may thus be unobservable experimentally. Since this treatment neglects effects such as coherency stress that tend to stabilize the solid against spinodal decomposition, the "unstable" solutions have been included in the event that some of them might be realized experimentally. Representative tie-lines are shown to aid the reading of the diagrams. As the interface speed is increased further to 3 v_D, or 0.03 v_0 (Fig. 6(c)), the kinetic liquidus drops below T_0 and k approaches unity. At this point the solute drag term is rapidly disappearing and most of the free energy dissipation is due to the interface friction term, RT $\ln(1 - v/v_0)$, which is now almost constant across the interface condition diagram. This conclusion can be drawn by observing the undercooling of pure Ag and pure Cu at the two ends of the diagram, where the solute drag term is identically zero. Note that at this speed a partitionless transformation is thermodynamically possible but, however, does not occur until still greater speeds, where the kinetic solidus and liquidus merge and become essentially indistinguishable (Fig. 6(d)).

Although the existence of "solute drag" has been established in grain boundary migration (42), there is the possibility that the idea does not apply to alloy solidification. There are treatments similar to that presented here except that they neglect solute drag (11,43). It is useful to see how the interface condition diagrams presented here are altered if the free energy dissipation due to solute drag is set equal to zero. In Fig. 7 is shown the numerical solutions of the same equations with $\Delta G_D = 0$. In this case, we see significantly less undercooling, in general, than in the former case. The difference between the undercoolings predicted in Figs. 6 and 7 is most evident at $v/v_D = 0.1$ (Fig. 7b). In the latter case, the kinetic solidus rises with increasing v toward T_0 much more rapidly than the liquidus drops. Consequently, the intersection of the two liquidus lines moves rather little as v is increased. Note also how close the kinetic solidus lies to T_0 at $v/v_D = 0.7$ when solute drag is neglected, compared to the proximity of the kinetic liquidus and T_0 at $v/v_D = 0.3$ when solute drag is included.

An instability in the steady-state solution occurs where a tie line connects a kinetic solidus and a kinetic liquidus having opposite slopes, such as in the central regions of Figs. 6(b) and 7(a). A fluctuation whereby a small amount of crystal grows with more solute than the steady-state amount will make the liquid at the interface leaner in solute, thus shifting the steady-state solution to yet greater solute concentrations. Similarly, a fluctuation in the opposite direction will also tend to grow. This effect could give rise to oscillatory behavior when a constant solidification velocity is imposed on material with a bulk liquid composition in this region. Consider, for example, a liquid of bulk composition 40 at.% Cu, undergoing solidification with a speed of $v = 0.01\ v_D$ (Fig. 7(a)) imposed by heat flow. Two solid compositions that can grow from this liquid are approximately 94 and 15 % Cu. If the system chooses the latter, rejected Cu will build up in the liquid at the interface and its composition as well as that of the growing solid will migrate toward the Cu-rich end of the diagram. Normally, this process would continue until the solid was growing with a composition equal to that in the <u>bulk</u> of the liquid, whereupon the enriched layer in the liquid at the interface would cease to grow or shrink. In the 40 % Cu case, however, such a final steady state is impossible because the liquidus reaches a maximum composition of 90 % Cu when the solidus is at only 28 % Cu. Even at this point the solid is continuing to reject excess Cu into the liquid and the buildup of Cu in the liquid at the interface continues. The only solution is to jump up to the other root with a liquidus at 90 % Cu, namely that of the Cu-rich phase with a solidus at 97 % Cu. Along this branch, the solid will reject Ag and the root will move to the Ag-rich side of the diagram until the same problem is encountered. In fact, the system may jump to the other branch at any time before being forced to do so. Factors that will tend to discourage such behavior include a gradient energy and a strain energy associated with depositing a layer of composition and equilibrium lattice parameter significantly different from those of the substrate. Strictly speaking, this theory yields steady-state response functions only, and any oscillatory behavior cannot be described in terms of such roots. However, we expect the time it takes for the interface to reach a "steady-state" configuration to be rather short, on the order of ℓ^2/D_i, where ℓ is the width of the interface. Thus, the steady state theory may adequately describe unsteady behavior also. If the interface temperature varies significantly along a branch or between branches during such oscillatory behavior, we might also expect oscillations in v about the value "imposed" by heat flow. A full treatment coupling the interface response functions presented here to the temperature field and the concentration field in the bulk of the liquid has not been attempted. When the variations in T_i are not too large, then the idea of oscillations in the composition of the growing solid at a velocity "imposed" by heat flow may be a reasonable approximation.

Such an effect may account in part for the so-called "banded" regions (41,44) observed in scanned laser or e-beam welds at speeds of ~1 m/s on Ag-Cu alloys of composition ~15-40 at.% Cu. It appears in this case that completion of a full cycle described above might be interrupted by the periodic cellular breakdown of the interface.

During rapid regrowth from pulsed laser melting of semiconductors, the interface has been observed to slow down upon encountering a heavily doped region (45,46). The interface velocity was measured in these experiments by the transient electrical conductance technique. Whether this effect is mainly due to solute drag, a reduced driving force for solidification, or a reduced interface mobility (v_0) remains to be seen. This interface velocity measurement technique has recently been successfully applied to metallic thin films (47) as well. In the near future, experiments of this type may enable us to make a quantitative test of the theory for the interface velocity vs. undercooling and composition response function.

Summary

The transition from diffusion-controlled to diffusionless solidification follows simply and directly from the collision-limited growth model of Turnbull and coworkers, wherein the atomic motions responsible for interface advancement can be much more rapid than those necessary for solute escape at the interface. The continuous growth solute trapping model best describes the measured velocity dependence of trapping of Bi in Si. The measured orientation dependence of solute trapping in this system is understood in terms of the lateral motion of {111} ledges. The interfacial undercooling has been predicted for the continuous growth model by equating the driving free energy with the free energy dissipated by the irreversible processes at the interface, assuming interface friction to be the same in alloys as in pure systems. The resulting interface condition diagrams reproduce the equilibrium phase diagram at zero velocity, and depict interfacial undercooling and suppressed solute partitioning as the interface speed increases. The results are sufficiently different if solute drag is neglected that experiments may be possible to determine whether the solute drag effect is present in alloy solidification.

Acknowledgments

Much of the work reported here has benefited from the invaluable contributions of my collaborators C.W. White, J.Y. Tsao, M.O. Thompson, and P.S. Peercy, on the experimental work, and T. Kaplan on the theoretical work.

This research was sponsored by the Division of Materials Sciences, U.S. Department of Energy under contract DE-AC05-84OR21400 with Martin Marietta Energy Systems, Inc., and in part by the Harvard Materials Research Laboratory under contract NSF-DMR-83-16979.

References

(1) J.C. Baker and J.W. Cahn, Solidification (Metals Park, OH: American Society of Metals, 1970), 23-58.

(2) J.C. Baker and J.W. Cahn, Acta Metallurgica, 17 (1969) 575.

(3) C.W. White, S.R. Wilson, B.R. Appleton and F.W. Young, Jr., Journal of Applied Physics, 51 (1980) 738.

(4) P. Baeri, J.M. Poate, S.U. Campisano, G. Foti, E. Rimini, and A.G. Cullis, Applied Physics Letters, 37 (1980) 912.

(5) R.N. Hall, Journal of Physical Chemistry, 57 (1953) 836.

(6) A.A. Chernov, Growth of Crystals, vol. 3, eds. A.V. Shubnikov and N.N. Sheftal' (New York: Consultants Bureau, 1962), p. 35.

(7) J.C. Brice, The Growth of Crystals from the Melt (Amsterdam: North-Holland, 1965).

(8) J.W. Cahn, S.R. Coriell and W.J. Boettinger, Laser and Electron Beam Processing of Materials, eds. C.W. White and P.S. Peercy (New York: Academic Press, 1980), p. 89.

(9) M. Hillert and B. Sundman, Acta Metallurgica, 25 (1977) 11.

(10) K.A. Jackson, Surface Modification and Alloying by Laser, Ion and Electron Beams, eds. J.M. Poate, G. Foti and D.C. Jacobson (New York: Plenum Press, 1983), p. 51.

(11) M.J. Aziz, Journal of Applied Physics, 53 (1982) 1158.

(12) M.J. Aziz, Applied Physics Letters, 43 (1983) 552.

(13) G.H. Gilmer, Materials Research Society Symposia Proceedings, 13 (1983) 249.

(14) D. Turnbull, Journal of Physical Chemistry, 66 (1962) 609.

(15) D. Turnbull, Contemporary Physics, 10 (1969) 473.

(16) D. Turnbull, Solidification (Metals Park, OH: American Society of Metals, 1970), pp. 1-22.

(17) D. Turnbull, Journal de Physique, Colloque C-4 (1974).

(18) D. Turnbull and B.G. Bagley, Treatise on Solid State Chemistry, vol. 5, ed. by N.B. Hannay (New York: Plenum Press, 1975), pp. 513-514.

(19) F. Spaepen and D. Turnbull, Rapidly Quenched Metals, 2nd International Conference, eds. N.J. Grant and B.C. Giessen (Cambridge, MA: MIT Press, 1976), pp. 205-29.

(20) F. Spaepen and D. Turnbull, Laser Annealing of Semiconductors, eds. J.M. Poate and J.W. Mayer (New York: Academic Press, 1982), pp. 15-42.

(21) D. Turnbull, Metallurgical Transactions A, 12 (1981) 693.

(22) S.R. Coriell and D. Turnbull, Acta Metallurgica, 30 (1982) 2135.

(23) J.Q. Broughton, G.H. Gilmer, and K.A. Jackson, Physical Review Letters, 49 (1982) 1496.

(24) C.A. MacDonald, M.A. Malvezzi and F. Spaepen, Materials Research Society Symposia Proceedings, vol. 51, eds. H. Kurz, G.L. Olson, and J.M. Poate (1986), 277-282.

(25) C.W. White, B.R. Appleton, B. Stritzker, D.M. Zehner and S.R. Wilson, Materials Research Society Symposia Proceedings, vol. 1, eds. J.F. Gibbons, L.D. Hess and T.W. Sigmon (1981), 59.

(26) P. Baeri, G. Foti, J.M. Poate, S.U. Campisano and A.G. Cullis, Applied Physics Letters, 38 (1981) 800.

(27) J.M. Poate, Materials Research Society Symposia Proceedings, vol. 4, eds. B.R. Appleton and G.K. Celler (1982) 121.

(28) C.W. White, D.M. Zehner, S.U. Campisano and A.G. Cullis, Surface Modification and Alloying by Laser, Ion and Electron Beams, eds. J.M. Poate, G. Foti and D.C. Jacobson (New York: Plenum Press, 1983), p. 94.

(29) M.J. Aziz, Ph.D. thesis, Harvard University (1983), pp. 76-82.

(30) M.J. Aziz, J.Y. Tsao, M.O. Thompson, P.S. Peercy and C.W. White, Physical Review Letters, 56 (1986) 2489.

(31) M.O. Thompson, G.J. Galvin, J.W. Mayer, P.S. Peercy and R.B. Hammond, Applied Physics Letters, 42 (1983) 445.

(32) M.J. Aziz, J.Y. Tsao, M.O. Thompson, P.S. Peercy, and C.W. White, Materials Research Society Symposia Proceedings, vol. 35, eds. D.K. Biegelsen, G. Rozgonyi, C. Shank (1985) 153.

(33) M. Hillert and B. Sundman, Acta Metallurgica, 25 (1971) 11.

(34) T. Papa, F. Scudieri, M. Marinelli, U. Zammit, and G. Cembali, Journal de Physique, Colloq. C-5 (1983) 73.

(35) M.J. Aziz and C.W. White (manuscript in preparation).

(36) L. Onsager, Physical Review, 37 (1931) 405; 38 (1931) 2265.

(37) J.W. Christian, Theory of Transformations in Metals and Alloys, Part I, 2nd ed. (Oxford: Pergamon, 1975), p. 479.

(38) M. Hillert and B. Sundman, Acta Metallurgica, 24 (1976) 733.

(39) J.W. Cahn, Acta Metallurgica, 10 (1962) 789.

(40) J.L. Murray, Metallurgical Transactions A, 15 (1984) 261.

(41) D.G. Beck, S.M. Copley, and M. Bass, Metallurgical Transactions A, 13 (1982) 1879.

(42) P. Gordon and R.A. Vandermeer, Recrystallization, Grain Growth, and Textures (Metals Park, OH: American Society of Metals, 1966), pp. 205-66.

(43) W.J. Boettinger, S.R. Coriell and R.F. Sekerka, Materials Science and Engineering, 65 (1984) 27.

(44) W.J. Boettinger, D. Shechtman, R.J. Schaefer, F. Biancaniello, Metallurgical Transactions A, 15 (1984) 55.

(45) G.J. Galvin, J.W. Mayer and P.S. Peercy, Materials Research Society Symposia Proceedings, vol. 23, eds. J.C.C. Fan and N.M. Johnson (1984) 111.

(46) P.S. Peercy and M.O. Thompson, Materials Research Society Symposia Proceedings, vol. 35 (1985) 53.

(47) J.Y. Tsao, S.T. Picraux, P.S. Peercy, and M.O. Thompson, Applied Physics Letters, 48 (1986) 278.

RAPID SOLIDIFICATION OF Al-6Fe-6Ni ALLOY POWDER

Young-Won Kim[a] and F. H. Froes[b]

[a]Metcut-Materials Research Group
P.O. Box 33511
Wright-Patterson AFB, OH 45433

[b]Air Force Wright Aeronautical Laboratories
Materials Laboratory, AFWAL/MLLS
Wright-Patterson AFB, OH 45433

Abstract

The solidification sequence occurring during atomization was studied in a hypereutectic Al-5.7Fe-6.1Ni aluminum alloy. Both single and multiple homogeneous nucleation events were observed in fine alloy powder (<15μm) which can be undercooled about 100 K below the eutectic temperature (640°C) or 310 K below the liquidus temperature (∿850°C) resulting in $0.3T_L$ undercooling. The homogeneous nucleation is followed by partitionless or planar solidification (PS) with the initial growth rate greater than 6m/s. The PS gradually becomes morphologically unstable as, with recalescence, the growth rate decreases and the diffusion rate of solute in the liquid increases. Fine-scale cellular solidification (CS) and relatively coarse dendritic solidification (DS) then develop, which eventually transforming to coupled eutectic (CE) growth with a final growth rate of ∿1cm/s. The stable τ (Al_9FeNi) compound is the only second phase observed after solidification and in PS, CS, and DS regions it is present as particles which increase in size from 5nm to 35nm in going from the PS to DS region. The compound particles in the PS region are precipitated during solid state cooling behind the solid/liquid interface, with the orientation relationship of $[012]_{Al}//[0\bar{1}1]_{\tau}$ and $(1\bar{2}1)_{Al}//(\bar{1}11)_{\tau}$. The particles in CS and DS regions are formed mainly by segregation during the solidification itself. The whole process of solidification is explained in terms of undercooling with respect to the T_0 curve and metastable phase transitions, determined by DTA and microstructural observations.

Introduction

The Al-Fe-Ni system at the Al-rich corner has been of great interest both fundamentally and practically. Under equilibrium conditions, the ternary phase τ $[Al_9(Fe_xNi_{1-x})_2]$, isomorphous with monoclinic Al_9Co_2, exists over a range of x [1, 2]. Rapid solidification (RS) can widen the x range such that Al_9Fe_2 may form instead of Al_3Fe or Al_6Fe on the Al-Fe binary side, or Al_9Ni_2 may form instead of Al_3Ni on the Al-Ni side [3]. The τ phase particles are resistant to thermal coarsening up to 370°C, making aluminum alloys with a fine dispersion of τ phase attractive for elevated temperature use, as has been shown in atomized Al-5.7Fe-6.1Ni [4-6]. Upon RS, the τ phase in low solute (Fe+Ni) content alloys is present. For example, in atomized Al-1.5Fe-3.7Ni alloy powder after hot pressing [7], interdendritic eutectic (<0.4μm), cellular eutectic (<0.1μm), and dispersed precipitates (<0.05μm) of τ are observed. The finely dispersed τ particles have also been observed in as-melt spun ribbons of this alloy, with definite orientation relationships with the matrix [8]. In the case of the Al-5.7Fe-6.1Ni alloy, the predominant microstructure in atomized alloy powder has been reported to be eutectic/coupled eutectic [5, 6], a recent observation [9] has confirmed that this alloy additionally shows regions containing fine τ phase particles having a preferred orientation relationships with the alpha-Al matrix. In this paper, the RS microstructure of atomized Al-5.7Fe-6.1Ni alloy powder is characterized and explained in terms of the undercooling and metastable phase relations developed by analyzing the microstructure.

Experimental

The ternary alloy with the composition in Table I was received as alloy powder that was rapidly solidified (RS) by the air atomization technique. The alloy powder having a size range of 1 to 37μm and an average powder diameter (APD) of ∿10μm was consolidated at 315°C. Both as-RS and consolidated powder particles were examined in the scanning electron microscope (SEM) and transmission electron microscope (TEM) and also characterized using the x-ray diffractometer technique. Crystallographic orientation relationships between the alpha-Al matrix and second phases were determined from the electron diffraction patterns (EDP). Differential thermal analyses (DTA) were conducted on the consolidated material to determine both equilibrium and non-equilibrium phase transitions.

Table I. Alloy Composition

	Al	Fe	Ni	Fe+Ni	Si+Cu+Mn+Cr+Zn
Weight %	88.02	5.70	6.07	11.77	0.21
Atomic %	93.94	2.94	2.98	5.92	0.14

Results

Rapidly Solidified Microstructure

Figure 1 shows SEM micrographs of (a) as-atomized alloy powders and (b) consolidated powders after polishing and etching with Keller's etch. As is shown in Figure 1b, alloy powders finer than 10μm are generally etch insensitive and featureless. Although it is not shown here, the shape change of individual powders after consolidation is generally limited to the surface area [6]. The interior of powder particles remains undeformed resulting in the retention of the RS microstructure.

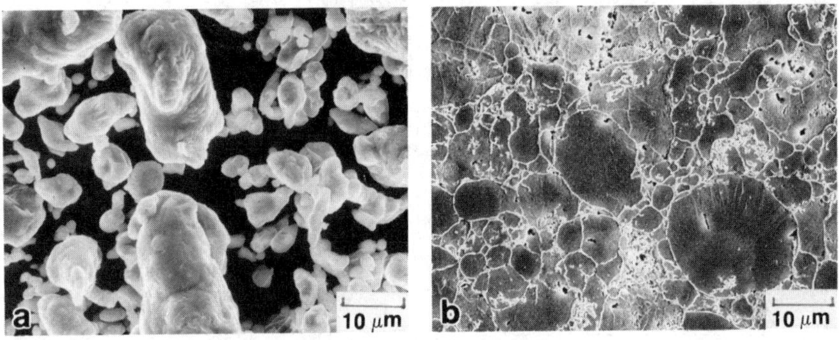

Figure 1 - Rapidly solidified alloy powder particles of alloy Al-5.7Fe-6.1Ni in (a) the as-atomized condition and (b) after consolidation.

Figure 2 shows TEM micrographs taken from consolidated powder particles featuring homogeneous nucleation events. In Figure 2a powder particles with sizes ranging roughly from 2 to ∿15μm reveal single nucleation events. The fine (2-4μm) powder particles exhibit only a featureless microstructure, but larger particles additionally show a well defined, fine eutectic microstructure. Occasionally, multiple homogeneous nucleation events are observed in relatively large powder particles, as in Figure 2b where a 16μm size particle contains at least four nucleation sites (N1, N2, N3, and N4). Solidification proceeds radially from each nucleation site until the boundary is formed between the solidification fronts advancing from different sites. The boundaries are outlined by fine compound particles (boundary C) or coupled eutectic families (boundary E) of different orientation. The boundary compounds are formed when the featureless zones meet. The regions solidified following N1, N2, and N3 nucleations are large enough to develop coupled eutectic (CE) solidification; the N4 region is small and contains mostly a featureless area. The featureless regions around the nucleation site frequently show the radially oriented band contrasts under certain tilt conditions (see N1 and N2 regions in Figure 2b).

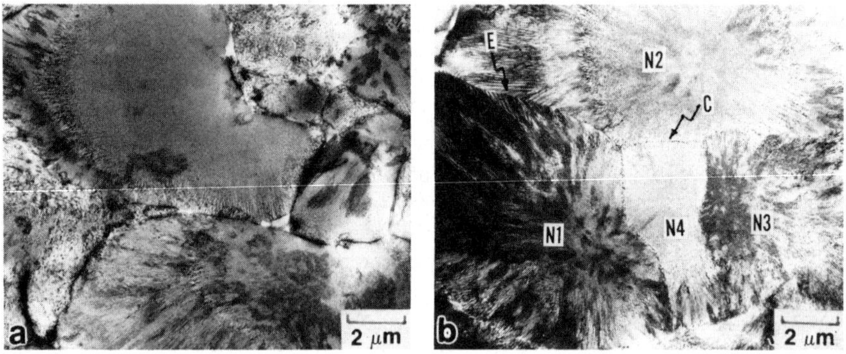

Figure 2 - Rapidly solidified alloy droplets showing (a) single nucleation events and (b) a multi-nucleation event.

Figure 3a shows details of the microstructure observed in a 15μm alloy powder particle that was apparently solidified by a single nucleation site. Area 1, very close to the nucleation site, shows randomly distributed ultrafine compound particles. As solidification proceeds the compound particles are gradually aligned into arrays which are briefly disrupted (Area 3) before transforming to a CE solidification. The microstructure of Area 1 at a higher magnification, shown in Figure 3b, reveals ultrafine (<30nm) compound particles that are distributed randomly. Their size appears to gradually increase with increasing distance from the nucleation site (bottom left). Figure 3c, which is a higher magnification of Area 3 in Figure 3a, shows that the transition region to CE appears to be dendritic with segregated interdendritic compounds. The compound size ranges from ∿5 to 50nm before the transition and 10 to ∿70nm in the interdendritic regions.

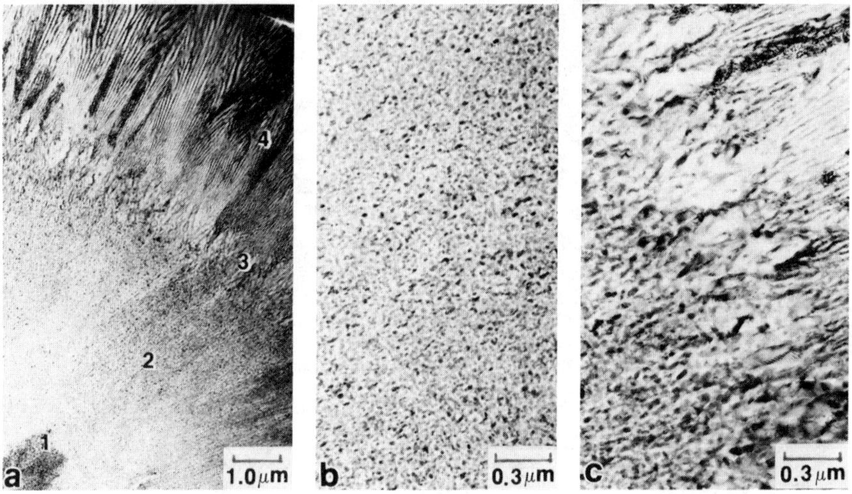

Figure 3 - (a) The sequence of rapid solidification (1-2-3-4) observed in medium size alloy droplets, (b) Area 1 near the nucleation site, and (c) the transition Area 3 between 2 and 4 in (a).

Figure 4 shows details of a coupled eutectic microstructure (in the 15μm size powder) in longitudinal direction (Figure 4a) and transverse direction (Figure 4b). The coupled eutectic consists of α-Al and rod-type Al_9FeNi [9]. The eutectic rod diameter and the eutectic spacing increase as solidification proceeds. The diameter and the spacing are respectively ∿0.015 and ∿0.03μm in the beginning, and 0.05 and 0.1μm when solidification is complete. The initial eutectic spacing appears to be the same regardless of the powder particle size, however, the final spacing generally increases with particle size. The eutectic rods appear to grow with the same crystallographic orientation (see EDP), and they have a preferred crystallographic orientation relationship with the matrix [9].

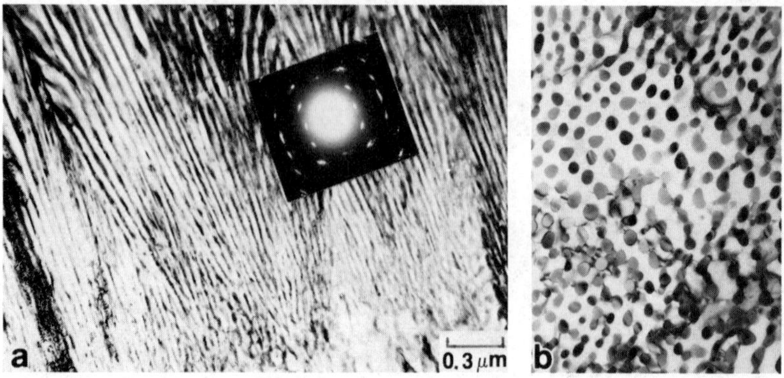

Figure 4 - Coupled eutectic microstructure developed as the last step of rapid solidification, showing (a) longitudinal and (b) cross-sectional morphologies.

Crystallographic Orientation Relationship

Details of the microstructure adjacent to the nucleation sites (Areas 1 and 2 in Figure 3a) are shown in Figure 5 with corresponding electron diffraction patterns (EDP) and dark field (DF) images. The bright field (BF) image (Figure 5a), EDP, and DF image (Figure 5c) obtained from the reflections from compounds reveal some important features: (1) the compounds have slightly elongated or round shape with sizes ranging from 3nm to 25nm, (2) they appear to be randomly distributed but most of them are actually oriented in the same crystallographic direction (see EDP), and (3) the number of compounds observed is roughly the same in the BF and DF images indicating that the majority of compounds have the same crystallographic orientation.

The BF image (Figure 5b), EDP, and DF (Figure 5d) obtained from Area 2 in Figure 3a reveal the following features: (1) the compound particles are more or less spheroidal with sizes ranging between 3nm and 40nm, (2) they become gradually aligned as solidification progresses, with increased separation, (3) however, their preferred orientation is reduced considerably (transition from spot to ring pattern), and (4) the number density of compound particles in DF is much lower than in BF, indicating that the compound particles now show much less preference for a single orientation.

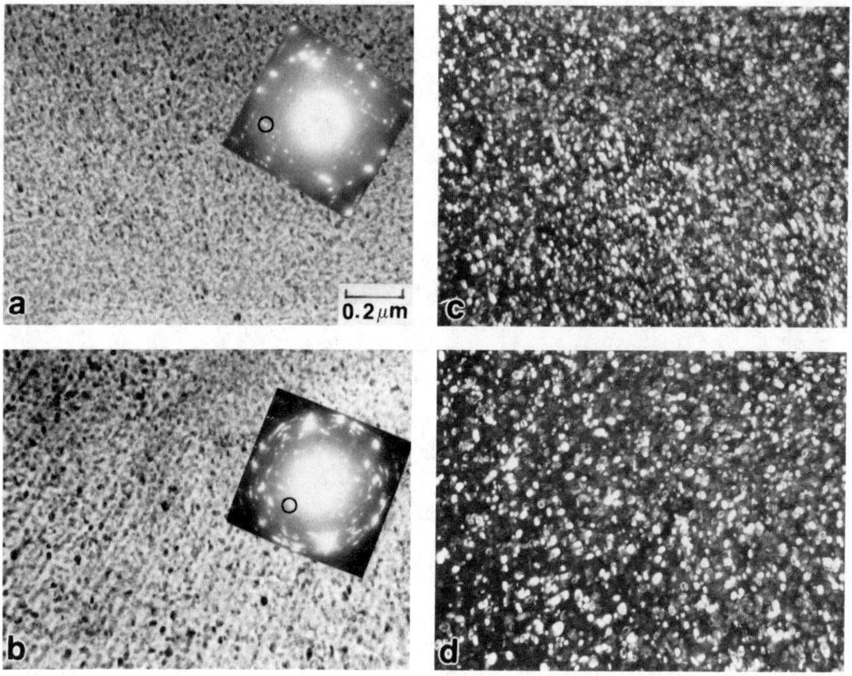

Figure 5 - Microstructure variations from Area 1 (Fig. 3) (a and c) around the nucleation site, (b and d) to the next stage of compound alignment showing the size and distribution of τ (Al_9FeNi) compound particles. (a) and (b) are bright field images and (c) and (d) are corresponding dark field images obtained from the circled reflections in the electron diffraction patterns.

Figure 6a shows the microstructure around the nucleation site (Figure 5a) obtained by tilting the specimen. Corresponding EDP (Figure 6b) and DF image (Figure 6c) show that the τ phase (Al_9FeNi) compound particles have a definite crystallographic relationship with the alpha-Al matrix. The selective indexing (Figure 6d) of the EDP indicates that with the zone axes $[012]_{Al}//[0\bar{1}1]_{\tau}$, the α-matrix and the phase particles have orientation relationships: $(1\bar{2}1)_{Al}//(\bar{1}11)_{\tau}$, $(0\bar{2}1)_{Al}//(011)_{\tau}$, $(100)_{Al}//(1\bar{0}0)_{\tau}$, etc. Figure 6e shows a DF image of Figure 6a after tilting, revealing that the compound particles are disk-shaped, with $(\bar{1}00)_{\tau}$ as the disk plane.

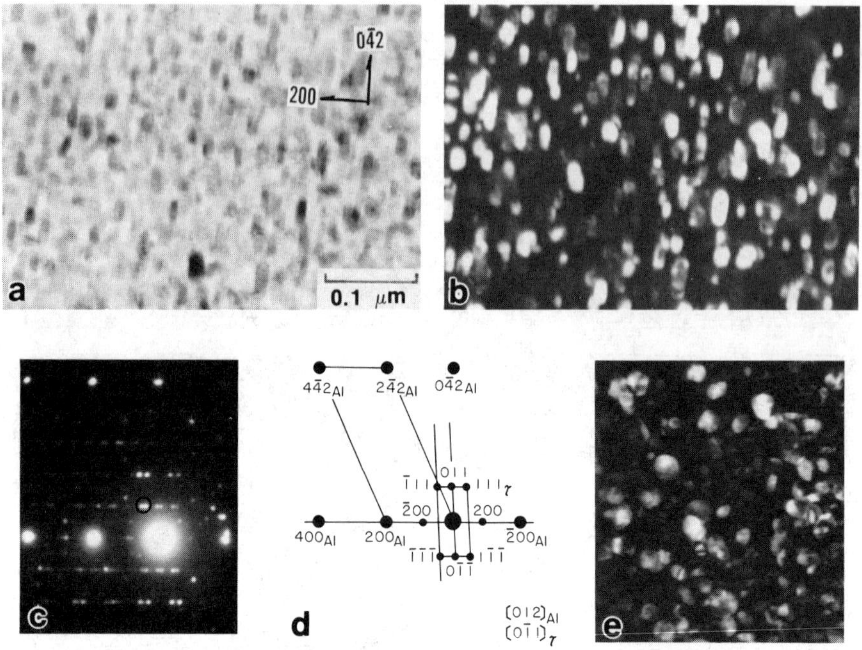

Figure 6 - (a) The microstructure of the area around the nucleation site in bright field mode, (c) after tilting to reflect the diffraction patterns, (d) with the indices. The dark field image (b) obtained from the encircled reflections in (c) shows that the compound particles are elongated perpendicularly to $[100]_{Al}$. The dark field image (e) obtained after arbitrary trials shows that the compounds have a disk-shape.

Table II lists the crystallographic relations (with zone axes $[012]_{Al}//[0\bar{1}1]_{\tau}$) observed in Figure 6, with the misfits calculated using the d-spacings and misorientations measured from the stereographic projections [9]. Also included are the orientation relationships obtained when the stereographic projections are rotated such that the zone axes $[001]_{Al}$ and $[100]_{\tau}$ are parallel [9]. It is clear from Table II that the most probable orientation relationship with the smallest misfit and misorientation is $[001]_{Al}//[100]_{\tau}:(100)_{Al}//(031)_{\tau}$. The same relationship has been suggested in melt-spun Al-1.5Fe-3.7Ni ribbons previously [8], but actual observations have yet to be made.

Table II. Orientation Relationships and Mismatches Between τ (Al_9FeNi) and the Al-Matrix

Orientation Relationship		d_{Al}(nm)	nd_{Al}(nm)	d_τ(nm)	Δ^a	$\Delta\phi^b$
$[012]_{Al}//[0\bar{1}1]_\tau$	$(1\bar{2}1)_{Al}//(\bar{1}11)_\tau$	d_{242}=0.827	5d=4.136	$d_{\bar{1}11}$=4.060	1.8	∿1
	$(0\bar{2}1)_{Al}//(011)_\tau$	$d_{0\bar{4}2}$=0.906	5d=4.530	d_{011}=4.446	1.8	<1.5
$[001]_{Al}//[100]_\tau$	$(100)_{Al}//(031)_\tau$	d_{200}=2.025	d=2.025	d_{031}=2.004	1.0	∿1.5
	$(210)_{Al}//(011)_\tau$	d_{420}=0.906	5d=4.530	d_{011}=4.446	1.8	∿3

$^a\Delta = |(nd_{Al}-d_\tau)/nd_{Al}|$ (%).
$^b\Delta\phi$ = misorientation between $(hkl)_{Al}$ and $(hkl)_\tau$ (in degree).

X-ray and DTA Observations

Figure 7 shows X-ray diffraction patterns from as-RS powder (Figure 7a) and from hot consolidated powder (Figure 7b). Both patterns are consistent with those from Al_9FeNi (τ phase) previously identified in the same alloy [9], which is monoclinic (isomorphous with $CoAl_9$) with lattice parameters of a = 0.87nm, b = 0.635nm, and c = 0.625nm and β (between a and c axes) = 95 ±0.5°. The peak heights of the strongest peaks (P's) of the τ phase in the two conditions are about the same magnitude in dictating that the amounts of τ phase are about the same. The same intensities of the same aluminum peaks (compare: $(220)_{Al}$ peaks) in both conditions confirms this observation. There are, however, some differences in X-ray patterns between the two conditions, that is, changes in sharpness and/or peak heights of some peaks. For example, peaks 1, 2, 3, and 5 are sharpened somewhat with a slight increase in peak heights and peaks 3 and 4 reverse their relative intensities after hot consolidation. These observations suggest that there may be a slight coarsening of ultrafine compounds reducing the line broadening without increasing the amount of the compounds.

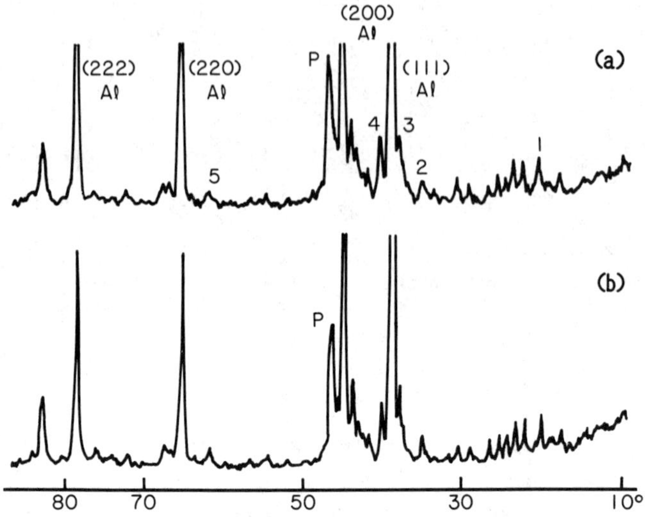

Figure 7 - X-ray diffraction patterns of atomized Al-5.7Fe-6.1Ni alloy powder particles in (a) as-rapidly solidified condition and (b) after consolidation at 315°C.

Figure 8 shows DTA curves of hot consolidated material following two heating/cooling cycles between 600°C and 890°C with both heating and cooling rates of 10°C/minute. During the first heating a major melting endotherm is observed around 670°C with an onset of 650°C, followed by a shallow, broad melting endotherm around 745°C with an onset of 720°C. The determination of liquidus (T_L) from the last endotherm is quite ambiguous but it appears to be around 850°C. During the first cooling, the liquid starts to solidify at 835°C forming Θ (Al_3Fe) phase followed by a small exotherm at 715°C which likely results in τ phase formation. The major exotherms occur at 635°C with an onset of 645°C and then shortly at 625°C with an onset of 630°C. The major exotherms are results of the formation of both alpha and τ phases. The second heating of the same specimen results in quite different results. The first and major melting endotherm occurs at 665°C with an onset of 640°C and having a shoulder at 655°C. The second endotherm occurs at around 745°C with an onset of 735°C and the T_L occurs at the same temperature as in the first heating around 850°C. The second cooling results in an identical DTÀ curve to that of the first cooling, except that the first two small exotherms occur at a few degrees higher temperatures.

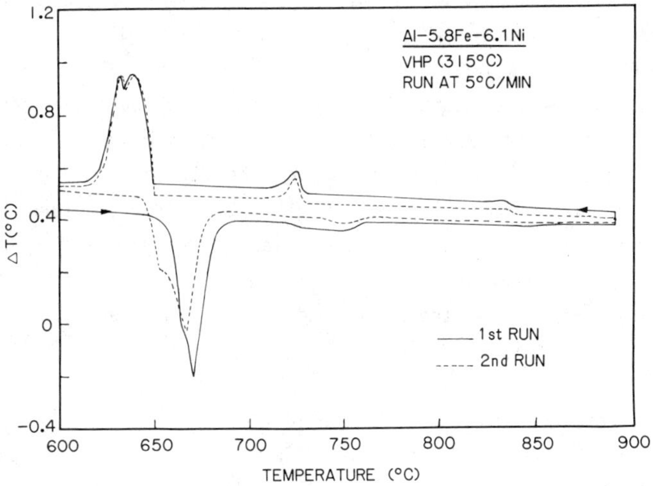

Figure 8 - DTA thermograms for hot consolidated (315°C) Al-5.7Fe-6.1Ni alloy powders. The second run was made using the same sample as for the first run.

Discussion

RS Microstructure

When solidified by homogeneous nucleation, atomized droplets of Al-5.7Fe-6.1Ni alloy typically exhibit four steps of RS microstructure. Step 1 microstructure (Figure 3a) formed around the nucleation site is featureless at a low magnification but actually consists of ultrafine Al_9FeNi compound particles (Figures 3b and 5). The particles are in fact precipitates that form by solid state decomposition during post-solidification cooling and/or during hot consolidation [9]. This conclusion is based on the fact that the particles have a definite crystallographic relationship with the matrix (Figure 6 and Table II). This means that the solid at the solid/liquid interface during solidification was a solid solution with complete solute trapping as the result of partitionless or planar solidification (PS).

The PS solidification front gradually breaks down, as solidification proceeds, because of recalescence and a reduction in the interface velocity. This leads to formation of the step 2 microstructure as a result of cellular solidification (CS) with intercellular regions outlined by fine (3-30nm) Al_9FeNi particles (region 2 in Figures 3a and 4b). Many of the particles in this region, however, are crystallographically random (Figure 4b), indicating that they are formed by segregation during the solidification event itself. The intercellular spacing increases from ∿0.03μm initially to ∿0.06μm just before the transition to step 3 dendritic solidification (see Figures 3 and 5b). The dendritic solidification (DS) results in a 0.5-1μm wide band of dendritic microstructure (Figure 3c). The interdendritic compound particles are generally larger than intercellular compounds and they are crystallographically random, indicating again that the interdendritic Al_9FeNi particles are formed during solidification.

The coupled eutectic (CE) solidification is the final step that follows the DS solidification and features rod type Al_9FeNi eutectic compounds with eutectic spacing (λ_e) ranging from 0.03μm initially up to ∿0.1μm at the end of solidification.

Figure 9 shows a schematic of the RS microstructure development in a medium size (∿15μm) alloy powder particle, including the measurements described above. Here only those features developed during solidification are shown, thus no precipitate particles, formed from the solid state, are shown in the PS region. Some quantitative measurements made from rapidly solidified particles are shown in Figure 9, and are discussed below.

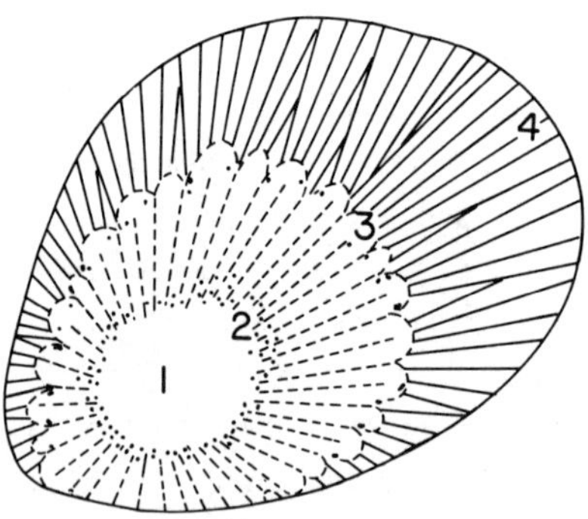

τ ∿0.03μm at Area 2
∿0.06μm at Area 3

τ_e ∿0.03μm at Area 3
∿0.1μm at Area 4

Figure 9 - Schematic of a 15μm size Al-5.7Fe-6.1Ni alloy droplet solidified by homogeneous nucleation at 1 showing the measured values of inter-cellular spacing (λ) and eutectic spacing (λ_e).

The solid/liquid interface velocities (R) during PS and CS solidification can be derived by calculating first the R value at the boundary between the PS and CS regions in Figure 9 where the first perturbation or solute segregation occurs. This breakdown takes place when the interface velocity R becomes slower than the diffusive velocity (V_D) which is expressed as [12]:

$$V_D = D_L/\xi \qquad (1)$$

where D_L is the diffusivity of the solute in the liquid solvent and ξ is the interatomic distance. Since D_L for Fe or Ni in Al is roughly $10^{-10} m^2 s^{-1}$ [10] and ξ of Al is $\sim 3 \times 10^{-10}$m, the velocity R at the onset of instability is $R = V_D$ or $R \sim 3ms^{-1}$. At this time, the perturbation wave length [11] or the intercellular spacing (λ) is $\sim 0.03\mu m$ as shown in Figure 9. When compared to the Al-4.5Cu alloy [12], the present alloy particle maintains a stable S/L interface to a slower interface velocity, consistent with the slower diffusion of Fe, as contrasted to Cu, in Al.

The cooling rate (ε) at the end of CS solidification is roughly estimated to be 10^8 K/s according to $\lambda = 0.06\mu m$ (Figure 9) and experimental ε-λ data [13] on Al-5Cu and Al-10Si alloys. By taking the temperature gradient (G) of 70 K/μm in the liquid at the S/L interface [12], the interface velocity (R) at the end of CS solidification is estimated to be roughly $R = 1.4ms^{-1}$ using:

$$\varepsilon = G \cdot R \qquad (2)$$

It is important to note that the cooling rate (ε) in this case is actually the heating rate since the solidification takes place during recalescence. If R is assumed to be inversely proportional to the distance from the nucleation site up to the end of CS, $R \sim 6.5ms^{-1}$ is estimated at the onset of planar solidification around the nucleation sites, Figure 9. The cooling, or the heating rate, at this time should approach 10^9 Ks^{-1} since G>100 K/μm [12] and actually $R > 6.5ms^{-1}$.

At the onset of CE solidification, the interface velocity (R) for the CE structure and the undercooling ($\Delta T_e = T_e - T_L$, where T_e is the eutectic temperature) at the solidification front can be estimated using [11, 13]:

$$\Delta T_e \sim (1\mu m\ K)/\lambda_e \quad \text{and} \quad R \sim \frac{100\mu m^3\ s^{-1}}{\lambda_e^2} \tag{3}$$

to be $\Delta T_e \sim 30$ K and R = 11cm s^{-1} since λ_e = 0.03μm (Figure 9). At the end of CE solidification in Figure 9 where λ_e = 0.1μm, $\Delta T_e \sim 10$ K and R = 1cm s^{-1}, indicating that the solidification is complete 10 K below the eutectic temperature and that the solidification rate is considerably reduced, compared with that at the onset of solidification, to almost 1 x 10^{-3}. Considering the slower recalescence at the later stage of solidification and the relative length of PS, CS/DS, and CE solidification, it does not appear unreasonable that the undercooling (ΔT_e) at the nucleation area is estimated to be 100 K or more.

Because of recalescence and slow heat dissipation, the solid behind the S/L interface should experience a relatively high temperature (up to 630°C) for a short time. The minimum time at 600°C can be estimated by calculating the total elapsed time from the nucleation to the completion of solidification. Since the average growth rate of CE solidification is ∼4cm/sec for the distance of ∼6μm (see Figures 3 and 9), the minimum elapsed time (t) at around 600°C may be 3 x 10^{-4}s when the relatively slow heat extraction effect associated with the air atomization is considered. By taking the diffusivity of Fe or Ni in Al, D $\sim 1 \times 10^{-13} m^2 s^{-1}$ [10], the diffusion distance is estimated using $x = \sqrt{Dt}$ to be ∼6nm. Fine precipitates with the size comparable to the diffusion distance may form during this solid state cooling, and this is evident in the planar solidification region containing 2-25nm precipitates.

Phase Diagram for RS Microstructure

Figure 10 shows a pseudobinary (Al-τ) equilibrium phase diagram consisting of the isopleth of the Al-rich corner of Al-Fe-Ni phase diagram [1] constructed by sectioning at constant Al, using the alloy composition and the Al_9FeNi intermediate phase composition in conjunction with the results of DTA analysis shown in Figure 8. The details of the phase diagram construction including the phase designations developed using DTA results will be published elsewhere [14]. Also shown in Figure 10 are non-equilibrium phase relations including the coupled eutectic (CE) zone which is asymmetrical and the T_0 curve. The asymmetrical CE zone is consistent with observations in RS Al-transition metal binary alloys like

Al-Fe [13]. If the CE zone was symmetrical, then the present alloy composition would not show the PS-CS-DS-CE solidification sequence, but rather perhaps all eutectics.

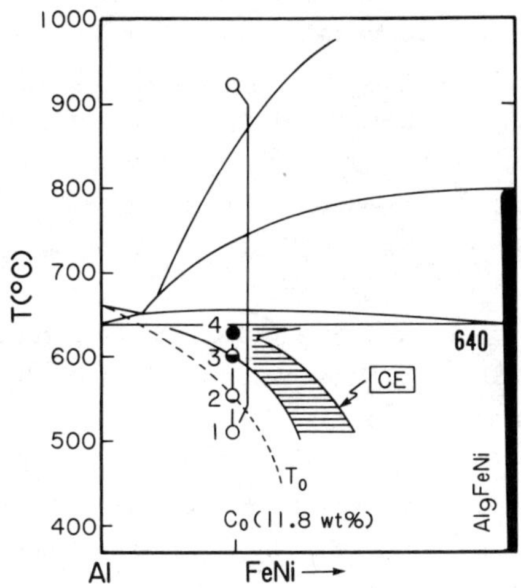

Figure 10 - An isopleth of the Al-rich corner of the Al-Fe-Ni phase diagram containing Al, the alloy composition (C_o) and the τ phase (Al_9FeNi). Also included are the sequence of solidification occurring after the alloy melt undercools to the position 1, with respect to the T_o curve and coupled eutectic (CE) zone.

If the present alloy with the total solute content C_o = 11.8wt% (Fe+Ni) is extremely slowly cooled from the melt, first the Θ phase ($Al_{13}Fe_4$) will start to form at the liquidus temperature, T_L ∿850°C (also see Figure 8), and then the second phase transition forming τ phase starts at T = 735°C. At T = 654°C, alpha-Al starts to form giving a ternary phase region (L + τ + Al) which reaches down to the ternary eutectic temperature (T_e) of 640°C.

When rapidly solidified, for example, at a cooling rate of 10^9 K/sec as in Figure 9, the Al-5.7Fe-6.1Ni melt powder undercools below the T_o curve (position 1 in Figure 10) before homogeneous nucleation occurs. The undercooling (ΔT_e) below T_e was estimated to be 100 K or greater, which means that the undercooling from the T_L, $\Delta T_L = T_L - T_N$ ∿310 K, can be achieved in a Al-5.7Fe-6.1Ni droplet of ∿15μm diameter. This undercooling results in $\Delta T_L/T_L$ ∿0.3 which is much greater than that of pure aluminum (∿0.2) [15].

Upon nucleation and subsequent growth and recalescence, the planar solidification transforms into a cellular solidification (position 2 in Figure 10), and then dendritic solidification before transforming to CE solidification at ΔT_e ∿30 K (position 3 in Figure 10). The formation of a CS/DS microstructure between the T_0 curve and the lower limit of CE zone can be understood if this transition phase region is considered as a mixture of eutectic and supersaturated alpha-Al, the latter forming below or to the left-hand side of the T_0 curve. The intercellular and interdendritic compounds observed in Figures 3 and 5 are the same as the interdendritic eutectic formed in slowly cooled hypoeutectic alloys.

The formation of CE requires a great deal of solute diffusion which is possible when the S/L interface velocity is relatively slow. This is observed, and the drastic decrease of R from 1.4m/s at the end of CS (near position 3, Figure 10) to 11cm/s at the onset of CE (right above position 3) occurs during the short dendritic solidification (around position 3). This can be explained by the coarse interdendritic spacing in DS region compared to that observed in CS regions (Figures 3c and 9), the former being characterized by much lower cooling rates. This behavior, when combined with the reduced temperature gradient at the solidification front, should yield a greatly reduced growth rate.

Phases

The τ phase (Al_9FeNi) appears to be the only second phase that forms when the alloy melt is undercooled below T_e at least up to ΔT_e ∿100 K, suggesting that no metastable phase form up to an undercooling of ∿$0.3T_L$. This makes it difficult to understand why: (1) the onset temperature (650°C) of the major endotherm of RS material is higher than that (640°C) of the bulk material and (2) the endotherm is a single event (at the heating rate of 10°C/minute) while the bulk material shows an endotherm with a shoulder indicating two consecutive transformation events, Figure 8. Further studies including interrupted DTA experiments are required to understand the phenomena.

Conclusions

1. Atomized alloy powders of Al-5.7Fe-6.1Ni were studied to determine the influence of undercooling and subsequent recalescence on the evolution of the microstructures.

2. As determined by DTA analysis and the existing liquidus surface derived from the phase diagram, the alloy has a melting (liquidus) temperature (T_L) of ∿850°C, a eutectic temperature (T_e) of 640°C for complete solidification, and two additional transformation temperatures between T_L and T_e.

3. The alloy melt powder smaller than 16μm can be undercooled down to 540°C or lower, which is ΔT_L ∿310 K (ΔT_L ∿$0.3T_L$) or ΔT_e ∿100 K. When this happens, the melt powder temperature is below T_0 (with $540<T_0<610$°C) and generally homogeneous nucleation of the solid takes place. The microstructure that develops during subsequent recalescence is sequentially characterized by planar solidification (PS), cellular solidification (CS), dendritic solidification (DS), and coupled eutectic (CE) solidification.

4. The solid/liquid interface velocity decreases gradually during PS from 7 to 3m/s which decreases further down to 1.4m/s at the end of CS. The growth rate decreases significantly during the short period of DS that develops after CS. The velocity is roughly 11cm/s at the onset of CE solidification and reduced to ∿1cm/s at the end of CE which is also the completion of solidification.

5. The only second phase observed in the RS alloy powder is the stable τ (Al_9FeNi) compound that is formed either during solidification in CS/DS and CE regions or during solid state cooling after solidification in PS/CS regions. The compounds are present in particle form in PS, CS, and DS regions, with gradually increasing size (average) from 5 to 35nm. The compound particles formed by the solid state reaction have the orientation relationship of $[012]_{Al}//[0\bar{1}1]_\tau$ and $(1\bar{2}1)_{Al}//(\bar{1}11)_\tau$, while the simpler orientationship of $[001]_{Al}//[100]_\tau$ and $(100)_{Al}//(031)_\tau$ is also possible.

Acknowledgments

This work was supported by Air Force Materials Laboratory Contract No. F33615-82-C-5078. The authors wish to thank Ms. Karen A. Sitzman for her assistance in typing the manuscript.

References

1. V. G. Rivilin and G. V. Raynor, "Critical Evaluation of Constitution of Aluminum-Iron-Nickel System," International Metallurgical Review, 25(3)(1980), 79-93.

2. L. F. Mondolfo, Aluminum Alloys: Structure and Properties (London: Butterworths, 1979), 532.

3. J. H. Perepezko and W. J. Boetinger, "Use of Metastable Phase Diagrams in Rapid Solidification," in Alloy Phase Diagrams, vol. 19, ed. L. H. Bennett, T. B. Massalski, and B. C. Giessen (New York: North Holland, 1983), 223-240.

4. H. G. Paris, J. W. Mullins, and T. H. Sanders, Jr., "The Effect of Solidification Microstructure on the Strength, Ductility and Toughness of Dispersion-Hardened Al P/M Alloys," in High-Strength Powder Metallurgy Aluminum Alloys, ed. M. J. Koczak and G. J. Hildeman (Warrendale, PA: The Metallurgical Society, 1982), 277-296.

5. M. J. Koczak and A. Lawley, "A Fundamental Study of P/M Processed Elevated Temperature Aluminum Alloys" (Annual Report, AFOSR Contract No. 82-0010, Drexel University, Philadelphia, PA, 1983).

6. Y-W. Kim and W. M. Griffith, "Annealing Behavior and Tensile Properties of Elevated-Temperature PM Aluminum Alloys," in Rapidly Solidified Powder Aluminum Alloys, ASTM STP 890, ed. M. E. Fine and E. A. Starke, Jr. (Philadelphia, PA: American Society for Testing and Materials, 1985), in print.

7. G. J. Hildeman, D. J. Lege, and A. K. Vasudevan, "Influence of Powder Morphology on the Mechanical Properties of Rapidly Solidified Aluminum Alloys," in High-Strength Powder Metallurgy Aluminum Alloys, ed. M. J. Koczak and G. J. Hildeman (Warrendale, PA: The Metallurgical Society, 1982), 249-276.

8. L. Bendersky, "Orientation Relationship Between Precipitated $Al_9(Fe,Ni)_2$ Phase and α-Aluminum," Met. Trans. A, 16A (1985), 683-686.

9. Y-W. Kim and A. G. Jackson, "Phases and Orientation Relationships in a Rapidly Solidified Al-6Fe-6Ni Alloy," Scripta Met., 20(5)(1986), 777-782.

10. M. Bishop and K. E. Fletcher, "Diffusion in Aluminum," International Metallurgical Review, 17 (1972), 203-225.

11. W. Kurz and D. J. Fisher, Fundamentals of Solidification (Switzerland-Germany-UK-USA: Trans. Tech. Publications, 1984), 58.

12. R. Mehrabian, "Rapid Solidification," International Metallurgical Review, 27(4)(1982), 185-208.

13. H. Jones, Rapid Solidification of Metals and Alloys, Monograph No. 8 (London: Institute of Metals, 1982), 40-47.

14. Y-W. Kim, unpublished results (1986).

15. J. H. Perepezko, this proceedings.

NUCLEATION IN THE PRESENCE OF A METASTABLE LIQUID MISCIBILITY GAP IN THE Al-Be SYSTEM

D.C. Van Aken and H.L. Fraser
Department of Metallurgical Engineering
University of Illinois
Urbana, IL 61801

Binary alloys of Al-Be were rapidly solidified by laser surface remelting and melt-spinning. Alloy compositions of up to 20 atomic percent Be were investigated and the microstructures were characterized by analytical electron microscopy. Formation of the microstructure in the rapidly solidified material was related to the liquid undercooling at the solid-liquid interface and the presence of a metastable liquid miscibility gap. The mechanism for the formation of microstructure is believed to involve a metastable monotectic reaction near the Al side of the phase diagram. A range of alloy compositions was observed to produce a uniform micro-structure of spherical Be particles (5-20 nm in diameter) in an Al matrix. A thermodynamic analysis of the metastable liquid miscibility gap was used to define a region of cooperative growth for the monotectic reaction.

Introduction

Thermodynamic considerations for nucleation in systems which exhibit immiscible liquids have been presented by Cahn (1) and two distinct reactions are possible when a second liquid product forms. The first is a liquid separating into two liquids of different composition, e.g. $L \rightarrow L_1 + L_2$, and the second reaction is a cooperative growth of a solid and a new liquid, e.g. $L \rightarrow S_1 + L_2$. Transformations involving this second reaction are called monotectics. Monotectic solidification is similar to an eutectic transformation, since both reactions may produce composite structures. A schematic diagram of a liquid miscibility gap and the monotectic reaction is shown in fig. 1 where solid A is S_1. Spinodal and extended metastable liquidus lines have been included in the figure. The extended liquidus for S_1 depicts the equilibrium between an undercooled liquid of uniform composition and the first solid to nucleate. Transformation of L_2 is not shown, but in this type of system liquid L_2 usually transforms as a divorced eutectic reaction.

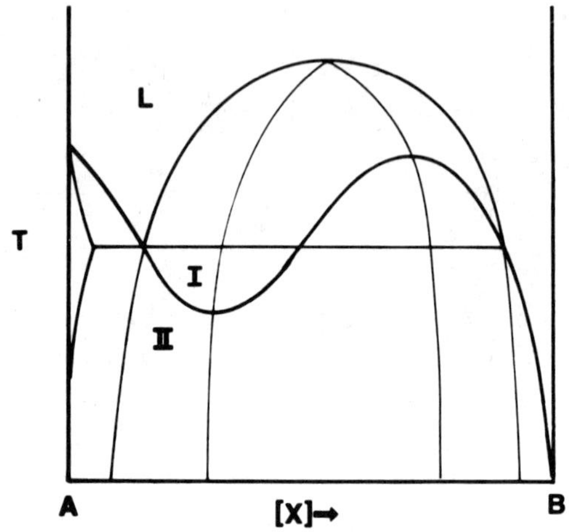

Fig. 1 - A schematic diagram illustrating a liquid miscibility gap and the associated monotectic reaction. Extensions of the metastable liquidus and spinodal lines for the liquid miscibility gap are shown with the extended liquidus for the solid. The portion of the later curve is unstable within the region of spinodal decomposition. Regions I and II are defined in the text and in Fig. 2.

Liquid phase separation within the liquid miscibilty gap (and above the monotectic temperature) may occur by either spinodal decomposition or nucleation and growth. A single phase liquid which is undercooled within the liquid miscibility gap, but outside the spinodal region would be metastable. Liquids undercooled within the bounds of the spinodal are thermodynamically unstable and these liquids decompose after minor compositional fluctuations and subsequent mass transport. Experimental evidence for spinodal decomposition is abundant in ceramic systems (e.g. 2). The high liquid viscosity permits extensive undercooling in the melt and spinodal decomposition is encountered commonly. In comparison to ceramics, metallic systems exhibit very low liquid viscosity and faster diffusional transport in the liquid state. Thus, nucleation within the metastable region (between the spinodal and the liquidus) may be more difficult to suppress.

Nucleation of the solid phase can be more complex and a metastable region exists below the monotectic isotherm where the nucleation of the solid is restricted. This is shown in fig. 1 as the region bounded by the monotectic isotherm, the extended S_1 liquidus, and the liquid spinodal. This metastable region was constructed from the free energy curves for the liquid and solid and a schematic diagram for the free energy curves, at a temperature below the monotectic temperature, is shown in fig. 2.

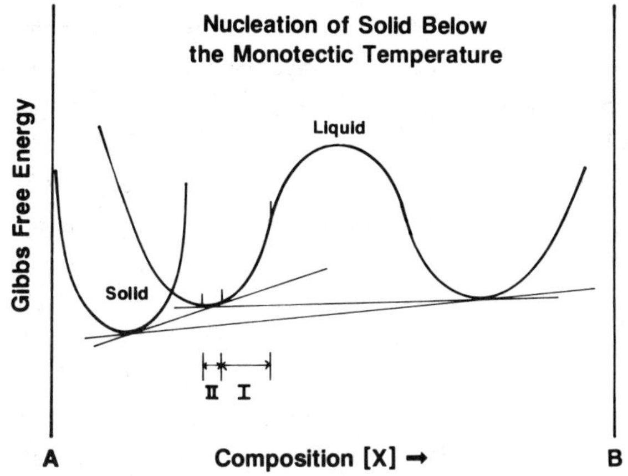

Fig. 2 - A schematic diagram illustrating free energy curves for the solid and the liquid at a temperature below the monotectic reaction. Region I is defined by the spinodal and the liquid composition (dilute in solute B) in equilibrium with solid. Equilibrium tangents for metastable liquids within this region fall below the free energy curve of the solid and nucleation of the solid is not permitted. Region II defines a composition range where either solid or liquid or both may form and coupled growth is possible.

Tangents drawn between the free energy curves of the liquid and solid define two important regions for consideration of undercooled liquids. When a homogeneous liquid is undercooled into region I, nucleation of the second liquid is required before the solid phase becomes thermodynamically stable. This can be demonstrated by drawing a tangent to the liquid free energy curve within region I, and observing that the tangent falls below the free energy of the solid for all compositions. However, if L_2 is present, and in equilibrium with L_1, a tangent drawn between the two liquids is above the free energy of the solid. Nucleation of the solid then becomes possible. Thus, a precursor reaction ($L \rightarrow L_1 + L_2$) would be expected for an hypermonotectic composition undercooled into region I. Alternatively, solid nucleation is possible for liquid compositions within region II for the same temperature and the greatest driving force for the nucleation of S_1 will exist if L_2 is already present. Region II also provides a thermodynamic criterion for cooperative growth from an undercooled liquid, since the tangent drawn to the liquid free energy curve lies above both product phases. Regions I and II are schematically shown in fig. 1.

Nucleation of the solid phase in the presence of a metastable liquid miscibility gap is governed by the same thermodynamic restrictions. For ceramic systems, metastable diagrams involving submerged liquid immiscibilty are well established. Glass formation in the SiO_2-Al_2O_3 system is complicated by the presence of a metastable liquid miscibility gap (3). As a result, liquid separation produces a crystalline phase. Metastable liquid immiscibility has also been reported for metallic alloys of Cu-Co and Cu-Fe by Nakagawa (4). In these studies, significant undercooling was required to cause liquid phase separation and these effects were observed near the critical point of the miscibility gap. A more recent investigation involving the rapid solidification of hypereutectic Al-Be alloys revealed microstructures consistent with a metastable monotectic reaction (5). These microstructural effects resulted from the large undercooling in the liquid induced by rapid solidification.

Rapid solidification processing (RSP) may indeed produce significant undercooling in the liquid prior to nucleation; for example, the results of melt-spinning (MS) and laser surface remelting (LSR) of hypereutectic Al alloys are consistent with liquid undercooling's of ≈ 150-200 K (6). The large undercooling produced during RSP provides a means to further investigate the effects of a metastable liquid miscibility gap on the formation of microstructure in the Al-Be metallic system. The purpose of this paper is to relate the concepts of undercooling and formation of microstructure in the Al-Be system with the thermodynamic criteria for nucleation in the presence of the metastable liquid miscibility gap.

Experimental Procedure

Al-Be alloys with nominal compositions 4.4, 5.8, 11, and 20 atomic percent Be were prepared by induction heating under an atmosphere of argon. Purity of the Al and Be was 99.97% and 99.7%, respectively.

Rectangular blocks (40 mm x 50 mm x 18 mm) for LSR experiments were cast into carbon chill-molds. Surfaces to be treated were first milled, to remove 2-9 mm per side, and then finish ground (wet) with 220 grit SiC paper. Machined substrates measured 12-14 mm in thickness. An AVCO/15 kW continuous CO_2 laser was used for surface remelting experiments. Rapidly solidified filaments were produced by scanning two sides of the alloy substrate at a rate of 6 cm/s or 8 cm/sec with the laser operating between 3.5-5.0 kW. Laser passes were spaced 5-6 mm apart with the laser beam focused to a 3 mm spot. LSR filaments measured approximately 1 mm wide with a remelted zone 150-250 μm deep.

Additional material was prepared and reduced to rapidly solidified ribbons using conventional melt-spinning technology. The melt-spinning chamber, containing the alloy and the wheel, was capable of being evacuated to 2.0 x 10^{-6} torr. Prior to melting the alloys, the chamber was back-filled with He gas. Melt-spun ribbon was produced by induction melting and ejecting the molten alloys onto a 15 cm (diameter) Cu wheel having a surface velocity of 35 m/s. Alloys were contained in either carbon or silica-glass crucibles and ejected through a .06-.08 mm orifice by pressurized Ar gas. Alloys were super heated (100-200°C) prior to ejection to ensure a homogeneous melt. Temperatures were measured by a thermocouple contained in a silica-glass tube immersed in the molten alloy. Ribbons produced were 2-4 mm wide with an average thickness of less than 40 μm. The measured thickness (40 μm) represents the distance from the wheel side of the ribbon to the asperity height on the opposite side.

Microstructural characterization was performed on thin foil (electron transparent) specimens. Specimens were prepared by electro-chemical jet polishing 3 mm discs of the rapidly solidified material. A solution 60% methanol, 8% perchloric, 12% butylcellosolve, and 20% H_2O, operating at 50 V and -25°C, produced acceptable thin foils. Thin foil observations were performed on Philips EM 420 and EM 430 transmission electron microscopes.

Results

Results from LSR and MS experiments are combined according to the type of microstructure produced. For compositions greater than 11 atomic percent Be only MS material was considered, since experimental difficulties were encountered in obtaining uniform liquid compositions for the LSR experiments. Metallographic examination of the remelted zone for

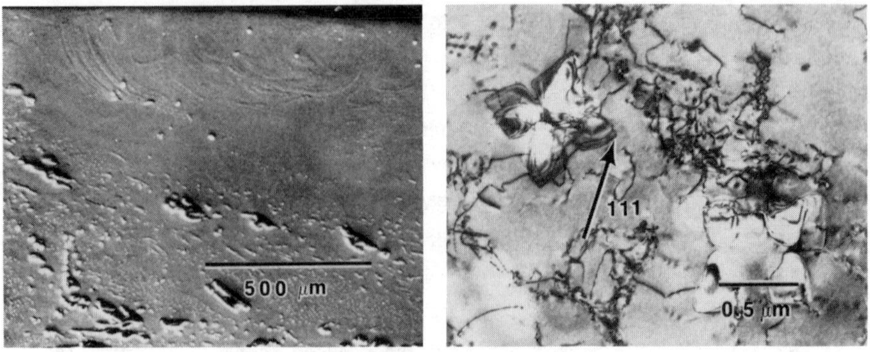

Fig. 3 - (a) A secondary electron image of a transverse section through the LSR filament. Swirls in the melt pool indicate compositional inhomogeneities. (b) Unmelted Be dendrites within the LSR filaments. These particles were associated with the swirls shown in (a).

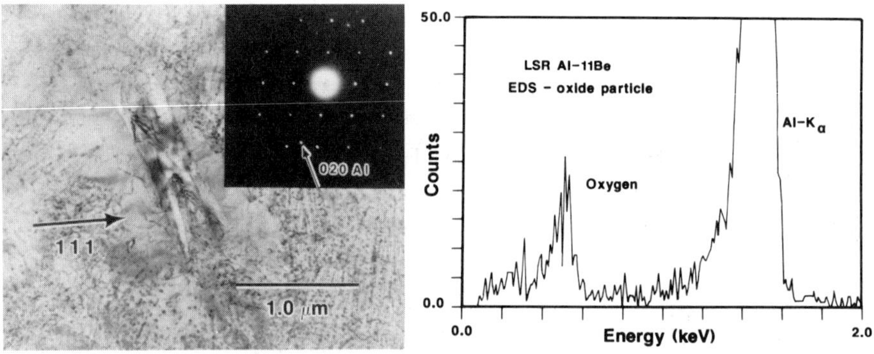

Fig. 4 - (a) An unmelted particle believed to be an oxide. These observations were recorded from an LSR Al-11Be filament. (b) Windowless EDS measurements indicating the presence of oxygen in this particle.

the alloys with higher Be contents indicated compositional inhomogeneities evidenced by swirls in the melt pool (see fig. 3). The presence of unmelted Be dendrites and oxide particles was evident in these areas (see figs. 3b and 4a). Oxygen was detected in the particle shown in fig. 4a using both windowless energy dispersive spectroscopy (EDS) (see fig. 4b) and electron energy loss spectroscopy (EELS) techniques.

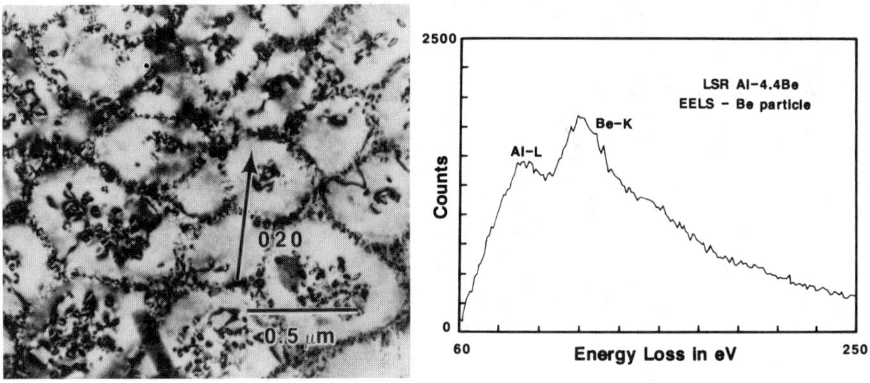

Fig. 5 - (a) A cellular microstructure observed in a LSR Al-4.4Be alloy. The electron beam was aligned parallel with the [100] direction in Al and the image contrast shows little if any variation in the orientation of the individual cells. (b) EELS recorded from a small intercellular particle showing the presence of Be.

Alloys of 4.4 atomic percent Be, when laser surface remelted, exhibited a uniform cellular morphology based on primary cells of Al with Be rich particles present in the cell boundaries (see fig. 5). Large groups of cells were aligned with a parallel crystal orientation which produced large grains containing many cells with an <100> Al growth direction. Dislocation loops (vacancy in nature) produced the dark contrast within the Al cell structure and the defect analysis showing the vacancy nature of the loops has been previously reported (7). Microstructural variations were more pronounced in the melt-spun ribbons. Electron transparent regions of unpolished ribbon exhibited a microstructure consisting of Al grains with a uniform dispersion of Be particles (see fig. 6a). This morphology will be discussed in more detail later for the higher Be composition alloys, although it should be noted that the shape of some of the Be particles suggests either spheroidization of a rod or the coalescence of two particles. Observations made of regions in electropolished ribbon revealed a cellular morphology (see fig. 6b). Two types of contrast features were observed, which have been reported previously (5); intracellular regions contained pure-edge and prismatic dislocation loops (vacancy in nature) and hcp Be particles were observed within the individual grains forming the cellular structure, as well as, at the boundaries between grains.

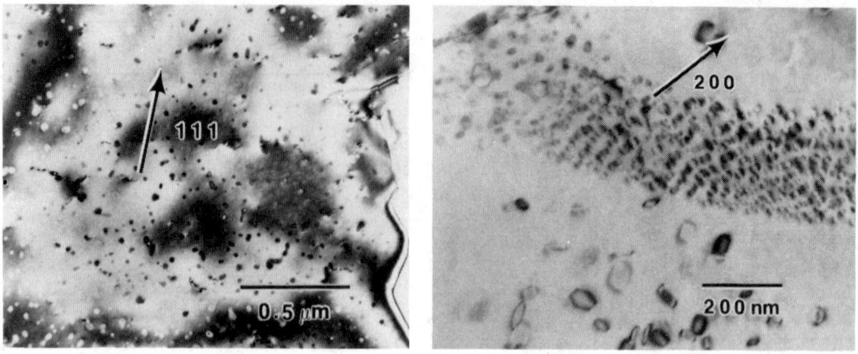

Fig. 6 - (a) An unpolished Al-4.4Be ribbon microstructure showing a cooperative growth morphology in an electron transparent region. (b) A preferentially polished Al-4.4Be ribbon showing the microstructure near the wheel-side. The microstructure consists of primary Al cells with an intercellular precipitate.

As the concentration of Be was increased, the extent of the precipitation at the cell boundary also increased. At a nominal composition of Al-5.8Be the microstructure consisted of a uniform dispersion of Be particles in the Al matrix and this microstructure will be referred to as a cooperative growth morphology. The spherical Be particles measured 5-10 nm in diameter with an interparticle distance of approximately 20 nm. The size of the Be dispersion could be varied by changes in the traverse speed of the laser. Thus, faster speeds produced a finer particle size and spacing.

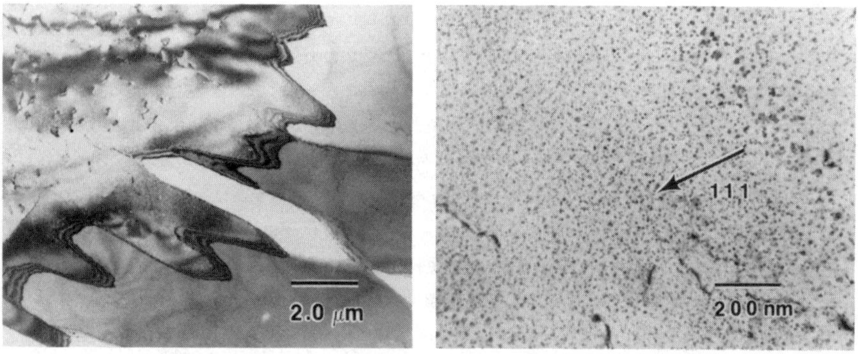

Fig. 7 - An Al-5.8 Be alloy processed by LSR at 8 cm/s; (a) grain structure observed within the LSR filament. (b) microstructure produced by cooperative growth.

Microstructures for LSR Al-5.8Be produced at a traverse speed of 8 cm/s are shown in figs. 7a and b. Alloy compositions up to 11 atomic percent Be produced the cooperative growth morphology when processed by LSR (see figs. 8a and b). The Be particles observed in the Al-11Be alloy were large enough (15-20 nm diameter) to obtain diffracted intensity in the SAD patterns; these particles produced rings of diffracted intensity. The smallest ring measured 0.174 nm and the second ring measured 0.123 nm (see fig. 9). Measurements were made relative to the Al matrix and this was justified by the x-ray diffraction work of Jordan on rapidly quenched Al-Be alloys (8). This work measured changes in the lattice parameter of Al to be less than 0.04% for RSP Al-Be alloys.

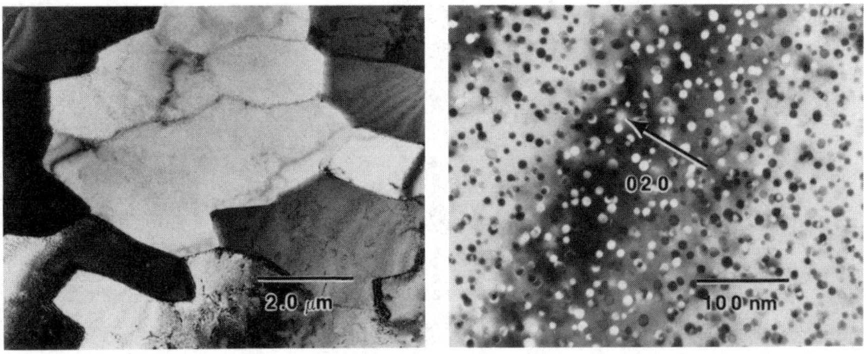

Fig. 8 - An Al-11Be alloy processed by LSR at 6 cm/s; (a) grain structure observed within the LSR filament. (b) microstructure produced by cooperative growth.

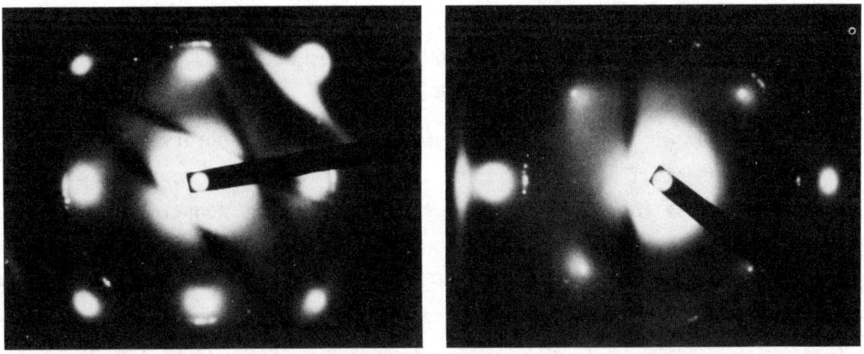

Fig. 9 - Selected area diffraction patterns from an Al-11Be alloy processed by LSR. In addition to the diffracted intensity from the Al matrix rings of diffracted intensity were observed which may be indexed using a bcc crystal structure for Be; (a) 100 zone axis of Al (b) tilted off zone along a 022 systematic row.

Convergent beam electron diffraction (CBED) patterns were also obtained from these Be particles (see fig. 10). The patterns were formed by using a small (20 μm) condenser aperture. The patterns may be indexed by assuming a bcc structure with a cubic lattice parameter of 0.246 nm and the CBED patterns suggest the following orientation relationship: $(100)_{Al} \parallel (100)_{Be}$ with $[010]_{Al} \parallel [010]_{Be}$. It should be noted that the patterns exhibit a slight misorientation between the matrix and particle, and the diffracted intensity in the SAD patterns indicate a predominant distribution of $\{110\}$ bcc Be planes nearly parallel to either $\{100\}$ or $\{110\}$ Al planes. This would indicate that more than one orientation relationship may exist.

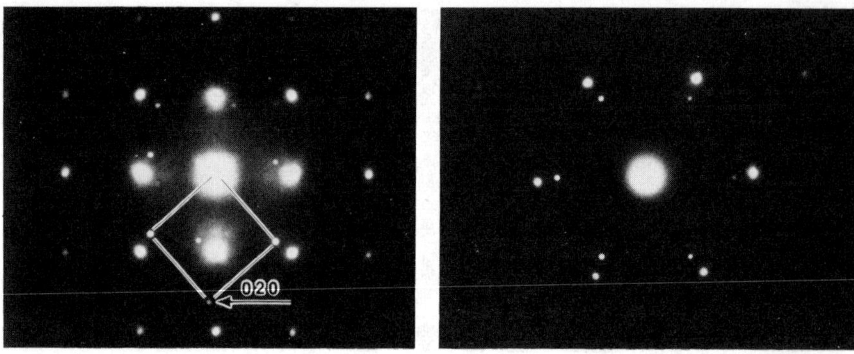

Fig. 10 - Convergent electron beam diffraction patterns of the Be rich particles observed in the Al-11Be LSR filaments. The two patterns shown were obtained from two different Be particles in the same Al grain. The diffraction results indicate an equivalent crystallographic orientation between the Be particles and the Al matrix. (a) 100 zone axes with $(010)_{Be} \parallel (010)_{Al}$ (b) 111 zone axes with $(110)_{Be} \parallel (110)_{Al}$.

Beryllium with an hcp crystal structure was also observed in the LSR processed material, but the Be particles exhibited a rod morphology (see fig. 11). These rod morphologies were found in areas containing unmelted Be rich particles as previously described. The rod orientation was determined by computer simulation of the experimentally obtained diffraction pattern (see fig. 11) and the following orientation relationship was determined: $[101]_{Al} \parallel [0001]_{Be}$ and $\{11\bar{1}\}_{Al} \parallel \{1\bar{2}10\}_{Be}$. As the solidification front advanced, the rod morphology was observed to degenerate into a series of aligned hcp Be particles (see fig. 12). The crystal orientation of the Be particles became less defined as solidification proceeded and rings of diffracted intensity were then observed as previously shown. DF images of the Be particles were recorded using the smallest reciprocal lattice spacing and only a small fraction of the Be particles showed any intensity in the recorded image (see fig. 13). As previously described these particles were characterized as having a bcc crystal structure. Transitions from the hcp to bcc crystal structure occurred over 1-2 μm, and could be observed within a single Al grain (see fig. 12a).

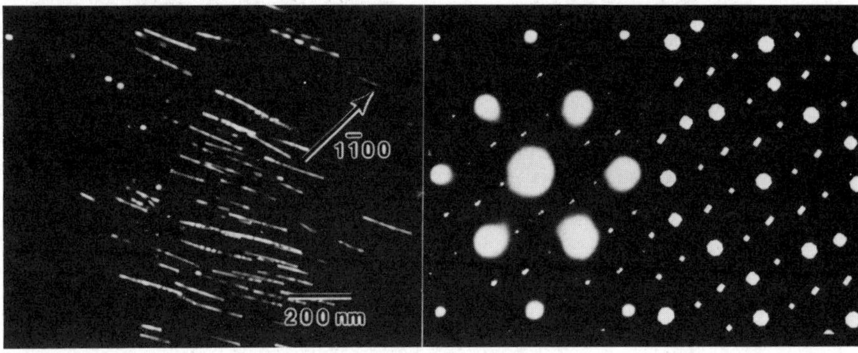

Fig. 11 - (a) A rod morphology observed near regions of unmelted substrate observed in the Al-11Be alloy processed by LSR.
(b) Experimental and computed electron diffraction patterns for the rod morphology. Calculations were based on the following orientation relationship: $[101]_{Al} \parallel [0001]_{Be}$ with $(11\bar{1})_{Al} \parallel (1\bar{2}10)_{Be}$. The simulated pattern is constructed from large, medium, and small spots representing the Al matrix, Be precipitate, and the possible double diffracted routes, respectively.

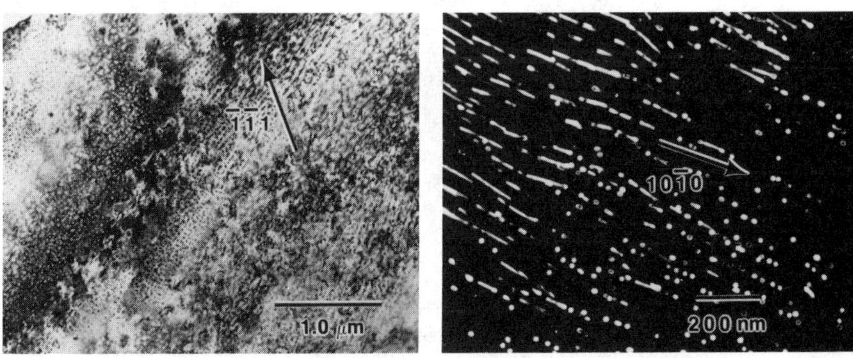

Fig. 12 - (a) A single Al grain exhibiting the transition from the rod to a spherical dispersion, top right to bottom left. (b) A dark-field image showing the morphological break-down of the rod structure. An equivalent crystal orientation is suggested by the intensity observed in the rods and the smaller spherical particles.

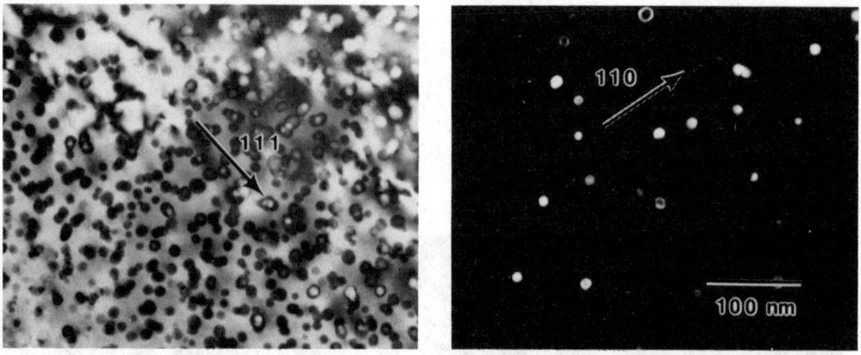

Fig. 13 - (a) A microstructure showing the random distribution of Be rich particles observed in LSR Al-11Be filaments. (b) Corresponding dark-field image produced by centering a portion of the smallest diffracted ring, i.e. 110_{Be}.

The microstructures produced by MS consisted only of hcp Be particles and the cooperatively grown microstructures were restricted to alloy compositions near 5.8 atomic percent Be. Transparent regions of the ribbon consisted of Al grains with a uniform dispersion of Be particles, and polished ribbons consisted of 1-2 μm Al grains showing a dispersion of spherical Be particles (see fig. 14). Ribbons produced with higher Be compositions showed evidence of a bimodal distribution of Be particles.

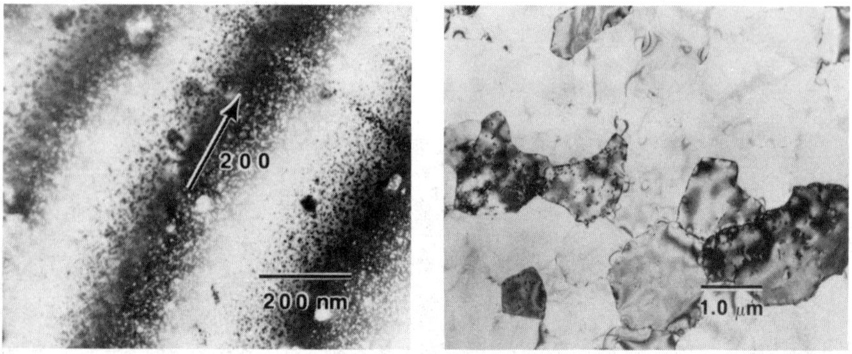

Fig. 14 - (a) An example of the microstructure observed in unpolished Al-5.8Be ribbons. (b) A preferentially polished Al-5.8Be ribbon showing the fine grain structure and cooperative growth morphology near the wheel side of the ribbon.

Representative microstructures for melt-spun Al-11Be (melted in silica-glass) and Al-20Be (melted in graphite) are shown in fig. 15. These images were recorded near the wheel-side of the ribbons. Previous studies have shown that these larger Be particles contain small precipitates believed to be Al (5,7). This Be particle distribution coarsened as a function of distance from the surface of the ribbon which had been in contact with the wheel, and only the larger Be particles were observed in the thicker regions of the ribbons (see fig. 16).

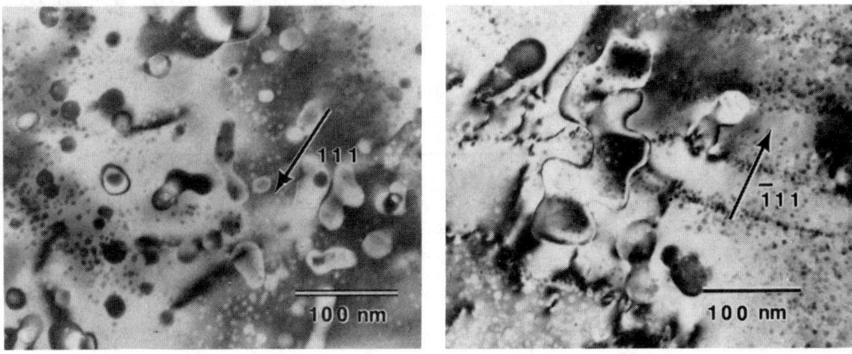

Fig. 15 - (a) A preferentially polished Al-11Be ribbon showing the wheel-side microstructure. The microstructure consisted of a cooperative growth morphology along with larger Be particles. (b) An unpolished Al-20Be alloy ribbon produced from a graphite crucible. The microstructure consisted of large Be rich particles with areas of cooperative growth.

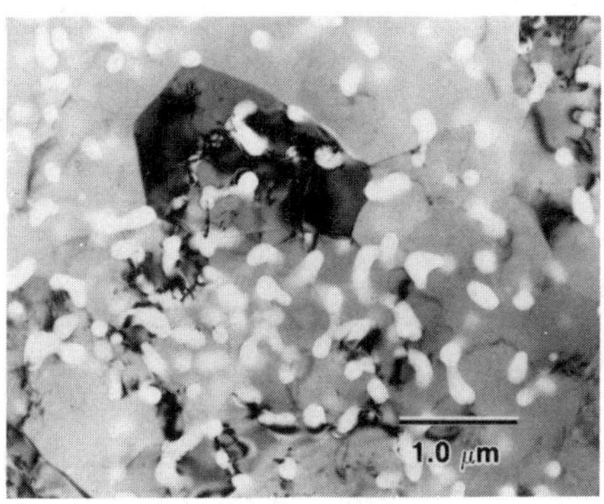

Fig. 16 - Melt-spun Al-20Be ribbon exhibiting large Be particles.

Discussion

Equilibrium phase reactions for the Al-Be system are defined by a simple eutectic diagram (see fig. 17) and the presence of a metastable liquid miscibility gap is suggested by the flattened hcp Be liquidus (9,10). Metastable phase diagrams for the Al-Be system have been reported by Perepezko and Boettinger (11). These diagrams were developed for

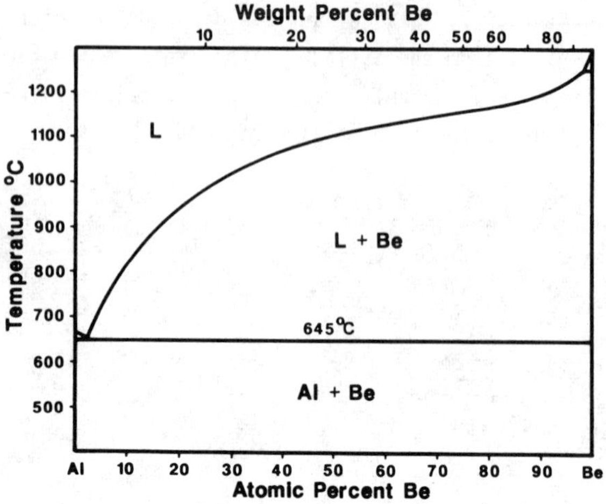

Fig. 17 - An equilibrium phase diagram for the eutectic Al-Be system. An eutectic transformation to a mixture of Al and hcp Be occurs at a composition of 2.5 atomic percent Be, and an allotropic transforma-tion from hcp to bcc occurs upon heating Be above 1254°C.

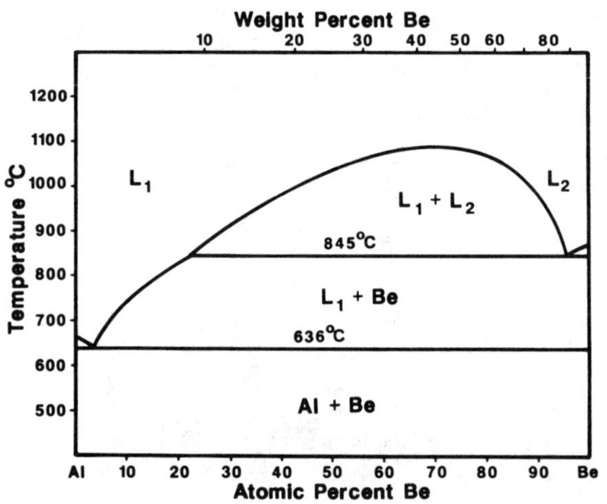

Fig. 18 - A metastable phase diagram for equilibrium with fcc Be. The critical point of the miscibility gap occurs at 1100°C and at a composition of 70 atomic percent Be. A monotectic reaction involving an Al rich liquid and fcc Be occurs at a composition 96 atomic percent Be. The Al rich liquid transforms by an eutectic reaction to Al and fcc Be at a composition of 4.1 atomic percent Be.

various equilibria with metastable Be solids which possess either a bcc or fcc crystal structure. However, the presence of the liquid miscibility gap was only evident for equilibria established with the fcc crystal structure of Be (see fig. 18). An alternative metastable phase diagram involving a monotectic reaction between Al and a second Be rich liquid was used previously to interpret the microstructures of rapidly solidifed hypereutectic alloys (5). A monotectic temperature (625°C) was determined by extending the Al liquidus to a composition which exhibited a cooperative growth morphology, i.e. Al-5.8Be. A schematic drawing which incorporates the shape of the miscibility gap is shown in fig. 19. The hcp liquidus has been included in this diagram for reference. Transformation of L_2 has not been defined in the present metastable phase diagram, since transformation to solid Be would not be thermodynamically possible if the metastable equilibrium were maintained. However, metastable equilibrium can be altered by diffusion of Al from L_2 to the surrounding matrix and equilibrium with either bcc or hcp Be established.

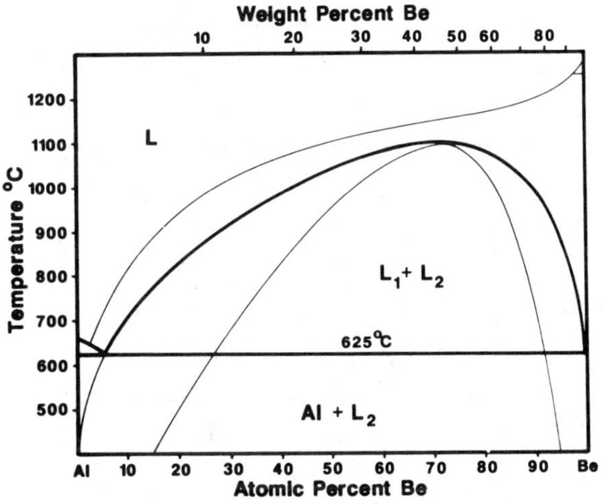

Fig. 19 - A metastable phase diagram for equilibrium between Al and two liquids. A monotectic reaction occurs at a composition of 5.8 atomic percent Be. The equilibrium β-liquidus is shown for reference.

Cooperative growth between Al and L_2 was demonstrated over a large range of composition and this region of cooperative growth may be defined within the present metastable phase diagram. This is shown schematically in fig. 20. The utility of this diagram is demonstrated by examining the relationship of the observed microstructures with the magnitude of undercooling (estimated from the diagram) required to produce the same result. For example, electron transparent areas of the Al-4.4Be ribbon exhibited a cooperative growth morphology while thicker regions of the ribbon were cellular in nature. These microstructures may be interpreted using the concepts of undercooling and the proposed metastable phase diagram. Thus, initial solidification of the Al-4.4Be alloy may occur within the region of cooperative growth for an undercooling of approximately 115 K. As stated earlier, as much as 200 K undercooling has been reported for the solidification techniques used in this investigation. The magnitude of undercooling was estimated from the temperature difference between the equilibrium Be liquidus and the region of cooperative growth. Recalescence during solidification would then reduce the degree of undercooling and result in the observed cellular growth. In way of contrast, phase reactions based on the equilibrium phase diagram would predict an eutectic growth at a lower Be composition and less undercooling, but this was not experimentally observed. This holds true even if an asymmetrically disposed region of cooperative growth is assumed. This latter case would require that the electron transparent areas of the Al-4.4Be ribbon exhibit an initial cellular structure leading into a eutectic solidification in regions away from the wheel-side of the ribbon (6).

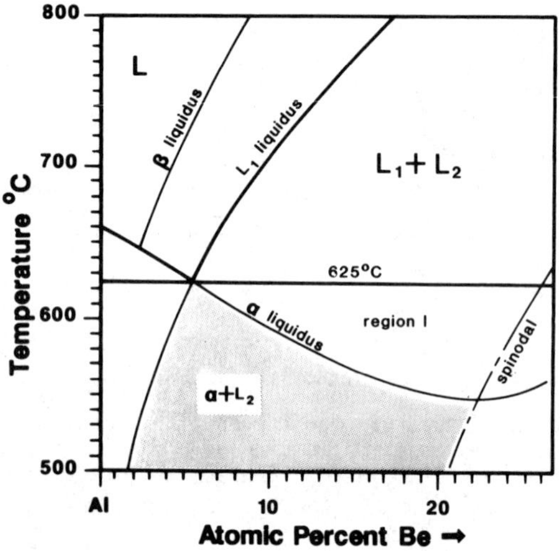

Fig. 20 - A schematic diagram for the region of cooperative growth for the metastable monotectic reaction.

An equivalent degree of undercooling (120 K) predicts a cooperative monotectic growth for the Al-5.8Be alloy and a uniform distribution of Be particles was observed in both melt-spun ribbon and LSR filaments. The magnitude of undercooling required for coupled growth for higher compositions depends on the extended Al liquidus. This would be true for coupled growth by either eutectic or metastable monotectic reactions. Graphical measurements indicate that undercoolings on the order of 230 K, for the Al-11Be composition, are required for cooperative growth. However, inhomogeneities in the melt pool may have resulted in liquid compositions deficient in Be and thus less undercooling would be required for cooperative growth. In addition, LSR filaments of Al-11Be exhibit both a rod morphology and random dispersion of Be particles and associated with this change in morphology was a transition from hcp to bcc Be crystal structure. An eutectic reaction is suggested by the common crystallographic orientation of the hcp Be rods and this rod morphology was observed to break down into a linear array of discrete particles. Similar observations have been observed in the $Cu-Cu_2S$ eutectic system (12). However, the decomposition of a rod eutectic does not explain the apparent random orientation and spatial distribution observed for the bcc Be particles. Although a metastable phase diagram involving a eutectic reaction between bcc Be and Al may provide an alternative explanation, the random spatial distribution suggests the metastable monotectic case. Contrary to eutectc solidification, a monotectic reaction may produce all of the observed microstructures (13). The crystallographic orientation of the bcc Be particles also appears to be random, since rings of diffracted intensity were observed in the selected area diffraction patterns. However, studies involving the thermal decomposition of this microstructure to the equilibrium hcp morphology have determined 8 crystallographic orientations (with 58 possible variants) between the Al matrix and the hcp Be particles (7). These studies concluded that the orientations were governed by the sharing of symmetry elements, since the lattice mismatch was very large. The crystallographic orientation of the bcc Be particles may also be governed by symmetry, since the lattice mismatch for the orientation relationship shown in fig. 10 is also large. If a number of orientations are possible, as observed for the hcp case, rings of diffracted intensity may be expected from groups of the bcc Be particles.

Melt-spinning results for the Al-11Be and Al-20Be alloys indicate primary nucleation of a Be rich phase. An interpretation of this microstructure may be established on the basis of a precursor reaction within region I of the metastable diagram. Thus, prior to nucleation of Al (and subsequent cooperative growth) a second Be rich liquid is formed. Cooperative growth would then occur as the composition of L_1 enters region II. As a result of recalesence, the thicker regions of the ribbon would only exhibit the larger Be particles. The observed morphologies for these large Be particles also suggest a liquid phase separation rather than dendritic solidification, since the morphologies are not consistent with the unrestricted growth of a dendrite in an undercooled liquid. It should also be noted that for alloys with even higher Be contents that the degree of undercooling may be insufficient to maintain the metastable equilibrium. As a result, alloys undercooled into the liquid miscibility gap initially may show liquid phase separation, but solid Be would form prior to the metastable monotectic reaction. The morphology of the Be particles would still be reminiscent of the liquid phase separation and appear irregular in shape rather than dendritic. This was observed in the Al-20Be ribbons and documented in fig. 16.

Finally, the authors would like to make a qualitative observation concerning the difference in microstructure produced by melting in graphite versus silica-glass prior to melt-spinning. Ribbons produced from alloys melted in graphite produced a greater fraction of cooperative growth. Originally, this difference was attributed to cooling rate, since melt-spinning from graphite was performed under an atmosphere of He (5). However, the addition of a ternary element, such as silicon, may shift the position of the metastable miscibility gap, and promote the monotectic reaction at lower Be compositions. This would in turn promote the formation of the larger Be particles. Indeed, some Si has been detected in EDS studies of thin foils of the material produced from the silica-glass crucibles.

Summary

Experimental evidence suggests liquid immiscibility in undercooled Al-Be alloys. Microstructures produced during rapid solidification are consistent with a metastable phase diagram exhibiting a monotectic reaction between Al and a second Be rich liquid. Liquid phase separation is evident in melt-spun ribbons having compositions greater than 11 atomic percent Be. Transformation of the L_2 product appears to be controlled by the heat extraction rate after monotectic solidification. Thus, the faster heat extraction experienced during laser surface remelting results in a bcc Be morphology whereas, melt-spinning produces the hcp Be structure.

Acknowledgements

This work was supported in part by the U.S. Air Force Office of Scientific Research, Division of Electronics and Materials Science under contract no. AFOSR86-0191. The technical monitor was Dr. Alan Rosenstein. Electron microscopy studies were performed using the facilities of the Center for the Microanalysis of Materials, operated by the Materials Research Laboratory of the University of Illinois.

References

1. J.W. Cahn, "The Metastable Liquidus and Its Effect on the Crystallization of Glass", J. Am. Ceram. Soc., 52, 118 (1960).

2. R.J. Charles, "Phase Separation in Borosilicate Glasses", J. Am. Ceram. Soc., 47, 559 (1964).

3. J.F. MacDowell and G.H. Beall, "Immiscibility and Crystallization in Al_2O_3-SiO_2 Glasses", J. Am. Ceram. Soc., 52, 17 (1969).

4. Y. Nakagawa, "Liquid Immiscibility in Copper-Iron and Copper-Cobalt Systems in the Supercooled State", Acta Metall., 6, 704 (1958).

5. D.C. Van Aken and H.L. Fraser, "The Microstructures of Rapidly Solidified Hyper-Eutectic Al-Be Alloys", Acta Metall., 33, 963 (1985).

6. J.W. Zindel, J.T. Stanley, R.D. Field and H.L. Fraser, "Rapid Solidification and Subsequent Analysis of Some Hypereutectic Aluminum Based Alloys", Rapidly Solidified Powder Aluminum Alloys, eds. M.E. Fine and E.A. Starke, ASTM STP 890 (1986), p. 136.

7. D.C. Van Aken, Al-Be Composites Produced by Rapid Solidification, PhD thesis 1986 University of Illinois.

8. C. Jordan, "Investigation of the Effect of Ultra-rapid Quenching on Metallic Systems, Including Beryllium Alloys", Tech. Rep. ASD-TDR-62-181, Wright Patterson AFB, 16 (1962).

9. L.S. Darken and R.W. Gurry, Physical Chemistry of Metals, McGraw-Hill Publ., p. 330 (1953).

10. J.H. Perepezko and W.J. Boettinger, MRS Symposium on Alloy Phase Diagrams, vol. 19, (L.H. Bennett, T.B. Massalski and B.C. Giessen, eds.), Elsevier, North-Holland, p. 223 (1983).

11. W.J. Boettinger and J.H. Perepezko, "Fundamentals of Rapid Solidification", Rapidly Solidified Crystalline Alloys, eds. S.K. Das, B.H. Kear, and C.M. Adam, TMS-AIME 1985, pp. 21-58.

12. S. Marich and D. Jaffrey, "The Morphology and Thermal Stability of the Cu-Cu_2S Eutectic System", Met. Trans., 2, 2681 (1971).

13. B. Derby, "Monotectic Mictostructure at High Growth Rates," Scripta Metall., 18, 169 (1984).

INTERFACIAL SUPERCOOLING

S. D. Peteves and G. J. Abbaschian

Department of Materials Science and Engineering
University of Florida
Gainesville, Florida 32611

Solidification mechanisms of supercooled melts are commonly described based on the bulk supercooling. Although this quantity is useful in describing characteristics of the nucleation process and heat transfer conditions, only a portion of it at the solid/liquid (S/L) interface, "Interfacial Supercooling", is the most dominant factor in determining the solidification process, particularly the growth kinetics, interfacial segregation and growth anisotropy. Results of recent experiments on the solidification of Ga into the supercooled melt involving direct measurement of the bulk and interface supercooling and growth rates will be discussed. The study covers interfacial supercoolings up to about 4.5°C, corresponding to about 50°C bulk supercoolings. The bulk and interfacial growth kinetics will be analyzed and compared; for the latter a theoretical description of the atomistic interfacial processes will be also given.

Introduction

Proceeding nucleation is the spatially discontinuous growth process, as growth takes place only in a small portion of the volume occupied by the system, namely at the interface between the existing nucleus (crystal seed or substrate) and the liquid. The interface, strictly speaking, is not at local equilibrium conditions since a certain amount of driving force is necessary to cause interfacial migration. Growth of a solid from liquid involves two sets of processes; one on the atomic scale, and the other on the macroscopic scale. The first one deals with the atomistic processes occurring at the interface where fluid atoms rearrange their configuration and assume the order of the crystalline phase. The second set of processes involve the transport of matter (solute and/or impurity species) and heat away from the growth front. Information regarding the interfacial process, both from a theoretical and technical point of view, can be obtained from the interfacial growth kinetics. Growth kinetics in turn express the mathematical relationship between the growth rate (V) and the thermodynamic driving force, i.e. the supercooling (ΔT) or the supersaturation (ΔC, ΔP) at the interface.

For growth from a pure melt the macroscopic scale process is identified with the long range transport of the latent heat released upon solidification away from the interface through the crystal and/or through the undercooled liquid. As may be surmised already, discussions of the growth mechanism must consider both the interfacial and transport effects since usually both control the rate cooperatively and interactively with each other. This is particularly true for melt growth where, regardless of the thermal arrangements, the growth is a coupled heat flow-interface kinetics process. Under such circumstances the interfacial supercooling $\Delta T = T_m - T_i$ (where T_m and T_i are the equilibrium and the actual interface temperatures, respectively) may be a small fraction of other temperature differences within the system and most important of the bulk supercooling, ΔT_b, shown in Fig. 1a. Furthermore, as it will be explained later, the latter is not explicitly related to the interface supercooling.

The interfacial conditions are even more complicated when growth involves additional matter transport as for example during crystal growth from a supersaturated solution as schematically shown in Fig. 1b. Since the interfacial temperature (T_i) is higher than the bulk temperature (T_b), the growing interface has a higher equilibrium concentration (C_i^e) than that of the bulk (C_e). At the same time, solute diffuses from the supersaturated solution (C_∞) to the growing interface resulting in an interfacial composition of C_i. Although the magnitude of C_i depends among other factors on the growth rate

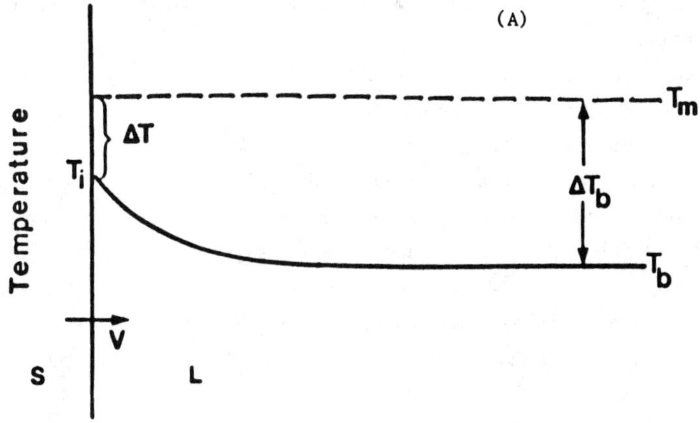

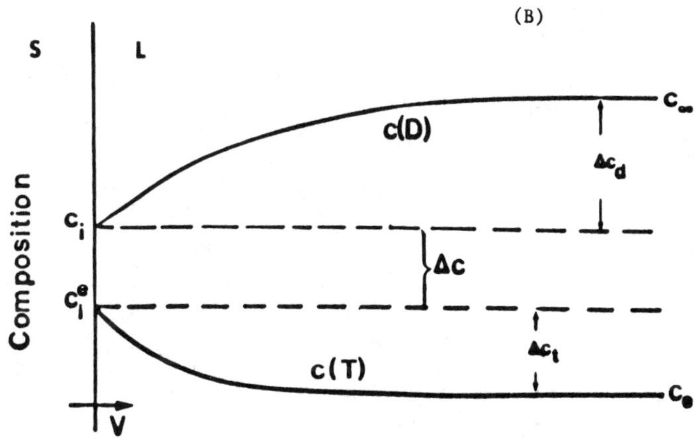

Fig. 1. A) Temperature profile with distance from the S/L interface during growth from the melt and form solution.

B) Concentration profile with distance form the interface during solution growth.

and convection, it is the difference $\Delta C = C_i - C_i^e$ which governs the interfacial processes and therefore, has to be evaluated in order to determine the growth kinetics (1).

For growth from a pure melt, a determination of the growth kinetics depends on a direct measurement of the interfacial supercooling. Unfortunately, measuring the interfacial supercooling accurately is a difficult task and the few proposed techniques of direct (2,3) or indirect (4,5) measurements have proved to be limited, and in most cases unreliable. In view of this difficulty, growth kinetics are commonly discussed in terms of the "Bulk Kinetics". As an example the bulk kinetics of Ni dendrites growing into pure undercooled melt as reported by Walker (6) are shown in Fig. 2 for bulk supercoolings up to about 245°C. The ΔT_b dependence of V in the range of 0-175°C supercoolings can be approximately described by a $K_b(\Delta T_b)^2$ law with K_b = .14 cm/sec.(°C)^2. In comparison, a growth law of .28 $(\Delta T_b)^{1.8}$ has been determined from another investigation (7) for ΔT_b's in the range of 20-200°C.

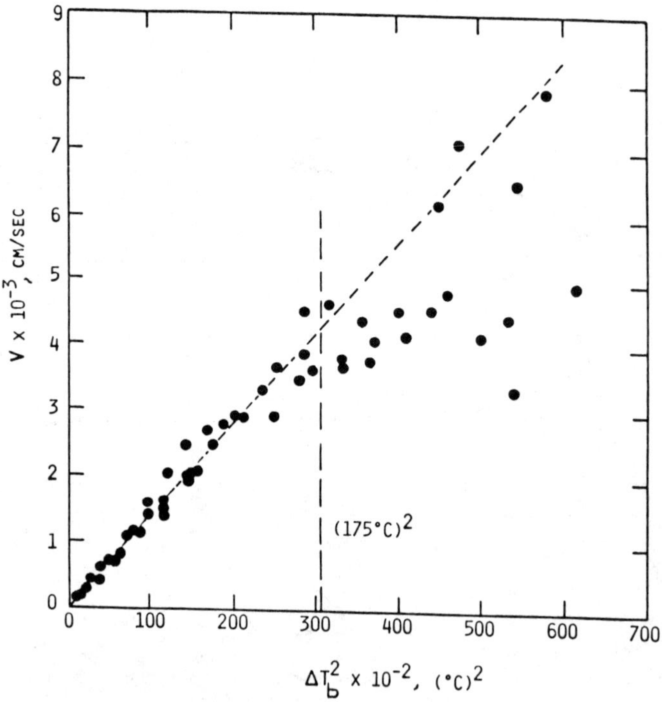

Fig. 2. Bulk growth kinetics of Ni in undercooled melt (after J. L. Walker (7)).

However, calculations based on heat transport limited growth rates (8) and an upper limiting kinetic coefficient based on the speed of sound in the melt indicate that the interfacial supercooling is only a small portion of the bulk supercooling, e.g. it is 26°C when the bulk supercooling is 175°C. Hence, the bulk kinetics relation in no way provides accurate information, even in a qualitative sense, as far as crystal growth kinetics is concerned, and may also be misleading (9-12) when a comparison is attempted with the existing theoretical treatments for crystal growth.

Recently we have conducted a detailed interfacial kinetics study during growth of high purity Ga single crystals grown into the supercooled melt (13,14). The study involves direct measurements of the interfacial and bulk supercoolings, and growth rate for the given interfacial conditions and orientation. In the present paper the results of this study with respect to interfacial and bulk supercooling will be given. For the latter the heat flow problem and its differentiation from the interfacial supercooling will be discussed.

Heat Flow Problem and Crystal Growth Theories

As noted earlier growth starts after an appropriate group of atoms (or molecules) has attained the crystalline structure through the nucleation process. The nucleus may then grow by incorporating liquid atoms at its interface provided that (i) the heat generated at the interface is transported away, and (ii) there is enough driving force at the interface to cause the interfacial migration. The rate at which the heat of fusion can be removed from the solidification front depends on the specifics of the experimental set up, and thermal properties of the solid and liquid. Whether or not this rate is greater than the interfacial kinetics determines whether the growth is limited by the interfacial kinetics (kinetics controlled) or by heat-transfer (heat-transfer controlled). The interfacial kinetics, on the other hand, depend on the nature of the interface, crystal anistropy and perfection. During steady state growth, the interface motion is such that both the heat transfer and interfacial kinetic conditions are simultaneously satisfied.

In this section we briefly analyze the heat-flow problem during "unconstrained" growth into a supercooled pure melt using the interfacial kinetic laws as boundary conditions for the transport (Stefan) problem. The term unconstrained is used to differentiate the current process from those in which the interface velocity is imposed by external means such as, for example, travelling furnace or crystal pulling. When the liquid is supercooled and the growth is unconstrained, the coupled boundary conditions at the interface can be described as

$$\Delta T = \Delta T_b - h_c V, \quad \Delta T_b \geq \Delta T \geq 0 \tag{1}$$
$$V = f(\Delta T)^*, \quad f(0) = 0 \tag{2}$$

where h_c is the parameter representing the interfacial heat transfer coefficient which depends on the experimental design, and physical and thermal properties of the materials involved such as the heat of fusion, thermal conductivities and densities. There are few developed heat transfer models (15-17) that allow for a calculation of h_c; however, the accuracy of these calculations cannot be validated as long as only one of the kinetic parameters (V) is measured directly. Furthermore, at fast growth rates and for rapid interfacial kinetics the problem of calculating the interface temperature is very complex (12,18). The nature of the function f depends upon the specific growth mechanism by which the growth proceeds for a given growth direction. For facet forming materials such as Ga, Bi, Si and most semiconducting compounds with sluggish atomic kinetics, the interfacial supercooling is appreciable, so that the assumption of local equilibrium ($T_i \approx T_m$) does not apply (13). On the other hand the approximation that $\Delta T_b \approx \Delta T$ (i.e., $h_c \to 0$) can hardly be true, even at very slow growth rates. For these materials at small supercoolings, growth is of the lateral (stepwise) type according to which the interface advances normal to itself layer by layer. The growth then can be either dislocation-assisted (SDG) or two-dimensional nucleation-assisted (2DNG) when the interface is free of dislocations. For non-faceted interfaces and also for faceted interfaces at high supercoolings, the growth is such that the interface advances continuously and normal to itself at any site, called continuous growth (CG). The kinetics relations associated with these mechanisms are:

$$V_1 = K_1 \Delta T \quad \text{(CG)} \tag{3}$$
$$V_2 = K_2 \Delta T^n \quad \text{(SDG)} \tag{4}$$
$$V_3 = K_3 \exp(-K_4/\Delta T) \quad \text{(2DNG)} \tag{5}$$

where the K's are approximately constant and n is around 1.7 - 2 (17). Note also that at low supercoolings, $V_1 > V_2 > V_3$, while at high supercoolings $V_1 \approx V_2 = V_3$. Based on these growth rate equations then the coupled

---

*Strictly speaking, eqn. (2) should be expressed as $V=f(\bar{n}, T_i, \Delta T)$ where $\bar{n}$ denotes growth direction.

interfacial kinetics and heat balance requirements, eqns. (1) and (2) can be rewritten as:

$$V = \frac{\Delta T_b - \Delta T}{h_c} = \begin{matrix} K_1 \ \Delta T \\ K_2 \ \Delta T^n \\ K_3 \ \exp(-K_4/\Delta T) \end{matrix} \quad (6)$$

Hence any change in the growth kinetics (e.g., the rate changes from V_3 to V_2 when a dislocation intersects the interface) alters the interfacial supercooling so that the balance between the external heat removal and the interfacial kinetics is maintained in accord with any of the equations in (6).

Experimental Results

The experimental investigation of the Ga interfacial growth kinetics mentioned earlier covers a range of 10^{-3} to 2×10^{4} μm/sec growth rates and interfacial supercoolings up to about 4.6°C, corresponding to the bulk supercooling up to 50°C. In addition, the kinetics have been determined as a function of crystal perfection (dislocation-free versus dislocation-assisted interface) and crystal orientation ([111] and [001]). The interfacial supercooling was measured directly using a technique that utilizes the Seebeck emf generated across the S/L interface with an accuracy of ± .005°C. The bulk supercooling was read via a thermocouple placed in the coolant medium (water or ethylene glycol solution) with an accuracy of ± .025°C. The crystal growth rates were measured optically and/or by utilizing the resistance change across the sample upon growth. The experimental apparatus and procedure have been described in references (13,14,19).

The bulk and interfacial supercoolings as a function of time are plotted in Fig. 3 during growth of the (111) interface. At low supercoolings, the faceted interfaces of Ga grow with either of the lateral growth mechanisms indicated earlier. The bulk supercooling is much higher than the interfacial one at any growth rate regardless of the operating growth mechanism. Furthermore, at a given bulk supercooling, the interfacial supercooling decreases drastically when the growth mechanism changes from 2DNG to SDG, causing the growth rate to suddenly increase. The latter resulted after dislocations were introduced by deforming the Ga crystal (14) and it could be easily detected by the Seebeck technique. This difference between the two

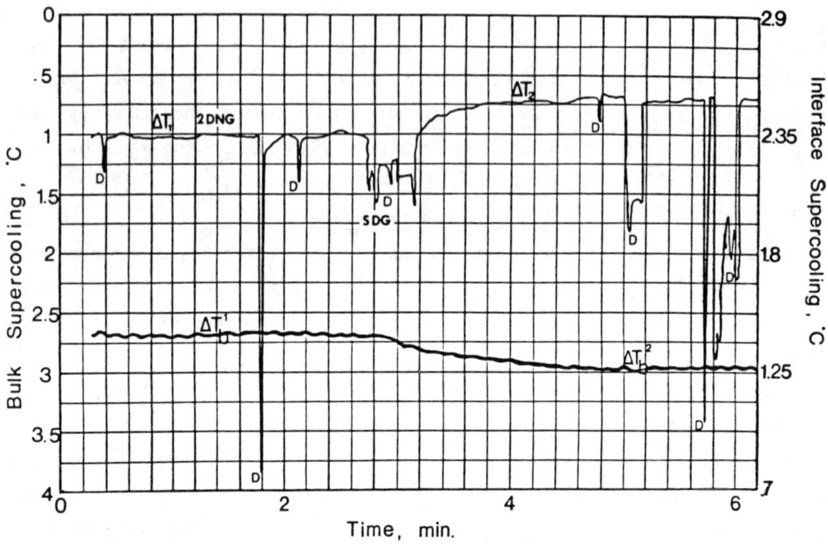

Fig. 3. Interfacial kinetics and heat transfer effects as recorded during unconstrained growth of a Ga S/L (111) interface into the supercooled melt; D denotes SDG growth mechanism.

supercoolings is better noted in Fig. 4 where the (111) results are given in a linear scale of V vs. ΔT and ΔT_b for growth rates in the range 10^{-3} to 2×10^4 μm/sec. As shown in this figure the faster the kinetics, the larger the difference $\Delta T_b - \Delta T$ is; e.g., the difference is 45°C at 1.9×10^4 μm/s**.

**The fact that the two supercoolings are nearly equal at low rates as revealed in Fig. 4, denotes that the assumption of $\Delta T_b \approx \Delta T$ may be used at small growth rates.

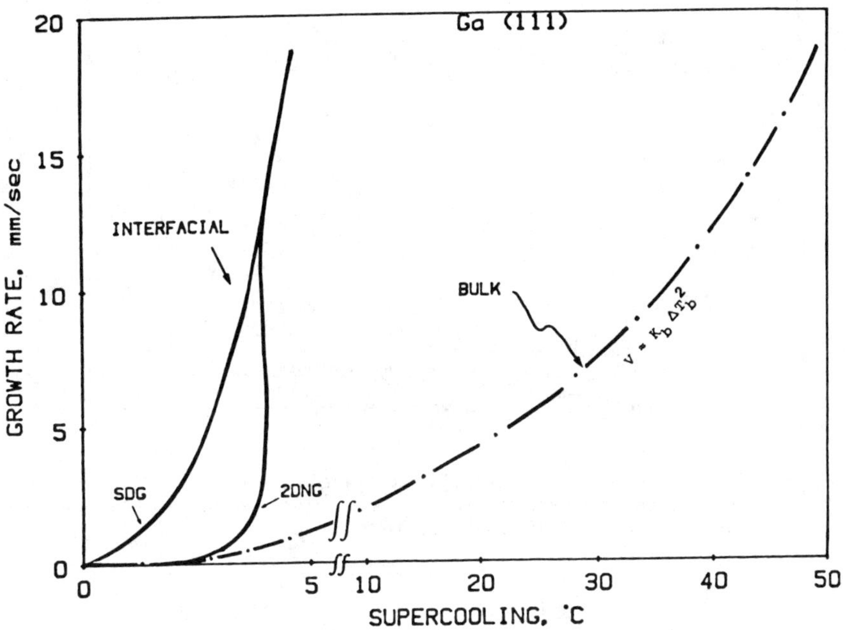

Fig. 4. Growth rates of the (111) interface as a function of interfacial and bulk supercooling.

The (111) growth rates are plotted on a linear scale versus the interface supercooling in Fig. 5. In the range of supercoolings of about 1.5 to 3.5°C it can be seen that the interface has a choice between two growth rates for the same ΔT; the fastest one belongs to the imperfect interface, in contrast with the slowest one which is for the defect-free interface. When the interface was free of dislocations, it was practically motionless ($V < 10^{-3}$ μm/s) for supercoolings less than 1.5°C. Beyond this supercooling the interface advanced with a rate that was strongly dependent on ΔT. This is apparent from the exponential kinetics involved on this growth mode as revealed from Fig. 6, where the data are plotted on a log (V) vs. $1/\Delta T$ scale. Also, note on this plot that the dislocation-free rates at low supercoolings (but higher than the critical threshold supercooling of 1.5°C) are dependent on the S/L interfacial area (A). The interfacial area for samples with the higher growth rates was 4.5 times that of the samples with the lower growth rates. As will be discussed later, the rates in this region are proportional to the interfacial area. However, this is not the case at higher

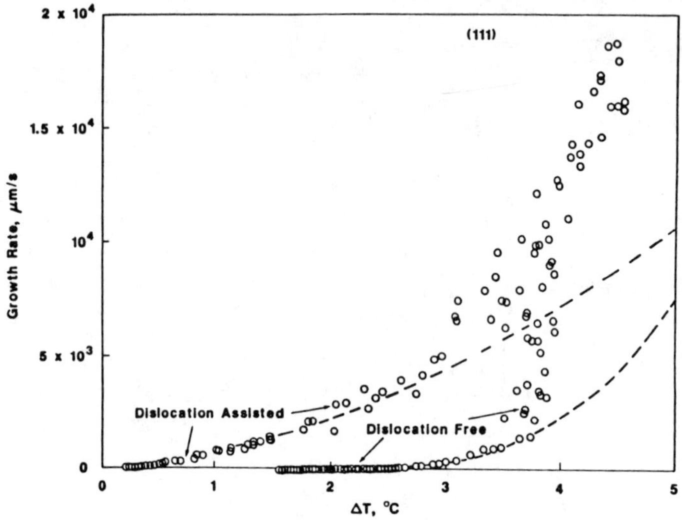

Fig. 5. Growth rates of the (111) interface as a function of interfacial perfection and supercooling; dashed curves represent the 2DNG and SDG rate equations as given in Table 1.

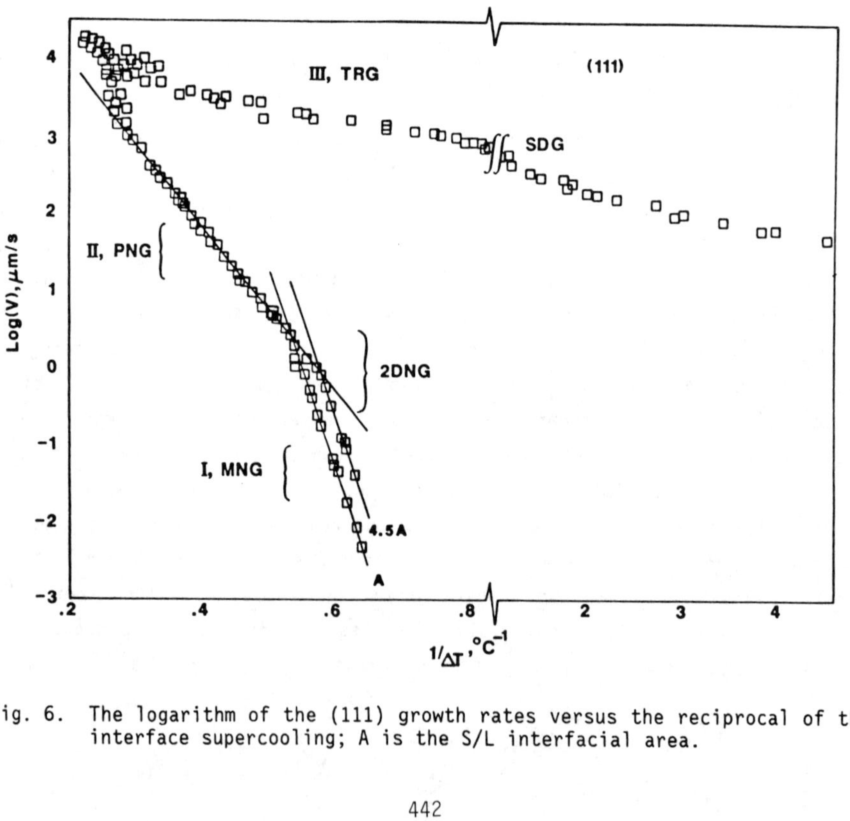

Fig. 6. The logarithm of the (111) growth rates versus the reciprocal of the interface supercooling; A is the S/L interfacial area.

supercoolings ($\Delta T > 2°C$) where the kinetics are independent of the area and become a "weaker" function of ΔT as reflected by the smaller slope of the line passing through the data points. Contrary with the dislocation-free interface, the dislocated one is mobile at all supercoolings even at negligibly small ones as indicated here and elsewhere (12).

The existence of a threshold supercooling and the exponential dependence of the dislocation-free growth rates upon $1/\Delta T$ are clearly indicative of the 2DNG mechanism, whereas those of the dislocated samples belong to the spiral growth (SDG). At higher than about 3.5°C supercoolings, the dislocation-free and -assisted growth rates become approximately equal and it is rather difficult to differentiate between them. Indeed for the highest supercooling regime, the data seem to fall within an upper and a lower rate bound that are linear functions of ΔT, as shown in Fig. 5. The growth behavior of the (001) Ga interface into the supercooled liquid is qualitatively similar to that of the (111) interface as illustrated in Fig. 7. The growth rates of the (001) interface at a given supercooling, however, are much faster than those of the (111) interface.

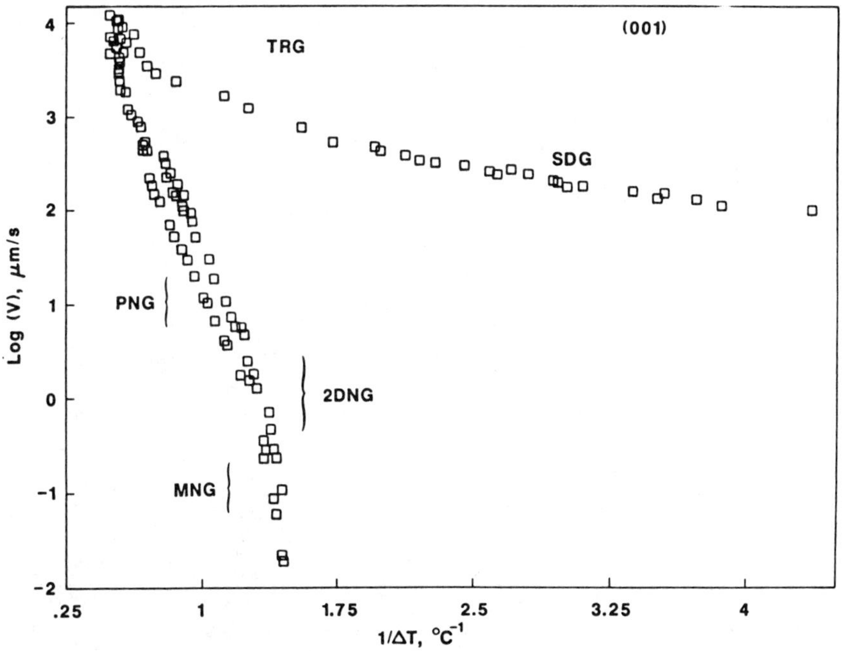

Fig. 7. The logarithm of the (001) rates versus the reciprocal of the interfacial supercooling.

The rate equations of the dislocation-assisted and dislocation-free growth data for both (111) and (001) interfaces up to supercoolings of about 3.5 and 1.5°C respectively are given in Table 1. The experimental growth kinetics for supercoolings higher than the above mentioned ones are quantitatively described elsewhere (20).

Table 1. Experimental Growth Rate Equations; V in μm/sec and A in μm^2.

Interface	Growth Mechanism		Supercooling Range, °C	Growth Kinetics
(111)	Dislocation-free, 2DNG	MNG	1.5 - 1.9	$V = 1.7 \times 10^9 A \exp(-\frac{58.76}{\Delta T})$
		PNG	2 - 3.5	$V = 9.5 \times 10^5 \exp(-\frac{23.9}{\Delta T})$
	Dislocation-assisted, SDG		.2 - 2	$V = 700\, \Delta T^{1.7}$
	Kinetic Roughening		> 3.5	(Ref. 20)
(001)	Dislocation-free, 2DNG	MNG	.6 - .8	$V = 2.95 \times 10^9 A \exp(-\frac{25.4}{\Delta T})$
		PNG	.8 - 1.45	(A) $V = 6 \times 10^5 \exp(-\frac{9.7}{\Delta T})$ (B) $V = 2.4 \times 10^5 \exp(-\frac{9.78}{\Delta T})$
	Dislocation-assisted, SDG		.2 - 1	$V = 1640\, \Delta T^{1.93}$
	Kinetic Roughening		> 1.5	(Ref. 20)

Discussion

Interfacial Kinetics Versus Bulk Kinetics

Since the beginning of this century, when the first crystal growth mechanisms were proposed (21), the importance of the interfacial supercooling has been realized as all theoretical treatments of growth deal with the form of the relation between the growth velocity and the supercooling at the interface. The interfacial supercooling is also important in describing various aspects of solidification of undercooled melts such as, for example, morphological stability and growth anisotropy. Notwithstanding, most experimental investigations (22) disregard the essential role of the interfacial supercooling and describe the growth process by the bulk supercooling and bulk kinetics. According to the present experimental results, as shown in Fig. 4, the Ga growth kinetics as a function of the bulk supercooling can be represented by:

$$V = K_b \, \Delta T_b^2$$

In view of such a parabolic relationship, that is very common for growth into undercooled melts (23), and in the absence of interfacial supercooling measurements, a false agreement between the theoretical growth laws and the experiment can be readily concluded. This is due to the fact that a $(\Delta T)^2$ growth law could arise in different ways. For example, in the case of dendritic growth, a simple solution of the governing heat flow equations (neglecting interfacial kinetics) predicts (24) a growth rate that has approximately a parabolic dependence on the bulk supercooling. Another example is the growth rate by the dislocation mechanism, as discussed previously, which is also a parabolic function of the supercooling.

Another difficulty in using the bulk kinetics is that the value of the coefficient K_b depends on the heat transfer conditions, sample and interface geometry, and the specifics of the experimental set-up. This complicates interpretation of the results, and may explain the contradictory conclusions reached by various investigators on the growth mode and kinetics. The disagreement between kinetic coefficients found from various investigations for the same material is very common indeed (see, for example, the reported kinetic laws for melt growth of Sn (22,23,25)). Certainly, when these results are applied to the theoretical models, the true nature of the growth characteristics of the material under study remains unexplained (22). An example of the effect of the sample size on the bulk kinetics is illustrated in Fig. 8, where the interfacial and bulk kinetics for the Ga (111) interface are given for samples of different S/L interfacial areas but grown under otherwise identical conditions. The heat transfer geometry for the

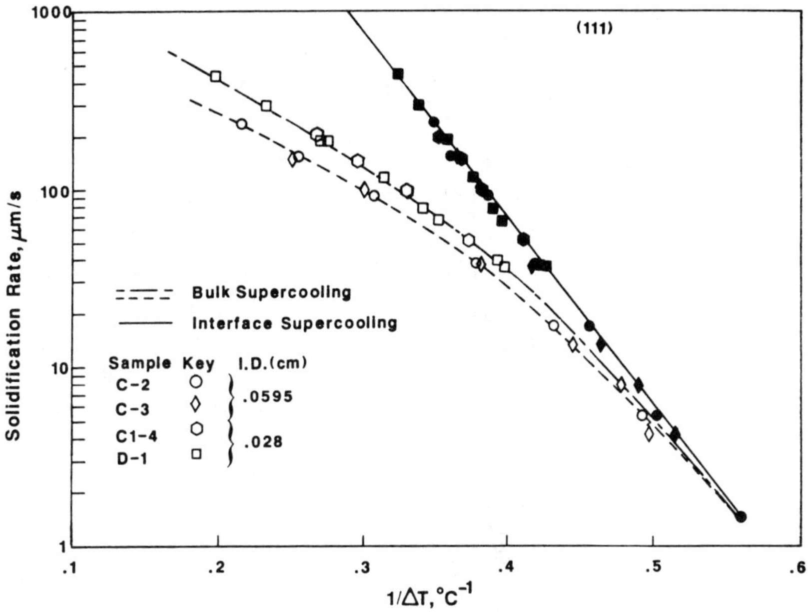

Fig. 8. The logarithm of the (111) rates versus the reciprocal of the interfacial (open symbols) and bulk supercooling (closed symbols) for two sample sizes.

experiments is schematically shown in Fig. 9. Heat transfer calculations (12), that have been verified during the present experiments (20) via the direct interfacial temperature measurements, show that the interface temperature is related to the bulk temperature by an overall heat transfer coefficient h_c. The coefficient is shown to increase approximately proportionally to the interfacial area. Therefore, at a given bulk temperature, the sample with the largest diameter exhibits the highest interface temperature, and consequently the slowest bulk kinetics. In contrast, the interfacial kinetics are the same for both cases in this example. On the other hand, as discussed earlier, there are situations where differences in the mechanism of growth of one and the same material really exist because of the unsimilar conditions in the crystal/melt interface. The problem with using bulk kinetics in this case is that, as can be seen in Fig. 4, it is not sensitive enough to allow for detection of these growth mechanisms.

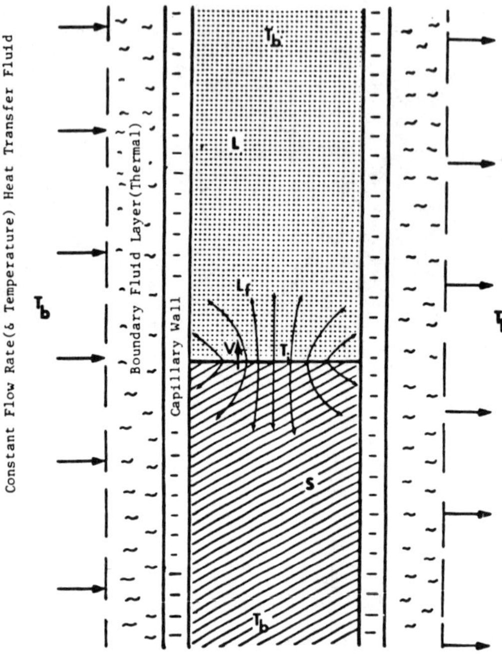

Fig. 9.
Geometry of the interfacial region of the Heat Transfer Analysis; L_f is the latent heat of fusion.

Interfacial Growth Kinetics

At low supercoolings the faceted Ga (111) and (001) interfaces grow by two-dimensional nucleation-assisted or screw dislocation-assisted lateral growth mechanisms. The qualitative features of the Ga dislocation-free growth kinetics are in excellent agreement with the predictions of the classical 2DNG models. These features are the threshold supercooling for a measurable rate and the exponential kinetics beyond this supercooling as given in Table 1. In addition it was shown that the rates are dependent on the S/L interfacial area at low supercoolings but not at higher supercoolings, where the slope of the log V vs. (1/ΔT) curve was found to be smaller. As predicted by the theoretical models (26), the 2DNG kinetics can be divided into two regimes depending on the relative time scales of 2D nucleation events and consecutive layer spreading. At low supercoolings, a single nucleus spreads laterally and completely covers the interfacial area (A) before another nucleation event happens on the same layer. According to this mononuclear (and monolayer) mechanism (MNG) the growth rate is directly related to the interfacial area as:

$$V = K_3' A \exp\left(-\frac{K_4}{\Delta T}\right) \qquad \text{(MNG)}$$

where K_3', and K_4 are constants. At increased supercoolings, however, the time between nucleation events is smaller and the interface advances through multi-nucleation and multilayer spreading. The growth rate for this polynuclear (and multilayer) growth mechanism (PNG) is

$$V = K_3 \exp\left(-\frac{K_4}{3\Delta T}\right) \qquad \text{(PNG)}$$

Hence, in this regime the growth rates is independent of the interfacial area (A) and the exponential constant is smaller than that of the MNG kinetics by a factor of 3. Therefore, the dislocation-free data for both (001) and (111) interfaces at low supercoolings correspond to the MNG mechanism while at higher supercoolings, the results and rate equations (see Table 1.) indicate that the PNG mechanism operates.

When dislocation(s) intersect the faceted interfaces the kinetic characteristics are entirely different than those of the 2DNG mechanism. The dislocated interfaces are mobile at all supercoolings and their growth rates, at a given supercooling, are several orders of magnitude higher than those of the dislocation-free interfaces. For example, at 1.5 and 2°C supercoolings the SDG rates of the (111) interface are higher than the 2DNG rates by 6 and 3 orders of magnitude, respectively. The growth rate equations indicate a nearly parabolic relationship between the rate and the interfacial supercooling. Theoretical analysis of the interfacial processes when screw dislocation(s) intersect the faceted interface, results in a growth law of the form (28)

$$V = K_2 \frac{\Delta T^2}{\Delta T_c} \tanh\left(\frac{\Delta T_c}{\Delta T}\right) \qquad \text{(SDG)}$$

where K_2 and ΔT_c are constants. This equation reduces to a parabolic growth law ($V \propto \Delta T^2$) when $\Delta T < \Delta T_c$, and to approximately a linear growth law when $\Delta T_c > \Delta T$. In between these extreme limits the growth law is in the form of $V = K \Delta T^n$ where the power n lies between 1 and 2. For dislocation-assisted growth from melt, the rate is commonly given by the parabolic rate equation (28). Indeed, the experimentally determined powers of the SDG rate equations for Ga are 1.7 and 1.9 for the (111) and (001) interfaces respectively.

Despite the qualitative agreement between the present results and the predictions of the phenomenological lateral growth theories, there is a substantial quantitative disagreement between the two. As discussed in detail elsewhere (20), the calculated PNG and MNG rates are about one and four orders of magnitude larger than the experimental values. Furthermore, the theories do not predict the observed deviation from the exponential rates of PNG as supercooling is increased beyond approximately 3.5°C for (111) and 1.5°C for (001). This region is designated as TRG (Transition Region Growth in Fig. 6. Neither is the deviation from the SDG rates predictable from the theories (28).

The above mentioned disagreements between the classical theories and experiments are due to the fact that the theories are based on an unrealistic model of the S/L interface; the interface for faceted materials is treated as being sharp such that the transition from the solid to liquid occurs in about one interatomic spacing. In reality, the S/L interface has a diffused nature (9,29), as seen in Fig. 10, which schematically shows the gradual transition from the liquid to solid. The diffuseness of the interface not only depends

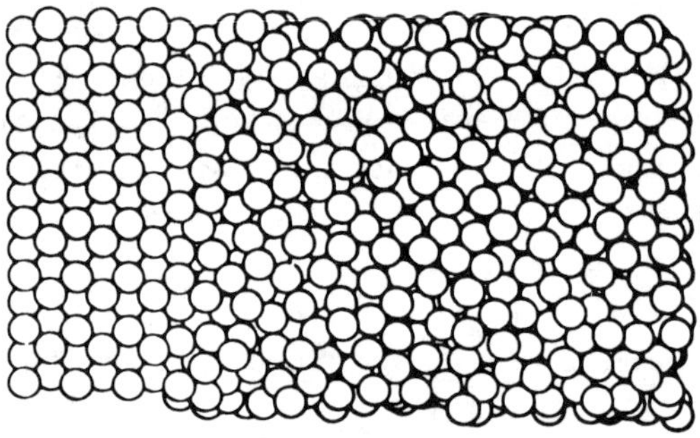

Fig. 10. Cross-sectional view of a model of the S/L interface (after A. Bonissent (28)).

on the material and orientation, but also on the driving force (supercooling) at the interface. Therefore, the activation energy for interfacial atomic transport and the step edge free energy, which are respectively incorporated in the rate coefficients K_3, K_3' and K_4, will depend on the supercooling as well as the position within the interface. Based on this model, the differences between the experimental results and theoretical predictions can be explained (20) in view of a decreased interfacial atom migration (compared with that in the bulk liquid) and a supercooling dependent step edge free energy. The latter vanishes as the interface gradually looses its faceted character and becomes kinetically rough. Indeed the kinetic roughening of the interface is responsible for the observed high supercoolings deviation from the exponential kinetics to approximately linear kinetics for both Ga (001) and (111) interfaces.

Conclusions

The experimental investigation of the Ga (111) and (001) interfacial and bulk growth kinetics has clearly highlighted the difficulties which may arise in utilizing the bulk kinetics to describe solidification behavior of supercooled liquids. In order to realize the macroscopic transport effects on the crystal growth process and differentiate them from the interfacial kinetics effects it was shown that the interfacial temperature has to be measured directly. It is concluded that the faceted Ga interfaces at low supercoolings grow with either of the lateral growth mechanisms, two-dimensional nucleation or screw dislocation-assisted, depending on the perfection of the interface. At high supercoolings the interfaces gradually become kinetically rough and the growth mode changes from lateral to normal. Furthermore, the experimental determination of the relation $V(\Delta T)$ allows us for a reliable quantitative comparison with the crystal growth theories. This comparison reveals that the theoretical rate equations describe the atomistic growth processes qualitatively, but not quantitatively. This is because the governing growth parameters are linked to the nature of the S/L interface as supercooling increases rather than to bulk properties, and the interface supercooling.

Acknowledgement

The support of this research by the National Science Foundation (Grant Number DMR-82-02724) is gratefully acknowledged.

References

1. P. Bennema, Phys. Stat. Sol., 17 (1966) 555.
2. J. C. Brice and P. A. C. Whiffin, Solid State Electron., 7 (1964) 183.
3. J. D. Verhoeven and E. D. Gibson, J. Crystal Growth, 5 (1969) 235.
4. J. J. Kramer and W. A. Tiller, J. Chem. Phys., 37 (1962) 841.
5. R. J. Schaefer and M. E. Glicksman, Acta Met., 16 (1968) 1009.
6. J. L. Walker, unpublished work in "Principles of Solidification," B. Chalmers, (John Wiley, NY, 1964) p. 114.

7. G. A. Colligan and B. J. Bayles, Acta Met., 10 (1962) 895.
8. S. R. Coriell and D. Turnbull, Acta Met., 30 (1982) 2135.
9. J. W. Cahn, W. B. Hilling, and G. W. Sears, Acta Met., 12 (1964) 1421.
10. K. A. Jackson, D. R. Uhlmann, and J. D. Hunt, J. Crystal Growth, 1 (1967) 1.
11. D. Turnbull, J. Phys. Chem., 66 (1962) 609.
12. G. J. Abbaschian and S. F. Ravitz, J. Crystal Growth, 44 (1978) 453.
13. S. D. Peteves, J. Alvarez, and G. J. Abbaschian, in: Rapidly Solidified Metastable Materials, B. H. Kear and B. C. Giessen, eds. (Elsevier, NY, 1984) p. 15.
14. S. D. Peteves and G. J. Abbaschian, Interfacial Kinetics and Roughening, Fall 1985 AIME-TMS Meeting, Toronto, Canada.
15. W. B. Hillig, in: Growth and Perfection of Crystals, R. H. Doremus, B. W. Roberts, and D. Turnbull, eds. (Wiley, NY, 1958) p. 350.
16. A. S. Michaels, P. L. T. Brian, and P. R. Sperry, J. Appl. Phys., 37 (1966) 4649.
17. P. R. Pennington, S. F. Ravitz, and G. J. Abbaschian, Acta Met., 18 (1970) 943.
18. D. E. Temkin, Sov. Phys.-Dokl., 5 (1960) 609.
19. S. D. Peteves, J. A. Sarreal, and G. J. Abbaschian, in: Semiconductor on Insulator and Thin Film Transistor Technology, A. Chiang et al., eds. (MRS Symp. Proc., Vol. 53, 1985) to be published.
20. S. D. Peteves and G. J. Abbaschian, to be published.
21. H. A. Wilson, Phil. Mag., 50 (1900) 238.
22. See for example: K. F. Kobayashi, M. I. Kumikawa, and P. H. Shingu, J. Crystal Growth, 67 (1984) 85.
23. D. E. Ovsienko and G. A. Alfinstev, in: Crystals, Vol. 2, H. C. Freyhardt, ed. (Springer-Verlag, Berlin, 1980) p. 119.
24. M. C. Flemings, Solidification Processing, (McGraw-Hill, NY, 1974) p. 167.
25. J. J. Favier, Dc. S. Thesis, University of Grenoble, 1977.
26. W. B. Hillig, Acta Met., 14 (1966) 1968.
27. W. K. Burton, N. Cabrera, and F. C. Frank, Phil. Trans. Roy. Soc. (London), A243 (1951) 299.
28. W. B. Hillig and D. Turnbull, J. Chem. Phys., 24 (1956) 914.
29. A. Bonissent, in: Interfacial Aspects of Phase Transformations, B. Mutaftschiev, ed. (D. Reidel Publ. Co., Dordrect, Netherlands, 1982) p. 143.

DENDRITIC GROWTH IN A SUPERCOOLED ALLOY MELT

by

V. Laxmanan

Department of Metallurgy and Materials Science, Case Western Reserve University, Cleveland, Ohio 44106. Concurrently, Visiting Scientist, NASA Lewis Research Center, Mail Stop 105-1, Cleveland, Ohio 44135.

ABSTRACT

A simple model which describes the growth of an "array" of dendrites into a supercooled, binary, alloy melt is presented. Solute diffusion is calculated by superposing the solutions given by Flemings and Zener, and also, by superposing the solutions given by Ivantsov and Flemings. A general expression for the transport solution is suggested from which all other dendrite growth models presented earlier may be obtained as special cases. It is shown that both "free" and "constrained" growth may be described by a single transport solution, which indicates that i) both thermal and solutal effects will be important during "free" growth in dilute alloys, ii) only solutal effects are predominant during "free" growth in concentrated alloys and during "constrained" growth. An examination of the relevant dimensionless parameters also suggests that all dendrite growth models, regardless of the assumptions used to determine the tip radius (marginal stability, minimum undercooling, maximum velocity, minimum entropy production) should predict the experimentally observed extrema in tip radius and growth velocity in dilute alloys, during "free" dendritic growth. Experimental data in binary H_2O-NaCl and succinonitrile-acetone solutions are shown to be in good agreement with the model.

Introduction

A binary alloy melt is considered to be supercooled or undercooled if the liquid phase can be maintained in metastable equilibrium at some temperature below the equilibrium liquidus temperature of the alloy, T_L. Dendritic growth occurs quite readily in such a supercooled bath when the solid phase nucleates either in an uncontrolled manner (heterogeneously due to the presence of a foreign substrate or even homogeneously if the bath supercooling is sufficiently large), or is nucleated in a controlled manner, as by insertion of a seed into the melt at some known and predetermined temperature below T_L. A number of techniques are available to supercool both small and large masses of liquid. These have been reviewed by the author in Ref. 1.

Let $\Delta T_b = T_L - T_\infty$ be the initial bath supercooling, where T_∞ is the bath temperature at which the solid phase, assumed here to be growing dendritically # , nucleated. The growing dendritic solid may be the stable equilibrium phase or some other metastable phase. Assume that T_∞ remains constant * as solidification proceeds, as in the rather precisely controlled experiments of Glicksman and co-workers [3]. The bulk liquid far away from the interface (tips of the dendrite) will thus remain at T_∞, fig.1. This is also a reasonably good assumption during the early stages of solidification in highly supercooled liquid alloy droplets, as discussed for example by Flemings, Mehrabian and their co-workers elsewhere in these proceedings. (More precisely, the assumption of a constant T_∞ is a good one when the Fourier number is very small [4]).

The dendrite tip temperature, T_t, is, however, much higher than the bath temperature, but is depressed below the equilibrium liquidus temperature T_L by an amount $\Delta T = T_L - T_t$, Fig. 1, due to constitutional, capillary and kinetic effects. The temperature gradient, G_L, in the bulk liquid ahead of the tips is thus negative [1-4]. Dendritic growth under these conditions is often called "free" growth [3,5], since for a fixed initial alloy composition and bath supercooling @, the dendrite tip growth rate R, although usually found to be a constant, is ***NOT KNOWN apriori.*** This situation is quite unlike that encountered during "constrained" dendritic growth [6]. In the latter instance both the growth rate and the temperature gradient are fixed and, moreover, can be controlled independently (as is the case in all directional solidification experiments or processes).

Thus, in a very simplified manner, it may be said that the differences between "free" and "constrained" growth arise from the fact that during "free" growth both R and G_L are unknown functions of the unknown tip radius r_t which in turn is a function of the bath supercooling, ΔT_b.

Other authors, for e.g. Levi and Mehrabian, assume the solid-liquid interface to be macroscopically planar but consider other important effects such as interface attachment kinetics, non-equilibrium partitioning of solute, etc. These effects may also be included in the analysis for dendritic growth, but will be neglected here as a first step.

* In many cases, such as during undercooling of liquid metal droplets during electromagnetic levitation or during "free flight" in a drop tube [2], it is very difficult to measure T_∞, which is then either unknown or only known approximately.

@ It is important to distinguish between the "undercooling" ΔT at the tip of the dendrite and the "undercooling" or "supercooling" in the bulk liquid

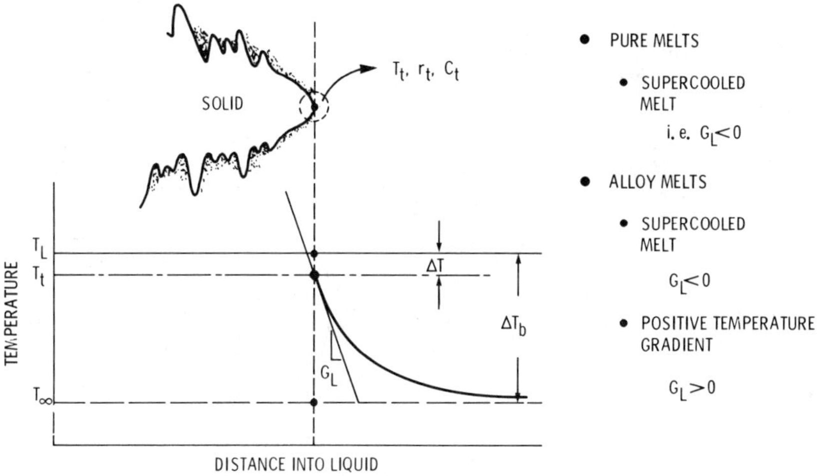

Figure 1. - Thermal conditions favoring dendritic growth.

The objective of all theoretical analyses for dendritic growth is to predict morphological characteristics such as the tip radius, r_t, the primary and secondary arm spacings as well as other observed dendrite characteristics such as the dendrite tip growth rate R (only in the case of "free" growth), the tip temperature T_t, the liquid composition C_t in equilibrium with the tips, the amount of interdendritic or intercellular eutectic formed, f_E, and, ultimately the detailed microsegregation profile in the finally solidified material. ** It may be noted here that hardly any measurements of tip temperature and/or tip composition have been reported in the literature for "free" growth in alloys in contrast to several good measurements available for "constrained" growth [7-10]. Even the measurements of morphological features such as the tip radius is relatively recent [11-14].

The main purpose of this paper is to present a simple model for dendritic growth in a supercooled alloy melt. The primary focus here will be the prediction of the dendrite tip radius and growth rate. These predictions will be compared with some of the recent experimental data of Glicksman and co-workers. The subject of tip temperature and composition predictions and detailed microsegregation analysis will be addressed seperately [15].

Figure 2 shows schematically the evolution of some solidification structures, such as observed in supercooled droplets, which may be analyzed

---

far away from the dendrite tips. The composition of the solid formed depends on the "tip undercooling" not the "bath supercooling".

** Please also see Levi & Mehrabian, Abbaschian and Kattamis elsewhere in these proceedings.

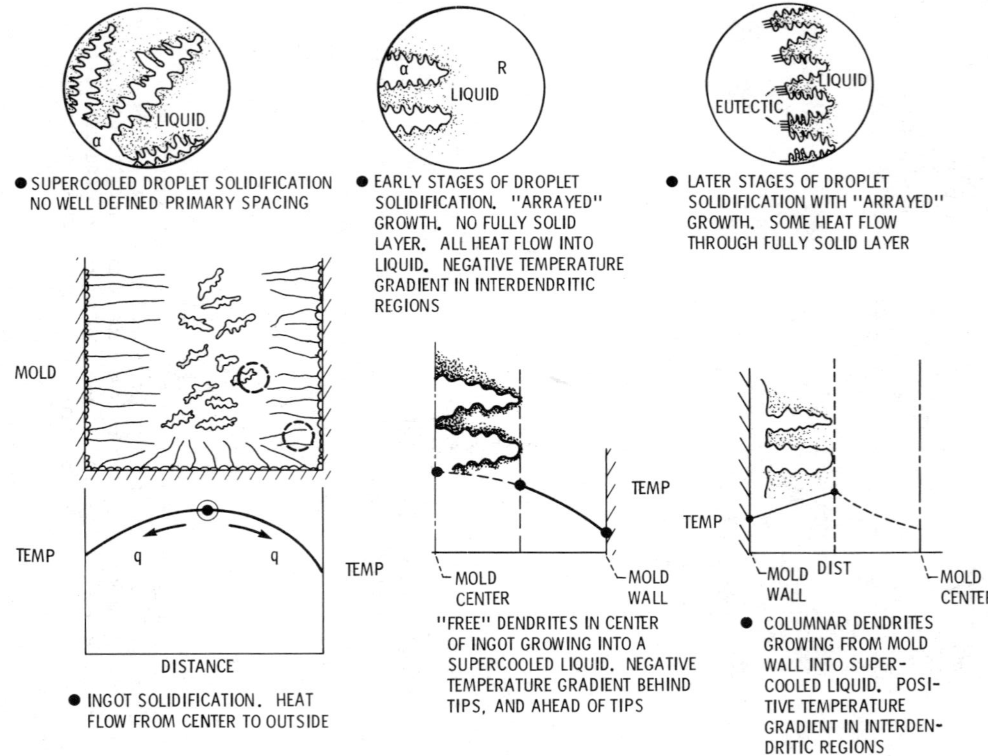

Fig. 2: Evolution of solidification structures in supercooled droplets and ingots.

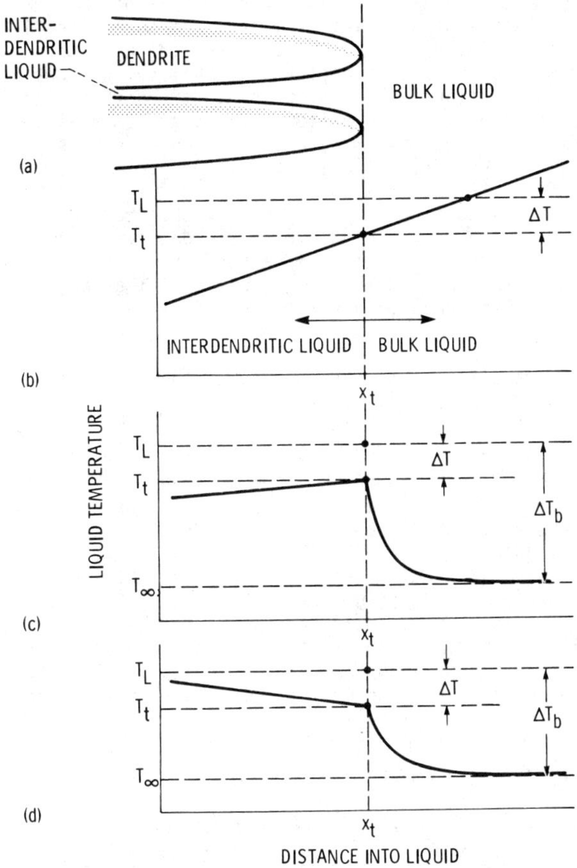

Fig. 3 - Thermal conditions during dendritic solidification in an alloy melt. (a) Schematic of dendrites, (b) "constrained" growth, (c) and (d) "free" growth.

using the model to be described here. A "dense array" of dendrites is often seen in the finally solidified alloy. Here we imagine the "dense" array to be advancing into a bath which is at a lower temperature than the tips. There may also be some heat flow to the surroundings through the fully solid layer, fig. 3. Liquid between the dendrites becomes enriched in solute due to segregation of alloying elements, and some eutectic or other second phases form giving rise to the observed segregation in the final alloy. There will also be a significant diffusive flux towards the tips because of the compositional changes in the interdendritic liquid. Fig. 2 also illustrates a group of dendrites radiating from a single "nucleus". The individual dendrites may be considered to be essentially isolated and growing independently of each other. The primary spacings are not very well defined [6] in this case. However, there will be a significant diffusive flux towards the tips even in this case due to the non-isothermal and non-isoconcentrate nature of the dendrite surface. It is suggested that an estimate of the diffusive flux may be obtained in this latter instance (although with less rigorous justification) from the model for growth of a dense array of dendrites, or other "constrained" growth models [16-22]

Figure 3 shows schematically the thermal conditions in both the interdendritic regions and in the bulk liquid ahead of the tips during "constrained" growth of a dendritic array ($G_i > 0$, $G_L > 0$) as well as during growth of a dendritic array into a supercooled melt. In the latter instance, the interdendritic thermal gradient, G_i, may be either positive (which will be the case if there is a significant amount of external heat extraction through the solid phase) or negative (if solidification is adiabatic, with the liquid phase providing the only available heat sink for the heat of fusion liberated. Adiabatic conditions will also be approached during the early stages of rapid solidification in highly supercooled droplets). Thermal conditions such as shown here will thus be encountered during solidification of initially supercooled ingots, as discussed in ref. 1, rapid solidification of atomized liquid droplets, solidification occurring in electron beam/laser beam melting processes, melt spinning or planar flow casting, and weld pool solidification.

TABLE I. - IMPORTANT LENGTH SCALES OF THE DENDRITE PROBLEM

LENGTH SCALE	PURE MELT	ALLOY	REMARKS
THERMAL DIFFUSION	$\ell = 2\alpha_L/R$	--------------	TYPICALLY 1 mm - 10 CM
SOLUTE DIFFUSION	--------------	$\ell_s = 2D_L/R$	TYPICALLY 10 μm - 1 mm
THERMAL CAPILLARY LENGTH	$d_0 = \dfrac{\gamma}{\Delta S(L_V/C_p)}$	--------------	TYPICALLY 1 - 10 nm
CHEMICAL CAPILLARY LENGTH	--------------	$\ell_c = \dfrac{\gamma}{k\,\Delta S\,\Delta T_0}$	FUNCTION OF ALLOY COMPOSITION; TYPICALLY 100 nm
THERMAL LENGTH	--------------	$\ell_t = \dfrac{k\,\Delta T_0}{G_L}$	1 mm TO 10 cm

γ = SURFACE TENSION

ΔS = ENTROPY OF FUSION

Important Dimensionless Groups Governing Dendritic Growth in Pure Melts and Alloys

The various physical mechanisms governing dendrite characteristics - the constitutional, capillary and kinetic effects - depend on a number of thermo-physical properties and the characteristics of the alloy itself, i.e. phase diagram properties. Table 1 lists several quantities which have been called the characteristic length scales of the dendrite problem. These length scales include among them essentially all the important thermo-physical properties influencing dendrite behavior [18, 23, 24]. These are briefly:

i) the thermal diffusion length and the solute diffusion length (which involve the thermal diffusivity and the chemical diffusivity respectively),

ii) the thermal capillary length and the chemical capillary length (which involve the surface tension, the entropy of fusion and a characteristic temperature). In the case of a pure melt the characteristic temperature is taken to be L_v/C_p, commonly called the hypercooling limit. For an alloy melt the characteristic temperature is taken to be the equilibrium freezing range.* (An expression for the hypercooling limit in alloy melts may be obtained from a simple re-examination of the basic notion of hypercooling, see ref. 1. The hypercooling limit is then found to depend also on the freezing range of the alloy, see also table 3).

One may also define the hypercooling limit for an alloy more simply by replacing the melting point (solidus temperature) of the pure melt, by the equilibrium solidus temperature, T_s, in the expression for hypercooling limit for a pure melt. Likewise, one may use the so-called "T_o" temperature instead of T_s for the same purpose. The "T_o" temperature is the temperature at which the free energies of the solid and liquid phases are equal for the given initial composition.

iii) Another length scale, simply called the thermal length here, involves the thermal gradient in the liquid, <u>at</u> the tip, the equilibrium partition ratio k, and the freezing range. This length scale is intimately related to the gradient of constitutional supercooling (GCS) at the tip.

Starting with the above length scales a number of important dimensionless groups may be constructed by considering the ratios of these length scales, Table 2. For example, the Chalmers number [22], symbol s, involves the ratio of solute diffusion length and the thermal length in (iii) above. It is closely related to the gradient parameters defined by Trivedi (symbol $\mathcal{G}$) and Flemings (symbol a). For s = 1 the solid-liquid interface will be planar and for very small values of s the interface will be dendritic.

The capillary number in the alloy case, symbol A, is the ratio of the chemical capillary length and the solute diffusion length. Capillary effects will be important when A > 1, i.e. when the growth rate is quite large. This ultimately leads to a stabilization of a planar solid-liquid

---

* The different characteristic temperatures appearing in the pure melt problem and the alloy problem are dictated simply by "mathematical convenience". Langer [24] defines the capillary length in a dilute alloy to be the capillary length ℓ_c here multiplied by (C_o/C_t).

Table II: Important dimensionless groups applicable during dendritic growth in pure melts and alloys.

Dimensionless Group	Pure Melt	Alloy Melt	Remarks
Chalmers Number*		$s = \frac{D_L G_L}{R \Delta T_o}$	$s \to 0$ AS $G_L/R \to 0$ $s \to 1$, PLANE FRONT
Gradient Parameters		$G = 2s/k = \ell_s/\ell_t$	Trivedi's Definition
		$a = \frac{D_L G_L}{m_L R C_o}$	Flemings' Definition
Capillary Number	$A_t = \frac{R d_o}{2\alpha_L}$	$A = \frac{R \ell_c}{2 D_L}$	$A = 2k^2 A^{\dagger}$

* $s = \frac{ak}{k-1} = \frac{1}{2} k G = \frac{R_p}{R}$ where $R_p = \frac{D_L G_L}{\Delta T_o}$ CS limit

† $A = \frac{R}{R_a}$ where $R_a = \frac{D_L}{\ell_c k^2}$ AS limit

CS = Constitutional supercooling AS = Absolute stability

TABLE II. - IMPORTANT DIMENSIONLESS PARAMETERS APPLICABLE DENDRITIC GROWTH IN PURE MELTS AND ALLOYS

DIMENSIONLESS GROUP	PURE MELT	ALLOY	REMARKS
THERMAL PECLET NUMBER	$p_t = \frac{R r_t}{2\alpha_L} = \frac{r_t}{\ell}$	---------------	$p \to 0,\ R \to 0$ $p \to \infty,\ R \to \infty$ $r_t \to \infty$
SOLUTE PECLET NUMBER	------------	$p = \frac{R r_t}{2 D_L}$	
TIP STABILITY PARAMETER	$\sigma^* = \frac{\ell d_o}{r_t 2} = \frac{2\alpha_L d_o}{R r_t 2}$	$\sigma_c = \frac{\ell_s \ell_c}{r_t 2} = \frac{2\ell_c D_L}{R r_t 2}$	$\sigma \to 0$ AS $r_t \to \infty$ $R \to \infty$ $R r_t^2$ = CONSTANT

TABLE III. - IMPORTANT DIMENSIONLESS PARAMETERS APPLICABLE TO DENDRITIC GROWTH IN PURE MELTS AND ALLOYS

DIMENSIONLESS GROUP	PURE MELT	ALLOY	REMARKS
BATH SUPERCOOLING	$\Delta\theta = \dfrac{(T_M - T_\infty)}{(L_v/C_p)}$	$\dfrac{(T_L - T_\infty)}{(L_v/C_p)}$	$\Delta\theta > 1$, HYPERCOOLING†
SOLUTE BUILDUP	--------------	$\dfrac{\Delta C}{\Delta C_0} = \dfrac{(C_t - C_0)}{C_0(1-k)/k}$	$\Delta C \to \Delta C_0$, PLANE FRONT
SUPERSATURATION	--------------	$\Omega = \dfrac{(C_t - C_0)}{C_t(1-k)}$	$\Omega \to 1$, PLANE FRONT
TIP UNDERCOOLING	--------------	$\dfrac{\Delta T}{\Delta T_0} = \dfrac{(T_L - T_t)}{(T_L - T_S)}$	$\Delta T \to \Delta T_0$, PLANE FRONT

† Hypercooling limit for alloy discussed further on next page.

interface in the rapid solidification regime as discussed by Mullins & Sekerka [25]. The dimensionless group A defined by them is related to A as follows: $A = 2k^2A$. Likewise, a capillary number may be defined for the pure melt case, by taking the ratio of the thermal capillary length d_0 and the thermal diffusion length. Although the capillary numbers are quite small in the low growth rate regime, capillary effects have a very important and subtle influence on the dendrite problem. Various physical mechanisms governing the selection of the dendrite tip radius - the maximum velocity hypothesis, the marginal stability hypothesis, the minimum undercooling assumption [3, 21] - all depend crucially on the seemingly small capillary effects. Dimensionless groups which determine the importance of kinetic effects may likewise be defined, see Levi & Mehrabian or Trivedi [4, 26].

Other dimensionless groups which characterize the dendrite problem involve either one of these length scales and the tip radius (the thermal Peclet number and the chemical Peclet number), or two of these length scales and the square of the tip radius (the tip stability parameter for the pure melt or the alloy melt). The tip stability parameter, σ^*, was first identified by Langer and Muller-Krumbhaar [23]. The present author has shown that the chemical analogue of σ^*, for which the notation σ_c has been proposed, is a very useful parameter in the analysis of dendritic growth in alloys [20-22]. σ_c is also equal to A/p^2. Hence σ_c will become zero at both very low and very large growth rates when $p \to \infty$ (planar interface) and will be a maximum for some p. Experimentally σ_c is found to be small and reasonably constant for small p [21]. Note that the chemical tip stability parameter, as defined here, will become infinitely large for a very dilute alloy and will not approach the value for the pure melt. This may be seen more clearly by considering their ratio:

$$B = \sigma_c/\sigma^* = (D_L/\alpha_L)/k\Delta\Theta_0 \qquad (1)$$

Table III (Cont'd) Hypercooling Limit for an Alloy Melt.

Dimensionless Group	Remarks
$\frac{T_L - T_\infty}{L_v/C_p} = \Delta\Theta$	Bath Supercooling
$\frac{T_L - T_s}{L_v/C_p} = \Delta\Theta_o$	Dimensionless freezing range
$\frac{T_s - T_\infty}{L_v/C_p} = \Delta\Theta_s$	Solidus temperature replaces T_L or T_M
$\frac{T_o - T_\infty}{L_v/C_p} = \Delta\Theta_{ef}$	"T_o" replaces T_L or T_M

For hypercooling in alloy melt we require

$\Delta\Theta > 1 + \Delta\Theta_o$, see ref. 1; this exceeds $\Delta\Theta_s$ and $\Delta\Theta_{ef}$.

Note: $\Delta\Theta = \Delta\Theta_s + \Delta\Theta_o$

$\Delta\Theta = \Delta\Theta_{ef} + 0.5\Delta\Theta_o$, assuming $T_L - T_o = 0.5\ \Delta T_o$

The dimensionless group identified here in equation (1) will be shown later in this paper to be an important parameter in the analysis of dendritic growth in a supercooled alloy melt. As the alloy concentration is reduced, $C_o \rightarrow 0$ and hence $\Delta\Theta_o \rightarrow 0$ so that $B \rightarrow \infty$ and $\sigma_c \neq \sigma^*$. A more complete discussion of the significance of this parameter may be found in Appendix A.

Table 3 defines some dimensionless groups which relate the dendrite tip temperature and tip composition to the corresponding values for a planar interface (dimensionless tip undercooling and dimensionless solute buildup). We also define a dimensionless bath supercooling $\Delta\Theta$ for both pure melts and alloy melts.

Table 3 also identifies a dimensionless group called the supersaturation, which is the ratio of the solute buildup at the tip to the solute rejected at the tip. The value of the supersaturation at the tip in the special case of an isothermal liquid will be denoted as Ω^0. In this limiting case the dendrite surface is isothermal, s = 0. There is no diffusion of solute or heat along the dendrite surface. All diffusion is strictly normal to the dendrite surface. This normal or "divergent" diffusion, fig.4d, depends only on the growth rate and the dendrite tip curvature [20], i.e. the "sharpness" or "pointedness" of the dendrite. Some authors refer to this as the "point" effect of diffusion. The "point" effect of diffusion depends on Peclet numbers, p or p_t, depending on whether solute flow or heat flow is being considered. Since the dendrite tip radius will become quite large in the low velocity regime, p becomes large close to the con-

TABLE IV. - GENERAL EXPRESSION FOR SUPERSATURATION IN ISOTHERMAL LIQUID

MODEL		SUPERSATURATION	REMARKS
IVANTSOV	(1947)	$\Omega^0 = \Phi = pe^p E_1(p)$	$\Omega^0 \rightarrow 1,\ p \rightarrow \infty$
ŻENER	(1946)	$\Omega^0 = 2p\lambda$	$\Omega^0 \rightarrow 1,\ p \rightarrow 1/2\lambda$ $\Omega^0 \rightarrow \infty,\ p \rightarrow \infty$
BURDEN AND HUNT	(1974)		$\lambda = 1$
FISHER KURZ AND FISHER	(1951) (1981)	$\Omega^0 = p$	$\lambda = 1/2$

● $\Omega^0 = \dfrac{\Delta C}{C_t^0(1-k)} = \dfrac{\text{SOLUTE BUILD-UP AT TIP}}{\text{SOLUTE REJECTED AT TIP}}$

● $\Omega^0 \rightarrow 1$

SOLID COMPOSITION $kC_t^0 \rightarrow C_0$

stitutional supercooling limit when s = 1. The "point" effect or "divergent" must then become negligibly small. Thus, it is clear that in the more general case when the dendrite surface is non-isothermal and a non-isoconcentrate, the "point" effect must depend on both p and s. Likewise, in the large growth rate regime (note s = 0, and $p \rightarrow \infty$, even in this case), when the interface again becomes planar, the "point" effect must be unimportant.

Table 4 summarizes the values of Ω^0, the supersaturation at the tips due to the "point" effect, for various isothermal dendrite growth models. These results may be generalized in terms of another important dimensionless parameter, λ, which has been called the Zener number or the Zener constant [21, 27]. In general $\Omega^0 = 2p\lambda$, where λ has different values for different dendrite growth models. By choosing different values of λ or Ω^0 a general expression for the supersaturation or dimensionless solute buildup and tip undercooling may be obtained which includes the results given by various investigators as special cases.

λ is also the ratio of the "effective" diffusion distance as defined by Zener [28] and the tip radius, appendix B, fig. 20. Hence, λ is also related to the concentration gradient at the tip and thus determines the GCS at the tip, and its variation with growth rate. It may be shown that the functional relationship for λ given by the Ivantsov solution [29] is such that the GCS predicted by this model becomes negative when s > k, i.e. long before the interface becomes planar (see appendix C). This is the fundamental reason for the motivation here to develop a model which does not depend only on the "Ivantsovian" description of heat flow and solute flow [3, 5, 12, 18, 24, 26].

Table 4 also defines a quantity called the thermal supersaturation, symbol Ω_t, which may be defined, by analogy with Ω^0, as the ratio of the enthalpy buildup at the tip to the heat of fusion liberated at the tip, L_v. The enthalpy buildup at the tip is simply the difference between the enthalpy of the liquid far away from the tip, C_pT_∞, and the enthalpy of the liquid at the tip itself, C_pT_t; assuming constant specific heat. The enthalpy buildup at the tip is thus $C_p \Delta T_H$. For the Ivantsov analysis $T_t = T_M$ so that $\Omega_t = \phi_t$, with $\phi_t = p_t \exp(p_t) E_1(p_t)$.

A thermal lambda may now be defined, by analogy with the solutal lambda: $\lambda_t = \Omega_t / 2p_t$.

Table IV (Cont'd) Thermal Supersaturation

Dimensionless Group	Remarks
$\Omega_t = \dfrac{\Delta T_H}{(L_v/C_p)}$	$\Omega_t = \Delta\Theta_H \quad \Omega_t \to 1, \Delta\Theta_H \to 1$ $\Delta\Theta_H = \Delta\Theta$ if $T_t = T_M$ or T_L

The Model for Dendritic Growth in a Supercooled Alloy Melt

General Background of the Present Model

We will assume that all heat flow and solute flow is unidirectional and occurs predominantly in the growth direction of the array, Fig. 4. We will distinguish between the thermal gradient in the interdendritic regions, G_i, and the thermal gradient in the bulk liquid ahead of the tips, G_L. The interdendritic concentration gradient is G_{cI}. Both G_i and G_{cI} are taken to be constant, Figs. 4b,c. The concentration gradient G_{cI} gives rise to a solute buildup, denoted in previous papers by the symbol ΔC_G, which may be estimated following the procedure described by Flemings and co-workers [16, 17]. The total solute buildup or supersaturation at the tip of the dendrite will depend on both this "unidirectional" or temperature gradient effect and the "point" effect of diffusion, Fig. 4d.

The "point" effect may be estimated from the Ivantsov solution or by adopting the simple idea first put forward by Zener to analyze the diffusion problem around a curved interface [27]. Since we seek a solution for both the temperature and concentration fields (both these fields must be evaluated to obtain an estimate of the "gradient of constitutional supercooling", appendix C, at the interface) the Zener approach thus involves the theoretical evaluation of two important dimensionless parameters, the solutal lambda, symbol λ, and a thermal lambda, symbol λ_t. These have been defined in appendix B.

The thermal and solutal lambdas may be lumped together to define an "effective lambda"; see eq. 11 in this section. The "effective lambda" so obtained is found to be a function of alloy composition. In very concentrated alloys, the "effective lambda", symbol λ_e, is approximately equal to the solutal lambda. In dilute alloys, the thermal lambda dominates.

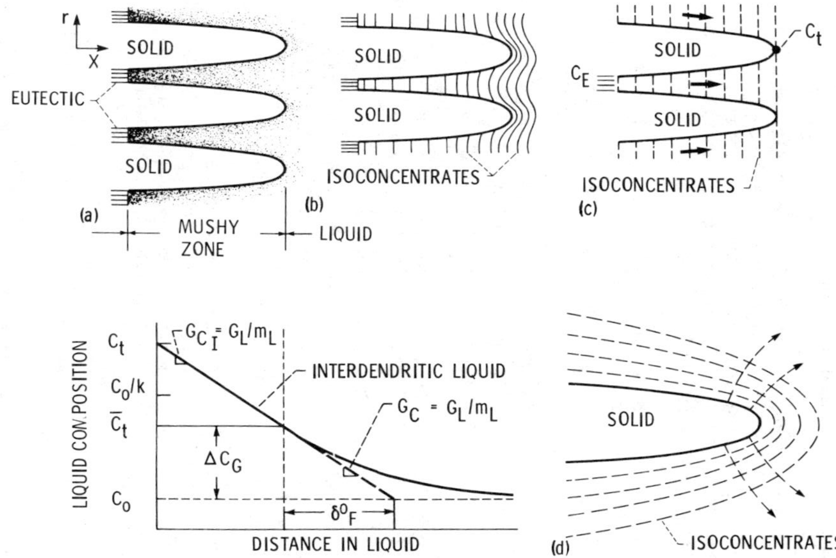

Fig. 4: Schematic illustration of basic solidification model for growth of an "array" of cells or dendrites. When the "point" effect of diffusion is neglected, isoconcentrates in the liquid phase will be planar right up to the tip, as in (c), and $G_{cI} = G_i/m_L = G_L/m_L$. When the "point" effect is included, $G_{cI} \neq G_L/m_L$, see refs (21) and (32).

A theoretical estimate for the solutal lambda has been obtained previously by the author when discussing the problem of "constrained" dendritic growth. (For "constrained" growth the "effective lambda" is simply equal to the solutal lambda). Here it is assumed that the solutal lambda has the same value even for "free" growth. This seems like a reasonable first approximation.

At this point there appear to be no fundamental physical principles, similar to those invoked to evaluate the solutal lambda, which could be used to obtain a theoretical estimate of the thermal lambda. For the moment, therefore, the behavior of the model will be discussed by treating the thermal lambda as an adjustable parameter.

Similar approaches, based on unknown fitting parameters (for example when kinetic effects, isotropic or anisotropic are discussed), have been proposed earlier in the literature. Also, the more detailed theoretical models for "pattern formation", which also discuss dendritic growth, contain a number of unknown parameters [26, 30, 31].

If the Ivantsovian description of heat and solute flow be accepted, however, the lambdas are no longer unknown. The solutal lambda is then simply given by $\lambda = \phi/2p$ and the thermal lambda is likewise given by $\lambda_t = \phi_t/2p_t$.

We thus have two models: a "Zenerian" model for which the solutal lambda $\lambda = 1/16$ and the thermal lambda λ_t is an unknown adjustable parameter, and, an "Ivantsovian" model for which the lambdas are as given above.

Tables 5 and 6 summarize the solute buildup obtained from these models. In general, the solute buildup is seen to be a function of both p and s. The results obtained by previous authors have also been included for comparison. The expressions in table 5 have been written in a form which allows easy comparison with the Flemings and the Zener solutions. The result obtained by the author may thus be viewed either as a modification of the Flemings solution to include the "point" effect or as a modification of the Zener solution to include the temperature gradient effect.

The expressions in table 6 have been written to allow easy comparison with the Ivantsov or Trivedi results. Again the expression obtained by the author may be viewed as a modification of the Ivantsov solution to include the temperature gradient effect. This result may be obtained by following the steps outlined in appendix B of ref. 20 with the exception that $\lambda = \phi/2p$ in the denominator of eq. (53). The final result so obtained differs from the earlier attempts at modifying the Ivantsov solutions, notably by Burden & Hunt [7], and Trivedi [18].

Comparison of the various expressions in table 6 indicates that the various models differ only in the value of the gradient function $N_g(p)$ which depends on p. Various models give various values of the gradient function which have been distinguished by the various superscripts: $N_g^o(p)$, $\hat{N}_g(p)$, etc. A general expression which includes the results obtained from both the "Zenerian" and "Ivantsovian" approaches is thus suggested and is given in the last row of table 6. Choosing different values of Ω^o and the gradient function gives the solute buildup given by all the models here. For the Flemings solution the gradient function is zero, or $\lambda = 0$. This also means $\Omega^o = 0$; see equation (4).

TABLE V. - COMPARISON OF SOLUTE BUILDUP EXPRESSIONS FOR VARIOUS DENDRITE GROWTH MODELS

MODEL	EXPRESSION FOR SOLUTE BUILDUP	REMARKS
FLEMINGS (1964) BOWER, BRODY, AND FLEMINGS (1966)	$\frac{\Delta C}{\Delta C_0} = s$	INDEPENDENT OF p $\Delta C \rightarrow \Delta C_0,\ s \rightarrow 1$ $\Delta C \rightarrow 0,\ s \rightarrow 0$
BURDEN AND HUNT (1974)	$\frac{\Delta C}{\Delta C_0} = s + 2pk$	$p \ll 1/2$ $\lambda = 1$ ASSUMED IMPLICITLY $\Delta C \rightarrow \infty$, AS $p \rightarrow \infty$ $\Delta C \rightarrow 2pk,\ s \rightarrow 0$
KURZ AND FISHER (1981) FISHER (1950)	$\frac{\Delta C}{\Delta C_0} = \frac{pk}{1 - p(1 - k)}$	$\lambda = 1/2$ ASSUMED IMPLICITLY, $\Delta C \rightarrow 0,\ p \rightarrow 0$ INTERDENDRITIC DIFFUSION NEGLECTED $\Delta C < 0$, AS $p \rightarrow \infty$
ZENER (1946)	$\frac{\Delta C}{\Delta C_0} = \frac{2p\lambda k}{1 - 2p\lambda(1 - k)}$	SUGGESTED $\lambda = 1$ $\Delta C < 0$, AS $p \rightarrow \infty$ $\Delta C \rightarrow 0,\ p \rightarrow 0$
LAXMANAN (1985)	$\frac{\Delta C}{\Delta C_0} = s + (1 - s)\left\{\frac{2p\lambda k}{1 - 2p\lambda(1 - k)}\right\}$	COMBINATION OF ZENER AND FLEMINGS SOLUTION, FOLLOWING BURDEN AND HUNT APPROACH. $\Delta C > 0$, AS $p \rightarrow \infty$

These results may also be expressed as a supersaturation, given by:

$$\Omega = \frac{\Omega^{o} k(1-s) + bs}{bs + k(1-s)} \quad (2)$$

$$\text{where} \quad b = 1 - \Omega^{o} (1-k) \quad (3)$$

When s = 0, i.e. when the temperature gradient effect is negligibly small, the supersaturation $\Omega = \Omega^{o}$, see figs. 5 and 6. On the other hand, when the "point" effect is negligible, $\Omega^{o} = 0$ and eq. (2) reduces to:

$$\Omega = \frac{s}{s + k(1-s)} \quad (4)$$

which is the same as the result obtained by Flemings [16, 17]. Note that both Ω and $\Delta C/\Delta C_0$ approach unity both when s = 1 and also when $\Omega^{o} = 1$. When $\Omega = 1$, the composition of the solid formed at the very tip of the dendrite approaches the initial alloy composition. This would lead to the formation of a segregation-free solid [32].

Fig. 5 plots the solute buildup given by the modified Ivantsov model of the present work as a function of p for various constant values of s. For low p, the solute buildup approaches asymptotically the Flemings value for all s. For large p, the solute buildup approaches the Ivantsov value for all s. Fig. 6 plots the same result as a supersaturation, versus p but for a single value of s for succinonitrile-1.3 wt% acetone (SCN-1.3 wt% Ace) alloy. This corresponds to a directional solidification experiment of Esaka & Kurz [8]. The experimental value of the supersaturation can be calculated from the measured values of the tip radius and tip temperature. This may be seen to fall almost exactly on the theoretical curve for the Flemings solution or the "Zenerian" model with $\lambda = 1/16$. The supersaturation predicted by the "Ivantsovian" modification of the present work is somewhat higher than the experimental value, whereas the "unmodified" Ivantsov solution predicts a supersaturation somewhat lower than the experiment value. The results here illustrate the importance of temperature gradient effects even during dendritic growth. Most authors have generally considered this effect to be unimportant in the dendritic regime, and assume a simple unmodified Ivantsov solution provides a satisfactory description: for example the model due to Lipton, Glicksman and Kurz [5] for "free" dendritic growth, the model due to Kurz, Giovanola and Trivedi for "constrained" growth [33], and a recent discussion by Trivedi and Somboonsuk [34].

Fig. 7 plots the solute buildup as a function of Peclet number as given by the Trivedi analysis for growth in a positive temperature gradient [18], for various values of s or G. When p becomes quite large, the Trivedi modification actually predicts a solute buildup which is less than the Ivantsov value. For sufficiently large values of s, in particular for $s > k$, the solute buildup becomes negative. The Trivedi dendrite thus behaves very much like the simpler Zener dendrite at large p in that $\Delta C < 0$ for both dendrites when p is large. These results are physically unrealistic since the solid composition should approach the initial alloy composition when both p and s become large. It has been suggested by the author [21] that this behavior of the Trivedi model may be related to the improper evaluation of the temperature gradient effect. (In reality, after the Trivedi result is combined with the marginal stability criterion, it is found that "difficulties" arise even for small p, if $s > k$. Thus tip radii

TABLE VI. - COMPARISON OF VARIOUS SOLUTE BUILDUP EXPRESSIONS FOR VARIOUS DENDRITE GROWTH MODLES

MODEL	EXPRESSION FOR SOLUTE BUILDUP	REMARKS
IVANTSOV (1947)	$\frac{\Delta C}{\Delta C_0} = \frac{\Phi k}{1 - \Phi(1 - k)}$	$\Phi = p \exp(p) E_1(p)$ $\Delta C \to 0,\ p \to 0$ $\Delta C \to \Delta C_0,\ p \to \infty$
TRIVEDI (1980)	$\frac{\Delta C}{\Delta C_0} = \frac{\Phi k}{1 - \Phi(1 - k)}\left[1 + \frac{1}{2}\mathscr{G} N_g(p) + \frac{2\mathscr{A}}{p^2} f(p)\right]$	FOR $\mathscr{A} \to 0$ MODEL YIELDS $\Delta C < 0$, AS $p \to \infty$, LIKE ZENER $N_g(p) \to -1$, AS $p \to \infty$
LAXMANAN (1985)	$\frac{\Delta C}{\Delta C_0} = \frac{2p\lambda k}{1 - 2p\lambda(1 - k)}\left[1 + \frac{1}{2}\mathscr{G} N_g^0(p)\right]$	$N_g^0(p) = \frac{1 - 2p\lambda}{2p\lambda}$ FOR $p \to \infty$, $\Delta C \to \Delta C_0$ IF $s \to 1$, OR $\mathscr{G} \to 2s/k$
PRESENT WORK (1986)	$\frac{\Delta C}{\Delta C_0} = \frac{\Phi k}{1 - \Phi(1 - k)}\left[1 + \frac{1}{2}\mathscr{G}\hat{N}_g(p)\right]$ $= s + (1 - s)\left\{\frac{\Phi k}{1 - \Phi(1 - k)}\right\}$	$\hat{N}_g(p) = \frac{1 - \Phi}{\Phi}$ COMBINATION OF IVANTSOV AND FLEMINGS
GENERAL EXPRESSION	$\frac{\Delta C}{\Delta C_0} = \frac{\Omega^0 k}{1 - \Omega^0(1 - k)}\left[1 + \frac{1}{2}\mathscr{G}\tilde{N}_g(p)\right]$	$\Omega^0 = 2p\lambda$ ZENER APPROACH $\Omega^0 = \Phi$ IVANTSOV APPROACH Ω^0 = SUPERSATURATION FOR ISOTHERMAL DENDRITE

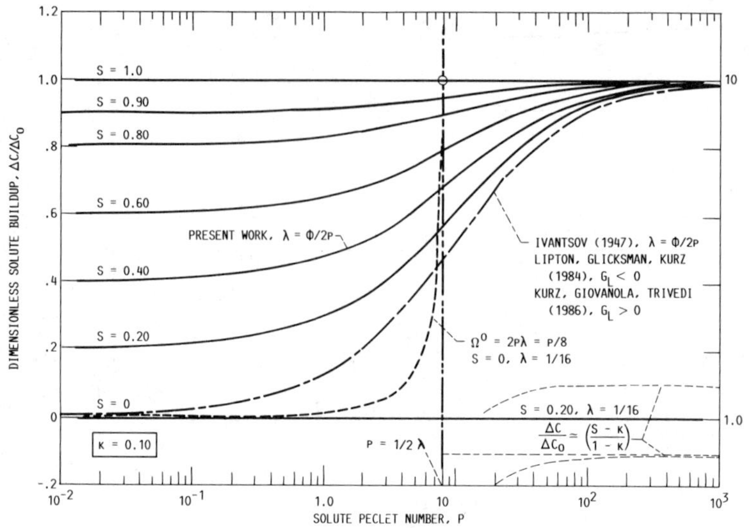

Fig. 5 : Dimensionless solute buildup $\Delta C/\Delta C_0$ versus solute Peclet number p for various fixed values of the Chalmers number, s according to the modified Ivantsov model of present work (variable λ-model, $\lambda = \phi/2p$). Also shown for comparison is the solute buildup given by the "Zenerian" model (constant $\lambda = 1/16$) of previous work by the author. Note that for large p, the $\Delta C > 0$ for the "Zenerian" model only if $s > k$. Large p behavior is plotted for this model for two values of s: $s = 0$ and $s = 0.20$. Partition ratio of $k = 0.10$ assumed in preparing these plots.

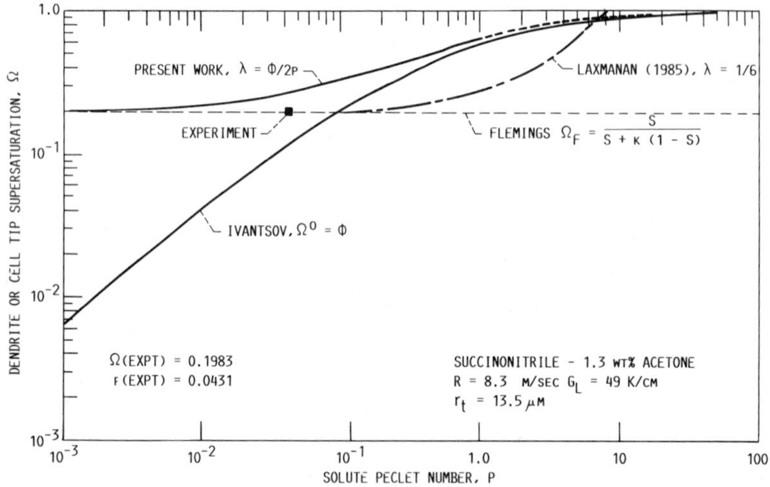

Fig. 6: Dendrite tip supersaturation Ω versus p for the "Zenerian" model ($\lambda = 1/16$) and the modified Ivantsov model ($\lambda = \phi/2p$) compared with the Flemings solution. Experimental data is from Esaka and Kurz [8].

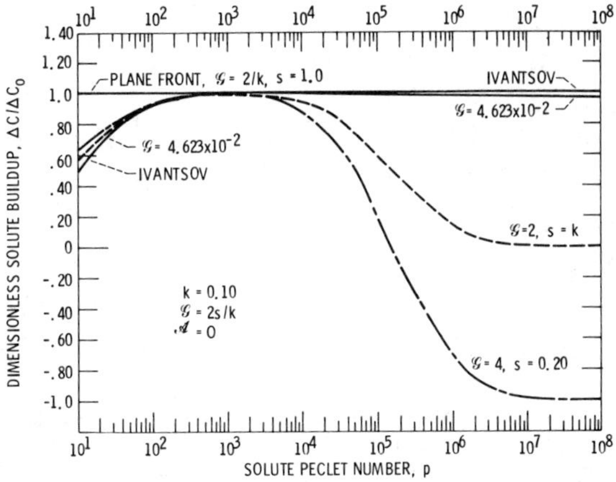

Fig. 7 : Dimensionless solute buildup versus p for the Trivedi analysis in the large p regime. For $s > k$, $\Delta C < 0$. Also for large p the solute buildup instead of approaching the Ivantsov value asymptotically is always less than the Ivantsov value.

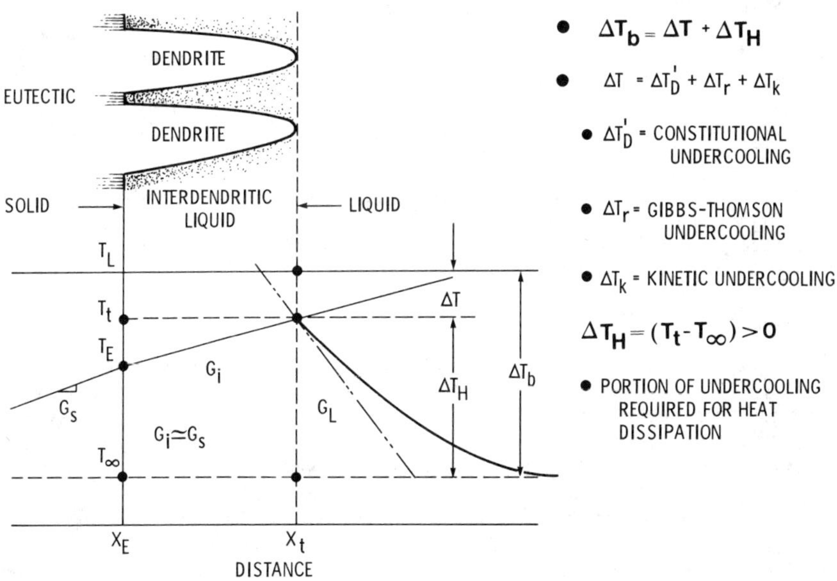

Fig. 8: "Arrayed" dendritic growth in a supercooled alloy melt.

predicted become "imaginary" and the gradient of constitutional supercooling becomes negative for $s > k$, see also appendix B).

It may be shown that the modification of the Ivantsov result as attempted by Burden & Hunt, eq. 32 in their paper, also leads to a behavior similar to that just described for the Trivedi model. Burden & Hunt's eq. (32) may be rewritten in a form remarkably similar to the Trivedi result:

$$\frac{\Delta T}{\Delta T_o} = \frac{\Omega^o k}{b}\left[1 + \frac{s}{k} N_g^{BH}(p) + \frac{2A}{p^2} f(p)\right] + \frac{2Ak}{p} \tag{5}$$

$$\text{where} \quad N_g^{BH}(p) = \frac{b-\phi}{\phi} \tag{6}$$

$\Omega^o = \phi$ and $f(p)$ is the function of Peclet number defined by Trivedi in ref. 18. Again this result differs from the results obtained by the author here,and the 1980 Trivedi result,only in the form of the gradient function. At least in the case of the Burden & Hunt result, it appears that some "algebraic problems" surface between their eq. (31) and the tip undercooling result of eq. (32). Retracing the algebra gives the modified Ivantsov result obtained by the author in table 6. ("Algebraic problems" also surface between their equations (15) and (16) and hence their result in eq. (17) must also be corrected. The corrected result is the same as the expressions for the "Zenerian" model in table 5 but with $\lambda = 1$. The significance of these modifications have been discussed in detail in ref. 20).

Finally, at least the results presented in eqs. (5) and (6) illustrate very clearly the advantage of expressing various theoretical results in dimensionless form. The last term in eq. (5), $2Ak/p$, gives the undercooling due to the Gibbs-Thomson effect. This term is usually quite small compared to the solutal undercooling represented by the first term. However, if the Gibbs-Thomson effect is excluded there will be no minimum in the tip undercooling for a fixed value of R and G_L, Fig. 5. Trivedi has ignored this contribution in the theoretical development of his model, see eqs. (32) and (33) in his 1980 paper [18], while retaining the term containing and $f(p)$ within the square brackets in eq. (5). This term gives the solute buildup at the tip due to the varying curvature of the dendrite surface. The solute buildup at the tips due to this effect is usually quite small. It is this term which may be neglected without any loss of accuracy. Terms involving the capillary number A are therefore absent in the expressions for the solute buildup obtained by the author in tables 5 & 6 or the expression for the supersaturation given in eq. (2).

The Transport Solution

The basic expression relating the dendrite tip undercooling, $\Delta T = (T_L - T_t)$, and the bath supercooling $\Delta T_b = (T_L - T_\infty)$ is called the transport solution. This may be easily obtained, as follows, in terms of the three dimensionless parameters p, s and σ_c, and, two additional dimensionless quantities - the solutal and thermal lambdas, λ and λ_t.

The total bath supercooling is simply the sum of the dendrite tip undercooling and the undercooling, ΔT_H, required to transport the latent heat of fusion away from the tip fig. 8. This may be expressed in dimensionless form:

$$\frac{\Delta T_b}{\Delta T_o} = \frac{\Delta\Theta}{\Delta\Theta_o} = \frac{\Delta T}{\Delta T_o} + \frac{\Delta\Theta_H}{\Delta\Theta_o} \tag{7}$$

The interdendritic thermal gradient, G_i is related to the Chalmers number, s by:

$$s = \frac{D_L G_i}{R\Delta T_o} = -\frac{D_L(G_i/m_L)}{R\Delta C_o} \tag{8}$$

where, G_i, may be either positive or negative. It will be assumed here that $G_i > 0$. Also, the thermal gradient in the fully solid region, $G_s \simeq G_i$. Hence a knowledge of the thermal gradient in the fully solidified material or the thermal gradient in the interdendritic regions will specify the value of the Chalmers number.

The total tip undercooling will be the sum of the solutal undercooling and the undercooling due to the Gibbs-Thomson effect. (It is assumed here that the kinetic undercooling is negligibly small). Next, the assumption of "local" equilibrium at the tip provides a relationship between the dendrite tip undercooling and the solute buildup at the tip in terms of the corresponding values for a planar interface, Fig. 9:

$$\frac{\Delta T}{\Delta T_o} = \frac{\Delta C}{\Delta C_o} + \frac{2\ell_c k}{r_t} = \frac{\Delta C}{\Delta C_o} + 2pk(\sigma_c) \tag{9}$$

Here it is assumed that both the liquidus and solidus lines are depressed uniformly, due to curvature, across the phase diagram [31]. In equation (9) above it will be recognized that only the term $\Delta C/\Delta C_o$ depends on the detailed assumptions made by the specific model used to describe dendritic growth. $\Delta C/\Delta C_o$ is the solutal or constitutional undercooling in dimensionless form: $\Delta T_D' = -m_L\Delta C$ and $\Delta T_o = -m_L\Delta C_o$.

Combining equations (7), (8), (9) above, with the general expression for $\Delta C/\Delta C_o$ yields:

$$\frac{\Delta T_b}{\Delta T_o} = \frac{\Delta\Theta}{\Delta\Theta_o} = s + 2pk(\sigma_c + \lambda_e) \tag{10}$$

where λ_e may be thought of as an "effective" lambda characteristic of dendritic growth in a supercooled alloy melt. λ_e includes both the solutal and thermal lambdas being given by:

$$\lambda_e = \lambda \left[\left(\frac{1-s}{b}\right) + \left(\frac{D_L/\alpha_L}{k\Delta\Theta_o}\right) \frac{\lambda_t}{\lambda} \right] \tag{11}$$

Note that λ_e also involves the dimensionless parameter, $\mathcal{B}$, discussed separately in Appendix A.

---

* The expression for $\Delta C/\Delta C_o$ is the more fundamental form of the transport solution.

Equation (10) above is a key result - the transport solution - a key element of this solution is the "effective lambda", λ_e, identified above. This will be discussed next.

The "Effective" Lambda

In general, one would expect the thermal and solutal lambdas to be a function of composition and the imposed growth conditions. However, it is entirely possible that lambdas are reasonably constant, as already demonstrated for "constrained" growth in alloys [20-22]. If the Ivantsov description of heat and solute diffusion is accepted then the ratio λ_t/λ is given by:

$$\frac{\lambda_t}{\lambda} = \frac{\phi_t}{\phi}\left(\frac{p}{p_t}\right) = \frac{\phi_t/\phi}{D_L/\alpha_L} \tag{12}$$

On the other hand, if the lambdas are constant, it may be easily seen from equation (11) above that for small values of the Peclet numbers (both thermal and solutal) and the Chalmers number s, λ_e is only a function of composition; $\lambda_e \simeq \lambda(1 + B)$ since $\phi_t \simeq p_t$, $\phi \simeq p$.

Figure 10 plots the effective lambda as a function of composition for SCN-Ace alloys for constant values of the ratio λ_t/λ. For concentrated alloys, the "effective" lambda is approximately equal to the solutal lambda. Hence thermal effects as well as solutal effects are important only in dilute alloys, whereas in concentrated alloys only solutal effects will be predominant. In extremely dilute alloys, thermal effects will completely dominate solutal effects; λ_e being essentially a function of the thermal lambda in this case. The results here clearly demonstrate the usefulness of the central concept of lambda introduced in this paper (actually goes back to Zener, see Appendices A and C).

For a fixed composition, the effective lambda is also a function of the Lewis number (the ratio of the chemical and thermal diffusivities). Hence in metallic alloys ($D_L << \alpha_L$, $\lambda_e \simeq \lambda$) solutal effects are more important, even for very small additions of solute elements, than in organic systems such as SCN-Ace. The extrapolation of results obtained in organic systems to metallic systems must therefore take into account these differences, as lumped together here in the form of the dimensionless parameter, B. This parameter includes the equilibrium partition ratio and so accounts for differences between solutes (large k solutes versus small k solutes). It also includes the dimensionless freezing range and thus includes thermal influences as well through the term L_v/C_p, see also Figure 19 in Appendix A.

Behavior of the Transport Solution

For fixed values of the bath supercooling, s and C_0 the "transport solution", equation (10) above, indicates that the tip stability parameter σ_c in general decreases with increasing values of the Peclet number p, Fig. 11. The "operating point" of the dendrite lies somewhere along this curve. The transport solution, however, provides only one relationship between these two unknowns - σ_c and p - which include between them the two key dendrite characteristics of interest, the tip radius r_t and the dendrite tip growth rate, R. An additional relationship between σ_c and p is clearly required to uniquely specify R and r_t. Such a relationship is provided by

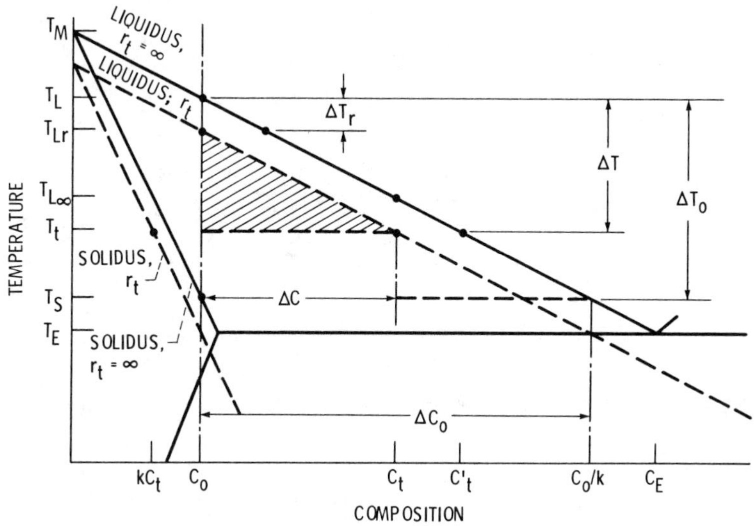

Fig. 9: Relationship between tip undercooling and solute buildup at tips.

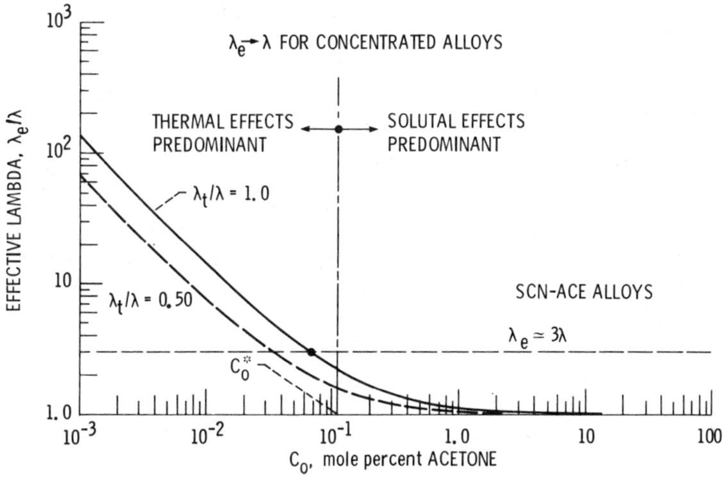

Fig. 10: Variation of "effective" lambda, λ_e for SCN-Ace alloys as a function of composition assuming $\lambda_t/\lambda = 1.0$ and $\lambda_t/\lambda = 0.50$.

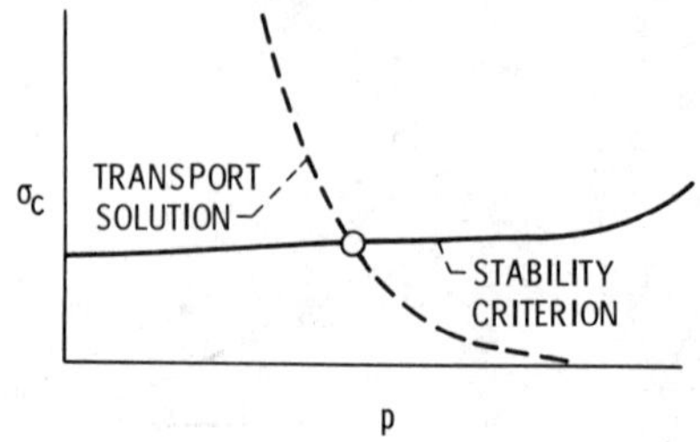

Fig. 11: Determination of dendrite tip radius from marginal stability hypothesis.

the marginal stability hypothesis. This will be discussed in the next section.

It may be noted here that other hypotheses such as minimum undercooling [20-21], minimum entropy production or maximum growth velocity [3] have also been used to determine the tip radius and growth rate.

Comparison of Transport Solutions for "Free" and "Unconstrained" Growth

The transport solution for the "constrained" growth problem obtained previously by the author [20] is given below in dimensionless form:

$$\frac{\Delta T}{\Delta T_o} = s + 2pk \left[\sigma_c + \frac{\lambda(1-s)}{b}\right] \qquad (13)$$

Comparing equations (10) and (13) it is seen that the "effective lambda" for the constrained growth problem involves only the solutal lambda, i.e. only the first term in equation (11) above. Stated differently, the basic equations describing "constrained" growth and "free" growth are remarkably similar in the case of concentrated alloys, with the "tip undercooling" playing exactly the same role as the "bath supercooling". However, since $\Delta T << \Delta T_b$, detailed predictions of dendrite features will clearly be quite different, even for the same alloy. Equations (10) and (13) highlight both the important similarities and the differences between "constrained" growth and "free" growth.

Determination of the Dendrite Tip Radius, Tip Temperature and Tip Composition: The Marginal Stability Hypothesis

Tip Radius

Kurz and Fisher [19] were the first to apply the marginal stability criterion (proposed by Langer and Muller-Krumbhaar for dendritic growth in a supercooled pure melt [23]) to the problem of "constrained" dendritic growth in an alloy. The Lipton, Glicksman and Kurz (LGK) analysis employs exactly the same approach to "free" growth in alloys. Both models, however, implicitly assumed that the Peclet number, p, is small. A more general stability criterion, without the small p approximation, is easily obtained and is given below, see also Ref. 21.

$$r_t = 2\pi \left[\frac{\Gamma}{(m_L G_c - \bar{G}) + m_L G_c g(\omega)} \right]^{1/2} \tag{14}$$

Equation (14) above follows directly from the basic expression for stability of planar interface given by Mullins and Sekerka, equation (20) in Ref. 25. Here it is assumed (following Langer and Muller-Krumbhaar) that the tip radius is smallest wavelength at the onset of instability at the planar interface. The function $g(\omega)$ in equation (14) depends only on the wave number, ω, of the perturbation and is exactly the same as defined by Mullins and Sekerka, equation (22) in Ref. 25. It is given below in terms of $p = Rr_t/2D_L = \pi R/\omega D_L$.

$$g(\omega) = - \frac{2k}{(2k-1) + (1 + 4\pi^2/p^2)^{1/2}} \tag{15}$$

For small p, $g(\omega) \rightarrow 0$ whereas for large p $g(\omega) \rightarrow -1$. Equation (14) here has also been employed recently by Lipton, Kurz and Trivedi for analyzing "free" growth and also by Kurz, Giovanola and Trivedi to analyze "constrained" growth [5, 33]. The assumptions leading to equation (14) and the results that follow from it, however, lead to some important fundamental issues which are beyond the scope of this paper and will be discussed elsewhere [22, 36].

Equation (14) may be rewritten as follows, in dimensionless form, to yield a second relationship between σ_c and p, i.e. R and r_t:

$$\sigma_c = \frac{1}{2\pi^2} \left[\frac{D_L}{Rk\Delta T_o} \{ (m_L G_c - \bar{G}) + m_L G_c g(\omega) \} \right] \tag{16}$$

Figure (11) also plots schematically the behavior of the stability criterion. Typically, σ_c obtained from equation (16) increases with p. The "operating point" of the dendrite is described uniquely by the intersection of the plots given by the transport solution and the stability criterion.

Appendix D provides a summary of some of the approximate expressions for the tip stability parameter, tip radius and tip growth rate obtained from equation (16). These approximate expressions will be shown to be in good agreement with available experimental data during "free" growth in alloys, in the next section.

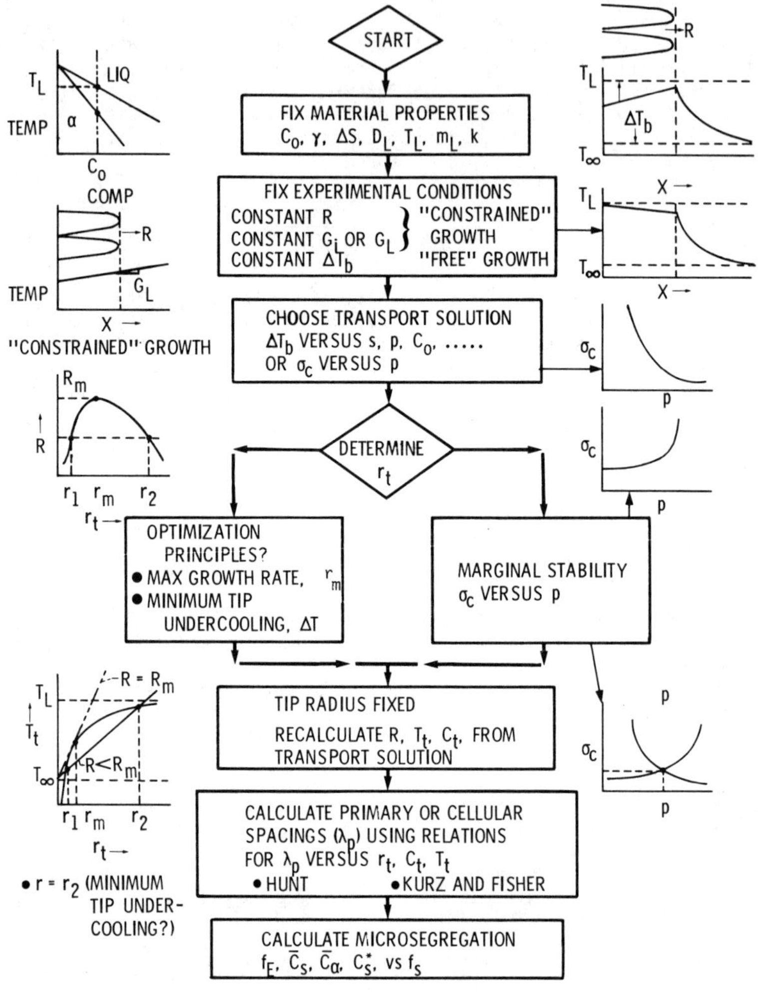

Fig. 12: Flow chart for calculation of various dendrite characteristics and microsegregation behavior during both "free" growth and "constrained" growth.

Tip Temperature and Tip Composition

Having determined the tip radius and the growth rate, other dendrite features such as the tip temperature and the tip composition may be obtained from the "transport solution". These may then be used to predict the primary and secondary arm spacings and the detailed microsegregation behavior. Figure 12 shows this calculation scheme in the form of a flow chart. In this paper only the tip radius and tip growth velocity predictions will be examined. As noted earlier in the introduction, although detailed microsegregation measurements have been reported for a number of metallic alloy systems [7, 37], no measurements of tip temperature and tip composition during "free" growth in alloys seem to have been made. Such measurements are required to verify completely the validity of the various transport solutions proposed, particularly since the tip radii predicted by various models (based on the marginal stability criterion) are very nearly indistinguishable, see also Refs. 3, 21 and 22.. Ref. 21 discusses the identical predictions of tip radii obtained from both minimum undercooling and marginal stability criteria during "constrained" growth.

Experimental Results: Comparison with Theory

In this section experimental data in binary SCN-ACE alloys will be compared briefly with the theoretical results summarized in the last section. A more detailed comparison will be presented in a later publication.

Figure 13 plots σ_c versus p as given by both the transport solution and the marginal stability criterion for SCN-0.40 mol% Ace alloy for a fixed bath supercooling of 0.50 K. Here it has been assumed that the interdendritic thermal gradient and concentration gradient are negligibly small, s = 0. Also, the solutal lambda and the thermal lambda are assumed to be given by the Ivantsov solution, $\lambda = \phi/2p$ and $\lambda_t = \phi_t/2p_t$. The present model then becomes identical to the LGK model

The theoretically predicted growth rate and tip radius, obtained from the intersection point are: R = 50.2 μm/sec and r_t = 7.5 μm. These are in good agreement with the experimentally measured values of R = 40 μm/sec and r_t = 8.5 μm [11]. Note that the tip velocity is slightly higher than the observed value. This is even more evident from figure 14 which plots σ_c versus p for the same alloy at a higher bath supercooling of 0.90 K. In this case the theoretically predicted growth rate R = 257.2 μm/sec, whereas the experimentally observed value is only about 100 μm/sec. The predicted tip radius of r_t = 3.24 μm is, however, again in very good agreement with the observed value of r_t = 3 μm.

Figures 15 through 16 take an entirely different approach to comparing theory and experiment. Here it is assumed that:

i) the solutal lambda is constant and has exactly the same value as obtained theoretically for "constrained" growth [20, 21]: $\lambda = 1/16$.

ii) the thermal lambda, λ_t, is unknown and at this point still needs to be theoretically evaluated. ***It is hypothesized here that the thermal lambda is perhaps also a constant.*** This hypothesis will be tested here.

Figure 15 plots the predicted tip radius versus bath supercooling for a dilute alloy, SCN-0.045 mol% Ace, for $\lambda = 1/16$ and various values of λ_t.

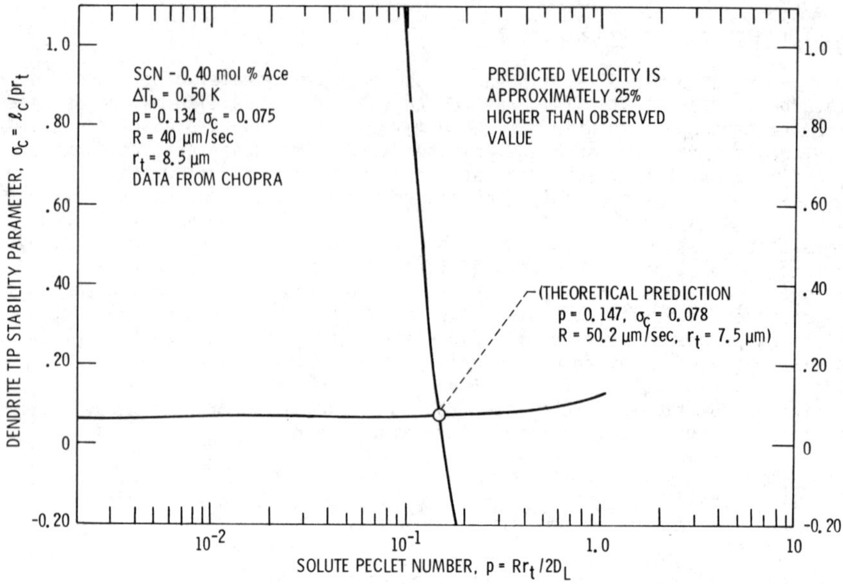

Fig. 13 - Operating point of dendrite determined by intersection of σ_c versus p curves from transport solution and marginal stability criterion, assuming $\lambda = \Phi/2_p$, $\lambda_t = \Phi_t/2p_t$ s = 0 and $g(\omega) = 0$. Present model is identical to LGK model with these assumptions. SCN - 0.4 mol% Ace, $\Delta T_b = 0.50$ K.

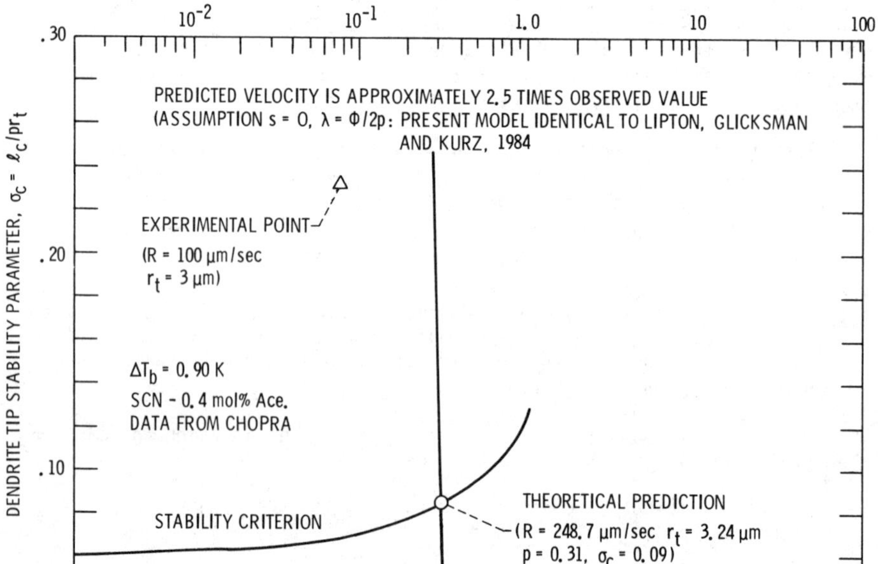

Figure 14 - Operating point of dendrite determined by intersection of σ_c versus p curves from transport solution and marginal stability criterion, assuming $\lambda = \Phi/2p$, $\lambda_t = \Phi_t \Phi 2p_t$, s = 0 and $g(\omega) = 0$. Present model is identical to LGK model with these assumptions. SCN - 0.4 mol% Ace, $\Delta T_b = 0.90$ K.

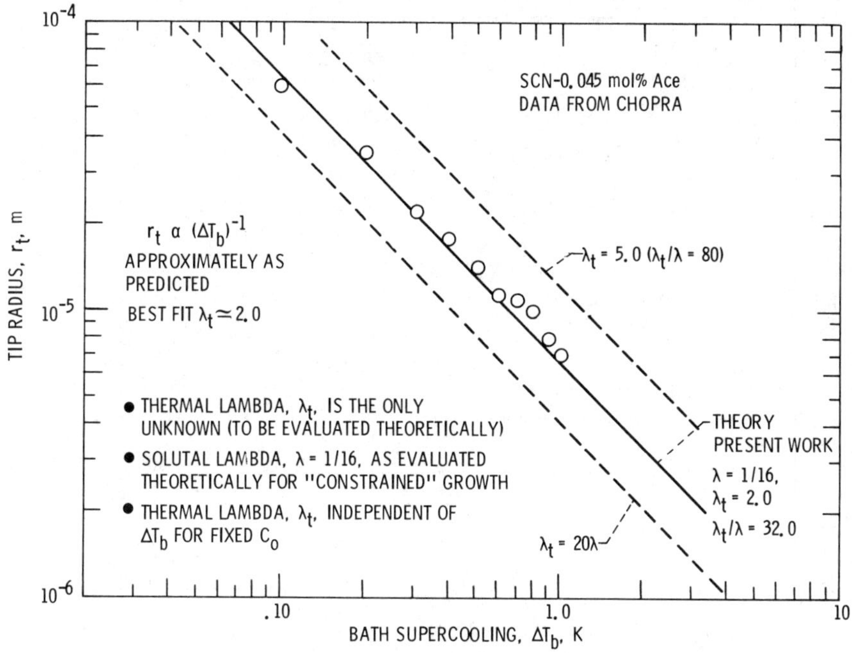

Figure 15 - Dendrite tip radius versus bath supercooling for SCN-0.045 mol% Ace. alloy. Marginal stability predictions based on approximate result of present model assuming constant λ and λ_t. Note $r_t \alpha (\Delta T_b)^{-1}$.

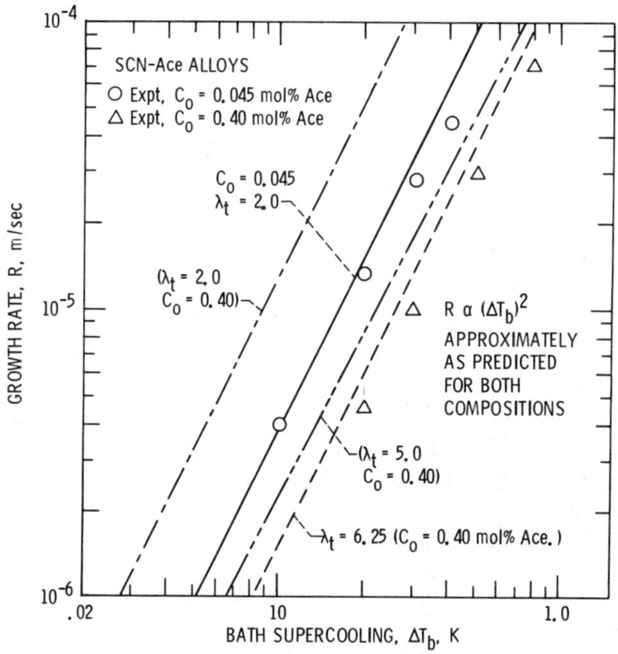

Fig. 16: Dendrite tip growth rate versus bath supercooling for binary SCN-0.4 mol% Ace. Marginal stability predictions of present model assuming constant λ and λ_t. Note: $R \alpha (\Delta T_b)^2$ for fixed composition.

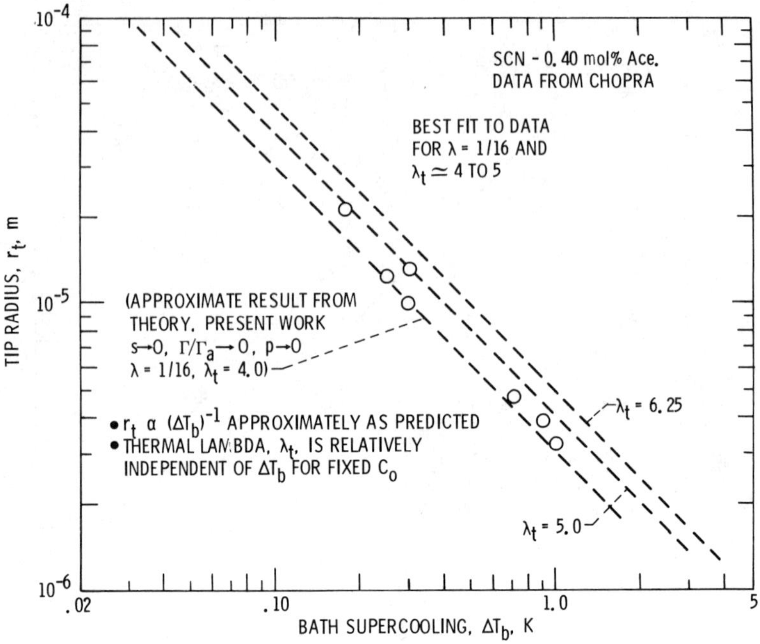

Figure 17 - Dendrite tip radius versus bath supercooling for binary succinonitrile-0.40 mol% Acetone alloy. Marginal stability predictions based on approximate result of present model. Constant λ and λ_t assumed. λ = 1/16. Note that $r_t \alpha (\Delta T_b)^{-1}$

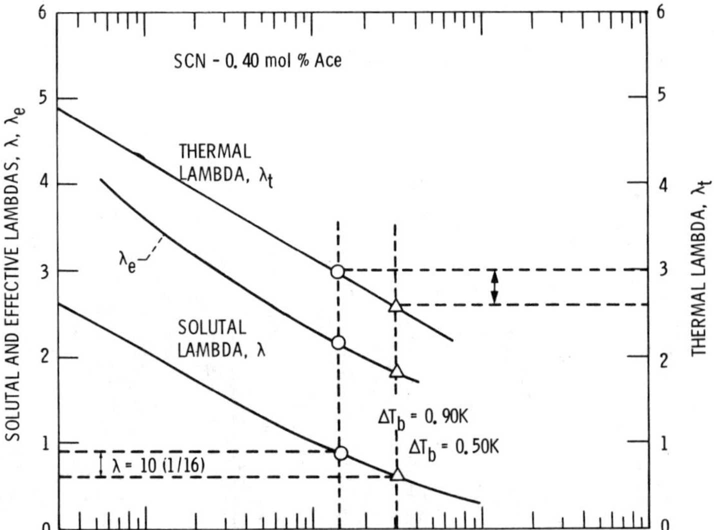

Figure 18 - Lambdas predicted for Ivantsov dendrite (present model) marginal stability calculations. Lambdas are reasonably constant, independent of ΔT_b, fixed C_o.

As expected from the approximate expression for the tip radius, equation (31) in Appendix D, the experimentally measured tip radii do vary approximately as the inverse of the bath supercooling. The best theoretical fit is obtained for $\lambda_t = 2.0$. For comparison, the theoretical predictions for $\lambda_t = 5.0$ or $\lambda_t/\lambda = 80$ and for $\lambda_t/\lambda = 20$ are also indicated here.

Figure 16 compares the predicted growth rates for the same alloy and for the same values of λ_t and λ. Again a thermal lambda of 2.0 and a solutal lambda of 1/16 may be seen to satisfactorily predict the measured growth velocities. Both these comparisons indicate that for a fixed alloy composition, both the solutal and thermal lambda's are approximately constant and independent of the bath supercooling. Also, as expected from the approximate result given by equation (30) in Appendix D it is seen that the measured growth rates do vary approximately as the square of the bath supercooling.

Figure 17 plots the predicted tip radii as a function of the bath supercooling for the concentrated alloy, SCN-0.40 mol% Ace. Again, the experimental data can be satisfactorily explained by assuming a slightly smaller value of the thermal lambda, $\lambda_t = 5.0$, than the value of $\lambda_t = 6.25$ which yields the best fit for the tip velocities in this alloy. It may be concluded again that the thermal lambda is reasonably constant for the concentrated alloy and independent of the bath supercooling.

Figure 18 plots the thermal, solutal and "effective" lambda's as given by the Ivantsov solution. The effective lambda here is calculated from equation (11). $\lambda = \phi/2p$ and $\lambda_t = \phi_t/2p_t$. The calculations are for a fixed composition, SCN-0.4 mol% Ace. At the operating point of the dendrite (as already determined in Figures 13 and 14), for a supercooling of 0.50 K and 0.90 K, the thermal lambda is again reasonably constant, being about 3 for the higher supercooling and about 2.6 at the smaller supercooling. The solutal lambda is more than an order of magnitude greater than the value $\lambda = 1/16$, varying between 0.9 and 0.625. These differences are simply because the Ivantsov solution involves a different set of assumptions from the Zener solution. In particular, the Ivantsov solution assumes the tip to be a paraboloid of revolution and thus yields the concentration gradient and solute buildup at the tip given by equations (20). The Zener solution makes no assumptions regarding the tip shape (only the tip radius or tip curvature is assumed) but assumes the concentration gradient and thermal gradient will be dissipated within a certain fixed multiple or fraction of the tip radius, equation (19) and (21).

Since both tip radii and growth velocities may be satisfactorily predicted using either approach, as discussed here, it is clear that further measurements of tip temperature and tip composition are required to distinguish between these two approaches. The theoretically predicted tip temperatures and compositions depend also on λ_e and so will be quite different for the Ivantsov and Zener solutions.

Finally, the theoretical analysis presented here will be complete only if the thermal lambda is also evaluated, following perhaps the procedure adopted earlier for the solutal lambda [20, 21].

Summary and Conclusions

1. A simple model for growth of a dendritic array into a supercooled alloy melt has been presented. Various models proposed in the literature for both "free" and "constrained" growth in alloys have been shown to follow as special cases of a general formulation attempted here.

2. A summary of the important length scales of the dendrite problem in pure melts and alloys is provided. A number of important dimensionless groups may be obtained by various combinations of these length scales. The model for dendritic growth is presented in terms of these dimensionless groups. An examination of the relationships between these dimensionless groups, as obtained by various models, also helps to verify the theoretical consistency of the models. This may be seen from the various tables provided here.

3. Two important dimensionless groups, λ_e and B, identified here are unique to the problem of the "free" dendritic growth in alloys. During "free" growth in concentrated alloys, the "effective lambda", $\lambda_e \rightarrow \lambda$. This is because the dimensionless parameter $B \rightarrow 0$ for concentrated alloys. Also, a comparison of the transport solutions for "free" growth and "constrained" growth indicates that $\lambda_e \simeq \lambda$ during "constrained" growth.

4. The B-number may also be used to define quantitatively the fundamental concept of infinite dilution. An alloy may be considered to be infinitely dilute if $D_L/\alpha_L << k\Delta\Theta_0$. Here k is the equilibrium partition ratio for the solute added, D_L and α_L are respectively the chemical diffusivity of the solute in the liquid and the thermal diffusivity of the liquid alloy in the presence of the solute.

5. The B-number is also exactly equal to the ratio of the tip stability parameter, σ_c, for the alloy and the tip stability parameter, σ^* for the pure melt. (This identity only applies if the ratio $\gamma/\Delta S$ is independent of alloy composition. Note the separate values of the surface tension and the volumetric entropy of fusion are not required to be a constant or independent of composition). A qualitative examination of the above relationship shows that all dendrite growth models, regardless of the assumptions used to determine the tip radius, are expected to predict the same behavior, as long as the analysis includes the various length scales of the dendrite problem. Thus, a maximum in the dendrite tip growth velocity and a minimum in the tip radius is predicted for dilute alloys, in agreement with experimental observations in binary succinonitrile-acetone and H_2O-NaCl solutions.

6. The dimensionless parameter B also indicates the important differences to be expected between metallic alloys and other organic and inorganic solutions such as succinonitrile-acetone and H_2O-NaCl. The ratio of the chemical and thermal diffusivities is quite different for these systems, thus yielding different values of and therefore, the "effective lambda". An extrapolation the results obtained from such systems to describe the behavior of commercial metallic (or semiconductor) alloy systems must take into account this difference in the behavior of B-number.

7. Approximate expression for the tip radius and tip velocity have been obtained, based on marginal stability considerations. These apply for small values of the dimensionless parameters s, B and the Peclet number, p. These approximate expressions indicate that the tip radius is expected to vary as the reciprocal of the bath supercooling for a fixed composition and that

the growth velocity is expected to vary as the square of the bath supercooling at a fixed composition. Both these predictions are shown to be in agreement with experimental observations.

8. A detailed comparison of the approaches taken by Zener and Ivantsov to describe the diffusion of solute or heat around a dendrite tip is provided. The dimensionless parameters λ and λ_t discussed above simply set the concentration gradient and the thermal gradient at the tip. (λ and λ_t may also be thought of as representing a dimensionless solute or thermal boundary layer thickness; in units of the tip radius). Choosing different values for λ and λ_t yields the results obtained by previous investigators for solute and heat diffusion around a dendrite tip.

9. "Free" growth, involves both the thermal and solutal lambda s, as lumped together in a quantity called "effective" lambda, λ_e . <u>The thermal lambda still needs to be theoretically evaluated</u>. Nevertheless, experimental data are shown to be consistent with the assumption of a constant value of the solutal lambda. (For the Ivantsov dendrite, both λ and λ_t decrease continuously with increasing Peclet numbers). The data also indicates that the thermal lambda is independent of the bath supercooling for a fixed composition. However, it appears that the thermal lambda, at a fixed supercooling, may be a function of composition. It is suggested that the maximum in the growth velocity and the minimum tip radius observed in dilute alloys may be related to this behavior of the thermal lambda (or a suggestive coupling of the thermal and solutal diffusive fluxes, particularly in dilute alloys).

<u>Acknowledgements</u>

The author is grateful for financial support provided by NASA Lewis Research Center under the "Microgravity Science and Applications Program". Special thanks are also due to Dr. Hugh R. Gray, Chief,Metals Science Branch, NASA Lewis, for his constant encouragement and support, to Dr. M.A. Chopra for kindly providing the experimental data used here and to Mrs. Carol J. Thomas for her meticulous typing and assistance in the preparation of the manuscript. Finally, the author would also like to thank the reviewers of this paper for their constructive comments, suggestions, in particular with regard to table 3, part 2, on hypercooling conditions in alloys.

APPENDIX A

Significance of The Dimensionless Parameter B

Figure 19 plots the dimensionless parameter B versus composition for a number of alloy systems. Here it has been assumed that the equilibrium partition ratio and both the liquidus and solidus slopes are constant.

Equation (1) and Figure 19 highlight the important differences to be expected from experimental measurements in metallic alloys and inorganic systems such as SCN-Ace. For the metallic alloys, the ratio D_L/α_L is much smaller than for the organic system. Hence for the same value of the dimensionless freezing range, $\Delta T_o/(L_v/C_p) = \Delta\theta_o$, B and hence the measured values of the tip stability parameter, σ_c, are expected to be much smaller in metallic alloys (when scaled with the corresponding pure melt value), than in inorganic systems. Reliable experimental measurement in both metallic and inorganic systems are, however, lacking to verify this important prediction.

Equation (1) also explains qualitatively that tip radii in very dilute alloys will be much sharper (smaller r_t) than in the pure melt case, since σ_c will be much larger than σ^*. Smaller values of r_t correspond to higher values of σ_c, but so do larger values of R. This, therefore, implies much faster growth rates in the infinitely dilute alloy compared to the pure melt. Both these observations have already been confirmed experimentally by Glicksman and co-workers [10, 11]. Also equation (1) implies that for very concentrated alloys σ_c will be much smaller than σ^*. Thus, tip radii will be much coarser, and correspondingly the tip growth rates much smaller, than for the pure melt case. Clearly, this implies a minimum in the tip radius and a maximum in the growth velocity with increasing compositions, an observation which is again in agreement with experiment [10-13].

More importantly, the above conclusions follow directly from a qualitative examination of the relevant length scales and dimensionless groups. Hence, these observations are very general and <u>do not depend</u> particularly, on any detailed assumptions or hypotheses (such as marginal stability) which govern dendrite behavior.

<u>All dendrite growth models</u>, regardless of the assumptions used to determine the tip radius (marginal stability, minimum undercooling, maximum velocity, minimum entropy production), <u>are therefore expected to predict qualitatively the same behavior</u>, as long as they include the above characteristic length scales in the analysis. They will differ only in <u>the quantitative details</u> such as the exact values of R, r_t, the composition at which the extremum values of R and r_t are attained, etc.

The differences in details of the predictions are mainly due to the different assumptions invoked to describe the transport of heat and solute, Ω^o (or λ and λ_t, as discussed in Appendix B) and the behavior of the various gradient functions. To iterate, it is shown from simple qualitative arguments that two important experimental observations during "free" dendritic growth in alloys (a maximum in the growth velocity and a minimum in the tip radius for dilute alloys) can be explained without necessarily resorting to marginal stability considerations. A satisfactory prediction of these trends has been considered to be a crucial ingredient of a satisfactory model for dendritic growth in supercooled alloy melts [3, 38].

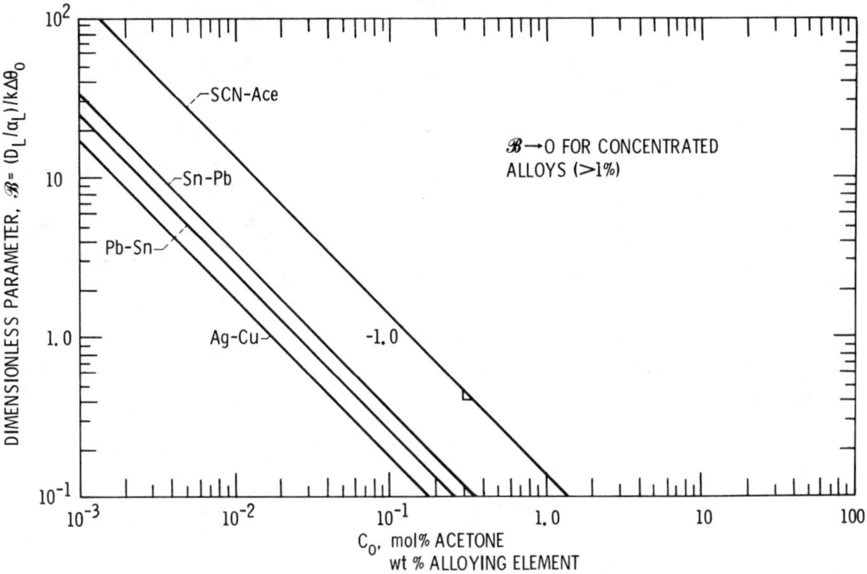

Figure 19 - 𝓑-number versus composition for various alloys.

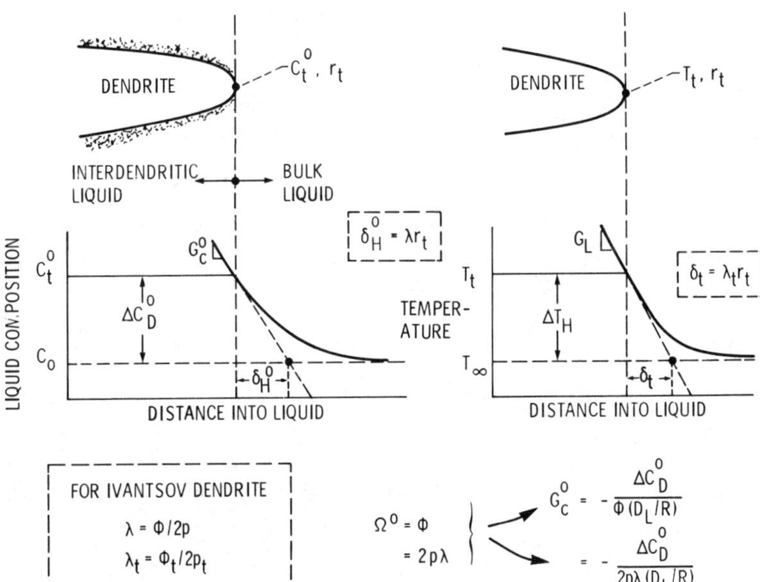

Figure 20 - Definition of solutal and thermal Lambdas. CS-86-0929

Finally, the dimensionless parameter may also be used to define quantitatively the concept of infinite dilution. An alloy is infinitely dilute if D_L/α_L is << $k\Delta\Theta_o$, i.e. if B is large, since for $C_o \to 0$, $\Delta\Theta_o \to 0$.

APPENDIX B

The Dimensionless Parameter λ. Definition of Solutal and Thermal Lambdas.

Consider the special case of an isothermal liquid. The dendrite surface is now also an isoconcentrate. A simple solute balance (assuming negligibly small diffusion in the solid phase) at the tip yields:

$$RC_t^o(1-k) = - D_L G_c^o \tag{17}$$

where the superscript o is a reminder of the isothermal conditions ($G_i = 0$) behind the tip, Fig. 20.

Following Zener [27, 28], the concentration gradient G_c^o may be expressed as the solute buildup ΔC_D^o divided by an "effective" diffusion distance, λr_t, when $\lambda > 0$. Thus,

$$G_c^o = - \frac{\Delta C_D^o}{\lambda r_t} \tag{18}$$

Combining (17) and (18) gives

$$G_c^o = - \frac{\Delta C_D^o}{\Omega^o(D_L/R)} = - \frac{\Delta C_D^o}{2p\lambda(D_L/R)} \tag{19}$$

To completely specify C_t^o and G_c^o Zener, however, subsequently suggested $\lambda = 1$, an assumption also retained by Burden and Hunt, and, Hunt [7, 39].

The concentration gradient at the tip of the Ivantsov dendrite may likewise be written as:

$$G_c^o = - \frac{\Delta C_D^o}{\Omega^o(D_L/R)} = - \frac{\Delta C_D^o}{\phi(D_L/R)} = - \frac{\Delta C_D^o}{(\phi/2p)r_t} \tag{20}$$

Comparing equations (18) & (20) it may be seen that different values of Ω^o, and thus λ, describe different modes of solute diffusion at the tip. Here for Ivantsov model $\lambda = \phi/2p$.

For heat diffusion, a thermal lambda, λ_t, may be defined, as illustrated in Figure 20b, which sets the thermal gradient at the tip. Note that the dendrite surface is again assumed to be isothermal but having a temperature $T_t < T_L$

$$G_L = - \frac{\Delta T_H}{\lambda_t r_t} = - \frac{(T_t - T_\infty)}{\lambda_t r_t} \tag{21}$$

Equation (21) follows from a simple heat balance at the tip, assuming negligible heat flow through the solid phase. A dimensionless expression relating the tip temperature and the bath temperature, T_∞, is (this is analogous to the dimensionless solute buildup

$$\Delta\theta_H = \frac{\Delta T_H}{(L_v/C_p)} = 2p_t\,\lambda_t \tag{22}$$

Again different values of the thermal lambda, λ_t yield different modes of heat diffusion. Fisher assumed $\lambda_t = 1/2$ [20, 40]. For the Ivantsov result, $T_t = T_M$, pure melt case, and hence $\lambda_t = \phi_t/2p_t$.

$$\Delta\theta = \frac{(T_M - T_\infty)}{(L_v/C_p)} = \phi_t \tag{23}$$

Comparing equations (21), (22) and (23) above with the solute buildup expressions it is seen that for $k \to 1$, (partitionless solidification, also pure melt) $\phi_t \to \phi$. Hence $\lambda_t \to \lambda$ as $k \to 1$.

APPENDIX C

Theoretical Inconsistencies Arising from the Application of the Simple, Unmodified Ivantsov Analysis.

The "gradient of constitutional supercooling" at the tip is defined as $S = (m_L G_c - \overline{G})$; see refs. 21 and 25 and eq. (38) in Appendix D.

For the Ivantsov dendrite, S may be obtained from eqns. (20) or (21) above. $\overline{G} \simeq G_i = G_L = sR\Delta T_o/D_L$ is the thermal gradient at the tip ("weighted" average value as defined in ref. 25), for "constrained" growth. This yields for the "constrained" growth problem:

$$S = (m_L G_c^o - \overline{G}) = \frac{Rk\Delta T_o}{D_L}\left[\frac{1}{b} - \frac{s}{k}\right] \tag{24}$$

where $b = 1 - \Omega^o(1-k) = 1-\phi(1-k)$

From eqn. (24) it is seen that $S < 0$, for $s > k$ if the Peclet number remains small ($b \simeq 1$). This is the situation under the experimental conditions reported by Somboonsuk et al, and, Esaka & Kurz (6, 8).

A negative value of S implies that the dendrite or cell tips experience no constitutional supercooling. In the experiments reported in refs. 6, 8, 34, a change in morphology from dendrites to cells was observed for $s > k$. Thus, the Ivantsov model for solute diffusion implies that constitutional supercooling is completely dissipated as the structure becomes cellular. This, however, contradicts the essential "physics" laid out by Chalmers and co-workers for breakdown of a planar interface.

As noted by Hillert [40], a theoretically "self-consistent" model must demonstrate that $S \to 0$ only as $s \to 1$, i.e. only at exactly the CS limit. Similar difficulties also arise with the Trivedi [18] Kurz & Fisher [19] and Kurz, Giovanola and Trivedi [33] models.

Including heat flow in the solid and taking $G_s = G_i = G_L$, $k_s = \beta k_L$, eqn. (24) will become

$$S = \frac{Rk\Delta T_o}{D_L} \left[\frac{1}{b} - \frac{s}{2k} \right] \qquad (25)$$

Allowing for heat flow through the solid will thus give $S > 0$, and hence predict instability (cellular or dendritic), only for $s < 2k$, with $b \simeq 1$.

The Ivantsov analysis will yield a theoretically consistent result of $S \to 0$ as $s \to 1$, if it can be shown that p increases rapidly long before the growth rate approaches the condition $s > k$. (This would yield $b \simeq k$ for $s > k$ since $\phi \to 1$ for $p \to \infty$). Presently available experimental data (very limited, though) does not support this possibility.

For the present model, however, the total concentration gradient, G_c, at the tip is (see ref. 21):

$$G_c = \frac{G_L}{m_L} + \frac{Rk\Delta T_o\,(1-s)}{m_L D_L[1-2p\lambda(1-k)]} \qquad (26)$$

which gives,
$$S = \frac{Rk\Delta T_o}{bD_L}\,(1-s) \qquad (27)$$

Eqn. (26) yields $S \to 0$ only for $s \to 1$. Thus, including the temperature gradient leads to a theoretically result for both the "Zenerian" model ($\lambda = 1/16$) and the "Ivantsovian" model ($\lambda = \phi/2p$).

Since a satisfactory model must yield $S \geq 0$ during both "free" dendritic growth and "constrained" growth, the results here raise some important theoretical questions about using the simple Ivantsov model without any allowance for the temperature gradient effect.

APPENDIX D

Approximate Expressions for the Tip Stability Parameter, Tip Radius and Dendrite Tip Growth Rate.

Tip Stability Parameter

When the Peclet number is small, $b = 1 - \Omega^o(1-k) \simeq 1$, and, $g(\omega) = 0$. Under these conditions, all models indicate that σ_c is only a function of G_i/R or s. For constant R and G_i and/or for $G_i/R \to 0$, σ_c = a constant, for a fixed composition. Only the value of this constant differs depending on the detailed assumptions made. The general expression for σ_c for small p is, therefore,

$$\sigma_c = \frac{1}{4\pi^2}\left[2(1-s) + \frac{s}{k} + B \right] \qquad (28)$$

For concentrated alloys, the dimensionless parameter $B \to 0$. Equation (28) then reduces to:

$$\sigma_c = \frac{(1-s)}{2\pi^2} + \frac{1}{4\pi^2}\left(\frac{s}{k} \right) \qquad (29)$$

The first term in equation (29) gives the tip stability parameter for "constrained" growth [21]. Thus σ_c for "free" growth is slightly higher than for "constrained" growth, depending on the value of s. This again implies sharper tip radii and/or faster growth rates during "free" growth.

Again, it is clear that "free" dendritic growth in concentrated alloys is remarkably similar to "constrained" growth. The predicted value of $\sigma_c = 1/2\pi^2$ is in very good agreement with experimental data in SCN-Ace alloys during "constrained" growth [21].

It is interesting to note that both equations (28) & (29) indicate that for small p, σ_c is independent of λ_e, although as shown below the separate values of R and r_t depend on λ_e.

Tip Radius and Growth Rate

The "Transport solution", equation (5), is a quadratic expression in r_t. For fixed values of R, ΔT_b and alloy composition, two tip radii are therefore predicted, being given by (and hence, also a maximum in growth rate as the tip radius increases. The maximum growth velocity is obtained when the roots of the quadratic become identical):

$$r_t = \left(\frac{D_L}{Rk}\right)\left(\frac{1}{2\lambda_e}\right)\left(\frac{\Delta T_b - s\Delta T_o}{\Delta T_o}\right)\left[1 \pm \left\{1 - \frac{8\lambda_e(\Gamma/\Gamma_a)}{(q-s)^2}\right\}\right] \tag{30}$$

where

$$\Gamma_a = (D_L/Rk)\ \Delta T_o;\ q = \Delta T_b/\Delta T_o \tag{31}$$

$$\Gamma = \ell_c k \Delta T_o = \gamma/\Delta S \tag{32}$$

Equation (30) above describes clearly the important factors determining the tip radius. Thus, the tip radius is some multiple or fraction of D_L/Rk, (which is the characteristic <u>transient</u> diffusion distance over which the solute buildup at a planar interface reaches its steady-state value, or D_L/R, the boundary layer thickness for a planar interface during steady-state growth).

The tip radius is also inversely proportional to the "effective" lambda and directly proportional to the bath supercooling. Also, the tip radius will become negative if the Chalmers number s exceeds a certain value, in other words, if the interdendritic gradient exceeds a certain value. The maximum value of s for which the tip radius remains positive is simply, $s_{max} = \Delta T_b/\Delta T_o$.

For $\Delta T_b/\Delta T_o > 1$ a positive tip radius is obtained only if $s < 0$ i.e. if $G_i < 0$. Thus, in deeply undercooled melts, heat flow must be entirely into the bulk liquid, with no heat flow through the solid phase (i.e. if the solid morphology is dendritic). Note that other solid morphologies, such as cylindrical dendrites and spherical "dendrite" elements have been reported in highly supercooled alloys [1, 41].

For $\Gamma/\Gamma_a = 0$ the <u>larger</u> of the two tip radii given by equation (30) is (assuming $s = 0$):

$$r_t = \left(\frac{D_L}{\lambda_e Rk} \right) \left(\frac{\Delta T_b}{\Delta T_o} \right) \tag{33}$$

and, correspondingly, the predicted Peclet number, p is given by:

$$p = \left(\frac{1}{2k\lambda_e} \right) \left(\frac{\Delta T_b}{\Delta T_o} \right) \tag{34}$$

Combining equation (34) with the approximate expression for the tip stability parameter (equation 28 with s = 0) yields the following approximate expressions for the tip radius and the growth rate (note these correspond to the larger of the two tip radii given by the transport solution):

$$R = M \left(\frac{2\ell_c D_L}{\Gamma^2} \right) (\Delta T_b)^2 \tag{35}$$

$$r_t = \left(\frac{\Gamma}{2M\lambda_e} \right) \left(\frac{1}{\Delta T_b} \right) \tag{36}$$

where $$M = \frac{2 + B}{(4\pi\lambda_e)^2} \tag{37}$$

Equation (35) predicts that the dendrite tip growth rate is expected to vary approximately as the square of the bath supercooling. This is in agreement with the experimental measurements of Huang and Barduhn [13] for ice crystals growing in a binary water-NaCl solution. The experimentally measured slope of R versus ΔT_b is 2 ± 0.2. Similar results have been obtained in SCN-Ace alloys.

Equation (36) predicts that the tip radius will vary approximately as the inverse of the bath supercooling at fixed composition. This is again confirmed by experimental measurements of Glicksman and co-workers in succinonitrile-alloys [10, 11].

It may be noted here details of assumptions such as the exact value of the "effective" thermal gradient at the tip, G, must be carefully specified. These assumptions can easily introduce factors of 2 or 1/2 or even 4 or 1/4 in the various approximate expressions above. The "effective thermal" gradient is defined as [3, 25]:

$$\overline{G} = \frac{k_S G_S + k_L G_L}{(k_S + k_L)} \tag{38}$$

In equations (28) through (38) above it has been assumed that $G_S = G_i$ and $k_S = k_L$. The LGK model assumes $G_S = 0$ and hence $G = G_L/2$.

Nomenclature

R	Dendrite tip growth rate
T_M	Melting point, pure melt
T_L	Liquidus temperature, alloy melt
T_s	Solidus temperature
T_o	Temperature at which liquid and solid, composition C_o, have equal free energies
T_∞	Supercooled bath temperature
T_t	Dendrite tip temperature
ΔT	Tip "undercooling", $(T_L - T_t)$
ΔT_b	Bath "supercooling", $(T_L - T_\infty)$
ΔT_H	Undercooling required for heat transport, $(T_t - T_\infty)$
$\Delta T_D'$	Solutal or constitutional undercooling, $-m_L \Delta C = -m_L(\Delta C_G + \Delta C_D)$
ΔT_r	Gibbs-Thomson undercooling
ΔT_k	Kinetic undercooling
C_o	Initial alloy composition
C_t	Liquid composition at tip
C_t^o	Liquid composition at tip, isothermal case
ΔC	Solute buildup at tip, $(C_t - C_o) = \Delta C_G + \Delta C_D$
ΔC_D	Solute buildup due to "divergent" or "point" effect, non-isothermal case
ΔC_D^o	Solute buildup, "point" effect, isothermal case
ΔC_G	Solute buildup due to temperature gradient effect, "unidirectional" component.
G_c	Concentration gradient at tip, non-isothermal case
G_c^o	Concentration gradient at tip, isothermal case
G_i	Interdendritic thermal gradient, in liquid
G_L	Thermal gradient in liquid, at tip
G_{cI}	Concentration gradient in interdendritic liquid
$\overline{G}$	Weighted average temperature gradient
G_s	Thermal gradient in solid
m_L	Slope of equilibrium liquidus line
S	Gradient of constitutional supercooling, $(m_L G_c - \overline{G})$
k	Partition ratio, assumed constant. May or may not be equilibrium value.
ΔC_o	Solute buildup for plane front, steady-state: $C_o(1-k)/k$

ΔT_o	Undercooling at planar interface, steady-state, $(T_L-T_s) = -m_L \Delta C_o$
D_L	Diffusivity of solute in liquid
α_L	Thermal diffusivity of liquid
L_v	Volumetric heat of fusion
C_p	Volumetric specific heat of liquid
$\Delta\Theta$	Dimensionless bath supercooling, $\Delta T_b/(L_v/C_p)$
$\Delta\Theta_o$	Dimensionless freezing range, $\Delta T_o/(L_v/C_p)$
Ω	Supersaturation at tip, non-isothermal case, $\Delta C/C_t(1-k)$
Ω^o	Supersaturation at tip, isothermal case, $\Delta C_D^o/C_t^o\ (1-k)$
Ω_t	Thermal supersaturation, $\Delta T_H/(L_v/C_p)$
$\Delta C/\Delta C_o$	Dimensionless solute buildup at tip
p	Solute Peclet number, $Rr_t/2D_L$
p_t	Thermal Peclet number, $Rr_t/2\alpha_L$
ϕ	Ivantsov function, solute case, $pe^pE_1(p)$
ϕ_t	Ivantsov function, thermal case $p_te^{p_t}E_1(p_t)$
λ	Solutal lambda, Zener number, $\Omega^o/2p$
λ_t	Thermal lambda, $\Omega_t/2p_t$
ℓ_c	Capillary length, alloy melt
A	Capillary number, solute case $R\ell_c/2D_L$
d_o	Capillary length pure melt
A_t	Capillary number, pure melt, $Rd_o/2\alpha_L$
σ^*	Tip stability parameter, Langer number, pure melt A_t/p_t^2
σ_c	Tip stability parameter, Langer number, alloy melt, A/p^2
$\mathcal{B}$	Dimensionless group, $\sigma_c/\sigma^* = (D_L/\alpha_L)/k\Delta\Theta_o$
s	Chalmers number, $D_LG_L/R\Delta T_o$; $D_LG_i/R\Delta T_o$
λ_e	"Effective" Lambda, includes thermal and solutal lambdas and alloy composition, eq. (11).
γ	Solid-liquid surface energy

REFERENCES

1. V. Laxmanan, Fundamentals of Alloy Solidification Applied to Solidification Processes, NASA Conference Publication No. 2337 (1984).

2. M. Barmatz, Materials Processing in the Reduced Gravity Environment of Space, Elsevier Publishing Co., (1982), Ed. G.E. Rindone, MRS Conf. Proc. p. 25.

3. M.E. Glicksman, Mat. Sci. & Eng. 65 (1984) 45-55.

4. R. Mehrabian, Int. Met. Reviews, 27, (1982) 185-208.

5. J. Lipton, M.E. Glicksman and W. Kurz, Mat. Sci. & Eng. 65 (1984) 57-63.

6. R. Trivedi and K. Somboonsuk, Mat. Sci. & Eng. 65 (1984) 65-74.

7. M.H. Burden and J.D. Hunt, J. Crystal Growth, 22 (1974) 99-108.

8. H. Esaka and W. Kurz, J. Crystal Growth, 72 (1985) 578-584.

9. H.A. Palacio, M. Solari and H. Biloni, 73 (1985) 369-378.

10. D.H. Kirkwood, Mat. Sci. & Eng. 65 (1984) 101-109.

11. M.A. Chopra, "Influence of diffusion and convective transport on dendritic growth in dilute alloys", Ph.D. thesis, Rensselaer Polytechnic Institute (1983).

12. M.E. Glicksman, N.B. Singh and M.A. Chopra, Materials Processing in the Reduced Gravity Environment of Space, (Elsevier Publishing Co., 1982) Ed. Guy E. Rindone, 461-477.

13. J.S. Huang and A.J. Barduhn AIChE Journal 31 (1985) 747-752.

14. J.P. Kallungal, "The Growth of Single Ice Crystals Parallel to the a-Axis in Subcooled Quiescent and Flowing Water", Ph.D. Thesis, Syracuse Univ., Syracuse, NY (1975).

15. V. Laxmanan, "Analysis of microsegregation during cellular and dendritic grwoth in a binary alloy", unpublished research (1984).

16. T.F. Bower, H.D. Brody and M.C. Flemings, Trans. AIME, 239 (1966) 624.

17. R.M. Sharp and M.C. Flemings, Metall. Trans. 5 (1974) 823.

18. R. Trivedi, J. Crystal Growth, 49 (1980) 219.

19. W. Kurz and D.J. Fisher, Acta Metall, 29 (1981) 11-20.

20. V. Laxmanan, Acta Metall, 33 (1985) 1023-1035, 1037-1049, 1475-1480.

21. V. Laxmanan, J. Crystal Growth, 75 (1986) 573-590.

22. V. Laxmanan, "Transition from a planar interface to cellular and dendritic structures during rapid solidification processing", Rapidly Solidified Alloys and their Mechanical and Magnetic Properties, (Elsevier Science Publishing Co., vol. 58, Mat. Res. Soc. Symp., 1986) Eds. B.C. Giessen, D.E. Polk and A.I. Taub 41-50.

23. J.S. Langer and H. Muller-Krumbhaar, J. Crystal Growth, 42 (1977) 11.

24. J.S. Langer, Rev. Mod. Phys. 52 (1980) 1.

25. W.W. Mullins and R.F. Sekerka, J. Appl. Phys. 35 (1964) 444.

26. R. Trivedi, H. Franke and R. Lacmann, J. Crystal Growth 47 (1979) 389.

27. C. Zener, Trans AIME 167 (1946) 550.

28. M. Hillert, J. Appl. Phys. 60 (1986) 1868

29. G.P. Ivantsov, Dokl Akad Nauk SSSR 132 (1947) 567.

30. D.A. Kessler, J. Koplik and H. Levine, Phys. Rev. A, 30 (1984) 3161, also Phys. Rev A 31 (1985) 1712.

31. E. Ben Jacob, N. Goldenfeld, J.S. Langer and G. Schon, Phys. Rev. A (1984) 330.

32. V. Laxmanan, "On Producing an alloy of uniform composition during rapid solidification processing", ASM Proc. Enhanced Mechanical Properties via Rapid Solidification Processing, ASM, Metals Park, Ohio (1987) in press, Eds. F.H. Froes and S.J. Savage.

33. W. Kurz, B. Giovanola and R. Trivedi, Acta Metall 34 (1986).

34. R. Trivedi and K. Somboonsuk, Scripta Met. (1986) in press.

35. M.C. Flemings, Solidification Processing, McGraw Hill (1974) 273.

36. V. Laxmanan, Unpublished research (1986).

37. Y. Miyata, T. Suzuki and J.I. Uno, Metall Trans A, 16A, (1985) 1799.

38. J.S. Langer, Mat Sci & Engr. 65 (1984) 37.

39. J.D. Hunt, Solidification & Casting of Metals (Metals Society, Book No. 192, 1979) 3.

40. M. Hillert, "Fundamentals of Aligned Growth", in Proc. Conf. on In-Situ Composites-III, Eds. J.L. Walter et al, pp. 1-15, Ginn Custom Publishing, Lexington (1979).

41. T.Z. Kattamis and M.C. Flemings, Trans. AIME 236 (1966) 1523.

Subject Index

Phase Diagram Index

Author Index